Lecture Notes in Computer Science 15016

Founding Editors

Gerhard Goos
Juris Hartmanis

AF167623

The series Lecture Notes in Computer Science (LNCS), including its subseries Lecture Notes in Artificial Intelligence (LNAI) and Lecture Notes in Bioinformatics (LNBI), has established itself as a medium for the publication of new developments in computer science and information technology research, teaching, and education.

LNCS enjoys close cooperation with the computer science R & D community, the series counts many renowned academics among its volume editors and paper authors, and collaborates with prestigious societies. Its mission is to serve this international community by providing an invaluable service, mainly focused on the publication of conference and workshop proceedings and postproceedings. LNCS commenced publication in 1973.

Michael Wand · Kristína Malinovská ·
Jürgen Schmidhuber · Igor V. Tetko
Editors

Artificial Neural Networks and Machine Learning – ICANN 2024

33rd International Conference on Artificial Neural Networks
Lugano, Switzerland, September 17–20, 2024
Proceedings, Part I

 Springer

Editors
Michael Wand ⓘ
IDSIA USI-SUPSI
Lugano, Switzerland

MeDiTech, SUPSI
Lugano, Switzerland

Jürgen Schmidhuber ⓘ
KAUST Center of Generative AI
Thuwal, Saudi Arabia

IDSIA USI-SUPSI
Lugano, Switzerland

Kristína Malinovská ⓘ
Comenius University
Bratislava, Slovakia

Igor V. Tetko ⓘ
Helmholtz Zentrum München
Neuherberg, Germany

BigChem GmbH
Unterschleißheim, Germany

ISSN 0302-9743   ISSN 1611-3349 (electronic)
Lecture Notes in Computer Science
ISBN 978-3-031-72331-5   ISBN 978-3-031-72332-2 (eBook)
https://doi.org/10.1007/978-3-031-72332-2

This Springer imprint is published by the registered company Springer Nature Switzerland AG
The registered company address is: Gewerbestrasse 11, 6330 Cham, Switzerland

If disposing of this product, please recycle the paper.

Preface

In recent years, Machine Learning has become more important than ever before. Large Language Models have revolutionized language-based tasks, with an impact far beyond the research community and IT-related industries: Artificial Intelligence for solving day-to-day tasks has become available for a wide range of end users across the world.

Machine Learning not only influences our daily lives, but also many fields of science and technology. As a specific example, we present Artificial Intelligence in organic chemistry and pharmaceutical research: a variety of tasks in this field are tackled with state-of-the-art Neural Network methods, leading to improved design and higher security of medical drugs, and to better solutions for chemical tasks in general, improving the quality of life of a large number of persons across the globe.

It is in this context that we proudly present the Proceedings of the 33th International Conference on Artificial Neural Networks (ICANN 2024). ICANN is the annual flagship conference of the European Neural Network Society (ENNS). This edition was co-organized by Istituto Dalle Molle di studi sull'intelligenza artificiale (IDSIA USI-SUPSI https://www.idsia.usi-supsi.ch) and by the Marie Skłodowska-Curie (MSC) Innovative Training Network European Industrial Doctorate "Advanced machine learning for Innovative Drug Discovery" (AIDD https://ai-dd.eu), supported by the MSC Doctoral Network "Explainable AI for Molecules" (AiChemist https://aichemist.eu). After two years of on-line and two years of hybrid conferences, ICANN 2024 was again organized as an in-person event, held on the premises of Università della Svizzera italiana (USI) and Scuola Universitaria Professionale della Svizzera italiana (SUPSI) in Lugano from September 17 to September 20, 2024.

ICANN 2024 featured three main conference tracks, namely Artificial Intelligence and Machine Learning, Bio-inspired Computing, and an Application Track. Dedicated members of the ICANN community also organized three workshops:

- AI in Drug Discovery
- Explainable AI in Human-Robot Interaction
- Reservoir Computing

  as well as three special sessions:

- Spiking Neural Networks and Neuromorphic Computing
- Accuracy, Stability, and Robustness in Deep Neural Networks
- Neurorobotics.

  Two tutorial sessions

- FEDn – A scalable federated machine learning framework for cross-device and cross-silo environments
- TSFEL - A Hands-on Introduction to Time Series Feature Extraction

were likewise proposed and organized by the community, as well as the

• Tox24 Challenge (prediction of toxicity of chemical compounds).

The proceedings of the conference are published as Springer volumes belonging to the Lecture Notes in Computer Science series. The conference had a total of 764 articles submitted to it. The papers went through a double-blind peer-review process supervised by experienced Area Chairs who suggested decisions to Program Chairs. In total, 564 Area Chairs, Program Committee (PC) members, and reviewers participated in the review process. The reviewers were on average assigned 3–4 articles each and submissions received on average 2.03 reviews each. A list of reviewers/PC Members who agreed to publish their names is included in the proceedings.

Based on the Area Chairs' and reviewers' comments, 310 articles (40.5% of initial submissions) were accepted, including 180 manuscripts selected for oral presentations. Out of the total number of accepted articles the majority (285 papers) were full articles with an average length of 15 pages, 20 manuscripts were short articles with an average length of 10 pages, and 5 were abstracts with an average length of 3 pages.

The accepted papers of the 33rd ICANN conference are published as 11 volumes, including one open-access volume with papers supported by the AIDD project.

The authors of accepted articles came from 29 different countries. As indicated by first author affiliation the largest number of articles came from China, followed by Germany, Japan, and Italy. While the majority of the articles were from academic researchers, the conference also attracted contributions from many industries including large pharmaceutical companies (Pfizer, Bayer, AstraZeneca, Johnson & Johnson), information and communication technology companies (Fujitsu and Baidu inc.), as well as multiple startups. This speaks to the increasing use of artificial neural networks in industry. Four keynote speakers were invited to give lectures on the timely aspects of advances in understanding the brain (Michael Reimann); new insights into cortical attention mechanisms and context-dependent gating and how they might inspire future developments in AI (Walter Senn); the current state of cognitive systems and how the full range of bio-signals can be utilized to further enhance human-robot interactions (Tanja Schultz); and a general overview of the past, present and future of machine learning (Jürgen Schmidhuber).

These proceedings provide comprehensive and up-to-date coverage of the dynamically developing field of Artificial Neural Networks. They are of major interest both for theoreticians as well as for applied scientists who are looking for new innovative approaches to solve their practical problems. We sincerely thank the Program and Steering Committee, Area Chairs, and the reviewers for their invaluable work.

September 2024

Michael Wand
Kristína Malinovská
Jürgen Schmidhuber
Igor V. Tetko

Organization

General Chairs

Jürgen Schmidhuber   KAUST Center of Generative AI, Saudi Arabia,
              and IDSIA USI-SUPSI, Switzerland
Igor V. Tetko   Helmholtz Munich, Germany and BigChem
              GmbH, Germany

Program Chairs

Michael Wand   IDSIA USI-SUPSI, Switzerland and MeDiTech,
              SUPSI, Switzerland
Kristina Malinovska   Comenius University Bratislava, Slovakia

Honorary Chair

Stefan Wermter   University of Hamburg, Germany

Organizing Committee Chairs

Katya Ahmad   Helmholtz Munich, Germany
Alessandra Lintas   University of Lausanne, Switzerland

Local Organizing Committee

Stefano van Gogh   IDSIA USI-SUPSI, Switzerland
Qinhan Hou   IDSIA USI-SUPSI, Switzerland
Nicolò La Porta   SUPSI, Switzerland
Alessandro Giusti   IDSIA USI-SUPSI, Switzerland
Vittorio Limongelli   USI, Switzerland
Cesare Alippi   IDSIA USI-SUPSI, Switzerland
Elena Invernizzi   IDSIA USI-SUPSI, Switzerland
Alessia Gianinazzi   IDSIA USI-SUPSI, Switzerland

Communication Chairs

Sebastian Otte	University of Lübeck, Germany
R. Omar Chavez-Garcia	IDSIA USI-SUPSI, Switzerland

Steering Committee

Stefan Wermter	University of Hamburg, Germany
Angelo Cangelosi	University of Manchester, UK
Igor Farkaš	Comenius University Bratislava, Slovakia
Chrisina Jayne	Teesside University, UK
Matthias Kerzel	University of Hamburg, Germany
Alessandra Lintas	University of Lausanne, Switzerland
Kristína Malinovská	Comenius University Bratislava, Slovakia
Alessio Micheli	University of Pisa, Italy
Jaakko Peltonen	Tampere University, Finland
Brigitte Quenet	ESPCI Paris, France
Ausra Saudargiene	Lithuanian University of Health Sciences, and Vytautas Magnus University, Lithuania
Roseli Wedemann	Rio de Janeiro State University, Brazil
Sebastian Otte	University of Lübeck, Germany

Area Chairs

Alessandro Antonucci	IDSIA USI-SUPSI, Switzerland
Alessandro Facchini	IDSIA USI-SUPSI, Switzerland
Alessio Micheli	University of Pisa, Italy
Anthony Cioppa	University of Liège, Belgium
Ausra Saudargiene	Lithuanian University of Health Sciences, and Vytautas Magnus University, Lithuania
Brigitte Quenet	ESPCI Paris PSL, France
Chen Zhao	King Abdullah University of Science and Technology, Saudi Arabia
Daniele Palossi	IDSIA USI-SUPSI, Switzerland
Davide Bacciu	University of Pisa, Italy
Fabio Rinaldi	IDSIA USI-SUPSI, Switzerland
Felix Putze	University of Bremen, Germany
Francesca Faraci	MeDiTech/BSP SUPSI-DTI, Switzerland
Gabriela Andrejková	P. J. Šafárik University in Košice, Slovakia
Hui Liu	University of Bremen, Germany

Igor Farkaš	Comenius University Bratislava, Slovakia
Kevin Jablonka	Friedrich Schiller University Jena, Germany
Marcello Restelli	Politecnico di Milano, Italy
Marco Forgione	IDSIA USI-SUPSI, Switzerland
Matthias Karlbauer	University of Tübingen, Germany
Michela Papandrea	ISIN, DTI, SUPSI, Switzerland
Mihai Andries	IMT Atlantique, France
Oleg Szehr	IDSIA USI-SUPSI, Switzerland
Rafael Cabañas de Paz	University of Almería, Spain
Silvio Giancola	King Abdullah University of Science and Technology, Saudi Arabia
Thang Vu	University of Stuttgart, Germany
Yibo Yang	King Abdullah University of Science and Technology, Saudi Arabia
Zuzana Černeková	Comenius University Bratislava, Slovakia

Workshop and Special Session Chairs

Workshop: AI in Drug Discovery

| Djork-Arné Clevert | Pfizer GmbH, Germany |
| Igor Tetko | Helmholtz Munich, Germany |

Workshop: Explainable AI in Human-Robot Interaction

Stefan Wermter	University of Hamburg, Germany
Angelo Cangelosi	University of Manchester, UK
Igor Farkaš	Comenius University Bratislava, Slovakia
Theresa Pekarek-Rosin	University of Hamburg, Germany

Workshop: Reservoir Computing

Alessio Micheli	University of Pisa, Italy
Gouhei Tanaka	Nagoya Institute of Technology, Japan
Claudio Gallicchio	University of Pisa, Italy
Benjamin Paassen	University of Bielefeld, Germany
Domenico Tortorella	University of Pisa, Italy

Special Session: Spiking Neural Networks and Neuromorphic Computing

Sander Bohté   CWI Amsterdam, Netherlands
Sebastian Otte   University of Lübeck, Germany

Special Session: Accuracy, Stability, and Robustness in Deep Neural Networks

Vera Kurkova   Institute of Computer Science of the Czech
                               Academy of Sciences, Prague Czech Republic
Ivan Tyukin   King's College, London, UK

Special Session: Neurorobotics

Igor Farkaš   Comenius University Bratislava, Slovakia
Kristína Malinovská   Comenius University Bratislava, Slovakia
Andrej Lúčny   Comenius University Bratislava, Slovakia
Pavel Petrovič   Comenius University Bratislava, Slovakia
Michal Vavrečka   Czech Technical University in Prague, Czechia
Matthias Kerzel   University of Hamburg, Germany
Hassan Ali   University of Hamburg, Germany
Carlo Mazzola   Italian Institute for Technology, Italy

Program Committee

Abraham Yosipof   CLB, Israel
Adam Arany   KU Leuven, Belgium
Adrian Mirza   Helmholtz Institute for Polymers in Energy
                               Applications, Germany
Adrian Ulges   RheinMain University of Applied Sciences,
                               Germany
Alan Anis Lahoud   Örebro University, Sweden
Albert Weichselbraun   University of Applied Sciences of the Grisons
                               (FHGR), Switzerland
Alessandra Roncaglioni   Istituto di Ricerche Farmacologiche Mario Negri,
                               Italy
Alessandro Giusti   IDSIA USI-SUPSI, Switzerland
Alessandro Manenti   USI, Switzerland

Alessandro Trenta	University of Pisa, Italy
Alessio Gravina	University of Pisa, Italy
Alex Doboli	Stony Brook University, USA
Alex Shenfield	Sheffield Hallam University, UK
Alexander Schulz	Bielefeld University, Germany
Alexandra Reichenbach	Heilbronn University of Applied Sciences, Germany
Ali Rodan	University of Jordan, Jordan
Alireza Raisiardali	Pragmatic Semiconductor Limited, UK
Aliza Subedi	Tribhuvan University, Nepal
Amir Mohammad Elahi	EPFL, Switzerland
Ana Claudia Sima	SIB Swiss Institute of Bioinformatics, Switzerland
Ana Sanchez-Fernandez	Johnson & Johnson Innovative Medicine, Belgium/JKU Linz, Austria
Andrea Licciardi	ICAR-CNR, Italy
Andreas Mayr	Johannes Kepler University Linz, Austria
Andreas Plesner	ETH Zurich, Switzerland
Andrej Lucny	Comenius University Bratislava, Slovakia
Aneri Muni	University of Montreal and Mila AI Institute, Canada
Angeliki Pantazi	IBM Research - Zurich, Switzerland
Angelo Moroncelli	IDSIA USI-SUPSI, Switzerland
Anne-Gwenn Bosser	Lab-STICC, ENIB, France
Anthony Strock	Stanford University, USA
Antonio Liotta	Free University of Bozen-Bolzano, Italy
Aparna Raj	BITS Pilani, Dubai Campus, United Arab Emirates
Ardian Selmonaj	IDSIA USI-SUPSI, Switzerland
Arnaud Gucciardi	University of Ljubljana, Slovenia
Artur Xarles	Universitat de Barcelona, Spain
Asma Sattar	University of Pisa, Italy
Aurelio Raffa Ugolini	Politecnico di Milano, Italy
Baohua Zhang	Beijing Institute of Technology, China
Baojin Huang	Wuhan University, China
Barbara Hammer	Bielefeld University, Germany
Bikram Kumar De	Texas State University, USA
Blerina Spahiu	University of Milan-Bicocca, Italy
Bo Li	Baidu Inc., China
Bogdan Kwolek	AGH University of Krakow, Poland
Bojian Yin	Innatera B.V., Netherlands
Brian Moser	German Research Center for Artificial Intelligence, Germany

Bulcsú Sándor	Babeş-Bolyai University, Romania
Cesare Donati	Politecnico di Torino, Italy
Chengeng Liu	Wuhan University, China
Chenxing Wang	Beijing University of Posts and Telecommunications, China
Chi Xie	Tongji University, China
Chong Zhang	Xi'an Jiaotong-Liverpool University, China
Chrisina Jayne	Teesside University, UK
Christoph Reinders	Leibniz University Hannover, Germany
Chrysoula Kosma	École Normale Supérieure Paris-Saclay, France
Cleber Zanchettin	Universidade Federal de Pernambuco, Brazil
Congcong Zhou	Sir Run Run Shaw Hospital, Zhejiang University, China
Coşku Can Horuz	University of Lübeck, Germany
Cunjian Chen	Monash University, Australia
Cyril Zakka	Stanford University, USA
Dania Humaidan	University Hospital Tübingen and Hertie Institute for Clinical Brain Research, Germany
Daniel Frank	University of Stuttgart, Germany
Daniel Nissani (Nissensohn)	Independent Research, Israel
Daniel Ortega	University of Stuttgart, Germany
Daniel Rose	University of Vienna, Austria
Daniele Angioletti	Università della Svizzera italiana, Switzerland
Daniele Castellana	Università degli Studi di Firenze, Italy
Daniele Malpetti	IDSIA USI-SUPSI, Switzerland
Daniele Zambon	IDSIA USI-SUPSI, Switzerland
Darío Ramos López	University of Almería, Spain
Davide Borra	University of Bologna, Italy
Dehui Kong	Sanechips; ZTE, China
Denis Kleyko	Örebro University, Sweden
Diana Borza	Babeş-Bolyai University, Romania
Dinesh Kumar	Bennett University, India
Dirk Väth	University of Stuttgart, Germany
Dongmian Zou	Duke Kunshan University, China
Doreen Jirak	Istituto Italiano di Tecnologia, Italy
Douglas McLelland	BrainChip, France
Duarte Folgado	Fraunhofer Portugal AICOS, Portugal
Dulani Meedeniya	University of Moratuwa, Sri Lanka
Dumitru-Clementin Cercel	Politehnica University of Bucharest, Romania
Dylan Muir	SynSense, Switzerland
Dylan R. Ashley	IDSIA USI-SUPSI, Switzerland

E. J. Solteiro Pires	Universidade de Trás-os-Montes e Alto Douro, Portugal
Elena Šikudová	Charles University, Czech Republic
Elia Cereda	IDSIA USI-SUPSI, Switzerland
Elia Piccoli	University of Pisa, Italy
Emmanuel Okafor	King Fahd University of Petroleum and Minerals, Saudi Arabia
Evaldo Mendonça Fleury Curado	Centro Brasileiro de Pesquisas Físicas and National Institute of Science and Technology for Complex Systems, Brazil
Evgeny Mirkes	University of Leicester, UK
Farhad Nooralahzadeh	Zurich University of Applied Sciences, University of Zurich, Switzerland
Fatemeh Hadaeghi	University Medical Center Hamburg-Eppendorf (UKE), Germany
Fatima Ezzeddine	Università della Svizzera italiana, Switzerland
Federico Errica	NEC Laboratories Europe, Germany
Fedor Scholz	University of Tübingen, Germany
Filipe Miguel Cardoso Micu Menezes	Helmholtz Munich, Germany
Flávio Arthur Oliveira Santos	Universidade Federal de Pernambuco, Brazil
Florian Lux	University of Stuttgart, Germany
Francesco Faccio	IDSIA USI-SUPSI, Switzerland/KAUST AI Initiative, Saudi Arabia
Francesco Landolfi	Università di Pisa, Italy
Francis Colas	Inria, France
Frédéric Alexandre	Inria, France
Gabriel Haddon-Hill	Keio University, Japan
Gabriela Sejnova	Czech Technical University in Prague, Czech Republic
Gabriele Lagani	ISTI-CNR, Italy
Gerrit A. Ecke	Mercedes-Benz AG, Germany
Gianvito Losapio	Politecnico di Milano, Italy
Giorgia Adorni	IDSIA USI-SUPSI, Switzerland
Giovanni Dispoto	Politecnico di Milano, Italy
Giovanni Donghi	University of Padua, Italy
Giuliana Monachino	University of Applied Sciences and Arts of Southern Switzerland (SUPSI), Switzerland
Gugulothu Narendhar	TCS Research, India
Guillaume Godin	BigChem, Switzerland
Habib Irani	Texas State University, USA
Hanno Gottschalk	TU Berlin, Germany
Haoran Yang	Sichuan University, China

Hasby Fahrudin	AIBrain, South Korea
Hicham Boudlal	Mohammed First University of Oujda, Morocco
Hitesh Laxmichand Patel	Oracle/New York University, USA
Houssem Ouertatani	IRT SystemX & Univ. Lille, CNRS, Inria, France
Huang Yifan	Northeast Electric Power University, China
Hubert Cecotti	California State University, Fresno, USA
Hugo Cesar de Castro Carneiro	Universität Hamburg, Germany
Huifang Ma	Northwest Normal University, China
Igor Tetko	Helmholtz Munich, Germany
Ivor Uhliarik	Comenius University Bratislava, Slovakia
Jan Kalina	Czech Academy of Sciences, Institute of Computer Science, Czech Republic
Jan Niehues	KIT, Germany
Jan Prosi	University of Tübingen/International Max Planck Research School for Intelligent Systems, Germany
Jan Wollschläger	Bayer Pharmaceuticals, Germany
Jannis Vamvas	University of Zurich, Switzerland
Jérémie Cabessa	University of Versailles Saint-Quentin, France
Jia Cai	Guangdong University of Finance and Economics, China
Jiahui Chen	Xiamen University, China
Jialiang Xu	Soochow University, China
Jian Zhang	Zhejiang University, China
Jing Han	University of Cambridge, UK
Jingzehua Xu	Tsinghua University, China
Jinlai Ning	King's College London, UK
Jiong Wang	Beijing Normal University, China
Jiwen Yu	Peking University, China
Jizhe Yu	Dalian University of Technology, China
João Ricardo Sato	Universidade Federal do ABC, Brazil
Johannes Kriebel	University of Münster, Germany
Johannes Zierenberg	Max Planck Institute for Dynamics and Self-Organization, Germany
Jorge Lo Presti	University of Pavia, Italy
Julian Cremer	Pfizer, Germany
Julie Keisler	EDF R&D, Inria, France
Julien Marteen Akay	Bielefeld University of Applied Sciences and Arts, Germany
Jun Zhou	Wuhan University, China
Junjie Zhou	Nanjing University of Aeronautics and Astronautics, China
Junzhou Chen	College of William and Mary, USA

Kai Mao	Xi'an Jiaotong University, China
Kevin Scheck	University of Bremen, Germany
Keyan Jin	Macao Polytechnic University, Macao SAR, China
Khoa Phung	University of the West of England, UK
Kiran Lekkala	University of Southern California, USA
Kohei Nakajima	University of Tokyo, Japan
Konstantinos Chatzilygeroudis	University of Patras, Greece
Krechel Dirk	RheinMain University of Applied Science, Germany
Kristína Malinovská	Comenius University Bratislava, Slovakia
Lapo Frascati	ODYS, Italy
Laura Azzimonti	IDSIA USI-SUPSI, Switzerland
Laurent Larger	FEMTO-ST Institute, Université Bourgogne-Franche-Comté, France
Laurent Mertens	KU Leuven, Belgium
Laurent Udo Perrinet	Institut des Neurosciences de la Timone, Aix Marseille Univ - CNRS, France
Lazaros Iliadis	Democritus University of Thrace, Greece
Lea Multerer	IDSIA USI-SUPSI, Switzerland
Lei Li	University of Copenhagen, Denmark
Lenka Tetkova	Technical University of Denmark, Denmark
Leon Scharwächter	University of Tübingen, Germany
Leonardo Olivetti	Uppsala University, Sweden
Lewis Mervin	AstraZeneca, UK
Lina Humbeck	Boehringer Ingelheim Pharma GmbH & Co. KG, Germany
Lindsey Vanderlyn	University of Stuttgart, Germany
Logofatu Doina	Frankfurt University of Applied Sciences, Germany
Lu Yang	Wuhan University, China
Lubomir Antoni	Pavol Jozef Šafárik University in Košice, Slovakia
Luca Butera	IDSIA USI-SUPSI, Switzerland
Luca Sabbioni	ML cube, Italy
Luís Gonçalves	Universidade Federal de Pernambuco, Brazil
Lyra Puspa	Vanaya NeuroLab, Indonesia and Canterbury Christ Church University, UK
Maëlic Neau	ENIB, France/Flinders University, Australia
Mahsa Abazari Kia	Northeastern University London, UK
Maksim Makarenko	Saudi Aramco, Saudi Arabia
Manas Mejari	IDSIA USI-SUPSI, Switzerland
Manon Dampfhoffer	Univ. Grenoble Alpes, CEA, List, France
Manuel Traub	University of Tübingen, Germany

Marco Paul E. Apolinario	Purdue University, USA
Marco Podda	University of Pisa, Italy
Marco Tarabini	Politecnico di Milano, Italy
Marcondes Ricarte da Silva Júnior	Federal University of Pernambuco, Brazil
Marek Suppa	Comenius University Bratislava, Slovakia
Marina Garcia de Lomana	Bayer AG, Germany
Markus Heinonen	Aalto University, Finland
Marta Lenatti	Consiglio Nazionale delle Ricerche, Italy
Martin Lefebvre	Université catholique de Louvain, Belgium
Martin Ritzert	Georg-August Universität Göttingen, Germany
Masanobu Inubushi	Tokyo University of Science, Japan
Matej Fandl	Comenius University Bratislava, Slovakia
Matej Pecháč	Tachyum s.r.o., Slovakia
Matteo Rufolo	IDSIA USI-SUPSI, Switzerland
Matthias Kerzel	Universität Hamburg, Germany
Matthias Rupp	Luxembourg Institute of Science and Technology, Luxembourg
Matus Tuna	Comenius University Bratislava, Slovakia
Maximilian Kimmich	University of Stuttgart, Germany
Maynara Donato de Souza	Federal University of Pernambuco, Brazil
Mengdi Li	University of Hamburg, Germany
Mengjia Zhu	IMT School for Advanced Studies Lucca, Italy
Michal Bechny	UNIBE/SUPSI, Switzerland
Michal Burgunder	Università della Svizzera italiana, Switzerland
Michal Vavrecka	CIIRC CTU, Czech Republic
Michela Sperti	Politecnico di Torino, Italy
Michele Fontanesi	University of Pisa, Italy
Mikhail Andronov	Università della Svizzera Italiana, Switzerland
Mingyang Li	Stanford University, USA
Mingyong Li	Chongqing Normal University, China
Miroslav Strupl	IDSIA USI-SUPSI, Switzerland
Moritz Wolter	Rheinische Friedrich-Wilhelms-Universität Bonn, Germany
Muhammad Arslan Masood	Aalto University, Finland
Muhammad Burhan Hafez	University of Southampton, UK
Mykhailo Sakevych	Texas State University, USA
Nabeel Khalid	German Research Center for Artificial Intelligence, Germany
Navdeep Singh Bedi	IDSIA USI-SUPSI, Switzerland
Nicolò La Porta	Università della Svizzera Italiana, Switzerland
Niklas Beuter	Technische Hochschule Lübeck, Germany
Niko Dalla Noce	University of Pisa, Italy

Oh-hyeon Choung   dsm-firmenich, Switzerland
Olivier J. M. Béquignon   Leiden University, The Netherlands
Omran Ayoub   University of Applied Sciences and Arts of
                                    Southern Switzerland, Switzerland
Oscar Mendez Lucio   Recursion, Spain
Osvaldo Simeone   King's College London, UK
Otto Brinkhaus   Spleenlab GmbH, Germany
Pascal Tilli   University of Stuttgart, Germany
Paul Czodrowski   JGU Mainz, Germany
Paul Kainen   Georgetown University, USA
Paula Štancelová   Comenius University Bratislava, Slovakia
Paula Torren-Peraire   Johnson & Johnson Innovative Medicine,
                                    Belgium
Pavel Denisov   University of Stuttgart, Germany
Pavel Kordík   Czech Technical University in Prague,
                                    Czech Republic
Pavel Petrovič   Comenius University Bratislava, Slovakia
Peiyu Liang   Temple University, USA
Peng Qiao   NUDT, China
Pengjie Liu   Southern University of Science and Technology,
                                    China
Pengyu Li   Yanshan University, China
Petia Koprinkova-Hristova   Institute of Information and Communication
                                    Technologies, Bulgarian Academy of Sciences,
                                    Bulgaria
Petra Vidnerová   Institute of Computer Science, Czech Academy of
                                    Sciences, Czech Republic
Philipp Allgeuer   University of Hamburg, Germany
Plinio Moreno   Instituto Superior Técnico/University of Lisbon,
                                    Portugal
Qinhan Hou   IDSIA USI-SUPSI, Switzerland
Quentin Jodelet   Tokyo Institute of Technology, Japan
Raphael Yokoingawa de Camargo   Universidade Federal do ABC, Brazil
Răzvan-Alexandru Smădu   National University of Science and Technology
                                    POLITEHNICA Bucharest, Romania
Reyan Ahmed   University of Arizona, USA
Ricardo O. Chávez García   IDSIA USI-SUPSI, Switzerland
Riccardo Massidda   Università di Pisa, Italy
Riccardo Renzulli   University of Turin, Italy
Robert Legenstein   Graz University of Technology, Austria
Robertas Damaševičius   Kaunas University of Technology, Lithuania
Robin Winter   Pfizer, Germany

Rodolphe Vuilleumier	École normale supérieure-PSL, Sorbonne Université, CNRS, France
Rodrigo Braga	NOVA School of Science and Technology, Portugal
Rodrigo Clemente Thom de Souza	Federal University of Paraná, Brazil
Roseli S. Wedemann	Universidade do Estado do Rio de Janeiro, Brazil
Roxane Jacob	University of Vienna, Austria
Ru Zhou	RuiJin Hospital LuWan Branch, Shanghai Jiaotong University School of Medicine, China
Ruinan Wang	University of Bristol, UK
Ruixi Zhou	Beijing University of Posts and Telecommunications, China
Rupesh Raj Karn	New York University Abu Dhabi, United Arab Emirates
Samuel Genheden	AstraZeneca R&D, Sweden
Sandra Mitrovic	IDSIA USI-SUPSI, Switzerland
Sankalp Jain	NCATS-NIH, USA
Sara Joubbi	University of Pisa, Italy
Seema Dilipkumar Aswani	BITS Pilani, Dubai Campus, UAE
Seiya Satoh	Tokyo Denki University, Japan
Semih Beycimen	Cranfield University, UK
Senhui Qiu	Ulster University, UK
Sergei Katkov	Free University of Bozen-Bolzano, Italy
Sergio Mauricio Vanegas Arias	LUT University, Finland
Shangchao Su	Fudan University, China
Sheng Xu	Chinese University of Hong Kong, Shenzhen, China
Shenyang Liu	University of Central Florida, USA
Sherjeel Shabih	Humboldt University, Germany
Shi Haoran	China Water Northeastern Investigation, Design & Research Co., Ltd., China
Shingo Murata	Keio University, Japan
Shinnosuke Matsuo	Kyushu University, Japan
Shiyao Zhang	University of Bremen, Germany
Sho Shirasaka	Osaka University, Japan
Simiao Zhuang	TUM Beijing, China
Simon Heilig	Ruhr University Bochum, Germany
Šimon Horvát	Slovakia
Simone Bonechi	University of Siena, Italy
Simone Lionetti	Hochschule Luzern, Switzerland
Siyu Wu	Central South University of Forests and Technology, China
Stefano Damato	IDSIA USI-SUPSI, Switzerland

Stéphane Meystre	MeDiTech/SUPSI, Switzerland
Steve Azzolin	University of Trento, Italy
Sudip Roy	Indian Institute of Technology Roorkee, India
Sujala D. Shetty	BITS Pilani, Dubai Campus, United Arab Emirates
Taoran Fu	Hunan University & Hunan Institute of Engineering, China
Teste Olivier	Université Toulouse 2, IRIT (UMR5505), France
Thierry Viéville	Inria, France
Tianyi Wang	Nanyang Technological University, Singapore
Tim Schlippe	IU International University of Applied Sciences, Germany
Tingyu Lin	TU Wien, Austria
Tuan Le	Pfizer, Germany
Valerie Vaquet	Bielefeld University, Germany
Vangelis Metsis	Texas State University, USA
Vani Kanjirangat	IDSIA USI-SUPSI, Switzerland
Varun Ojha	Newcastle University, UK
Veronica Lachi	Fondazione Bruno Kessler, Italy
Viktor Kocur	Comenius University Bratislava, Slovakia
Vincenzo Palmacci	University of Vienna, Austria
Wei Dai	Robo Space, China
Weiqi Li	Peking University, China
Weiran Chen	Soochow University, China
Wenjie Zhang	Shandong University, China
Wenwei Gu	Chinese University of Hong Kong, China
Wolfram Schenck	Bielefeld University of Applied Sciences and Arts, Germany
Xavier Hinaut	Inria, France
Xi Wang	National University of Defense Technology, China
Xiangxian Li	Shandong University, China
Xiangyuan Peng	Technical University of Munich, Germany
Xiaochen Yuan	Macao Polytechnic University, Macao SAR, China
Xiaomeng Fu	University of Chinese Academy of Sciences, China
Xiaowen Sun	University of Hamburg, Germany
Xiaoxiao Miao	Singapore Institute of Technology, Singapore
Xingda Yao	Zhejiang University of Technology, China
Xinxin Luo	Southeast University, China
XinZhi Lin	Beihang University, China
Xun Lin	Beihang University, China

Yan Jiang	Nanjing University of Information Science and Technology, China
Yang Cao	Shanghai University of Finance and Economics, China
Yangfan Zhou	Southwest University of Science and Technology, China
Yangxun Ou	East China Normal University, China
Yao Du	Beihang University, China
Yaxin Hu	University of Lübeck, Germany
Ye Hu	Pfizer, Germany
Yi Li	Lancaster University, UK
Yichi Zhang	Fudan University, China
Yiming Tang	Shanghai Lixin University of Accounting and Finance, China
Ying Tan	Key Laboratory for Computer Systems of State Ethnic Affairs Commission, Southwest Minzu University, China
Yiqing Shen	Johns Hopkins University, USA
Yixuan Xiao	University of Stuttgart, Germany
Yong Luo	Wuhan University, China
Yongtao Tang	National University of Defense Technology, China
Yuankun Chen	University of Science and Technology, China
Yuansheng Ma	Soochow University, China
Yuchen Guo	Institute of Information Engineering, Chinese Academy of Sciences, China
Yuichi Katori	Future University Hakodate, Japan
Yuji Kawai	Osaka University, Japan
Yusen Wu	Sichuan University, China
Yutaka Nakamura	Riken, Japan
Yuya Okadome	Tokyo University of Science, Japan
Zdravko Marinov	Karlsruhe Institute of Technology, Germany
Zeyao Liu	Key Institute of Information Engineering, Chinese Academy of Sciences, China
Zhang Ke	China University of Petroleum (Beijing), China
Zhenjie Yao	Institute of Microelectronics of the Chinese Academy of Sciences, China
Zheyan Gao	Tianjin University, China
Zhiheng Qiu	City University of Macau, China
Zhihuan Xing	Beihang University, China
Zuzana Berger Haladova	Comenius University Bratislava, Slovakia

Plenary Talks

Past, Present, Future, and Far Future of Machine Learning

Jürgen Schmidhuber

IDSIA USI-SUPSI, Switzerland, and KAUST AI Initiative, Saudi Arabia

I'll discuss modern Artificial Intelligence and how the principles of the G, P and T in Chat GPT emerged in 1991. I'll also discuss what's next in AI, and its expected impact on the future of the universe.

Dendritic Computations and Deep Learning in the Brain

Walter Senn

University of Bern, Institut für Physiologie, Computational Neuroscience Lab, Switzerland

Artificial Intelligence, through its working horse of neural networks, is inspired by the biological example of the brain. The unprecedented success of AI in modeling cognitive processes, in turn, inspires functional models of the brain. Yet, when looking into the brain, additional biological structures become apparent, such as dendritic morphologies, interneuron circuits, recurrent connectivity, error representations, top-down signaling and various gating hierarchies. I will give a review on these biological elements and show how they may integrate in an energy-based theory of cortical computation. Dendrites and cortical microcircuits turn out to implement a real-time version of error-backpropagation based on prospective errors. The theory is inspired by the least-action principle in physics from which all dynamical equations of motions are derived. We likewise derive the neuronal dynamics, including the synaptic dynamics with gradient-descent learning, from our Neuronal Least-Action (NLA) principle. The principle tells that the cortical activities and the real-time learning follows a path that minimizes prospective errors across all neurons of the network. Prospective errors in output neurons relate to behavioral errors, while prospective errors in deep network neurons relate to errors in the neuron-specific dendritic prediction of somatic firing. I will explain how these ideas relate to cortical attention mechanisms and context-dependent gating that link to, and potentially inspires, recent developments in AI.

Biosignal-Adaptive Cognitive Systems

Tanja Schultz

University of Bremen, Fachbereich 3 - Mathematik und Informatik, Cognitive Systems Lab, Germany

I will describe technical cognitive systems that automatically adapt to users' needs by interpreting their biosignals: Human behavior includes physical, mental, and social actions that emit a range of biosignals which can be captured by a variety of sensors. The processing and interpretation of such biosignals provides an inside perspective on human physical and mental activities, complementing the traditional approach of merely observing human behavior. As great strides have been made in recent years in integrating sensor technologies into ubiquitous devices and in machine learning methods for processing and learning from data, I argue that the time has come to harness the full spectrum of biosignals to understand user needs. I will present illustrative cases ranging from silent and imagined speech interfaces that convert myographic and neural signals directly into audible speech, to interpretation of human attention and decision making in human-robot interaction from multimodal biosignals.

A Model of Neocortical Micro- and Mesocircuitry and Its Applications

Michael Reimann

Blue Brain, Swiss Federal Institute of Technology Lausanne, Switzerland

We present a large-scale, biophysically detailed model of rat non-barrel somatosensory regions. Building upon an earlier version of such a model, we increased the spatial scale of the model and enhanced its biological realism. The most salient improvements are: First, construction of realistic synaptic connectivity as the union of two algorithms, one for local connections, and another for long-range connections. Second, introduction of methods to build a model inside a standardized voxel atlas. This, combined with the connectivity algorithms allows models of brain regions to be developed separately and then easily integrated. Third, improvements in the methods to compensate for missing extrinsic inputs and to validate an in-vivo-like activity regime.

We demonstrate several applications of the model that make use of its specific advantages over more simplified models: First, studying the rules of synaptic plasticity at the population level. Second, studying the effect of heterogeneous and non-random connectivity on circuit function and reliability. Third, studying the accuracy and biases inherent in spike sorting algorithms.

A Model of Neocortical Micro- and Mesocircuitry and Its Applications

Michael Reimann

Blue Brain, Swiss Federal Institute of Technology, Lausanne, Switzerland

Contents – Part I

Novel Neural Architectures

Neural Architecture Search

Self-Organization

Neural Processes

Novel Architectures for Computer Vision

Fairness in Machine Learning

Theory of Neural Networks
and Machine Learning

Multi-label Robust Feature Selection via Subspace-Sparsity Learning

Yunya Zhou[1], Bin Yuan[2], Yan Zhong[3(✉)], and Yuling Li[1(✉)]

[1] School of Statistics, Beijing Normal University, Zhuhai, China
202322011108@mail.bnu.edu.cn , liyuling@bnuz.edu.cn
[2] Trine University, Phoenix, AZ, USA
byuan22@my.trine.edu
[3] Peking University, Beijing, China
zhongyan@stu.pku.edu.cn

Abstract. Multi-label feature selection is crucial for managing feature redundancy and irrelevance in high-dimensional datasets. Existing methods reduce information redundancy through subspace dimensionality reduction but often suffer from instability due to high degrees of freedom and lack flexibility. This is because of the assumption of a shared subspace for features and labels, which leads to reduced performance. To address these problems, we introduce a novel multi-label feature selection approach. Specifically, we propose a dual subspace learning approach to capture both label correlations and feature correlations for feature selection. Therefore, our method can mitigate the adverse effects of noise, redundancy and imperfect features in the quest for discriminative features. Additionally, it reduces the sensitivity of the constructed model to the noise and outliers present in the data. Empirical experiments conducted on real-world datasets illustrate the efficiency and superiority of our proposed approach.

Keywords: Multi-label learning · Robust feature selection · Structure preservation · Sub-space learning

1 Introduction

Multi-label feature selection has found various applications in different domains, such as document classification [14,25], social media analysis [1,23], bioinformatics [19], and image annotation [9,24] due to the excessive numbers of features, which can reduce the classification performance of the employed model. Existing multi-label feature selection methods can be categorized into embedded, filter, and wrapper methods [21]. Embedded methods, in particular, have garnered considerable interest. These methods treat the multi-label feature selection (MLFS) problem as an optimization model [11,26], performing feature selection and model optimization simultaneously, and are claimed to be more efficient and informative with lower computational cost [12,15].

© The Author(s), under exclusive license to Springer Nature Switzerland AG 2024
M. Wand et al. (Eds.): ICANN 2024, LNCS 15016, pp. 3–17, 2024.
https://doi.org/10.1007/978-3-031-72332-2_1

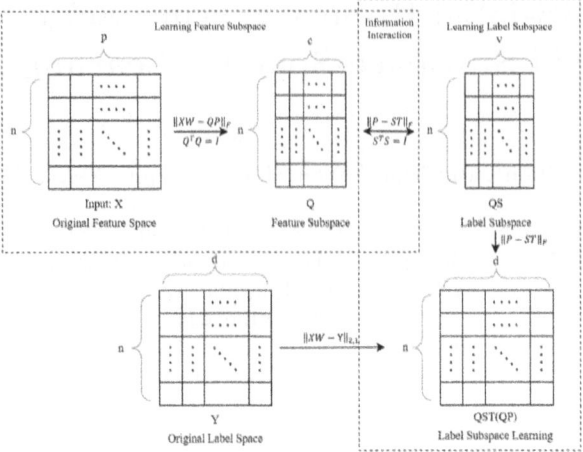

Fig. 1. The architecture of SS-MRFS. We learn the sparse subspace of X through Q. By imposing restrictions on the mapping matrix P, the model can learn the label subspace through QS and establish a relationship with the feature subspace to learn the shared information. Additionally, the degrees of freedom within the subspace can be restricted.

Many embedded methods tend to incorporate some sparsity-based approaches to exploit information correlations. The $\ell_{2,1}$-norm has shown effective performance and has been a subject of increasing studies [16,17,20], most of which aim to identify label or feature subspace that are more likely to effectively capture the relationships between features [7,13,17] and labels [5,10] prior to conducting feature selection. Classic methods include sub-feature uncovering with sparsity (SFUS) [17] and multi-label informed feature selection (MIFS) [13]. Obviously, they only use label correlations to select discriminative features, but ignore the relationship between features.

Methods [3,6,8,26] that assume identical subspaces for the feature and label spaces have been developed. Correlated multi-label feature selection method (CMFS) integrates sample similarities, label correlations, and feature correlations into an objective framework by embedding mode. It uses the Frobenius norm to measure the distance between subspace and sample spaces, which leads to the inability to induce sparsity effectively, making it less robust to noise and outliers compared to the $\ell_{2,p}$ norm. Zhang et al. [26] propose manifold regularized discriminative feature selection (MDFS), which makes use of the pseudo-label matrix to exploit the global and local, label and feature correlations. However, MDFS relies on the accuracy of the pseudo-label matrix **F**. By compressing multiple types of information into a single subspace, the method is prone to significant information loss, making it difficult to ensure the accuracy and reliability of the pseudo-label matrix. Recently, Hu et al. [12] utilize the shared common mode between features and labels (SCMFS) for MLFS, which ignores the integrity and underlying structured information.

To address the aforementioned challenges and issues, we propose a **M**ultilabel **R**obust **F**eature **S**election algorithm based on **s**ubspace **s**parsity learning (SS-MRFS). As shown in Fig. 1, original feature information is mapped to subspaces that exhibit more transparent and comprehensible structures. Rather than assuming an identical subspace, we incorporate information about the connections between the feature and label subspaces by imposing constraints on the mapping matrix during the subspace learning process. As a result, the subspace we learn encompasses both the intrinsic information that is specific to the original space and the common information that is shared between the feature and label spaces. Simultaneously, this strategy constrains the degrees of freedom within the subspace, which helps prevent overfitting and enhances numerical stability. Furthermore, by maintaining the orthogonality of feature and label subspaces, it enhances the effectiveness of subspace learning and preserves information integrity, making it adept at capturing intricate feature-label relationships while maintaining a certain level of robustness. The primary contributions of our work are as follows.

- We propose a novel multi-label robust feature selection algorithm based on subspace sparsity learning, named SS-MRFS. This algorithm can jointly clarify the structures of both feature and label spaces using sparse subspace learning, thus improving the performance of the feature selection process.
- Our innovative subspace learning method captures both distinct connections and shared information between feature and label spaces. By imposing constraints on the mapping matrix, we enhance the effectiveness of subspace learning and ensure information integrity and robustness.
- Extensive experiments conducted on multiple benchmark databases demonstrate the superior performance of the proposed method, highlighting its ability to effectively capture intricate feature-label relationships while maintaining robustness.

2   The Proposed SS-MRFS Framework

In this section, we first present a summary of the symbols used throughout this paper and then introduce the formulation of our SS-MRFS framework, along with a detailed approach for solving the objective problem.

2.1   Notations

For an arbitrary matrix $\mathbf{A}$, $\mathbf{a}_i$ denotes the i-th row of $\mathbf{A}$, $\mathbf{a}_{ij}$ is the (i, j)-th entry, $\mathbf{A}^{-1}$ is the inverse of A, $Tr(\mathbf{A})$ represents the trace of $\mathbf{A}$ and $\mathbf{A}^T$ is the transpose of $\mathbf{A}$. Suppose that in a multi-label dataset, our input data are $\mathbf{X} = (\mathbf{x}_1, \ldots, \mathbf{x}_n)^T \in \mathbb{R}^{n \times p}$, where p is the dimensionality of the feature space. The output data are $\mathbf{Y} = (\mathbf{y}_1, \ldots, \mathbf{y}_n)^T \in \{0, 1\}^{n \times d}$, where d stands for the number of class labels. $\mathbf{y}_{ij} = 1 (j = 1, \ldots, d)$ if and only if $\mathbf{x}_i$ is associated with the j-th label. For any matrix $\mathbf{A} \in \mathbb{R}^{n \times p}$, its Frobenius norm is defined as

$\|\mathbf{A}\|_F = \sqrt{\Sigma_{i=1}^n \Sigma_{j=1}^p \mathbf{a}_{ij}^2}$, and its $\ell_{2,1}$ norm is defined as $\|\mathbf{A}\|_{2,1} = \Sigma_{i=1}^n \sqrt{\Sigma_{j=1}^p \mathbf{a}_{ij}^2}$.
For any vector $\mathbf{a}_i$, its ℓ_p-norm is defined as $(\sum_{i=1}^n |\mathbf{a}_i|^p)^{\frac{1}{p}}$, where p ranges in the range of $(0, 1]$. $\mathbf{D} = diag\{\mathbf{d}_1, \ldots, \mathbf{d_n}\}$ denotes a diagonal matrix with its diagonal elements equal to $\mathbf{d_i}$.

2.2   Problem Formulation

In the multi-label feature selection process, we aim to identify features with high discriminative power and eliminate redundant features. This concept can be seen in the subfeature uncovering with sparsity (SFUS) framework, as given by Eq. (1).

$$arg \min_{W,P,Q} \|\mathbf{XW} - \mathbf{Y}\|_{2,1} + \alpha\|\mathbf{W}\|_{2,1} + \beta\|\mathbf{W\text{-}QP}\|_F^2$$
$$s.t.\quad \mathbf{Q}^T\mathbf{Q} = \mathbf{T} \tag{1}$$

$\mathbf{W} \in \mathbb{R}^{p \times d}, \mathbf{Q} \in \mathbb{R}^{n \times c}$ and $\mathbf{P} \in \mathbb{R}^{c \times d}$, where c is the dimensionality of the subspace.

SFUS employs sparse subspace learning to find the most relevant features in the given data. The structure of the objective function allows it to leverage orthogonality constraints to aid in exploring data correlations to a certain extent. However, the information mapping process operates indirectly on the feature space through a transformation matrix, significantly diminishing the auxiliary benefits derived from the discovered subspace. Motivated by this observation, Eq. (2) is proposed.

$$arg \min_{W,P,Q} \|\mathbf{XW} - \mathbf{Y}\|_{2,1} + \alpha\|\mathbf{W}\|_{2,1} + \beta\|\mathbf{XW\text{-}QP}\|_F^2$$
$$s.t.\quad \mathbf{Q}^T\mathbf{Q} = \mathbf{I} \tag{2}$$

In the above function, the first part regulates the information of each specific label, while $\beta\|\mathbf{XW\text{-}QP}\|_F^2$ can map the feature information to a low-dimensional subspace, which is denoted as $\mathbf{Q} \in \mathbb{R}^{n \times c}$. In this case, $\mathbf{P} \in \mathbb{R}^{c \times d}$ is a mapping matrix that projects the feature subspace onto the label subspace. It can be interpreted as clustering the original p features into c different clusters that explain the dependency structure of the feature space. Furthermore, an orthogonality constraint enables it to preserve as much information as possible within the mapping process.

This is a novel approach that retains the advantages of SFUS while directly exploring the correlation structure of the feature space and mitigating the adverse effects of incorrect and incomplete annotations on feature selection. However, two issues are worthy of further consideration. First, the resulting feature subspace $\mathbf{Q}$ exhibits a high degree of freedom, making our learning process vulnerable to noise and outliers. Second, while the primary focus of Eq. (2) is to clarify the information within the feature space, the underlying redundant noise within the label space still exerts an influence on the process, which makes it difficult to explore the feature-label correlations.

To address the above limitations, we integrate a constraint term into Eq. (3):

$$arg \min_{W,P,Q,S,T} \|\mathbf{XW} - \mathbf{Y}\|_{2,1} + \alpha\|\mathbf{W}\|_{2,1} + \beta\|\mathbf{XW\text{-}QP}\|_F^2 + \gamma\|\mathbf{P\text{-}ST}\|_F^2$$
$$s.t. \quad \mathbf{Q}^T\mathbf{Q} = \mathbf{I}, \quad \mathbf{S}^T\mathbf{S} = \mathbf{I} \tag{3}$$

As mentioned in Chap. 1, by incorporating this constraint into the mapping matrix $\mathbf{P}$, we can achieve a sparse decomposition of the label space while linking the feature subspace with the label subspace. This not only aids in the identification and interpretation of the latent structure within the original label space but also enables the subspaces to interact during the learning process, thereby simultaneously leveraging unique and shared information. Furthermore, this strategy imposes limitations on the degrees of freedom, which enhance the robustness and reliability of our proposed framework for feature selection.

2.3   Solution

As we know, the $\ell_{2,1}$-norm in Eq. (3) is nonsmooth and cannot be solved in a closed form, and for an arbitrary matrix $\mathbf{A}$, $\|A\|_F^2 = \mathrm{Tr}(A^T A)$. We rewrite this in the following form [20]:

$$arg \min_{W,P,Q,S,T} Tr((\mathbf{XW} - \mathbf{Y})^T \tilde{D}(\mathbf{XW} - \mathbf{Y}))) + \alpha Tr(\mathbf{W}^T D\mathbf{W})$$
$$+\beta Tr((\mathbf{XW\text{-}QP})^T(\mathbf{XW\text{-}QP})) + \gamma Tr((\mathbf{P\text{-}ST})^T(\mathbf{P\text{-}ST})) \tag{4}$$
$$s.t. \quad \mathbf{Q}^T\mathbf{Q} = \mathbf{I}, \quad \mathbf{S}^T\mathbf{S} = \mathbf{I}$$

Denoting $\mathbf{Z} = \mathbf{XW} - \mathbf{Y}$, we have $\tilde{D} = diag\{\frac{1}{2\|\mathbf{z_1}\|_2}, \dots, \frac{1}{2\|\mathbf{z_n}\|_2}\}$ and $\tilde{D} = diag\{\frac{1}{2\|\mathbf{w_1}\|_2}, \dots, \frac{1}{2\|\mathbf{w_p}\|_2}\}$. By sequentially setting the derivatives of Eq. (4) w.r.t T and P to zero and incorporating into the above function, we have

$$arg \min_{W,Q,S} Tr((\mathbf{XW} - \mathbf{Y})^T \tilde{D}(\mathbf{XW} - \mathbf{Y}))) + \alpha Tr(\mathbf{W}^T D\mathbf{W})$$
$$+\beta \mathbf{Tr}(\mathbf{W}^T\mathbf{X}^T\mathbf{XW}) - \beta^2 Tr(\mathbf{W}^T\mathbf{X}^T\mathbf{QN}^{-1}\mathbf{Q}^T\mathbf{XW}) + Tr(\Lambda(\mathbf{Q}^T\mathbf{Q\text{-}I})) \tag{5}$$
$$s.t. \quad \mathbf{S}^T\mathbf{S} = \mathbf{I}$$

The optimization problem in Eq. (5) is not convex and is unable to find the global optimum w.r.t three variables $\mathbf{W},\mathbf{Q},\mathbf{S}$ simultaneously. Fortunately, it becomes a convex optimization problem or can find the global optimum if we fix two of the model variables and update the third variable. Therefore, we propose using an efficient alternating optimization algorithm to solve Eq. (5) until the objective function converges.

Updating $\mathbf{S}$. Although the objective function in Eq. (5) is not convex for $\mathbf{S}$ when $\mathbf{W},\mathbf{Q}$ are fixed, we can obtain the optimal solution of $\mathbf{S}$ through the following process. The problem in Eq. (5) is equivalent to the following:

$$arg \max_{S} Tr(\mathbf{W}^T\mathbf{X}^T\mathbf{QN}^{-1}\mathbf{Q}^T\mathbf{XW})$$
$$s.t. \quad \mathbf{S}^T\mathbf{S} = \mathbf{I} \tag{6}$$

According to the Sherman-Woodbury-Morrison formula, $N^{-1} = (\beta I + \gamma(\mathbf{I}\text{-}\mathbf{SS}^T))^{-1} = \frac{1}{\beta+\gamma}I + \frac{\gamma}{\beta(\beta+\gamma)}\mathbf{S}^T$, the optimization problem becomes:

$$\max_{S} Tr(\mathbf{S}^T\mathbf{Q}^T\mathbf{X}\ \mathbf{WW}^T\mathbf{X}^T\mathbf{Q}\ \mathbf{S})$$
$$s.t.\quad \mathbf{S}^T\mathbf{S}=\mathbf{I} \tag{7}$$

Denoting $A = \mathbf{W}^T\mathbf{X}^T$, Eq. (7) can be easily solved via the eigen-decomposition of $\mathbf{A}^T\mathbf{A}$.

Updating Q. Similarly, we update the variable $\mathbf{Q}$ when $\mathbf{S}$ and $\mathbf{W}$ are fixed. By removing the terms that are irrelevant to $\mathbf{Q}$, we convert the optimization problem in Eq. (5) into the following problem:

$$arg\min_{Q} -\beta^2 Tr(\mathbf{W}^T\mathbf{X}^T\mathbf{Q}\mathbf{N}^{-1}\mathbf{Q}^T\mathbf{X}\mathbf{W})+Tr(\mathbf{\Lambda}(\mathbf{Q}^T\mathbf{Q}\text{-}\mathbf{I})) \tag{8}$$

The above optimization problem is a convex optimization problem with a differentiable objective function; thus, we can apply the gradient descent method in an alternating manner to update $\mathbf{Q}$. By taking the derivative of Eq. (8) w.r.t. the variables $\mathbf{Q}$ and $\mathbf{\Lambda}$ separately, we obtain the following formulations:

$$\begin{cases} \mathbf{Q}:=\mathbf{Q} - \lambda_Q(-\beta^2\mathbf{X}\mathbf{W}\mathbf{W}^T\mathbf{X}^T\mathbf{Q}\mathbf{N}^{-1} + 2\mathbf{Q}\mathbf{\Lambda}) \\ \mathbf{\Lambda} := \mathbf{\Lambda} - \lambda_\Lambda(\mathbf{Q}^T\mathbf{Q} - \mathbf{I}) \end{cases} \tag{9}$$

where λ_Q and λ_Λ are the step sizes for the gradient descent update rules. It is essential to choose suitable step sizes to accelerate the convergence rate, while the Armijo rule [2] can help speed up this process to reduce the total running time of SS-MRFS. Thus, we employ Armijo to adaptively determine the step sizes λ_Q and λ_Λ in each iteration.

Updating W. Finally, we discuss how to update the variable $\mathbf{W}$ when $\mathbf{Q}$ and $\mathbf{S}$ are fixed. The objective function can now be reformulated as follows:

$$arg\min_{W} Tr(\mathbf{W}^T\mathbf{Z}\mathbf{W}) - 2Tr(\mathbf{W}^T\mathbf{X}^T\tilde{\mathbf{D}}\mathbf{Y}) \tag{10}$$

where $\mathbf{L} = \mathbf{X}^T(\tilde{\mathbf{D}} + \beta)\mathbf{X} + \alpha\mathbf{D} - \beta^2\mathbf{X}^T\mathbf{Q}\ \mathbf{N}^{-1}\mathbf{Q}^T\mathbf{X}$. Equation (10) is convex w.r.t. $\mathbf{W}$. Hence, we take the derivative of Eq. (10) w.r.t. $\mathbf{W}$ and set it to zero. We can obtain the closed-form solution of W as follows:

$$2\mathbf{L}\mathbf{W} - 2\mathbf{X}^T\tilde{\mathbf{D}}\mathbf{Y} = 0 \Rightarrow \mathbf{W} = \mathbf{L}^{-1}\mathbf{X}^T\tilde{\mathbf{D}}\mathbf{Y} \tag{11}$$

With the above update rules, the pseudocode of the proposed SS-MRFS framework is summarized in Algorithm 1.

3   Experiments

To validate the effectiveness of our method on real-world data, we conduct extensive experiments on various types of multi-label datasets.

Algorithm 1: : Algorithm for solving the objective function of SS-MRFS.

Input : $\mathbf{X} \in \mathbb{R}^{n \times p}, \mathbf{Y} \in \mathbb{R}^{n \times d}$;
       Parameters α, β, γ.
Output: Optimized $\mathbf{W} \in \mathbb{R}^{p \times d}$.

1 Initialize $\mathbf{W}, \mathbf{Q}$;
2 **repeat**
3   Compute $\mathbf{Z} = \mathbf{XW} - \mathbf{Y}$;
4   Compute the diagonal matrices $\tilde{D} = diag\{\frac{1}{2\|\mathbf{z_1}\|_2}, \ldots, \frac{1}{2\|\mathbf{z_n}\|_2}\}$ and
       $\tilde{D} = diag\{\frac{1}{2\|\mathbf{w_1}\|_2}, \ldots, \frac{1}{2\|\mathbf{w_p}\|_2}\}$;
5   Update $\mathbf{S}$ by Eq. (7);
6   Compute matrix $\mathbf{N}$;
7   **while** *the objective function in Eq. (8) has not converged* **do**
8   Determine the step sizes λ_Q and λ_Λ via the Armijo rule;
9   Update $\mathbf{Q}$ with Eq. (9);
10   **end**
11   Compute matrices $\mathbf{L}, \mathbf{M}$;
12   Update $\mathbf{W}$ with Eq. (11);
13 **until** *convergence*;
14 Return $\mathbf{W}$

3.1   Comparison Methods

We compare six advanced feature selection methods with SS-MRFS, including CSFS [4], SFSS [18], MIFS [13], SFUS [17], FSNM [20] and GLOCAL [28]. Afterward, the selected features are incorporated into the classic ML-KNN multi-label classification algorithm [27] to complete the process of multi-label learning and conduct comparisons among the results.

3.2   Datasets

We choose five publicly available benchmark datasets to conduct the experiments, and these datasets come from five different domains. Detailed information is presented in Table 1, in which each cardinality value represents the average number of labels per sample and each density value is the modified cardinality that can be calculated by dividing the cardinality by the number of labels. These quantities serve as indicators of the relative scale of each dataset. As we can see, two relatively small-sample datasets and three relatively large-sample datasets are included.

Table 1. Details of the five benchmark datasets

Dataset	Domain	#Training	#Test	#Features	#Labels	Cardinality	Density
Bibtex	Text	5000	2395	1836	159	2.402	0.015
CAL500	Music	450	52	68	174	26.044	0.150
Mediamill	Video	10000	1000	120	101	4.376	0.043
Yeast	Biology	1817	600	103	14	4.237	0.303
Scene	Image	1600	807	294	6	1.074	0.179

3.3 Experimental Setup and Performance Comparison

To comprehensively compare the experimental results, we select two sample-based multi-label learning evaluation metrics, i.e., the *Hamming loss* and the *Ranking Loss*, and two label-based metrics, i.e., the *macro-average* and *micro-average* metrics, to assess the performance of each method; higher values are better for the first two metrics, and while lower values are better for the other metrics. As previously mentioned, multi-label classification learning is performed using the ML-KNN algorithm. To maintain experimental fairness, the same set of parameters is utilized for each feature selection algorithm employed in ML-KNN, the default value of the nearest-neighbor parameter k equals 5, and the smoothing parameter is 1. To guarantee fairness, we tune these parameters for all methods with a grid search strategy by varying their values within the range of $\{10^{-3}, 10^{-2}, 10^{-1}, 1, 10, 10^2, 10^3\}$. For each method, we incorporate the optimal parameters and take the average of the outcomes of ten runs to ensure the representativeness of the results. Finally, we report the results obtained with feature selection proportion ranging from 2% to 30% with a step size of 2%. The results of the comparison between the proposed SS-MRFS approach and the other multi-label feature selection algorithms in terms of various metrics are shown in Fig. 2. Based on this, we arrive at the following conclusions.

- The proposed SS-MRFS method demonstrates consistently better performance than the competing approaches, especially when the selection proportion ranges from 5% to 15%. The main reason for this is that the $\ell_{2,1}$-norm regularization strategy used in Eq. (3) possesses the ability to perform sparse feature selection, which may be more effective than the Frobenius norm.
- The performance of SS-MRFS is more prominent on datasets characterized by higher label dimensions (i.e., *Bibtex, CAL500*, and *Mediamill*). This observation highlights its remarkable ability to handle high-dimensional label spaces, demonstrating that mapping the label space by incorporating constraint terms can help capture its latent structures, thereby enabling the more effective capture of the intricate relationship between the feature space and the label space.
- Our algorithm demonstrates consistent and top-ranking performance across various datasets and evaluation metrics, thereby substantiating the robust generalizability of the proposed approach.

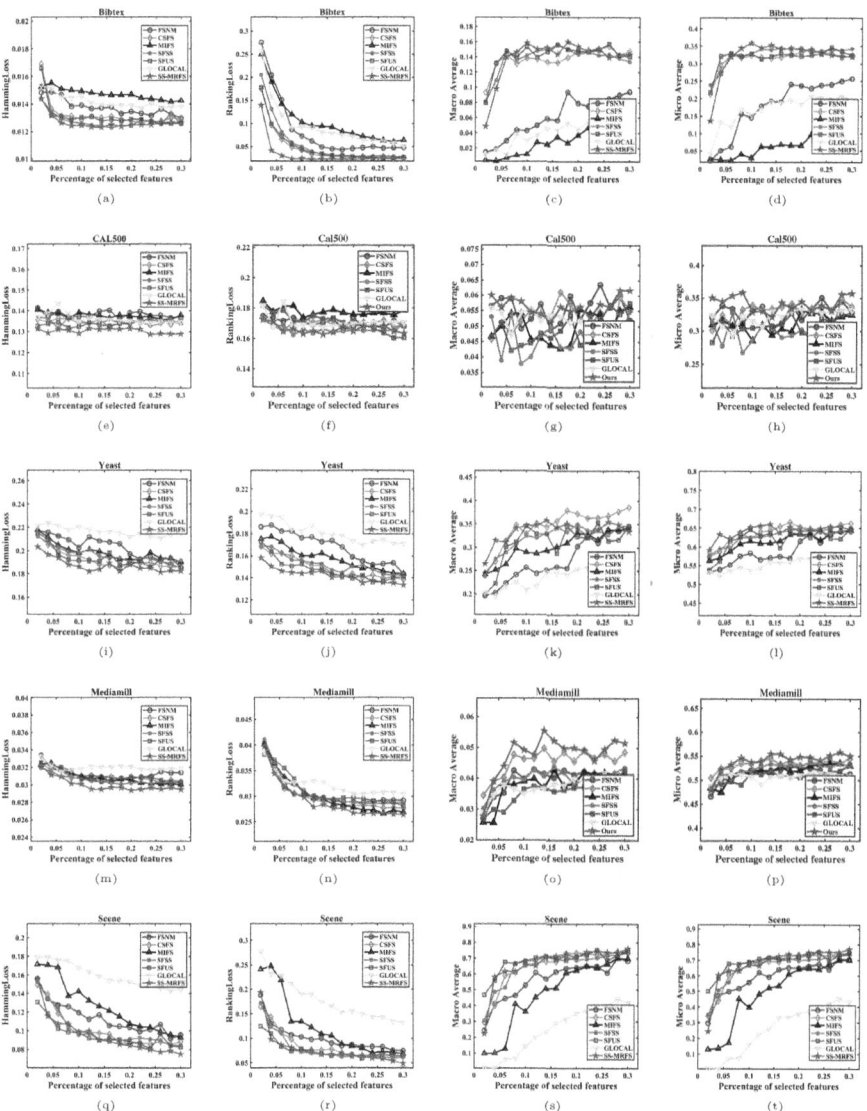

Fig. 2. The results of a comparison between SS-MRFS and other multi-label feature selection algorithms, including CSFS, SFSS, MIFS, SFUS, FSNM and GLOCAL. The Hamming loss, ranking loss, macro average and micro average are the four selected metrics.

3.4 Convergence Analysis

To demonstrate the efficiency of our proposed SS-MRFS method, we conduct an empirical experiment with the parameters fixed as $\alpha = 10$, $\beta = 4$, and $\gamma = 0.1$. We draw the produced convergence curves in Fig. 3, using the *Mediamill* and

Yeast datasets as examples. These datasets are characterized by having a large sample size and a relatively large scale, respectively. The following stopping criterion is used:

$$\frac{|\theta^t - \theta^{t-1}|}{\theta^{t-1}} < 10^{-7} \qquad (12)$$

where θ^t indicates the value of the objective function in the t-th iteration.

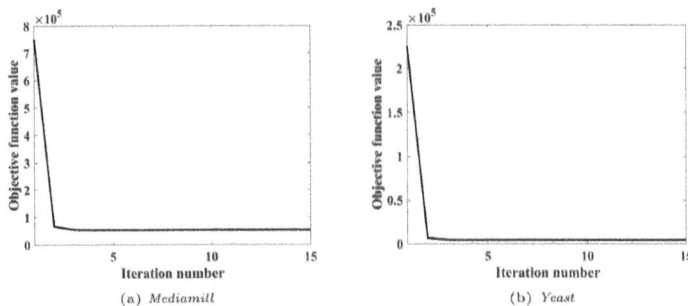

(a) *Mediamill*   (b) *Yeast*

Fig. 3. Convergence curves produced on the *Mediamill* and *Yeast* datasets.

Figure 3 shows that the objective function in Eq. (3) converges within 15 iterations on both datasets. It is worth noting that when setting the convergence criterion to be less than 10^{-5}, convergence is achieved within 5 iterations. Similar conclusions can be drawn on other datasets.

3.5   Parameter Sensitivity Analysis

SS-MRFS consists of three important parameters: α, β, and γ. The α parameter regulates the penalty strength for feature selection. The β and γ parameters control the preservation levels of the informational structures in the original input and output spaces, respectively, during the low-dimensional sparse subspace learning process.

To study the parameter sensitivity of SS-MRFS, we tune α and β with a fixed γ. Due to space limitations, we only present the results obtained on the *CAL500* dataset with a 30% feature selection proportion using another example-based metric (*average precision*) for a more persuasive demonstration. Figure 4 shows that the classification performance is not very sensitive to parameter changes, and the same conclusion holds true for the other four datasets.

3.6   Subspace Efficiency

To further assess the performance of SS-MRFS in the efficiency of subspace learning for label relevance, we compare the correlations in the predicted label space, label subspace and original label space in this section. To ensure the

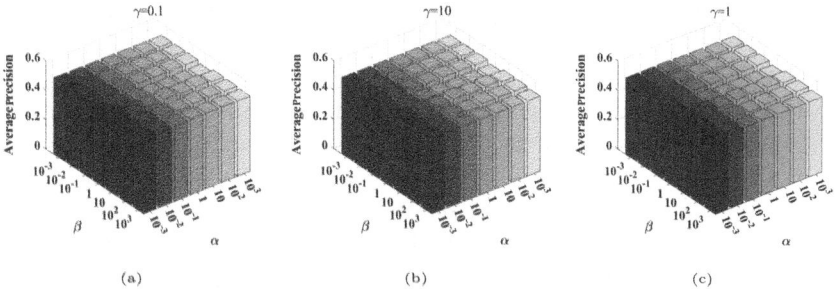

Fig. 4. Average precision rates achieved by SS-MRFS on the *CAL500* dataset with different α, β, and γ values.

accuracy of feature selection and prediction, it is imperative that the predicted label space maintains a high level of consistency with the original label space. Additionally, the label subspace should be designed to learn and amalgamate strongly correlated labels, thereby enhancing the model's ability to extract the underlying label structure more effectively.

Figure 5 shows correlations in the predicted label space $\mathbf{XW}$, original label space $\mathbf{Y}$ and label subspace $\mathbf{QS}$ on the Mediamill dataset calculated by cosine similarity. From the figure, we observe that label correlations extracted from $\mathbf{XW}$ and $\mathbf{Y}$ exhibit similar characters, which reveals the correlation structure learned from predicted label space and original label space is preserved consistent by SS-MRFS. The figure also illustrate that the correlations within the label subspace $\mathbf{QS}$ are notably weak, diverging from the correlation characteristics observed between $\mathbf{XW}$ and $\mathbf{Y}$. This underscores the efficacy of SS-MRFS in learning the label correlation structure. The above techniques promises capturing appropriate label correlations to guide feature selection and predict unknown labels.

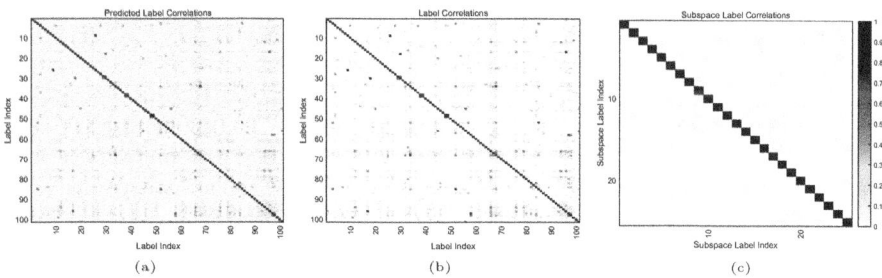

Fig. 5. Multi-label data correlations (calculated by cosine similarity) in the predicted label space original label space and label subspace constructed by SS-MRFS with $\alpha = 0.1$, $\beta = 0.1$, $\gamma = 10^3$.

3.7   Comparison Experiment

Nie et al. [20] have proved that $\ell_{2,1}$-norm based models can simultaneously exhibiting sparsity and robustness. However, many studies have demonstrated that ℓ_p-norm $(0<p<1)$ can detect sparser solution than ℓ_1-norm and Sheikhpour et al. [22] use $\ell_{2,p}$-norm $(0<p<1)$ instead of $\ell_{2,1}$-norm to conduct feature selection. Therefore, experimental evaluation is required to determine the suitable p $(0<p \leq 1)$ for the proposed SS-MRFS. We extend $\ell_{2,1}$-norm to $\ell_{2,p}$-norm $(0<p \leq 1)$ then solved and applied it to experiments on multiple datasets, where parameter P is tuned in the range of $\{0.25, 0.5, 0.75, 1\}$ and the α, β, γ parameters use a grid search strategy by varying their values within the range of $\{10^{-3}, 10^{-2}, 10^{-1}, 1, 10, 10^2, 10^3\}$. For each p, we take the optimal parameters and report the results in Tabel 2 with 30% feature selection proportion. Due to the space limitation, we only show the results of one dataset.

Table 2. The results of SS-MRFS($\ell_{2,p}$) on Scene datasets

Method	Hamming loss	Ranking loss	macro-average	micro-average
SS-MRFS ($\ell_{2,0.25}$)	0.094	0.069	0.692	0.697
SS-MRFS ($\ell_{2,0.5}$)	0.099	0.072	0.688	0.692
SS-MRFS ($\ell_{2,0.75}$)	0.098	0.077	0.685	0.687
SS-MRFS ($\ell_{2,1}$)	**0.085**	**0.059**	**0.743**	**0.742**

As observed from Tabel 2, the proposed SS-MRFS using $\ell_{2,1}$-norm significantly have better performance than other SS-MRFS ($\ell_{2,p}$) method with p tune in the range of $\{0.25, 0.5, 0.75, 1\}$. Thereby using $\ell_{2,1}$-norm is appropriate for the proposed SS-MRFS method.

3.8   Interpretability

To verify the accuracy of data structure capture by SS-MRFS, an assessment was conducted from the perspective of interpretability. We implement SS-MRFS on Yeast data set and analyse the relationship between features and labels. The Yeast data set contains 14 labels and 103 features, due to space constrains, only part of them are shown in Fig. 6.

From Fig. 6 we can conclude that the labels in label pairs (L_1, L_2), (L_7, L_8), (L_{10}, L_{11}) and (L_{12}, L_{13}) have the similar feature structure, and therefore stronger correlations can be seen within them than other pairs of label. Also, from the distribution of the shaded cell, we observe that feature F_{29} and F_{44} carry the information of similar labels, and F_{94} is a label specific feature of label L_1. Above conclusions are consistent with results obtained by SFUS [17].

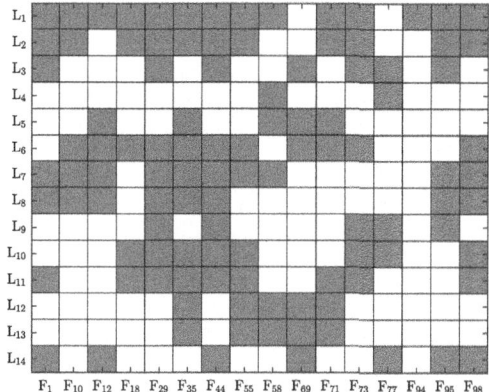

Fig. 6. The relation between each selected feature and each selected label in Yeast, where a feature corresponding to a column and a label corresponding to a row. The shaded cells represent the corresponding features have effect on the corresponding labels.

3.9   Significance Analysis

We examine the performance of the feature subspace by setting the parameters as $\beta = 0$ and $\beta > 0$, and we evaluate the performance of the label subspace and the between-subspace interaction obtained by imposing constraints on the mapping matrix $\mathbf{P}$ in SS-MRFS by setting the parameters as $\gamma = 0$ and $\gamma > 0$ in the experiment. Due to space limitations, we only present the results obtained on the *CAL500* and *Yeast* datasets using the average precision metric.

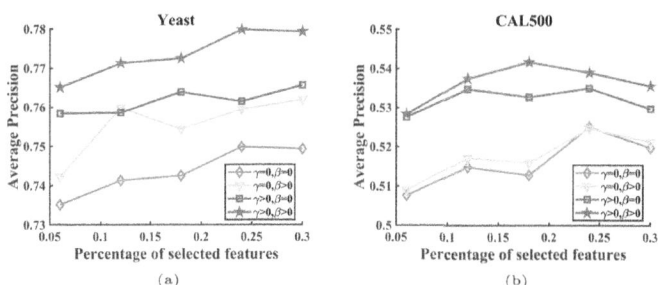

Fig. 7. Average precision achieved by SS-MRFS on the *CAL500* and *Yeast* datasets with different parameter sets.

Figure 7 illustrates that when $\beta = 0$ or $\gamma = 0$, the average precision is lower than that attained when $\beta > 0$ and $\gamma > 0$, supporting the effectiveness of the feature subspace, the label subspace and the interaction between them. Particularly, when $\beta = 0$ and $\gamma = 0$, the average precision drops to its lowest value, which also underscores the necessity of constructing the subspaces.

4 Conclusion

In this paper, we propose a novel multi-label robust feature selection algorithm based on subspace sparsity learning (SS-MRFS). The proposed method has two appealing properties. First, it utilizes subspace sparsity to jointly clarify the structures of both the feature and label spaces while sufficiently leveraging abundant information. Second, it imposes limitations on the degrees of freedom of each subspace, which enhance the robustness and reliability of our proposed feature selection framework. We conduct extensive experiments on various real-world datasets, demonstrating the efficiency and efficacy of the proposed framework. Therefore, we conclude that our method is a robust and efficient feature selection method that can be used in multi-labeled learning processes.

References

1. Agrawal, R., Gupta, A., Prabhu, Y., Varma, M.: Multi-label learning with millions of labels: recommending advertiser bid phrases for web pages. In: Proceedings of the 22nd International Conference on World Wide Web, pp. 13–24 (2013)
2. Bertsekas, D.P.: Nonlinear programming. J. Oper. Res. Soc. **48**(3), 334 (1997)
3. Braytee, A., Liu, W., Catchpoole, D.R., Kennedy, P.J.: Multi-label feature selection using correlation information. In: Proceedings of the 2017 ACM on Conference on Information and Knowledge Management, pp. 1649–1656 (2017)
4. Chang, X., Nie, F., Yang, Y., Huang, H.: A convex formulation for semi-supervised multi-label feature selection. In: Proceedings of the AAAI Conference on Artificial Intelligence, vol. 28 (2014)
5. Dai, J., Huang, W., Zhang, C., Liu, J.: Multi-label feature selection by strongly relevant label gain and label mutual aid. Pattern Recogn. **145**, 109945 (2024)
6. Fan, Y., Chen, B., Huang, W., Liu, J., Weng, W., Lan, W.: Multi-label feature selection based on label correlations and feature redundancy. Knowl. Based Syst. **241**, 108256 (2022)
7. Fan, Y., Liu, J., Tang, J., Liu, P., Lin, Y., Du, Y.: Learning correlation information for multi-label feature selection. Pattern Recogn. **145**, 109899 (2024)
8. Faraji, M., Seyedi, S.A., Tab, F.A., Mahmoodi, R.: Multi-label feature selection with global and local label correlation. Expert Syst. Appl. **246**, 123198 (2024)
9. He, K., Zhang, X., Ren, S., Sun, J.: Deep residual learning for image recognition. In: Proceedings of the IEEE Conference on Computer Vision and Pattern Recognition, pp. 770–778 (2016)
10. He, Z., Lin, Y., Wang, C., Guo, L., Ding, W.: Multi-label feature selection based on correlation label enhancement. Inf. Sci. **647**, 119526 (2023)
11. Hu, J., Li, Y., Gao, W., Zhang, P.: Robust multi-label feature selection with dual-graph regularization. Knowl. Based Syst. **203**, 106126 (2020)
12. Hu, L., Li, Y., Gao, W., Zhang, P., Hu, J.: Multi-label feature selection with shared common mode. Pattern Recogn. **104**, 107344 (2020)
13. Jian, L., Li, J., Shu, K., Liu, H.: Multi-label informed feature selection. In: IJCAI, vol. 16, pp. 1627–33 (2016)
14. Kim, Y.: Convolutional neural networks for sentence classification. arXiv preprint arXiv:1408.5882 (2014)

15. Li, X., Zhang, H., Zhang, R., Liu, Y., Nie, F.: Generalized uncorrelated regression with adaptive graph for unsupervised feature selection. IEEE Trans. Neural Netw. Learn. Syst. **30**(5), 1587–1595 (2018)
16. Li, Z., Liu, J., Yang, Y., Zhou, X., Lu, H.: Clustering-guided sparse structural learning for unsupervised feature selection. IEEE Trans. Knowl. Data Eng. **26**(9), 2138–2150 (2013)
17. Ma, Z., Nie, F., Yang, Y., Uijlings, J.R., Sebe, N.: Web image annotation via subspace-sparsity collaborated feature selection. IEEE Trans. Multimedia **14**(4), 1021–1030 (2012)
18. Ma, Z., Nie, F., Yang, Y., Uijlings, J.R., Sebe, N., Hauptmann, A.G.: Discriminating joint feature analysis for multimedia data understanding. IEEE Trans. Multimedia **14**(6), 1662–1672 (2012)
19. Mayr, A., Klambauer, G., Unterthiner, T., Hochreiter, S.: DeepTox: toxicity prediction using deep learning. Front. Environ. Sci. **3**, 80 (2016)
20. Nie, F., Huang, H., Cai, X., Ding, C.: Efficient and robust feature selection via joint 2, 1-norms minimization. In: Advances in Neural Information Processing Systems, vol. 23 (2010)
21. Pereira, R.B., Plastino, A., Zadrozny, B., Merschmann, L.H.: Categorizing feature selection methods for multi-label classification. Artif. Intell. Rev. **49**, 57–78 (2018)
22. Sheikhpour, R., Sarram, M.A., Gharaghani, S., Chahooki, M.A.Z.: A robust graph-based semi-supervised sparse feature selection method. Inf. Sci. **531**, 13–30 (2020)
23. Tan, Z., Wang, M., Xie, J., Chen, Y., Shi, X.: Deep semantic role labeling with self-attention. In: Proceedings of the AAAI Conference on Artificial Intelligence, vol. 32 (2018)
24. Uricchio, T., Ballan, L., Seidenari, L., Del Bimbo, A.: Automatic image annotation via label transfer in the semantic space. Pattern Recogn. **71**, 144–157 (2017)
25. Yang, Z., Yang, D., Dyer, C., He, X., Smola, A., Hovy, E.: Hierarchical attention networks for document classification. In: Proceedings of the 2016 Conference of the North American Chapter of the Association for Computational Linguistics: Human Language Technologies, pp. 1480–1489 (2016)
26. Zhang, J., Luo, Z., Li, C., Zhou, C., Li, S.: Manifold regularized discriminative feature selection for multi-label learning. Pattern Recogn. **95**, 136–150 (2019)
27. Zhang, M.L., Zhou, Z.H.: ML-KNN: a lazy learning approach to multi-label learning. Pattern Recogn. **40**(7), 2038–2048 (2007)
28. Zhu, Y., Kwok, J.T., Zhou, Z.H.: Multi-label learning with global and local label correlation. IEEE Trans. Knowl. Data Eng. **30**(6), 1081–1094 (2017)

Nullspace-Based Metric for Classification of Dynamical Systems and Sensors

Dominique Martinez[1]([✉]) and Mohamed Boutayeb[2]

[1] Aix-Marseille Université, CNRS, ISM, 13009 Marseille, France
`dominique.martinez@univ-amu.fr`
[2] Lorraine Université, CNRS, INRIA, CRAN, 54506 Nancy, France
`mohamed.boutayeb@univ-lorraine.fr`

Abstract. A growing field in machine learning is to develop computationally friendly algorithms optimized for classification of sensor data at the edge. As many sensors respond to different stimuli with different dynamics, the sensor dynamics may provide valuable information for classification. The problem of determining which of the dynamics has generated a particular observation vector is refered to as a multiclass discrimination problem. A discriminative metric can be computed in a Kronecker space provided the different dynamics are represented with state-space models having distinct observability matrices. In this paper, we express the metric in original space rather than in Kronecker space. One of the main contributions is to show theoretically that the metric for a given system is provided by the norm of the matrix-vector product between the left nullspace of the observability matrix and the observation vector. In practice, when the system is unknown and noisy, an "approximate" nullspace is obtained with a data-driven approach using eigenvalue or singular value decomposition. We tested the classifier on synthetic and real datasets. Results demonstrate the applicability of the method. The new formulation of the metric in original space leads to a classifier with reduced complexity so that learning can be implemented with a few lines of code. This low complexity is suitable for machine learning on low-cost, low-power, and resource-constrained devices like smartphones and sensors.

Keywords: Classification · Dynamical systems · machine Learning · State-space models · Nullspace · Observability · Time series · Edge computing · TinyML

1 Introduction

With the multiplication of sensors and IoT devices in our daily life, there is a growing interest for lightweight machine learning (ML) or tiny ML capable of performing sensor data classification at the edge, thus avoiding the need to transfer data over long distances [1]. Deep learning [2] is not the ideal candidate for classification on embedded systems as it requires a large number of training

© The Author(s), under exclusive license to Springer Nature Switzerland AG 2024
M. Wand et al. (Eds.): ICANN 2024, LNCS 15016, pp. 18–29, 2024.
https://doi.org/10.1007/978-3-031-72332-2_2

data combined with lots of computational and memory resources to train and store a huge number of parameters. In contrast, traditional ML is less expensive in terms of memory and computation. This is particularily the case for distance-based classifiers, such as k-nearest neighbors [3], k-means [5] or Parzen windows [4]), which are the simplest classification algorithms. The classification rule during the inference process is based on the dissimilarity, as measured by a distance function, between input samples and prototypes in the training set. Distance-based methods have been developped originally for static data using the Euclidean or Mahalanobis distance. Yet, sensor data are usually not static but evolves in time and the sensor dynamics may provide valuable information for classification. Chemical sensors, for example, react with different dynamics to different odorants and the sensor dynamics can be exploited to discriminate among odorants [6].

Here, one considers that sensor data from a given class are generated by some underlying dynamical system represented with a state-space model. State-space modeling is an effective tool to represent dynamical systems. A variety of artificial and natural systems have been described precisely by state-space models, e.g. the control of an aircraft that depends on the flight conditions [7], the response of gas sensors to different chemical stimuli [8] or the behaviour of animals relative to their environmental conditions [9]. In a state-space model, it is assumed that the state of the system evolves in time but cannot be observed directly. There are two equations. One is the transition equation which predicts the future state given the current state and the other is the measurement equation which provides the observations conditional on the current state. The system is said to be observable in a control-theoric sense if all its states can be retrieved from the sensor data, a condition that is satisfied when the observability matrix has full rank.

Recently, it has been shown that a classifier can be designed to discriminate between different state-space models from sensor data, even when the systems are not observable, provided their observability matrices are all different [10]. We also proposed an appropriate metric to quantify whether sensor data result from a particular state-space model or class. Yet, the distance is computed in a high-dimensional Kronecker space with quadratic complexity which may prevent its use for embedded classification on small devices. In this paper, we revisited the minimum distance classifier for dynamical systems to limit complexity. We show theoretically that the distance can be written in original space as a sum of dot products with a set of weight vectors that corresponds to the left nullspace of the observability matrix. This new formulation leads to a distance-based classifier with reduced learning complexity which makes it suitable for machine learning on embedded devices.

The rest of the paper is organized as follows. The classification problem is specified and described in Sect.2 for dynamical systems. Classifier inference, that is distance computation in original space, is presented in Sect. 3 with main result provided by Theorem 2. Classifier learning using eigenvalue or singular value decomposition in case of unknown dynamical systems is presented Sect. 4.

To demonstrate its simplicity, we implemented the learning process as one line of code in matlab (see Note 3). Results obtained on synthetic and real datasets are presented in Sect. 5. Finally, the main conclusions are presented in Sect. 6.

2   Problem Statement

Throughout this paper, the following notation is used. All vectors are column vectors unless transposed to a row vector by a prime notation $'$. We consider a dynamical system represented as the following state-space model:

$$x(k + 1) = Ax(k)$$
$$y(k) = Cx(k) \tag{1}$$

with state $x \in \mathbb{R}^{n \times 1}$, output $y \in \mathbb{R}^{m \times 1}$, transition matrix $A \in \mathbb{R}^{n \times n}$ and output matrix $C \in \mathbb{R}^{m \times n}$.

Let us further consider a vector of N consecutive observations written as $Y \in \mathbb{R}^{l \times 1}$ with $l = Nm$:

$$Y = \begin{pmatrix} y(k) \\ y(k + 1) \\ \dots \\ y(k + N - 1) \end{pmatrix} \tag{2}$$

The problem is to find an appropriate distance $d(Y)$ between Y and system (1) so that it can be applied for the classification of dynamical systems. For example, in one-class classification, (1) would describe the normal operating mode of the system. Unusual events (e.g. novelty, anomaly or fault events) would be detected whenever $d(Y)$ exceeds a predetermined threshold. In multi-class classification, one has different dynamical models (1) and the classifier would assign the observation vector Y to the one with the smallest distance.

3   Inference (Distance Computation)

If Y was generated by system (1) then we have:

$$Y = O_b \, x(k) \tag{3}$$

with O_b the observability matrix defined as:

$$O_b = \begin{pmatrix} C \\ CA \\ \dots \\ CA^{N-1} \end{pmatrix} \tag{4}$$

Even if system (1) is not observable in a control-theoric sense (i.e. O_b is not full rank), $x(k)$ can be determined from (3) by using the pseudo-inverse of O_b, denoted O_b^+. The entire set of solutions is represented by:

$$x(k) = O_b^+ Y + (I - O_b^+ O_b)\Gamma \tag{5}$$

with the arbitary vector Γ varying over all possible values [14].

The distance between any observed sequence Y and system (1) writes as follows [10]:

$$
\begin{aligned}
d^2(Y) &= ||Y - O_b x(k)||^2 \\
&= ||Y - O_b O_b^+ Y - (O_b - O_b O_b^+ O_b)\Gamma||^2 \text{ (from 5)} \\
&= ||Y - O_b O_b^+ Y||^2 \text{ (as } O_b O_b^+ O_b = O_b) \\
&= [(I - Q)Y]'[(I - Q)Y] \text{ (with } Q \triangleq O_b O_b^+) \\
&= Y'(I - Q' - Q + Q'Q)Y \\
&= Y'(I - Q)Y \text{ (as } Q = Q' \text{ and } Q'Q = Q)
\end{aligned}
$$

Thus the distance between any observed sequence Y and system (1) writes

$$d^2(Y) = Y'MY \tag{6}$$

With $M = (I - O_b O_b^+) \in \mathbb{R}^{l \times l}$. The matrix equation (6) is reshaped into a vector equation by using the vectorization operator $vec()$ and Kronecker product $\otimes$.

Theorem 1. Distance in Kronecker space.

$$d^2(Y) = Y'MY \Leftrightarrow d^2(Y) = w'(Y \otimes Y)$$
$$\text{with } w = vec(M) \in \mathbb{R}^{l^2 \times 1}$$

Proof.

$$
\begin{aligned}
d^2(Y) &= Y'MY \\
&= vec(Y'MY) \\
&= (Y' \otimes Y')vec(M) \\
&= vec(M)'(Y \otimes Y) \\
&= w'(Y \otimes Y) \tag{7}
\end{aligned}
$$

Note 1. In Matlab, the vec operation can be computed as **w=M(:)** and the Kronecker product is implemented as **kron(Y,Y)**.

Although (7) is a simple scalar product, the Kronecker product requires the computation of l^2 products between all pairs of the l components of Y. Yet, Theorem 2 allows us to limit complexity by computing the distance in original space rather than in Kronecker space.

Theorem 2. Distance in Original space

$$d^2(Y) = Y'MY \Leftrightarrow d(Y) = ||V'Y||$$
$$\text{with } V = Null(O'_b) \in \mathbb{R}^{l \times k}$$

where Null denotes the nullspace and $k = l - rank(O_b)$.

Proof. To avoid computing the distance in Kronecker space, w is rewriten as a sum of Kronecker products:

$$w = vec(M) = \sum_{i=1}^{k} v_i \otimes v_i$$

with weight vectors $v_i \in \mathbb{R}^{l \times 1}, i = 1 \cdots k$.
The distance then becomes as follow

$$d^2(Y) = w'(Y \otimes Y)$$
$$= \sum_{i=1}^{k} (v'_i \otimes v'_i)(Y \otimes Y)$$
$$= \sum_{i=1}^{k} (v'_i Y) \otimes (v'_i Y)$$
$$= \sum_{i=1}^{k} (v'_i Y)^2$$
$$= ||V'Y||^2 \text{ with } V = (v_1 \cdots v_k) \tag{8}$$

For Y generated by system (1), $d^2(Y) = 0$ leads to

$$\sum_{i=1}^{k} (v'_i Y)^2 = 0$$
$$\Leftrightarrow v'_i Y = 0 \ \forall i$$
$$\Leftrightarrow Y' v_i = 0 \ \forall i$$
$$\Leftrightarrow x'(k) O'_b v_i = 0 \ \forall i \ \forall x(k) \neq 0$$
$$\Leftrightarrow O'_b v_i = 0 \ \forall i$$
$$\Leftrightarrow v_i \in Null(O'_b)$$
$$\text{for } i = 1 \cdots k = l - rank(O_b)$$

Note 2. The distance is now expressed as the norm of a matrix-vector product in the original space Y, as stated by (8), and not in the Kronecker space as in (7). In Matlab, it can be implemented simply as **d=norm($V'*Y$)** with **V=Null(O'_b)**

4   Learning (Approximate Nullspace)

When computing the distance in original space, the weight vectors v_i's can be obtained from the left nullspace of the observability matrix (Theorem 2). A prerequisite, however, is that the dynamical system is known exactly. In case it is unknown or partially known, the v_i's have to be estimated from training data. Let $Y_{train} = (Y^1, \cdots, Y^p) \in \mathbb{R}^{l \times p}$ be the training set corresponding to the concatenation of p observation vectors Y^μ, $\mu = 1 \cdots p$, for system (1). Ideally, $d(Y^\mu)$ should be zero for all μ so that $Y'_{train} v_i$ is a null vector for all i. Thus, v_i belongs to the nullspace of Y'_{train}. In practice, however, the observations are corrupted by noise so that the nullspace of Y'_{train} does not necessarily exist. We therefore look for an "approximate" nullspace by considering the span of the left singular vectors associated with the k smallest singular values, with $k = l - rank(O_b)$. The solution can be obtained through singular value decomposition (SVD) or eigenvalue decomposition as the left singular vectors of Y_{train} correspond to the eigenvectors of the sample autocorrelation matrix $Y_{train} Y'_{train} = \sum_{\mu=1}^{p} Y^\mu (Y^\mu)'$. There are a number of algorithms that can be apply to extract the k minor components of the sample autocorrelation matrix [11], including online algorithms to update the v_i's as new observations become available [12].

Note 3. In Matlab, learning can be implemented as one line of code : $[\mathbf{V}, \sim] = $ **eigs**$(Y_{train} * Y'_{train}, \mathbf{k}, \text{'smallestabs'})$ or $[\mathbf{V}, \sim, \sim] = $**svds**$(Y_{train}, \mathbf{k}, \text{'smallest'})$.
   Alternatively, one might consider all data for training and not only those of the class under consideration. Let $\overline{Y}_{train}$ the training data of the other classes. Ideally, $d(Y)$ should be small for Y_{train} and large for $\overline{Y}_{train}$. This strategy leads to minimizing the Rayleigh quotient defined as $||V'Y_{train}||^2 / ||V'\overline{Y}_{train}||^2$. The solution is obtained by solving a generalized eigenvalue problem with matrices $(Y_{train} Y'_{train})$ and $(\overline{Y}_{train} \overline{Y}'_{train})$ in which V is the set of the k generalized eigenvectors that corresponds to the smallest eigenvalues.

5   Experiments

The nullspace-based classifier is tested on synthetic and real datasets. We compare the data-driven approach using SVD learning, as in Note 3, to the model based approach using system identification or a priori knowledge. When the exact knowledge of the system is available, the ground truth is the best possible classifier using the left nullspace of the observability matrix as in Note 2.

5.1   XOR Problem of Dynamical Systems

In this illustrative example, one considers the dynamical version of the XOR classification problem. Let $x(k+1) = x(k)$ (class 1) and $x(k+1) = -x(k)$ (class 2) be the two dynamical systems depicted in Fig. 1, with $y(k) = x(k)$ for the two classes. We consider observation vectors with $N = 2$, i.e. $Y = (y(k)\ y(k+1))'$, so that the observability matrices are $O_1 = (1\ 1)'$ (class 1) and $O_2 = (1\ -1)'$ (class 2). We further consider $p = 2$ training samples for each class such that:

$$Y_{train,1} = \begin{pmatrix} -1 & 1 \\ -1 & 1 \end{pmatrix} \qquad Y_{train,2} = \begin{pmatrix} -1 & 1 \\ 1 & -1 \end{pmatrix}$$

As seen in Fig. 1, the classification problem is not linearily separable, i.e. $Y_{train,1}$ and $Y_{train,2}$ cannot be separated with a straight line. The nullspaces of O'_1 and O'_2 resume to single vectors V_1 and V_2. Without observation noise, the data-driven approach leads to the same solution as the ground truth, that is $V_1 = (-0.7071\ 0.7071)'$ and $V_2 = (0.7071\ 0.7071)'$, which correspond to the crosslines in Fig. 1 (left). In the presence of noise, the solution obtained with $p = 10$ training samples is depicted in Fig. 1 (right).

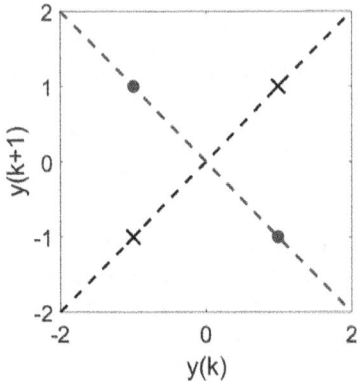 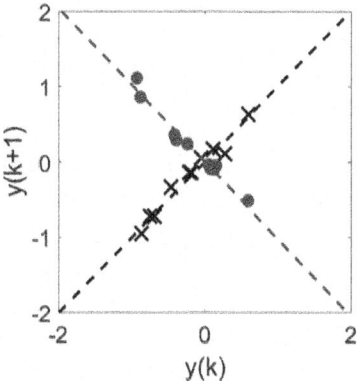

Fig. 1. Nullspace-based classification on the XOR example without noise (left) and with noise (right). The solution obtained with SVD (data-driven approach) is represented by the crosslines. The training data are depicted as crosses for class 1 and circles for class 2.

5.2   Aircraft Simulation

The aircraft simulation is derived from [7] and has been used in [10,13]. The dynamical equation for class $i = 1 \cdots 4$ which predicts the future state $x_i(k+1)$ given the current state $x_i(k)$ and input $u_i(k)$ is given by

$$x_i(k+1) = \mathcal{A}_i x_i(k) + \mathcal{B}_i u_i(k)$$

and

$$\mathcal{A}_i = \begin{pmatrix} a_{i,11} & a_{i,12} & a_{i,13} \\ a_{i,21} & a_{i,22} & a_{i,23} \\ 0 & 0 & a_{i,33} \end{pmatrix} \qquad \mathcal{B}_i = (b_{i,1}\ b_{i,2}\ b_{i,3})'$$

The entries of matrices $\mathcal{A}_i$ and $\mathcal{B}_i$ depend on the flight condition (speed and altitude). They are given in Table 1. They have been obtained by mapping the continuous-time models in [7] to discrete-time models using zero-order hold discretization and sampling time $dt = 0.1$ s. The state x_i has three component: normal acceleration, pitch rate and pitch angle. The input signal u_i is the input to the elevon servo. As the system is unstable for subsonic speeds, a linear quadratic controller is designed using Matlab function *dlqr.m* to improve stability. The feedback controller is written as $u_i(k) = K_i x_i(k)$ with $K_i = (k_{i,1}\ k_{i,2}\ k_{i,3})$ given in Table 1. Thus, the system rewrites as in Eq. 1, i.e. $x_i(k+1) = A_i\ x_i(k)$ with $A_i = \mathcal{A}_i + \mathcal{B}_i K_i$. The problem is to build a classifier that discriminates the different classes $(i = 1 \cdots 4)$ from noisy observations $y_i(k) = C_i x_i(k) + \mu$ (μ is a zero-mean gaussian noise).

Table 1. Parameters of the airplane state model for different flight modes (speed and altitude conditions)

Class	$i = 1$	$i = 2$	$i = 3$	$i = 4$
Mach	0.5	0.85	0.9	1.5
Altitude	5000	5000	35000	35000
a_{11}	0.9268	0.8915	0.9424	0.8648
a_{12}	1.6002	4.4207	1.6990	2.3959
a_{13}	4.3075	8.28	3.7561	6.4622
a_{21}	0.0243	0.0192	0.0077	−0.0613
a_{22}	0.9396	0.9162	0.9432	0.8018
a_{23}	−0.5090	−1.4456	−0.5389	−1.8297
a_{33}	0.2466	0.2466	0.2466	0.2466
b_1	−5.4001	−16.4777	−4.7666	−9.9568
b_2	−0.5931	−1.5991	−0.5101	−0.9536
b_3	0.7534	0.7534	0.7534	0.7534
k_1	0.1556	0.0524	0.1752	0.0788
k_2	0.3710	0.2767	0.4547	0.2664
k_3	0.6478	0.4530	0.6109	0.5204

We trained the nullspace-based classifier using SVD on $p = 500$ training data for each system, where each training data consists in $N = 4$ consecutive observations. The error rate is estimated over 1000 test data generated the same way as the training data. As seen in Fig. 2, the data driven approach leads to performance similar to the ground truth using the entire nullspace of the observation matrix and to support vector machines (SVMs) [15]. Moreover, the data driven approach outperforms a classifier that uses a single vector of the nullspace, and a model-based classifier built from system identification.

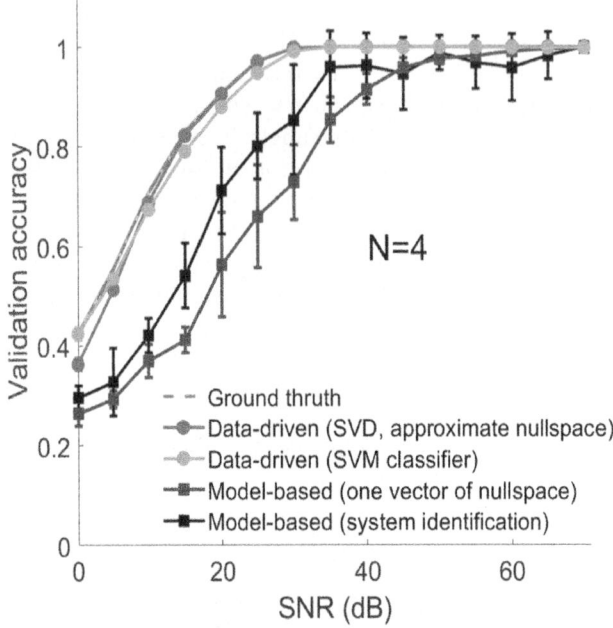

Fig. 2. Classification on the aircraft example. The observation vector consists in $N = 4$ consecutive sensor data. Noise is added to the sensor data. Classifier accuracy is represented versus signal-to-noise ratio (SNR) as mean±S.D estimated over 100 trials. The nullspace-based classifier is built from training data using SVD (note 3). Performance is compared to an SVM classifier and to model-based classifiers (Theorem 2) built from system identification or from the exact knowledge of the system (using the entire nullspace or a single vector). The ground truth (best possible classifier) uses the entire nullspace derived from the exact knowledge of the system.

In theory, the nullspace classifier can be built when the systems are not observable, provided their observability matrices are all different [10]. This condition is statisfied for $N \geq 2$ as the pair (A, C) in (1) is different for different systems. To check the validity of this theoretical condition, we trained the nullspace-based classifier with training data that consist of only $N = 2$ consecutive observations, i.e. $Y = (y(k), y(k+1))'$. As seen in Fig. 3, for $N = 2$, the nullspace-based classifier built using SVD has similar performance to the ground truth and outperforms all other classifiers. It is also worth noting that the system identification approach fails as $N = 2$ is not enough to identify the systems whose state dimension is 3.

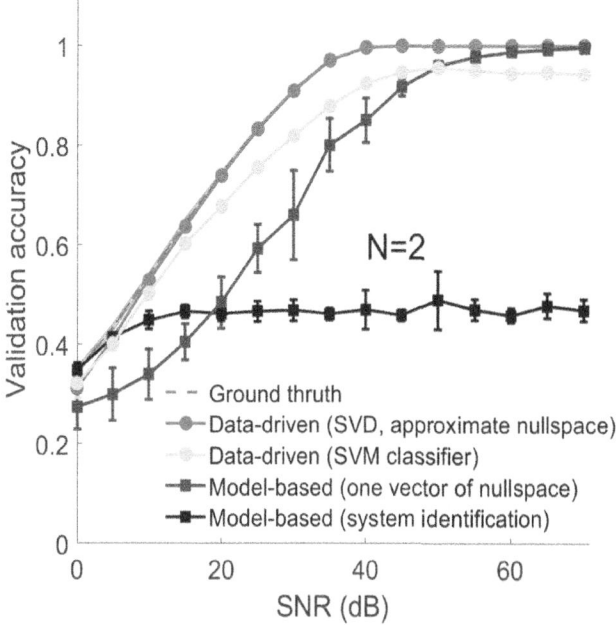

Fig. 3. Nullspace-based classification on the Aircraft example. Same conventions as in Fig. 2 except that the observation vector consists in $N = 2$ consecutive sensor data.

5.3 Handwriting Character Recognition from Pen Trajectory

We consider a real dataset from the UCI Machine learning archive[1]. The dataset consists of 2858 handwriting characters (20 classes 'a,b,c,d,e,g,h,l, m,n,o,p,q,r,s,u,v,w,y,z') collected using a WACOM pen tablet. The sensor data consists in pen trajectories with three components recorded at 200 Hz: pen velocity in x-position (mm/s), pen velocity in y-position (mm/s) and rate of the pen tip force (N/s). Figure 4 shows examples of characters reconstructed using the first $N = 10$, 50 and 100 pen trajectory data. We trained the nullspace-based classifier using observation vectors with $N=10$, 50 and 100. The training set consisted of 10% of the recordings and the validation set consisted of the remaining 90%. Training and validation were repeated ten times. The validation accuracy, plotted in Fig. 5 for different k (number of vectors considered for spanning the nullspace), indicates that optimal performance can be achieved by carefully choosing the dimension of the nullspace.

---

[1] https://archive.ics.uci.edu/dataset/175/character+trajectories.

Fig. 4. Reproduction of characters with the first N = 10 (plots in red), N = 50 (plots in blue) and N = 100 (plots in black) sensor data sampled during the pen trajectory. The data are the pen velocity in x-position (mm/s), pen velocity in y-position (mm/s) and rate of the pen tip force (N/s) recorded at 200 Hz. (Color figure online)

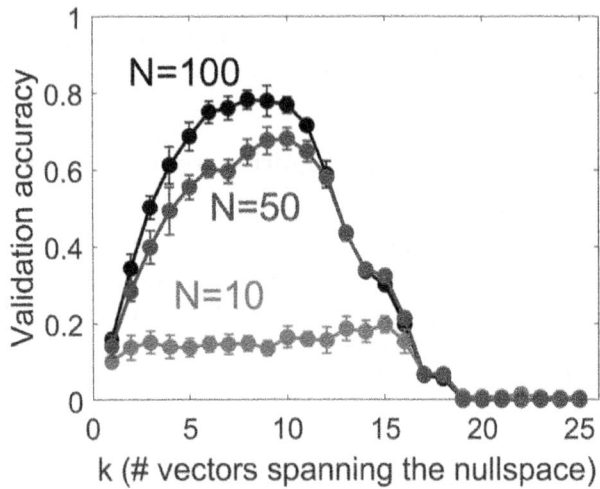

Fig. 5. Nullspace-based classification accuracy with N = 100 (plot in dark), N = 50 (plot in blue) or N = 10 (plot in red) sampled points in the pen trajectory. k is the number of vectors considered for (that span) the nullspace. Results are represented as means±S. D estimated over 10 trials. (Color figure online)

6   Conclusion

In this paper we propose a nullspace-based classifier for dynamical systems. Computing the metric in Kronecker space vs original space has similar complexity: $\mathcal{O}(l^2)$ vs $\mathcal{O}(kl)$. Indeed, it involves computing a single scalar product of dimension l^2 in Kronecker space vs k scalar products of dimension l in original space. In contrast, learning is more difficult to perform in Kronecker space than in original space as it involves singular or eigenvalue decomposition of a $(l^2 \times l^2)$ sample autocorrelation matrix in Kronecker space vs $(l \times l)$ in original space. For example, in the handwritten character recognition with $N = 100$, one has $l^2 = 90{,}000$ in Kronecker space vs $l = 300$ in original space. Still, for a large, near-singular, $(l \times l)$ sample autocorrelation matrix, the "approximate" nullspace is not well defined and can be computationally difficult to extract as it corresponds to eigenvectors associated with very small, but not necessarily zero, eigenvalues. Future work will concentrate in reducing computational requirements on large-scale problems to make it suitable for embedded learning at the edge.

Acknowledgements. This work was funded by France 2030 under the investment program ANR-20-PCPA-0007.

References

1. Ray, P.P.: A review on TinyML: state-of-the-art and prospects. J. King Saud Univ. Comput. Inf. Sci. **34**(4), 1595–1623 (2022)
2. LeCun, Y., Bengio, Y., Hinton, G.: Deep learning. Nature **521**, 436–444 (2015)
3. Cover, T., Hart, P.: Nearest neighbor pattern classification. IEEE Trans. Inf. Theory **13**(1), 21–27 (1967)
4. Parzen, E.: On estimation of a probability density function and mode. Ann. Math. Stat. **33**, 1065–1076 (1962)
5. Least squares quantization in PCM. IEEE Trans. Inf. Theory **28**(2), 129–137 (1982). https://doi.org/10.1109/TIT.1982.1056489
6. Bermak, A., Belhouari, S.B., Shi, M., Martinez, D.: Pattern recognition techniques for odor discrimination in gas sensor array. **10**, 1–17 Encyclopedia Sens. (2006)
7. Åström, K.J., Wittenmark, B.: Adaptive Control Reading. Addison-Wesley (1989)
8. Martinez, D., Burgués, J., Marco, S.: Fast measurements with MOX sensors: a least-squares approach to blind deconvolution. Sensors **19**(18) (2019). https://doi.org/10.3390/s19184029,
9. Patterson, T.A., Thomas, L., Wilcox, C., Ovaskainen, O., Matthiopoulos, J.: State-space models of individual animal movement. Trends Ecol. Evol. **23**(2), 87–94 (2008)
10. Martinez, D., Boutayeb, M.: Minimum-distance classification for dynamical systems and sensors. IEEE Sens. J. **23**(22), 28454–28461 (2023)
11. Golub, G.H., Van Loan, C.F.: Matrix Computations, 3rd edn. Johns Hopkins University Press, Baltimore (1996)
12. Cichocki, A., Amari, S.-I.: Adaptive Blind Signal and Image Processing. Wiley, Chichester (2002)
13. Battistelli, G., Tesi, P.: Classification for dynamical systems: model-based and data-driven approaches. IEEE Trans. Autom. Control **6**, 1741–1748 (2021)
14. James, M.: The generalised inverse. Math. Gazette **62**, 420 (1978). https://doi.org/10.1017/S0025557200086460
15. Burges, C.J.: A tutorial on support vector machines for pattern recognition. Data Min. Knowl. Disc. **2**, 121–167 (1998). https://doi.org/10.1023/A:1009715923555

On the Bayesian Interpretation of Robust Regression Neural Networks

Jan Kalina$^{(\boxtimes)}$ and Petra Vidneroá

Institute of Computer Science of the Czech Academy of Sciences,
182 07 Pod Vodárenskou věží 2, Prague 8, Prague, Czech Republic
{kalina,petra}@cs.cas.cz
http://www.cs.cas.cz/staff/kalina/en

Abstract. The aim of this work is to search for intuitive interpretations of regularized regression procedures within the framework of Bayesian inference. First, the paper considers Bayesian estimation of parameters of the linear regression model. Second, regularized neural networks are explained to correspond to the Bayesian approach obtained under specific assumptions. The contribution is a unique compact look on training neural networks with available prior information, i.e. a likelihood-based perspective of training neural networks. Attention is also paid to very recently proposed regularized versions of robust neural networks; as a novelty, these are expressed by means of quasi-likelihood and their connection to Bayesian reasoning is discussed as well.

Keywords: Bayesian estimation · Regularization · Regression · Neural networks · Robust estimation

1 Introduction

Bayesian reasoning represents an intuitive approach to considering conditional probabilities [10] and Bayesian estimation represents an important approach applicable to regression modeling, which is especially useful for solving difficult regression problems [6]. From the theoretical point of view, Bayesian estimates have been proven to dominate standard (e.g. maximum likelihood) estimates as overviewed in the monograph [19] for a variety of statistical methods. Bayesian neural networks have also found many recent applications revealing their advantages compared to standard (non-Bayesian) neural networks [5]. Bayesian procedures are often exploited for evaluating and controlling uncertainty of measurements within metrology as the science about measurement [3]. In fact, Bayesian methods are expected to be incorporated in later editions of the Guide to the Expression of Uncertainty in Measurement (GUM) methodology in the next 5 years [28]. Still, the methods of Bayesian linear regression or Bayesian neural

The research was supported by the project 22-02067S ("Approximate Neurocomputing") of the Czech Science Foundation.

networks [13] are often considered in practice as specific self-standing tools, without acknowledging their narrow connections to intuitive thinking and regularized (but non-Bayesian) estimation.

Regularized versions of neural networks are particularly appealing for their ability to avoid overfitting [21, 26]. L_2-regularization remains to be the most common regularization type used for shallow neural networks [9]; L_2-regularization was repeatedly found as the most effective choice for small data [20]. Regularization may be recommended also for robust neural networks, which are based on a robust loss function [24] aimed to yield resistant results also for data contaminated by outliers. Still, only very few studies have considered regularized versions of robust neural networks for data that are not too simplistic [7] and we are not aware of any papers on the connections between Bayesian thinking and regularization in the context of robust neural networks.

Bayesian thinking is important also for the field of approximate computing, where it is namely important to consider the estimated (trained) parameters of various models of statistics or machine learning with a possible contamination by small errors [18]. This corresponds to the desire to study the effect of rounding the parameters (say $\theta \in \mathbb{R}^d$ for now in some arbitrary model) or computing its estimate within a low-precision arithmetic [25]. If modeling the probabilistic distribution of the error (e.g. of the effect of rounding) to be specified, then estimating θ corresponds to the Bayesian estimation in the original (non-contaminated) model using the distribution about the errors as the prior distribution about θ. Thus, it remains useful to study the Bayesian connections of the regularized robust versions of neural networks of the previous paragraph.

The contribution of this work is a unique compact look on training neural networks with available prior information, a likelihood-based perspective of training neural networks. Regularized versions of robust multilayer perceptrons proposed recently in [17] are studied here and as the main contribution, Bayesian estimation of their parameters is investigated. Section 2 discusses interesting connections related to the Bayesian estimation of parameters in the linear regression model. Section 3 extends the results to nonlinear regression performed by multilayer perceptrons and Sect. 4 finds new connections e.g. for very recent robust neural networks, which are resistant to the presence of outliers in the data. Section 5 concludes the paper.

2 Bayesian Estimation in the Linear Regression

This section is devoted to Bayesian estimation of parameters of linear regression and its slightly generalized extension; important concepts are explained here before their using in the nonlinear model in Sect. 3.

2.1 Location Model

Principles of Bayesian estimation will be explained on the one-dimensional location model

$$X_i = \mu + e_i, \quad e_i \sim N(0, \sigma^2), \quad i = 1, \ldots, p, \tag{1}$$

with $\mu \in \mathbb{R}$. The model (1) implies that $X_i \sim N(\mu, \sigma^2)$ for every $i = 1, \ldots, n$.

Using the prior distribution

$$\mu \sim N(\theta, \gamma^2) \quad \text{with} \quad \theta \in \mathbb{R}^p, \quad \gamma > 0, \tag{2}$$

we may express the expectation of the posterior distribution as

$$\hat{\mu} = (1 - \delta)\bar{X} + \delta\theta, \quad \text{where} \quad \delta = \frac{\gamma^{-2}}{n\sigma^{-2} + \gamma^{-2}} = \frac{\sigma^2}{n\gamma^2 + \sigma^2} \in (0, 1). \tag{3}$$

This is an L_2-regularized version of $\bar{X}$, where the latter denotes the arithmetic mean of $X_1, \ldots, X_n$. In the special case with $\theta = 0$, we obtain a univariate version of the ridge regression estimator known from the linear regression model [9], which has the form $\hat{\mu} = (1 - \delta)\bar{X}$ here.

Using Laplace distribution as the prior for μ leads to the univariate version of the lasso (least absolute shrinkage and selection operator) estimator [8]. More formally, we assume that μ follows Laplace distribution with expectation 0 and variance $2/\tau^2$ for $\tau > 0$. The mode of the posterior distribution is the lasso, which may be expressed as

$$\min_{\hat{\mu} \in \mathbb{R}} \left\{ \sum_{i=1}^{n} (X_i - \hat{\mu})^2 + \delta|\hat{\mu}| \right\}, \tag{4}$$

or equivalently as

$$\hat{\mu} = \begin{cases} \bar{X} + \delta, & \text{if } \bar{X} < -\delta, \\ 0, & \text{if } -\delta \le \bar{X} \le \delta, \\ \bar{X} - \delta, & \text{if } \bar{X} > \delta, \end{cases} \tag{5}$$

where δ has the explicit form $\delta = 2\tau\sigma^2$. This estimator at the same time represents the solution of the optimization task

$$\hat{\mu} = \text{sgn}(\bar{X}) \left(|\bar{X}| - \delta \right)_+ = \text{sgn}(\bar{X}) \max \left\{ |\bar{X}| - \delta, 0 \right\} \tag{6}$$

and can be perceived as an L_1-regularized version of $\bar{X}$.

The ridge and lasso estimators are however not useful for $p = 1$, because the justification of the shrinkage estimation studied in [11] requires $p > 2$. The estimators are thus useful only when the specific prior information about μ is available indeed. Let us additionally note that the Jeffreys (non-informative) prior corresponds to the uniform distribution across $\mathbb{R}$ in (1); the corresponding Bayesian estimator as the mean of the posterior distribution of μ is precisely the sample mean.

2.2 Bayesian Linear Regression: Basic Model

We consider the standard linear regression model

$$Y_i = \beta_1 X_{i1} + \cdots + \beta_p X_{ip} + e_i, \quad i = 1, \ldots, n, \tag{7}$$

which may be expressed in the matrix notation as $Y = X\beta + e$. Here, we consider p fixed regressors (predictors) available for n measurements (observations). In our notation, $\beta = (\beta_1, \ldots, \beta_p)^T$ is the vector of parameters, where the i-th row of X will be denoted as $X_i = (X_{i1}, \ldots, X_{ip})^T$ for $i = 1, \ldots, n$. Under the assumption $e_i \sim N(0, \sigma^2)$, we have $Y_i \sim N(X_i^T \beta, \sigma^2)$ for $i = 1, \ldots, n$ and the likelihood can be expressed as the product of Gaussian (normal) densities

$$L(\text{data}|\beta, \sigma^2) = \left(\frac{1}{\sqrt{2\pi\sigma^2}}\right)^n \exp\left\{-\frac{\sum_{i=1}^n (Y_i - X_i^T \beta)^2}{2\sigma^2}\right\}. \tag{8}$$

Let us consider the prior distribution about the parameter $\beta \in \mathbb{R}^p$ to be p-variate Gaussian $\beta \sim N_p(0, \Psi)$ with a known positive definite matrix Ψ. The Bayes theorem requires the prior to be formulated by means of the joint density $f(\beta, \sigma^2)$, while σ^2 represents a nuisance parameter here. The prior in the form of the joint density is simply obtained as

$$f(\beta, \sigma^2) = f(\sigma^2) f(\beta|\sigma^2), \quad \beta \in \mathbb{R}^d, \quad \sigma^2 > 0, \tag{9}$$

under the usual assumption that $f(\sigma^2)$ is the density of the inverse gamma distribution [8]. Here, the prior for β does not actually depend on σ^2 and its density is obtained according to

$$f(\beta) = (2\pi)^{-k/2} \left(\det(\Psi)\right)^{-1/2} \exp\left\{-\frac{1}{2}\beta^T \Psi^{-1} \beta\right\}, \quad \beta \in \mathbb{R}^p. \tag{10}$$

Assuming the expectation of β equal to 0 simplifies the results; without loss of generality, let us assume the data to be transformed in such a way that it holds $\beta = 0$. Let us also remark that it is common in this context to denote all the densities by f so that f plays the role of different functions with different parameters.

The posterior distribution of β conditioned on the given data and on σ given by the Bayes theorem is the Gaussian distribution and the Bayesian estimator of β is given by the mean of the posterior distribution. This estimator

$$\hat{\beta} = \left(X^T X + \sigma^2 \Psi^{-1}\right)^{-1} X^T Y. \tag{11}$$

depends on σ. The special case with $\Psi = \mathcal{I}$ is commonly denoted as the ridge regression estimator [9]. The least squares estimator represents a special case of the Bayesian estimator (11) with $\lambda = 0$. An appropriate value of the regularization parameter is typically selected in a cross-validation.

If the expectation of the prior distribution is $\beta_0 \in \mathbb{R}^p$ rather $0 \in \mathbb{R}^p$, the posterior distribution gets more complex and its expectation has the form

$$\hat{\beta} = \left(X^T X + \sigma^2 \Psi^{-1}\right)^{-1} \left(X^T Y + \sigma^2 \Psi^{-1} \beta_0\right) \tag{12}$$

rather than just (11). The special case with the limiting case $\sigma^2 = 0$ reduces (12) to the least squares estimate. For the other extreme with $\sigma^2 \to \infty$, we obtain the convergence $\hat{\beta} \to \sigma^{-2} \Psi(\sigma^2 \Psi^{-1} \beta_0)$, i.e. $\hat{\beta} \to \beta_0$, which corresponds to estimating β just by the expectation of the prior distribution.

2.3   Prior Information in the Form of Additional Measurements

Let us have a set of the total number of R previously performed measurements and let the estimate of β computed from them by means of the least squares be denoted by $b_1, \ldots, b_R$. Further, let us assume that these R previous measurements were measured under the same conditions as the currently available data, i.e. each of the values b_r for $r = 1, \ldots, R$ was obtained in the linear model with the same regressors and the same number of measurements as in (7). It is now natural to choose $\Psi^{-1} = (X^T X)^{-1}/R$ and thus $\hat{\beta}$ (11) has the form

$$\hat{\beta} = \left(X^T X + R X^T X \right)^{-1} X^T Y = \frac{1}{R+1} b_{LS}, \tag{13}$$

where b_{LS} denotes the least squares estimator $b_{LS} = (X^T X)^{-1} X^T Y$. The expression (13) corresponds to the intuition, because it is a convex linear combination of the least squares estimate and 0, i.e. a shrinkage estimator obtained by shrinking the classical estimator towards 0.

3   Bayesian Estimation of Parameters of Neural Networks

The considerations of Sect. 2 for the linear model will now be extended to regression neural networks. Attention is paid also to very recent robust and at the same time regularized versions of neural networks.

3.1   Bayesian Interpretation of Regularized Multilayer Perceptrons

To introduce the nonlinear regression model, a continuous random variable $Y_i \in \mathbb{R}$ is considered as a response of the vector of regressors (independent variables) $X_i = (X_{i1}, \ldots, X_{ip})^T \in \mathbb{R}^p$, which are p-dimensional random vectors available for the total number of n measurements. The nonlinear regression model is formally expressed as

$$Y_i = \varphi(X_i) + e_i, \quad i = 1, \ldots, n, \tag{14}$$

with an unknown function φ and random errors $e_1, \ldots, e_n$, which are independent on the regressors. The matrix X will denote the matrix of all the regressors with the i-th measurement $(X_{i1}, \ldots, X_{ip})^T$ in the i-th row.

Let the function φ be fully specified up to a $\theta = (\theta_1, \ldots, \theta_d)^T \in \mathbb{R}^d$ and the aim of the training is reliable estimation of θ. In contrary to the linear model of Sect. 2, the dimension d of θ does not have to coincide with p. The loss function of multilayer perceptrons (MLPs) plays the role of φ here. Training MLPs is typically based on the optimization task

$$\min u_i^2(\hat{\theta}) \quad \text{over} \quad \hat{\theta} = (\hat{\theta}_1, \ldots, \hat{\theta}_d)^T \in \mathbb{R}^d, \tag{15}$$

where $u_1(\hat{\theta}), \ldots, u_n(\hat{\theta})$ are residuals computed for a given estimate $\hat{\theta}$ of θ, i.e.

$$u_i(\hat{\theta}) = Y_i - \varphi(X_i, \hat{\theta}), \quad i = 1, \ldots, n. \tag{16}$$

Most commonly, no probabilistic assumptions on the random noise are needed for training MLPs. Our likelihood-based considerations however need the assumption that $e_i \sim N(0, \sigma^2)$ for $i = 1, \ldots, n$. Then, we have

$$Y_i \sim N(\varphi(X_i, \theta), \sigma^2) \tag{17}$$

for $i = 1, \ldots, n$. Using the notation $\propto$ to denote proportionality, the likelihood is expressed (up to a normalizing constant) as

$$L(\text{data}|\theta, \sigma^2) \propto \exp\left\{ -\frac{1}{2\sigma^2} \sum_{i=1}^{n} (Y_i - \varphi(X_i, \theta))^2 \right\}. \tag{18}$$

Remark 1. The training of a particular model of an MLP based on (15) corresponds to maximizing the likelihood expressed as (18), i.e. to computing the maximum likelihood estimates of θ for Gaussian random errors $e_1, \ldots, e_n$ in (14). This is evident, because the optimization task does not depend on the multiplicative factor $2\sigma^2$ in (18) and also on the exponential function, which performs a monotone transformation.

We assume the prior information in the form $\theta \sim N_d(0, \delta^2 \mathcal{I}_d)$ i.e. a d-variate Gaussian distribution, where $\delta^2 > 0$ and $\mathcal{I}_d$ denotes the unit matrix of size $d \times d$. Just like in Sect. 2.2, the assumption of a zero expectation is without loss of generality and without limiting the applicability of the results. This prior can be alternatively expressed as a product density of individual Gaussian densities corresponding to $\theta_j \sim N(0, \delta^2)$ for every $j = 1, \ldots, d$, where $\theta_1, \ldots, \theta_d$ are mutually uncorrelated (and thus independent thanks to the normality) random variables. We stress that the prior information assumes not only normality of the errors but also homoscedasticity in the model. Using this prior information, the posterior density given by the Bayes formula has the form

$$f(\theta|\text{data}) \propto \exp\left\{ -\frac{1}{2\sigma^2} \sum_{i=1}^{n} u_i^2 \right\} \exp\left\{ -\frac{1}{2\delta^2} \sum_{j=1}^{d} \theta_j^2 \right\}$$

$$\propto \exp\left\{ -\frac{1}{2\sigma^2} \left[\sum_{i=1}^{n} u_i^2 + \frac{\sigma^2}{\delta^2} \sum_{j=1}^{d} \theta_j^2 \right] \right\}$$

$$\propto \exp\left\{ -\frac{1}{2\sigma^2} \left[\sum_{i=1}^{n} u_i^2 + \lambda \sum_{j=1}^{d} \theta_j^2 \right] \right\}, \quad \theta \in \mathbb{R}^d, \tag{19}$$

with the notation for the regularization parameter $\lambda = \sigma^2/\delta^2$. The maximum likelihood estimation of θ is based on solving the optimization task

$$\min\left\{ \sum_{i=1}^{n} \left(Y_i - \varphi(X_i, \hat{\theta}) \right)^2 + \lambda \sum_{j=1}^{d} \hat{\theta}_j^2 \right\} \tag{20}$$

over $\hat{\theta} \in \mathbb{R}^d$ and the network with the obtained estimate of θ is exactly a regularized MLP using the L_2-regularization.

Remark 2. The value of λ is indirectly proportional to δ^2, which means that increasing λ corresponds to the prior belief about smaller variability of θ, i.e. larger certainty (stronger prior belief) about θ being close to its expected value.

4   Robust Multilayer Perceptrons

Robust training of MLPs has the aim to perform resistant (reliable) training under the presence of possible outliers in the data. Particularly the versions inspired by highly robust statistical estimators [14] yield promising results in numerical experiments [24]. Recently proposed regularized versions of robust MLPs [17] are recalled in this section. As the research problem, it remains open how to show that their regularized versions may be derived in the Bayesian setting. Using a quasi-likelihood approach, these robust MLPs will be justified by the Bayesian reasoning in this section. Particularly, we are interested in formulating an analogy to Remark 1 for robust versions of MLPs.

We consider the class of robust MLPs denoted as LWS-MLP, which is inspired by the least weighted squares (LWS) estimator [15,27]. In linear regression, the LWS estimator (7) is able to combine efficiency, global robustness (with respect to outliers), and local robustness with respect to small changes of all data points. Let us refer to p. 30 of [14] for a discussion of local and global concepts of statistical robustness. MLP-LWS is formally defined by minimizing the loss

$$\arg\min_{\hat{\theta}\in\mathbb{R}^d} \sum_{i=1}^{n} \Psi\left(\frac{i-1/2}{n}\right) u_{(i)}^2(\hat{\theta}) \tag{21}$$

with the notation $u_{(1)}^2(\hat{\theta}) \leq \cdots \leq u_{(n)}^2(\hat{\theta})$. We may formulate a quasi-likelihood function for the LWS-MLP model. Such function is minimized to obtain the estimates of the parameters, i.e. it plays the role of the likelihood, although it does not correspond to the actual probability distribution (cf. [23] for the context of robust linear regression). The quasi-likelihood has the form

$$L(\text{data}|\hat{\theta}, \hat{\sigma}^2) \propto \exp\left\{ -\frac{1}{2\hat{\sigma}^2} \sum_{i=1}^{n} w_i u_{(i)}^2(\hat{\theta}) \right\} \tag{22}$$

with the weights

$$w_i = \Psi\left(\frac{i-1/2}{n}\right), \quad i = 1, \ldots, n, \tag{23}$$

where $\hat{\sigma}^2$ is an estimate of σ^2, which has however no influence on the maximization of (22). Minimizing the loss function in (21) yields the same solution as maximizing the quasi-likelihood over $\hat{\theta} \in \mathbb{R}^d$. If Bayesian estimation is performed with the prior $\theta \sim N_d(0, \delta^2 \mathcal{I}_d)$, the maximum of the posterior density is achieved for the $\hat{\theta}$ obtained by

$$\arg\min\left\{ \sum_{i=1}^{n} w_i u_{(i)}^2(\hat{\theta}) + \lambda \sum_{j=1}^{d} \hat{\theta}_j^2 \right\} \tag{24}$$

considered over all $\hat{\theta} \in \mathbb{R}^d$. Thus, the L_2-regularized LWS-MLP [17] is obtained.

If the loss of the least trimmed squares (LTS) estimator [22] is used instead of the LWS, the optimization task solved within the training of the robust network denoted as LTS-MLP [24] has the form

$$\arg\min_{\hat{\theta} \in \mathbb{R}^d} \sum_{i=1}^{h} u_{(i)}^2(\hat{\theta}) \qquad (25)$$

for a chosen trimming constant $h \in [n/2, n)$. Using the prior $\theta \sim N_d(0, \delta^2 \mathcal{I}_d)$ as above, the maximum of the posterior density is obtained by solving

$$\min \left\{ \sum_{i=1}^{h} u_{(i)}^2(\hat{\theta}) + \lambda \sum_{j=1}^{d} \hat{\theta}_j^2 \right\} \qquad (26)$$

considered over all $\hat{\theta} \in \mathbb{R}^d$. The network with estimates found by this optimization task is known as the L_2-regularized LTS-MLP [17].

4.1   Prior Information in the Form of Additional Measurements

The Bayesian estimation for LWS-MLP will now be discussed to correspond to the intuition for the specific situation with the total number of R additional measurements. We assume an LWS-MLP model with a known architecture that is fully specified up to the knowledge of a parameter $\theta = (\theta_1, \ldots, \theta_d)^T \in \mathbb{R}^d$. Based on given training data, the parameters are estimated by minimizing the sum of squared residuals. The assumption of a common variance

$$\operatorname{var} \hat{\theta}_1 = \cdots = \operatorname{var} \hat{\theta}_d =: \varepsilon^2 \qquad (27)$$

corresponds to the idea of performing the measurements leading to estimating θ under the same conditions, which are common across the measurements. Just like above, the prior information is considered in the form $\theta \sim N_d(0, \delta^2 \mathcal{I})$. The residuals of the R additional measurements evaluated for the trained LWS-MLP will be denoted by $t_1, \ldots, t_R$. Choosing the prior for θ as

$$f(\theta) \propto \exp \left\{ -\frac{\sum_{r=1}^{R} t_r^2}{2\varepsilon^2} \right\}, \quad \theta \in \mathbb{R}^d, \qquad (28)$$

is natural, because θ is estimated from the task to solve the optimization task $\min \sum_{r=1}^{r} t_r^2$. This prior corresponds to the assumption that $(t_1(\theta), \ldots, t_R(\theta))^T$ has R-variate normal distribution $N_R(0, \sigma^2 \mathcal{I}_R)$. The posterior distribution in such a situation is obtained as

$$f(\theta|\text{data}) \propto \exp \left\{ -\frac{\sum_{i=1}^{n} w_i u_{(i)}^2}{2\sigma^2} - \frac{\sum_{r=1}^{R} t_r^2}{2\varepsilon^2} \right\}, \quad \theta \in \mathbb{R}^d. \qquad (29)$$

In a special situation with $\varepsilon^2 = \sigma^2$, the posterior density has the form

$$f(\theta|\text{data}) \propto \exp\left\{ -\frac{1}{2\sigma^2}\left[\sum_{i=1}^{n} w_i u_{(i)}^2 + \sum_{r=1}^{R} t_r^2 \right] \right\}, \quad \theta \in \mathbb{R}^d, \tag{30}$$

so the resulting $\hat{\theta}$ exploits the previous measurements in a very intuitive way.

Remark 3. The additional measurements may have different uncertainty (variability) than current measurements. Then, still considering an LWS-MLP model, let us assume the prior knowledge $\theta_j \sim \mathsf{N}(0, \delta^2)$ to be replaced by

$$\theta_j \sim \mathsf{N}(0, k\delta^2) \quad \text{for a given } k > 0 \quad \text{for } j = 1, \ldots, n, \tag{31}$$

i.e. for the p-variate parameter θ in the form $\theta \sim \mathsf{N}_d(0, k\delta^2 \mathcal{I}_d)$. Then, the maximizing the quasi-likelihood to obtain an estimate of θ requires to solve

$$\min_{\hat{\theta} \in \mathbb{R}^d}\left\{ \sum_{i=1}^{n} w_i u_{(i)}^2 + k\lambda \sum_{j=1}^{d} \hat{\theta}_j^2 \right\}. \tag{32}$$

In other words, λ is replaced by $k\lambda$ and the effect of regularization is k-times stronger compared to the effect in (24).

5   Conclusion

The aim of this work is to understand the connection between recently proposed regularized versions of robust neural networks and Bayesian statistical estimation. Bayesian thinking allows to incorporate previously performed measurements in a very intuitive way. Some of the presented results have been discussed in the literature previously but often without keeping in mind the necessary assumptions on normality and homoscedasticity. As a novelty, we perceive here the training of robust MLPs as a maximization of certain quasi-likelihood functions, i.e. as an optimization task very analogous to likelihood-based statistical estimation problems. The presented considerations are valid even in the situation that no explicit formulas for the robust estimates of θ are available.

The Bayesian reasoning is applicable in metrological scenarios either with available prior knowledge or under errors in the estimation computations (e.g. approximate computing in a low-precision arithmetic). On the other hand, the Bayesian methodology is its large skepticism against hypothesis testing about the parameters, as admitted in Sect. 16.4.3 of [6].

This work is theoretical; extensive experiments with robust regularized neural networks were presented e.g. in [12]. They reveal that neural networks with a highly robust loss function outperform standard versions for contaminated data; the effect is remarkable with the L_2 regularization, particularly if the number of samples is not very large. Also, the results are not harmed by a possible multicollinearity due to a high dimension. However, previous experiments revealed the pre-processing such as cleaning the data and transformations of the variables to have a stronger effect compared to the choice of the particular regularization [17].

From the practical point of view, it deserves to be stressed that the robustness considered here is focused on the presence of outliers in the data. The idea of the LWS is to be efficient for normally distributed errors and to be robust for contaminated models. In other words, this type of robustness is tailor-made for distributions that belong to a certain neighborhood of the normal distribution [2], rather than for other distributions including those with heavy tails (e.g. Laplace distribution). Furthermore, the robust procedures are not intended to be robust to heteroscedasticity, which may have severe consequences on the variability of individual parameters [4]. At the same time, it remains an open problem whether the robust training is robust (or on the other hand sensitive) to small changes of the architecture or to changes of the weight function, which has to be chosen by the user, typically without any available prior knowledge.

Connections between regularization and Bayesian estimation bring theoretical arguments in favor of the commonly used L_2-regularization and justify the common usage of regularization in scenarios with small samples. A connection between Bayesian regularized regression and data imputation for scenarios with limited data availability was investigated in [1]. As future work, it is planned to extend the considerations to deep learning, to classification tasks [16], or to develop (non-Bayesian) likelihood-based hypothesis testing for neural networks.

References

1. Al-Jamali, N.A.S., Al-Saedi, I.R.K., Zarzoor, A.R., Li, H.: A new imputation technique based a multi-spike neural network to handle missing data in the internet of things network (IoT). IEEE Access 11, 112941–112850 (2023)
2. Berenguer-Rico, V., Johansen, S., Nielsen, B.: A model where the least trimmed squares estimator is maximum likelihood. J. Roy. Stat. Soc. 85, 886–912 (2023)
3. Cheng, Y.B., et al.: Analysis and comparison of Bayesian methods for measurement uncertainty evaluation. Math. Probl. Eng. 2018, 7509046 (2018)
4. Deka, B., Nguyen, L.H., Goulet, J.A.: Analytically tractable heteroscedastic uncertainty quantification in Bayesian neural networks for regression tasks. Neurocomputing 572, 127183 (2024)
5. Goan, E., Fookes, C.: Bayesian neural networks: an introduction and survey. In: Mengersen, K.L., Pudlo, P., Robert, C.P. (eds.) Case Studies in Applied Bayesian Data Science. LNM, vol. 2259, pp. 45–87. Springer, Cham (2020). https://doi.org/10.1007/978-3-030-42553-1_3
6. Greene, W.H.: Econometric Analysis, 8th edn. Pearson Education Limited, Harlow (2018)
7. Gupta, D., Hazarika, B.B., Berlin, M.: Robust regularized extreme learning machine with asymmetric Huber loss function. Neural Comput. Appl. 32, 12971–12998 (2020)
8. Hans, C.: Bayesian lasso regression. Biometrika 96, 835–845 (2009)
9. Hastie, T., Tibshirani, R., Wainwright, M.: Statistical Learning with Sparsity: The Lasso and Generalizations. CRC Press, Boca Raton (2015)

10. Hooten, M.B., Hefley, T.: Bringing Bayesian Models to Life. CRC Press, Boca Raton (2019)
11. James, W., Stein, C.: Estimation with quadratic loss. Estimation with quadratic loss. In: Proceedings of the Fourth Berkeley Symposium on Mathematical Statistics and Probability, vol. 1, pp. 361–379 (1961)
12. Jiang, H.: Forecasting global solar radiation using a robust regularization approach with mixture kernels. J. Forecast. **42**, 1989–2010 (2023)
13. Jospin, L.V., Laga, H., Boussaid, F., Buntine, W., Bennamoun, M.: Hands-on Bayesian neural networks–a tutorial for deep learning users. IEEE Comput. Intell. Mag. **17**, 29–48 (2022)
14. Jurečková, J., Picek, J., Schindler, M.: Robust Statistical Methods with R, 2nd edn. CRC Press, Boca Raton (2019)
15. Kalina, J.: Highly robust statistical methods in medical image analysis. Biocybern. Biomed. Eng. **32**(2), 3–16 (2012)
16. Kalina, J., Matonoha, C.: A sparse pair-preserving centroid-based supervised learning method for high-dimensional biomedical data or images. Biocybern. Biomed. Eng. **40**(2), 774–786 (2020)
17. Kalina, J., Tumpach, J., Holeňa, M.: On combining robustness and regularization in training multilayer perceptrons over small data. In: Proceedings IJCNN 2022, pp. 1–8 (2022)
18. Klauenberg, K., Martens, S., Bošnjaković, A., Cox, M.G., van der Veen, A.M.H., Elster, C.: The GUM perspective on straight-line errors-in-variables regression. Measurement **187**, 110340 (2022)
19. Maruyama, Y., Kubokawa, T., Strawderman, W.E.: Stein Estimation. Springer, Singapore (2023). https://doi.org/10.1007/978-981-99-6077-4
20. Piotrowski, A.P., Napiorkowski, J.J., Piotrowska, A.E.: Impact of deep learning-based dropout on shallow neural networks applied to stream temperature modelling. Earth Sci. Rev. **201**, 103076 (2020)
21. Poggio, T., Smale, S.: The mathematics of learning: dealing with data. Notices Am. Math. Soc. (AMS) **50**, 537–544 (2003)
22. Rousseeuw, P.J., Van Driessen, K.: Computing LTS regression for large data sets. Data Min. Knowl. Disc. **12**, 29–45 (2006)
23. Ruli, E., Sartori, N., Ventura, L.: Robust approximate Bayesian inference. J. Stat. Plan. Infer. **205**, 10–22 (2020)
24. Rusiecki, A.: Robust learning algorithm based on LTA estimator. Neurocomputing **120**, 624–632 (2013)
25. Sze, V., Chen, Y.H., Yang, T.J., Emer, J.S.: Efficient Processing of Deep Neural Networks. Morgan & Claypool Publishers, San Rafael (2020)
26. Vang-Mata, R.: Multilayer Perceptrons: Theory and Applications. Nova Science Publishers, New York (2020)
27. Víšek, J.Á.: Consistency of the least weighted squares under heteroscedasticity. Kybernetika **47**, 179–206 (2011)
28. Wübbeler, G., Marschall, M., Elster, C.: A simple method for Bayesian uncertainty evaluation in linear models. Metrologia **57**, 065010 (2020)

Probability-Generating Function Kernels for Spherical Data

Theodore Papamarkou[1]([⊠]) [iD] and Alexey Lindo[2] [iD]

[1] Department of Mathematics, The University of Manchester, Manchester, UK
theo.papamarkou@manchester.ac.uk
[2] School of Mathematics and Statistics, University of Glasgow, Glasgow, UK
alexey.lindo@glasgow.ac.uk

Abstract. The class of probability-generating function (PGF) kernels is introduced, which consists of kernels supported on the unit hypersphere, for the purposes of spherical data analysis. PGF kernels generalize RBF kernels in the context of spherical data. The properties of PGF kernels are studied. A semi-parametric learning algorithm is introduced to enable the use of PGF kernels with spherical data.

Keywords: Compositional kernels · Gaussian processes · deep kernel learning · kernels · probability-generating functions · spherical data

1 Introduction

This paper contributes to spherical data analysis by developing a new class of kernels, called probability-generating function (PGF) kernels. By construction, PGF kernels are supported on the unit hypersphere, making them a natural choice for spherical data.

PGF kernels are dot-product kernels. Furthermore, PGF kernels generalize radial basis function (RBF) kernels, and therefore the former are expected to have better predictive performance on spherical data than the latter.

Contribution. A summary of contributions follows. PGF kernels are introduced as compositions of PGFs. This way, another connection between machine learning and probability is drawn. It is shown that PGF kernels generalize RBF kernels. A semi-parametric learning algorithm is introduced to fit PGF kernels to spherical data. PGF kernels are equipped with a notion of depth and width, both of which are explored via an ablation study. Three examples are presented to demonstrate the applicability of PGF kernels, namely two Gaussian process (GP) regression examples on circular and spherical data, and a deep kernel learning (DKL) classification example on higher-dimensional spherical data.

© The Author(s), under exclusive license to Springer Nature Switzerland AG 2024
M. Wand et al. (Eds.): ICANN 2024, LNCS 15016, pp. 41–59, 2024.
https://doi.org/10.1007/978-3-031-72332-2_4

Comparison with Other Works. [2] have studied theoretical aspects of compositional kernels based on generating functions (GFs), with an emphasis on the duality between GF kernels and (neural network) activation functions. In contrast, this paper focuses on PGF kernels (rather than GF kernels), studying their properties and relation to other kernels, and introducing a learning algorithm that enables the use of PGF kernels in practice. [11] have focused on a subclass of PGF kernels, based on compositions of a single PGF, and have introduced a learning algorithm for this subclass by drawing a link between kernels and activation functions. In contrast, this paper defines PGF kernels in their generality, allowing compositions of different PGFs, and puts forward a learning algorithm that allows the use of PGF kernels as a standalone entity, without necessitating a link with activation functions.

Paper Structure. Sect. 2 highlights related work in the literature. Section 3 introduces PGF kernels, studies their properties and relations with other kernels, and provides examples of kernels based on known PGFs. Section 4 introduces a learning algorithm to fit PGF kernels to spherical data, and uses this algorithm for GP regression and DKL classification. The paper concludes with a discussion, providing directions for future work (Sect. 5).

2   Related Work

Spherical Data Analysis. Spherical data analysis refers to the analysis of data supported on a hypersphere. Numerous methods have been developed for spherical data analysis. [10] have introduced a neural network architecture and a spherical convolutional kernel, the latter being used as a component of the former. The approach of [10] is designed for point cloud data in $\mathbb{R}^3$, which are typically processed using spherical regions. [12] have applied GP regression to approximate incident radiance functions on the sphere in $\mathbb{R}^3$. Several works have focused on the development of kernels on a hypersphere. [6] have studied the differentiability properties of kernels on a hypersphere. [14] have introduced a method to approximate polynomial kernels for data on the unit sphere using random spherical Fourier features. [21] have proposed a power series expansion of the RBF kernel for heat diffusion on a hypersphere.

PGFs. A PGF is defined as

$$g(s) := \sum_{i=0}^{\infty} p_i s^i, \tag{1}$$

where $p_i \in [0, 1]$, $i = 0, 1, 2, \ldots$, is a sequence of probabilities. It is assumed that $\sum_{i \geq 0} p_i \leq 1$, and therefore the power series in Eq. 1 converges absolutely for all $|s| \leq 1$ [5]. PGFs were introduced by De Moivre in 1730 [3,17]. Nowadays, PGFs are widely used in branching processes, theory of random walks, renewal theory, and analytic combinatorics.

Compositional Kernels. Composition of kernels is a common operation in practice. Python packages, such as `GPyTorch` [4], provide functionality and examples of kernel composition. The term 'compositional kernel' is also linked to the more topical class of kernels investigated by [2,11], establishing a duality between kernels and activation functions. This paper has its methodological origin in the topical notion of compositional kernels [2,11], but it shifts the attention away from their duality with activation functions. In doing so, this paper is thus focused on the general operation of kernel composition, introducing a methodological and computational framework to compose kernels using PGFs.

GPs and DKL. Section 4 relies on GPs and DKL. The log-marginal likelihoods used for GP regression and GP classification (as part of DKL) are available in [15, equation 2.30] and [13], respectively. For DKL, the reader is referred to [20].

3   PGF Kernels

This section introduces PGF kernels, their relations to other kernels, and their properties. It also provides examples of PGF kernels based on known PGFs. Hereafter, $\rho(x, z) := \langle x, z \rangle$ denotes the correlation between $x \in \mathbb{S}^h$ and $z \in \mathbb{S}^h$, where $\langle \cdot, \cdot \rangle$ is the dot product and $\mathbb{S}^h := \{x \in \mathbb{R}^{h+1} : \|x\| = 1\}$ is the unit h-sphere with $h \geq 1$.

3.1   Definition

Definition 1 introduces PGF kernels. PGF kernels apply compositions of PGFs on the correlation of their inputs.

Definition 1. *Let $g_1, \ldots, g_n$ be PGFs and $\rho(x, z)$ the correlation between $x \in \mathbb{S}^h$ and $z \in \mathbb{S}^h$. The n-depth PGF kernel $\mathcal{K}[g_1, \ldots, g_n]$ is defined as*

$$\mathcal{K}[g_1, \ldots, g_n](x, z) := g_n \circ \cdots \circ g_1(\rho(x, z)). \tag{2}$$

Setting $g = g_1 = \cdots = g_n$ in Eq. 2 yields the special case

$$\mathcal{K}[(g)_n](x, z) := \mathcal{K}[\underbrace{g, \ldots, g}_{n \text{ times}}](x, z). \tag{3}$$

Equation 3 yields the n-fold composition of g, which is also known as the n-fold iteration of g and is commonly denoted by $g_{(n)}$. $\mathcal{K}[(g)_n]$ represents the PGF kernel defined by iterations of a single PGF, previously referred to as compositional kernel in [11]. The composition $g_n \circ \cdots \circ g_1$ of different PGFs $g_1, \ldots, g_n$ is known in the literature on branching processes as an iterated function system in varying or random environment [1,8,9], but it has not been used in machine learning problems. Definition 1 introduces PGF kernels as compositions of different PGFs $g_1, \ldots, g_n$.

Definition 2 introduces the notions of layer and layer width for PGF kernels. This way, PGF kernels have layers, depth and width, similarly to neural networks.

Definition 2. *Let $\mathcal{K}[g_1,\ldots,g_n]$ be a PGF kernel. The i-th PGF g_i is called the i-th compositional layer or, shortly, the i-th layer of the kernel. The number l_i of non-zero terms in the series expansion of PGF g_i (see Eq. 1) is called the width of the i-th layer.*

An n-depth PGF kernel $\mathcal{K}[g_1,\ldots,g_n]$ is parameterized by the coefficients of its PGFs $g_1,\ldots,g_n$. Note that each layer width l_i can be finite or infinite. If all layer widths $(l_1,\ldots,l_n)$ are finite, then the kernel contains $\sum_{i=1}^{n} l_i$ parameters. If at least one of the layer widths is infinite, then the kernel contains an infinite number of parameters.

3.2   Relations to Common Kernels

PGF kernels are dot-product kernels. A dot-product kernel $g(\rho(x,z))$ entails an arbitrary function g [18], while a PGF kernel employs a composition $g_n \circ \cdots \circ g_1$ of PGFs as g. In fact, a composition $g_n \circ \cdots \circ g_1$ of PGFs is a PGF. So, a PGF kernel is a dot-product kernel for which g is a PGF. This remark clarifies that a PGF kernel can alternatively be defined as $g(\rho(x,z))$, where g is a PGF; instead, Definition 1 is chosen to emphasize the role of compositional depth in machine learning problems.

RBF kernels are PGF kernels. In other words, each RBF kernel can be expressed as $g(\rho(x,z))$, where g is a PGF.

Proposition 1. *RBF kernels are PGF kernels.*

Proof. Consider the RBF kernel

$$K(x,z) = \exp\left(-\frac{1}{2\sigma^2}\|x-z\|^2\right),\tag{4}$$

where σ is the lengthscale. Recall the polarization identity:

$$\|x-z\|^2 = \|x\|^2 + \|z\|^2 - 2\langle x,z\rangle.\tag{5}$$

Since $\|x\| = \|z\| = 1$, it follows from Eq. 5 that

$$\|x-z\|^2 = 2(1-\langle x,z\rangle).\tag{6}$$

Combining equations 4 and 6 yields

$$K(x,z) = \exp\left(\frac{1}{\sigma^2}(\rho(x,z)-1)\right).\tag{7}$$

By expanding the exponential function as a power series around the correlation, Eq. 7 becomes

$$K(x,z) = \sum_{i=0}^{\infty} p_i \rho^i(x,z),\tag{8}$$

$$p_i = \exp\left(-\frac{1}{\sigma^2}\right)\frac{1}{i!\sigma^{2i}}.\tag{9}$$

Equation 8 is a PGF kernel if for any $j \in \mathbb{N} \cup \{0\}$ it holds that $0 \leq p_i \leq 1$ and $\sum_{i=0}^{\infty} p_i = 1$. It is obvious from Eq. 9 that $p_i > 0$. Moreover, combining equations 7 and 8 gives

$$\sum_{i=0}^{\infty} p_i \rho^i(x, z) = \exp\left(\frac{1}{\sigma^2}(\rho(x, z) - 1)\right),$$

and setting $\rho(x, z) = 1$ leads to $\sum_{i=0}^{\infty} p_i = 1$, whence it follows that $p_i \leq 1$.

Polynomial kernels with positive coefficients summing up to one are PGF kernels. In reverse, PGF kernels with a finite number of positive coefficients are polynomial kernels. It follows directly from these observations that neither polynomial kernels nor PGF kernels are subsumed by each other.

3.3 Properties

Proposition 2 establishes the fact that PGF kernels are positive-definite and expresses their eigensystem in closed form. Proposition 2 is a special case of the corresponding proposition of [18] for the eigensystem of dot-product kernels.

Proposition 2. *Let $\mathcal{K}[g_1, \ldots, g_n] : \mathbb{S}^h \times \mathbb{S}^h \to [-1, 1]$ be a PGF kernel.*

1. *$\mathcal{K}[g_1, \ldots, g_n]$ is positive-definite on $\mathbb{S}^h \times \mathbb{S}^h$.*
2. *The eigenfunctions of $\mathcal{K}[g_1, \ldots, g_n]$ are the spherical harmonic functions Y_{ij} of degree $i \in \{0\} \cup \mathbb{N}$ and of order $j = 1, \ldots, r(h, i)$, with associated eigenvalues*

$$\lambda_{ij} = \frac{2p_i \pi^{h/2}}{\Gamma(h/2) r(h, i)}$$

of multiplicity

$$r(h, i) = \frac{2i + h - 2}{i} \binom{i + h - 3}{i - 1}.$$

p_i are the probabilities of the PGF $g_n \circ \cdots \circ g_1$ and Γ is the Gamma function.

Proof. Since $\mathcal{K}$ is a PGF kernel, it is a dot-product kernel. This allows to invoke [18], which completes the proof.

As stated in proposition 3, PGF kernels are rotationally stationary, which means that they are stationary with respect to rotations about the origin. In the present context, rotations are bijections that preserve the origin and spherical distance. PGF kernels are rotationally stationary due to the fact that the correlation $\rho(x, z)$ between two points x and z corresponds to the cosine of the angle between x and z, as elaborated in the proof of proposition 3.

Proposition 3. *PGF kernels are rotationally stationary.*

Proof. Let $\mathcal{K}$ be a PGF kernel. The dot product between the inputs x and z of $\mathcal{K}$ is given by $\langle x, z \rangle = \|x\| \|z\| \cos \theta$, where θ denotes the angle between x and z. Since $\|x\| = \|z\| = 1$, it follows that $\rho(x, z) = \cos \theta$. Let x' and z' be rotations of x and z. Due to the applied rotation, the angle between x' and z' is θ. Thus, $\rho(x', z') = \rho(x, z)$, and therefore $\mathcal{K}(x, z) = \mathcal{K}(x', z')$ according to Eq. 2.

Proposition 3 generalizes to dot-product kernels. More generally, rotational stationarity is a common property for other kernel families operating on spherical data.

PGF kernels based on PGFs that map perfectly correlated spherical data to one are correlation kernels (see proposition 4). In other words, such kernels map any two points into $[-1, 1]$, and any two perfectly correlated points to one.

Proposition 4. *Let $g_1, \ldots, g_n$ be PGFs. If for each $i = 1, \ldots, n$, the PGF g_i satisfies $g_i(1) = 1$, then the PGF kernel $\mathcal{K}[g_1, \ldots, g_n]$ is a correlation kernel.*

Proof. Let x and z be two points on the unit sphere $\mathbb{S}^h$. For any PGF g_i, $i = 1, \ldots, n$, it holds that $|g_i(s)| \leq 1$, $s \in [-1, 1]$. Furthermore, the correlation function satisfies $|\rho(x, z)| \leq 1$. Thus, $|\mathcal{K}[g_1, \ldots, g_n](x, z)| \leq 1$. It is assumed that $g_i(1) = 1$ for all $i = 1, \ldots, n$. Since $\rho(x, x) = 1$, it follows that $\mathcal{K}[g_1, \ldots, g_n](x, x) = 1$. $\square$

3.4   Kernels Based on Known PGFs

A PGF can be expressed as an infinite series according to Eq. 1 or as a closed-form function. Consequently, the resulting PGF kernel is expressed as an infinite series or as a closed-form function. In this subsection, common PGFs are used to provide examples of PGF kernels in closed form. Such examples are provided to spark interest for future research, as PGF kernels arising from common PGFs have not been studied.

Two possible ways to classify PGFs and the resulting PGF kernels are based on the existence of explicit PGF iterations and the number of non-zero terms in the series expansion of Eq. 1. First, two classes of PGFs are induced, depending on whether the PGFs have explicit iterations. The resulting PGF kernels are said to satisfy the closure property if the corresponding PGFs have explicit iterations. Second, PGFs without explicit iterations can have a finite or infinite series expansion, whereas PGFs with explicit iterations have an infinite series expansion. Thus, PGF kernels satisfying the closure property are not polynomial kernels.

Two examples of PGF kernels that do not satisfy the closure property are based on binomial and Poisson PGFs, hereafter termed binomial and Poisson kernels. The lack of closure property implies that binomial and Poisson kernels can not be expressed in closed form, and are therefore approximated numerically.

In what follows, PGF kernels that arise from compositions of Harris, linear fractional and θ kernels are discussed. As a prerequisite, Appendix A recaps on the definitions of Harris, linear fractional and θ PGFs. Appendix B states and proves the closure and other properties of some of these kernels.

Definition 3. *Let $\rho(x, z)$ be the correlation between $x \in \mathbb{S}^h$ and $z \in \mathbb{S}^h$. The Harris kernel with parameters c and r is defined as*

$$\mathcal{H}_{c,r}(x, z) := \left(c\rho(x, z)^{-r} - (c - 1)\right)^{-\frac{1}{r}}, \tag{10}$$

where $r \in \mathbb{N}$ and $c > 1$.

Definition 4. *Let $\rho(x, z)$ be the correlation between $x \in \mathbb{S}^h$ and $z \in \mathbb{S}^h$. The linear fractional kernel with parameters a and b is defined as*

$$\mathcal{F}_{a,b}(x, z) := 1 - \left(a(1 - \rho(x, z))^{-1} + b\right)^{-1}, \tag{11}$$

where $a > 0$, $b > 0$, and $a + b \geq 1$.

Table 5 in Appendix B shows all nine possible cases of PGF kernels that arise from n-fold iterations of the nine cases of θ PGFs (see Definition 7 in Appendix A). Note that the PGF kernels based on θ PGFs generalize linear fractional kernels, since the latter are derived from the first three cases of Table 5 by setting $\theta = 1$. A PGF kernel that arises from the n-fold composition of θ PGFs is considered in Propositions 5 and 6. This kernel and its infinite-depth version can be expressed in closed form according to Propositions 5 and 6, respectively (the proofs are available in Appendix B). The PGF kernel appearing in Propositions 5 and 6 is qualitatively different from Harris and linear fractional kernels, because it arises from a composition, rather than an iteration, of θ PGFs.

Proposition 5. *Consider the n-depth PGF kernel $\mathcal{T}_{\theta,c}[g_1, \ldots, g_n]$ that arises from the n-fold composition of θ PGFs $g_i(s) = 1 - ((1 - s)^{-\theta} + c_i)^{-1/\theta}$, where $c_i > 0$ for $i = 1 \ldots, n$. This kernel experesses as*

$$\mathcal{T}_{\theta,c}[g_1, \ldots, g_n](x, z) = 1 - ((1 - \rho(x, z))^{-\theta} + c)^{-1/\theta}, \tag{12}$$

where $c = \sum_{i=1}^{n} c_i$.

Proposition 6. *The infinite-depth limit of the PGF kernel of Proposition 5 is given by*

$$\lim_{n \to \infty} \mathcal{T}_{\theta,c}[g_1, \ldots, g_n](x, z) = 1 - \left((1 - \rho(x, z))^{-\theta} + c\right)^{-\frac{1}{\theta}}, \tag{13}$$

where $c = \sum_{i=1}^{\infty} c_i$.

All the PGF kernels mentioned in Subsect. 3.4 are correlation kernels. However, arbitrary PGF kernels, as used in learning tasks (Sect. 4), are not necessarily correlation kernels.

4   Learning with PGF Kernels

This section introduces a learning algorithm to fit PGF kernels to spherical data (Subsect. 4.1). The proposed algorithm is put in use to run GP regression (Subsect. 4.2) and DKL classification (Subsect. 4.3).

4.1 Learning Algorithm

Fitting a model equipped with a PGF kernel to a dataset, involves learning the kernel parameters, which are the PGF coefficients. In practice, the number of PGF parameters is reduced by introducing a numerical approximation to ensure numerical stability and computational tractability. Consider an n-depth PGF kernel $\mathcal{K}[g_1, \ldots, g_n]$ with layer widths $(l_1, \ldots, l_n)$. Each PGF g_i is numerically approximated by a series truncation. More concretely, the first $m_i < \infty$ terms are maintained in the series expansion of PGF g_i, where $m_i \leq l_i$. So, the numerical approximation of PGF kernel $\mathcal{K}[g_1, \ldots, g_n]$ contains $m = \sum_{i=1}^{n} m_i$ parameters. In this paper, $(m_1, \ldots, m_n)$ are referred to as (truncation) widths.

Typically, kernels are used in Bayesian non-parametrics. While PGF kernels are suitable for non-parametric modeling, they contain more parameters than other kernels. From this perspective, numerical approximations of PGF kernels introduce a semi-parametric setup.

To learn the m PGF kernel parameters, stochastic optimization is run on a set of corresponding real-valued parameters. More concretely, at each iteration of the optimization algorithm, a gradient descent step is taken in the space of real-valued parameters, which are then transformed to approximate PGF coefficients via the softmax function in order to evaluate the loss function.

4.2 GP Regression

Circular Von Mises Density. This subsubsection provides examples based on the circular von Mises density $f : \mathbb{S}^1 \to [0, \infty)$ given by

$$f(x|\kappa, \mu) = \frac{\exp\left(\kappa \cos\left(x - \mu\right)\right)}{2\pi I_0(\kappa)},$$

where κ and μ correspond to the shape and location parameters, and $I_0(\kappa)$ denotes the modified Bessel function of the first kind of order zero. In the experiments, the von Mises density parameters are set to $\kappa = 2$ and $\mu = (0, 0)^T$. Two GP regression examples related to the circular von Mises density are presented, namely an ablation study of PGF kernel width and depth, and a performance comparison between PGF and other kernels.

The ablation study aims to demonstrate the role of width and depth in the predictive performance of PGF kernels. To this end, two sets of comparisons are run. To assess the role of width, three single-layer PGF kernels with increasing width $m_1 = 2$, $m_1 = 100$ and $m_1 = 200$ are compared. To assess the role of depth, three fixed-width PGF kernels with increasing depth $n = 1$, $n = 2$ and $n = 3$ are compared; more specifically, the widths are set to $m_1 = 10$, $(m_1, m_2) = (10, 10)$ and $(m_1, m_2, m_3) = (10, 10, 10)$.

A training set is simulated by drawing 250 samples from the circular von Mises density and adding Gaussian noise $\mathcal{N}(\mu = 0, \sigma = 0.5)$ to them. For each PGF kernel, a GP regression model is fitted to the training set by running the Adam optimizer for the same number of iterations (200). The standard log-marginal likelihood function for GP regression is used; see [15, equation 2.30]. A

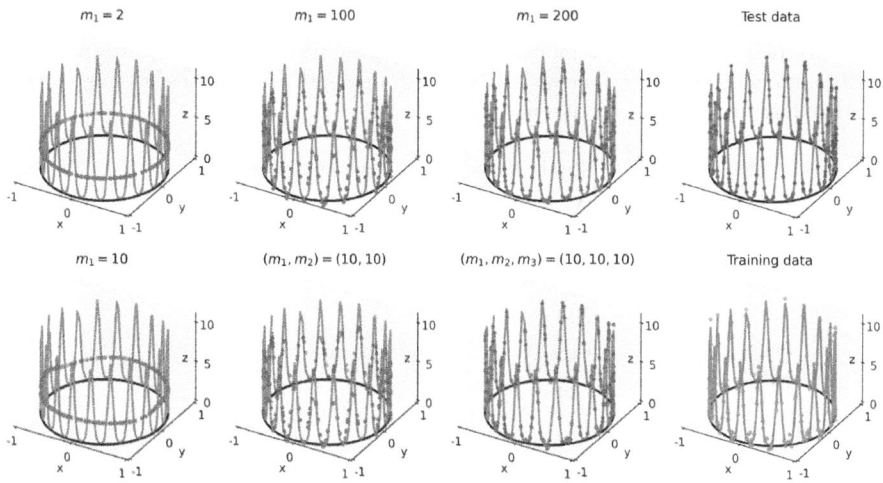

Fig. 1. Visual demonstration of the effect of PGF kernel width and depth on predictive performance for GP regression fitted to data drawn from the circular von Mises density. The blue line represents the circular von Mises density from which the data are simulated. The orange, pink and red points represent training data, test data and predictions, respectively. Increasing the width or depth brings the predictions (red points) closer to the test data (pink points) and therefore improves predictive performance. (Color figure online)

test set of 250 data points is drawn from the circular von Mises density. Using the estimated PGF kernel parameters obtained from training, predictions are made on the known ground truth, that is, on the noiseless test set.

Figure 1 visualizes the relative predictive performance of PGF kernels of varying width or depth. The blue line in each plot represents the underlying circular von Mises density. The orange points on the bottom right plot and the pink points on the top right plot correspond to the noisy training set and the noiseless test set. The red points in the other plots of Fig. 1 depict predictions. The first three plots in row one correspond to single-layer PGF kernels with widths $m_1 = 2$, $m_1 = 100$ and $m_1 = 200$, while the first three plots in row two correspond to fixed-width PGF kernels with depths $n = 1$, $n = 2$ and $n = 3$. As can be seen from the first and second rows of plots, increasing the width or depth produces predictions (red points) in closer proximity to the test set (pink points in the top right plot). In other words, Fig. 1 shows that increasing the width or depth of a PGF kernel results in higher predictive performance.

Table 1 summarizes numerically the predictive performance of the simulation visualized in Fig. 1. More specifically, two error metrics are reported, namely the mean absolute error (MAE) and the mean squared error (MSE). As observed, both MAE and MSE are strictly decreasing functions of width and depth.

Next, a predictive performance comparison is made between the PGF kernel and other well-known kernels. In this example, the PGF kernel with depth $n = 3$

Table 1. Error metrics (MAE and MSE) demonstrate the effect of PGF kernel width and depth on predictive performance for GP regression fitted to data drawn from the circular von Mises density. Increasing the width or depth reduces these error metrics and therefore improves predictive performance.

Single layer, varying width		
Width	MAE	MSE
2	2.8564	11.8405
100	1.3331	2.7496
200	0.5144	0.4973
Fixed width, varying depth		
Depth	MAE	MSE
1	2.8576	11.9041
2	1.4035	3.1663
3	0.2530	0.1082

and widths $(m_1, m_2, m_3) = (30, 30, 30)$ is used. Thus, the employed PGF kernel has $m = 90$ parameters. The comparison is carried out against the RBF, Matern, periodic and spectral mixture kernel.

Ten training sets, each consisting of 500 data points, are simulated following the sampling procedure described in the ablation study. For each kernel, a GP regression model is fitted to each training set by running the Adam optimizer for the same number of iterations $(1,000)$. The adopted log-marginal likelihood function is mentioned in [15, equation 2.30]. Ten test sets, each consisting of 500 data points, are drawn from the circular von Mises density. For each kernel, predictions are made and error metrics are computed on each test set.

Table 2 summarizes numerically the predictive performance based on the kernels under comparison. For each kernel, the mean MAE and mean MSE across the ten test sets are tabulated. For these means, standard errors (estimated standard deviations) are also reported. In this GP regression experiment, the PGF kernel attains lower MAE and MSE than the RBF kernel. This is an empirical confirmation of the fact that PGF kernels generalize RBF kernels (see Proposition 1). Besides, the PGF kernel yields the smallest means and standard errors for MAE and MSE in comparison to the other kernels used in the experiment.

Spherical Exponential-Cosine Function. This subsubsection provides an example based on spherical input data from $\mathbb{S}^2$ and output data simulated from an exponential-cosine (exp-cos) function. Points on the sphere are simulated in polar coordinates (θ, ϕ), where θ and ϕ denote the polar angle and azimuthal angle, respectively. The polar coordinates are transformed to Cartesian coordinates, and the latter constitute the input data. The output data are generated

Table 2. Mean MAEs and MSEs (accompanied by standard errors) for GP regression fitted to data, drawn from the circular von Mises density, using PGF, RBF, Matern, periodic and spectral mixture kernels. For each kernel, the error metric means and standard errors are obtained from ten runs (for details, see the main text). Metrics reported in bold indicate best performance.

Kernel	MAE	MSE
PGF	**0.1683±0.0181**	**0.0501±0.0122**
RBF	1.3144±1.3583	4.5370±5.7027
Matern	0.3766±0.0360	0.2621±0.0669
Periodic	0.3031±0.0265	0.1748±0.0289
Spectral	0.2553±0.0328	0.1170±0.0360

from the spherical exp-cos function $f : [0, \pi] \times [-\pi, \pi] \to [0, \infty)$ given by

$$f(\theta, \phi) = \exp\left(u \sum_{i=1}^{4} \cos\left(v(a_i\theta + b_i\phi + c_i)\right) \right), \tag{14}$$

where $(a_1, a_2, a_3, a_4) = (0, 1, 1, 1)$, $(b_1, b_2, b_3, b_4) = (1, 1, 0, 1)$, $(c_1, c_2, c_3, c_4) = (0, \pi/2, \pi, 3\pi/2)$, $u = 0.5$ and $v = 15$.

A predictive performance comparison is made between the PGF kernel and the RBF, Matern, periodic and spectral mixture kernel. In this example, the PGF kernel with depth $n = 3$ and widths $(m_1, m_2, m_3) = (20, 20, 20)$ is used, which has $m = 60$ parameters.

Four training sets, each consisting of 5,000 data points, are generated. For each training set, 5,000 input data points are simulated by sampling uniformly at random the polar angle θ and azimuthal angle ϕ. The corresponding output data points are generated by evaluating the exp-cos function of Eq. 14 at the input data points and by adding Gaussian noise $\mathcal{N}(\mu = 0, \sigma = 0.2)$. For each kernel, a GP regression model is fitted to each training set by running the Adam optimizer for the same number of iterations (1,000). The adopted log-marginal likelihood function is available in [15, equation 2.30]. Four noiseless test sets, each consisting of 5,000 data points, are generated similarly to the training sets, the only difference being that noise is not added to the output. The test sets are generated without noise to ensure that predictive performance is evaluated against the known ground truth given by Eq. 14. For each kernel, predictions are made and error metrics are computed on each test set.

Figure 2 demonstrates the training and test set generation. The sphere on the left-hand side displays the exp-cos function of Eq. 14, from which test points are drawn. The sphere on the right-hand side displays a noisy version of the exp-cos function, with noise drawn from $\mathcal{N}(\mu = 0, \sigma = 0.2)$; training points are simulated from such realizations.

Table 3 provides an empirical comparison of the kernels involved by showing numerical summaries of predictive performance. For each kernel, the mean MAE,

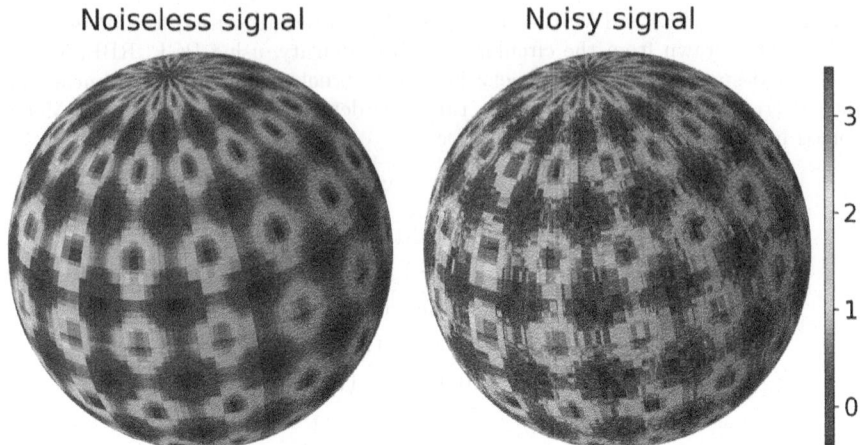

Fig. 2. Noiseless (left) and noisy version (right) of the spherical exp-cos function given by Eq. 14. For the noisy version, Gaussian noise $\mathcal{N}(\mu = 0, \sigma = 0.2)$ is added to the exp-cos function. Training sets are drawn from such noisy realizations, whereas test sets are drawn from the noiseless version of the exp-cos function.

Table 3. Mean MAEs and MSEs (accompanied by standard errors) for GP regression fitted to data, drawn from the spherical exp-cos function (Eq. 14), using PGF, RBF, Matern, periodic and spectral mixture kernels. For each kernel, the error metric means and standard errors are obtained from four runs (for details, see the main text). Metrics reported in bold indicate best performance.

Kernel	MAE	MSE
PGF	**0.1476**±0.0076	0.0443±0.0085
RBF	0.3094±0.1598	0.1865±0.1444
Matern	0.1504±0.0011	**0.0442**±0.0033
Periodic	0.2500±0.0088	0.1149±0.0078
Spectral	0.1549±0.0106	0.0533±0.0075

mean MSE and associated standard errors across the four test sets are tabulated. In this GP regression experiment, the PGF kernel outperforms the RBF kernel in terms of MAE and MSE. This observation confirms empirically Proposition 1, according to which PGF kernels form a superset of RBF kernels. Moreover, the PGF kernel attains the lowest MAE and second lowest MSE across all kernels. The observed advantage of the PGF kernel is supported by the different scaling of error metrics, indicated by the reported standard errors. Notably, only the PGF and Matern kernels exhibit comparable predictive performance, outperforming the RBF, periodic and spectral mixture kernel.

4.3   Deep Kernel Learning

As a last example, the supervised learning problem of multiclass classification on the hypersphere is considered. To this end, the hyperspherical Thomas process [19] on $\mathbb{S}^h$ is used. The hyperspherical Thomas process is conditioned on the number of clusters, given the supervised learning setup.

Data are generated from this doubly stochastic process. First, cluster centers are simulated on the hypersphere uniformly at random. Second, for each cluster, data points are simulated from a von Mises-Fisher distribution whose mean direction is set to be the cluster center. The number of data points per cluster is drawn from a Poisson distribution. Thus, three sets of parameters are involved in the data generation process, namely the intensity λ of the underlying Poisson process, and the mean directions μ and concentration parameters κ of the von Mises-Fisher distributions. Four datasets are simulated this way from the hyperspherical Thomas process. Each of the four datasets is split into two halves to generate four corresponding training and test sets. In this example, the multiclass classification problem involves four clusters, and the parameters are set to $h = 17$, $\lambda = 850$ and $\kappa = (20, 20, 20, 20)$. To give an indication of sample size, the mean number of data points across the four datasets (before being split into training and test sets) is 8147.

A DKL classification model is fitted to the data simulated from the hyperspherical Thomas process. The adopted model is a modification of the DKL framework introduced by [20]. More specifically, the multilayer perceptron of [11] is used as the neural network component of the DKL model. This choice of representation for the multilayer perceptron is made to map the input data on the unit hypersphere to embeddings on the unit sphere. The embeddings must be on the unit sphere to respect the input requirements for the PGF kernel. The Dirichlet-based GP of [13] is employed as the GP component of the DKL model. Thus, GP classification is performed through the latter part of the DKL model, using the embeddings as input to the GP. This modified DKL model has been chosen over standalone GP classification for two reasons. First, the DKL model reduces the computational cost of GP classification by embedding the input into a lower-dimensional latent spherical space. Second, these embeddings facilitate explainability, as they constitute latent features in $\mathbb{R}^3$ that can be visualized.

A predictive performance comparison is made between the PGF kernel and the RBF, Matern and periodic kernel in the context of this DKL example. The PGF kernel with depth $n = 3$ and widths $(m_1, m_2, m_3) = (20, 20, 20)$ is used.

For each kernel, two steps are taken. First, a DKL model is fitted to each training set by running the Adam optimizer for the same number of iterations (500). Second, predictions are made and error metrics are computed on the corresponding test set.

Figure 3 shows the embeddings generated by fitting the DKL model to one of the training sets. The embeddings produced by each kernel are visualized on the unit sphere. On each displayed sphere, the four colors of the embeddings represent the four classes of the classification problem. The PGF kernel yields

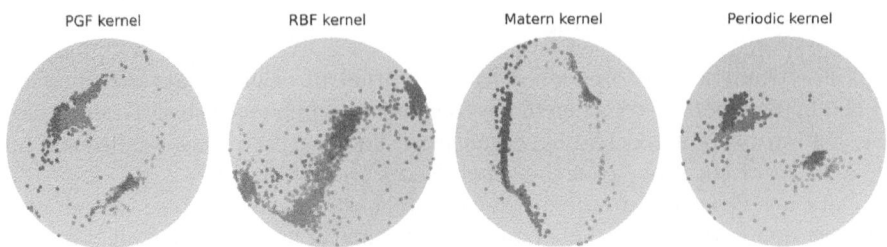

Fig. 3. Embeddings in $\mathbb{R}^3$ generated by training a DKL classification model on a dataset simulated from a hyperspherical Thomas process in $\mathbb{R}^{18}$. Each of the four spheres displays the embeddings generated by training the DKL model with a different kernel. On each sphere, the four colors of the embeddings represent the four classes of the classification problem.

Table 4. Mean predictive accuracies and standard errors for DKL classification fitted to data, drawn from the hyperspherical Thomas process, using PGF, RBF, Matern and periodic kernels. For each kernel, the mean predictive accuracy and standard error are obtained from four runs (for details, see the main text). The bold font indicates best performance.

Kernel	Accuracy (%)
PGF	**88.55**±4.8772
RBF	88.30±5.1280
Matern	88.10±5.0961
Periodic	88.27±4.9245

more distinct clusters of embeddings than the RBF kernel, and more generally than the other kernels.

Table 4 summarizes the predictive performance of the four compared kernels. For each kernel, the table shows the mean predictive accuracy and its standard error across the four test sets. In this DKL classification experiment, the PGF kernel outperforms the RBF kernel in terms of predictive accuracy. Moreover, the PGF kernel achieves the highest predictive accuracy among the four kernels.

So, the predictive performance of Table 4 and the clustering visualization of Fig. 3 demonstrate the superiority of the PGF kernel in this DKL example. However, the standard errors of Table 4 indicate that the observed differences in performance between the four kernels seem negligible.

5   Discussion and Future Directions

This paper has introduced the family of PGF kernels. Due to their parameterization, PGF kernels replace kernel hyperparameter optimization with a semiparametric learning task. Since PGF kernels generalize RBF kernels, the former provide an alternative to the latter in the case of spherical data.

Several directions for theoretical research arise. PGF kernels satisfying the closure property may be used to approximate arbitrary PGF kernels; characterizing the quality of such an approximation is an open problem. This approximation is beneficial in two ways. First, PGF kernels with or without the closure property lead to non-parametric or semi-parametric learning tasks, respectively. For example, the fewer parameters of a PGF kernel that satisfies the closure property can be treated as hyperparameters in a non-parametric setting. Thus, PGF kernels with the closure property, if used as proxies to PGF kernels without the closure property, can simplify the learning task. Second, the closure property substantially reduces the computational cost.

Another theoretical problem is to identify conditions on the parameters of a PGF kernel so that the kernel is optimal for certain classes of data or learning tasks. In other words, the question becomes how to select an optimal PGF for a given dataset under the PGF kernel setup.

Spherical data analysis gives rise to future applications of PGF kernels. For instance, simulations of partial differential equations on a sphere, such as Boussinesq convection on a spherical shell or viscous shallow-water motion on a sphere, are computationally expensive. GP emulators with PGF kernels can be used to reduce computational cost. As another example, PGF kernels can find applications in earth data science, including air temperature forecasting and ocean surface topography interpolation.

Acknowledgments. This material is based upon work supported by the Google Cloud Research Credits program with the award GCP19980904. The authors would like to thank Devanshu Agrawal, Serik Sagitov and Umberto Noe for useful discussions.

Compliance with Ethical Standards. The authors have no competing interests to declare that are relevant to the content of this article.

Software. This paper is accompanied by a Python package, called `pgfml`, to facilitate the use of PGFs in machine learning. `pgfml` (https://github.com/papamarkou/pgfml), which has been developed using `GPyTorch`, provides an implementation of PGF kernels and PGF activations. The source code for the experiments of Sect. 4 can be found in `pgf_kernel_experiments` (https://github.com/papamarkou/pgf_kernel_experiments), forming a separate Python package based on `pgfml`.

A   PGFs with Explicit Iterations

This appendix recaps on the definitions of Harrris, linear fractional and θ PGFs. Definition 7 reproduces the definition of θ PGFs, as introduced in [16].

Definition 5. *A Harris PGF with parameters c and r is defined as*

$$g(s) = \left(cs^{-r} - (c-1)\right)^{-\frac{1}{r}}, \ s \in [-1, 1], \tag{15}$$

where $r \in \mathbb{N}$ and $c > 1$.

Definition 6. *A linear fractional PGF with parameters a and b is defined as*

$$g(s) = 1 - \left(a(1-s)^{-1} + b\right)^{-1}, \ s \in [-1,1], \tag{16}$$

where a > 0, b > 0, and a + b ≥ 1.

Definition 7. *For θ ∈ (−1,0) ∪ (0,1], consider the PGF*

$$g(s) = r - (a(r-s)^{-\theta} + c)^{-1/\theta}, \theta \in (-1,0) \cup (0,1]. \tag{17}$$

It is assumed that the parameters θ, a, c and r satisfy one of the three options

$$\theta \in (0,1], \qquad\qquad a \geq 1, \qquad c > 0, \qquad\qquad\qquad\qquad r = 1,$$
$$\theta \in (-1,0) \cup (0,1], \ a \in (0,1), \ c = (1-a)(1-q)^{-\theta}, \ q \in [0,1), \ r = 1,$$
$$\theta \in (-1,0) \cup (0,1], \ a \in (0,1), \ c = (1-a)(r-q)^{-\theta}, \ q \in [0,1], \ r > 1.$$

For θ = 0, consider the PGF

$$g(s) = r - (r-q)^{1-a}(r-s)^a, \theta = 0. \tag{18}$$

It is assumed that parameters q and r satisfy one of the two options

$$q \in [0,1), r = 1,$$
$$q \in [0,1], r > 1.$$

For θ = −1, consider the PGF

$$g(s) = as + (1-a)q, \theta = -1. \tag{19}$$

It is assumed that a ∈ (0, 1) and q ∈ [0, 1].
    For each of the three cases θ ∈ (−1, 0) ∪ (0, 1], θ = 0 and θ = −1, the corresponding PGF given by Eq. 17, 18 and 19 is called a θ PGF.

For any fixed choice of θ, a, c and r, an n-fold iteration of the corresponding θ PGF yields a θ PGF. Thus, an n-fold iteration of a θ PGF is expressed in closed form. The families of θ PGFs and of Harris PGFs [7, p. 10] are the only two known PGF families with explicit n-fold iterations.

B   PGF Kernels Satisfying the Closure Property

This appendix establishes some properties of PGF kernels that arise from compositions of Harris, linear fractional and θ kernels. Proposition 7 states a symmetry property of the Harris kernel. Propositions 8, 9 and 10 state the corresponding closure property of the Harris kernel, linear fractional kernel and PGF kernels arising from n-fold iterations of θ PGFs. The appendix concludes with the proofs of Propositions 5 and 6.

Proposition 7. *For any even r and any $(x, z) \in \mathbb{S}^h \times \mathbb{S}^h$, the Harris kernel safisfies*

$$\mathcal{H}_{c,r}(x, z) = \mathcal{H}_{c,r}(-x, z) = \mathcal{H}_{c,r}(x, -z). \tag{20}$$

Proof. Eq. 20 follows from Eq. 10 for even r.

Proposition 8. *For any $n \in \mathbb{N}$, let $\mathcal{K}[g_1, \ldots, g_n]$ be an n-depth PGF kernel, where g_i, $i = 1, \ldots, n$, is a Harris PGF with parameters c_i and r. The n-depth PGF kernel $\mathcal{K}[g_1, \ldots, g_n]$ is the Harris kernel $\mathcal{H}_{w_n, r}$ where $w_n = \prod_{i=1}^{n} c_i$.*

Proof. Eq. 20 follows from Eq. 10 for even r.

Proof. The proposition will be proved by induction. For $n = 1$, according to Eqs. 2, 15 and 10, notice that

$$\mathcal{K}[g_1](x, z) = g_1(\rho(x, z)) = \mathcal{H}_{w_1, r}(x, z).$$

Assuming that

$$\mathcal{K}[g_1, \ldots, g_n](x, z) = \left(w_n \rho(x, z)^{-r} - (w_n - 1) \right)^{-\frac{1}{r}} = \mathcal{H}_{w_n, r}(x, z) \qquad (21)$$

holds for $n = j$, it will be shown that it also holds for $n = j + 1$. To this end,

$$
\begin{aligned}
\mathcal{K}[g_1, \ldots, g_{j+1}](x, z) &= g_{j+1} \circ g_j \circ \cdots \circ g_1 (\rho(x, z)) && \text{See equation 2}\\
&= g_{j+1} \circ \mathcal{K}[g_1, \ldots, g_j](x, z) && \text{See equation 2}\\
&= g_{j+1} \left(\left(w_j \rho(x, z)^{-r} - (w_j - 1) \right)^{-\frac{1}{r}} \right) && \text{See equation 21}\\
&= \left(w_{j+1} \rho(x, z)^{-r} - (w_{j+1} - 1) \right)^{-\frac{1}{r}}, && \text{See equation 15}
\end{aligned}
$$

which completes the proof.

Proposition 9. *For any $n \in \mathbb{N}$, let $\mathcal{K}[g_1, \ldots, g_n]$ be an n-depth PGF kernel, where g_i, $i = 1, \ldots, n$, is a linear fractional PGF with parameters a_i and b_i. The n-depth PGF kernel $\mathcal{K}[g_1, \ldots, g_n]$ is the linear fractional kernel $\mathcal{F}_{u_n, v_n}$, where $u_n = \prod_{i=1}^{n} a_i$ and $v_n = \sum_{i=1}^{n} u_{i-1} b_i$.*

Proof. The proposition will be proved by induction. For $n = 1$, according to Eqs. 2, 16 and 11, notice that $\mathcal{K}[g_1](x, z) = g_1(\rho(x, z)) = \mathcal{F}_{u_1, v_1}(x, z)$. Assuming that

$$\mathcal{K}[g_1, \ldots, g_n](x, z) = 1 - \left(u_n (1 - \rho(x, z))^{-1} + v_n \right)^{-1} = \mathcal{F}_{u_n, v_n}(x, z) \qquad (22)$$

holds for $n = j$, it will be shown that it also holds for $n = j + 1$. To this end,

$$
\begin{aligned}
\mathcal{K}[g_1, \ldots, g_{j+1}](x, z) &= g_{j+1} \circ g_j \circ \cdots \circ g_1 (\rho(x, z)) && \text{See equation 2}\\
&= g_{j+1} \circ \mathcal{K}[g_1, \ldots, g_j](x, z) && \text{See equation 2}\\
&= g_{j+1} \left(1 - \left(u_j (1 - \rho(x, z))^{-1} + v_j \right)^{-1} \right) && \text{See equation 22}\\
&= 1 - \left(u_{j+1} (1 - \rho(x, z))^{-1} + v_{j+1} \right)^{-1}, && \text{See equation 16}
\end{aligned}
$$

which completes the proof.

Table 5. PGF kernels $\mathcal{K}[(g)_n]$, where g is a θ PGF and $\rho(x,z)$ is the correlation between $x \in \mathbb{S}^h$ and $z \in \mathbb{S}^h$.

Case	$\mathcal{K}[(g)_n]$	θ	a	q, c	r
1	$1 - (a^n(1-\rho)^{-\theta} + (a^n - 1)c)^{-1/\theta}$	$(0,1]$	$(1,\infty)$	$(0,\infty)$	1
2	$1 - ((1-\rho)^{-\theta} + nc)^{-1/\theta}$	$(0,1]$	1	$(0,\infty)$	1
3	$1 - (a^n(1-\rho)^{-\theta} + (1-a^n)(1-q)^{-\theta})^{-1/\theta}$	$(0,1]$	$(0,1)$	$[0,1)$	1
4	$1 - (1-q)^{1-a^n}(1-\rho)^{a^n}$	0	$(0,1)$	$[0,1)$	1
5	$1 - (a^n(1-\rho)^{|\theta|} + (1-a^n)(1-q)^{|\theta|})^{1/|\theta|}$	$(-1,0)$	$(0,1)$	$[0,1)$	1
6	$a^n\rho + (1-a^n)q$	-1	$(0,1)$	$[0,1]$	1
7	$r - (a^n(r-\rho)^{-\theta} + (1-a^n)(r-q)^{-\theta})^{-1/\theta}$	$(0,1]$	$(0,1)$	$[0,1]$	$(1,\infty)$
8	$r - (r-q)^{1-a^n}(r-\rho)^{a^n}$	0	$(0,1)$	$[0,1]$	$(1,\infty)$
9	$r - (a^n(r-\rho)^{|\theta|} + (1-a^n)(r-q)^{|\theta|})^{1/|\theta|}$	$(-1,0)$	$(0,1)$	$[0,1]$	$(1,\infty)$

Proposition 10. *A PGF kernel that arises from the n-fold iteration of a θ PGF expresses in closed form according to one of the nine cases of Table 5 depending on the values of parameters θ, a, q, c, r of the θ PGF.*

Proof. Closed-form expressions for n-fold iterations of a θ PGF f are provided in [16, section 4]. Thus, Table 5 follows.

Proof (Proof of Proposition 5). The proposition will be proved by induction, starting from Definition 1. For $n = 1$, notice that $K[g_1](x,z) = g_1(\rho(x,z)) = T_{\theta,c_1}(x,z)$. Assuming that Eq. 12 holds for $n = j$, it will be shown that it also holds for $n = j + 1$. To this end,

$$\mathcal{K}[g_1,\ldots,g_{j+1}](x,z) = g_{j+1} \circ g_j \circ \cdots \circ g_1(\rho(x,z)) \qquad \text{See equation 2}$$

$$= f_{j+1} \circ \mathcal{K}_{1:j}(x,z) \qquad \text{See equation 2}$$

$$= g_{j+1}\left(1 - \left((1-\rho(x,z))^{-\theta} + \sum_{i=1}^{j} c_i\right)^{-\frac{1}{\theta}}\right) \qquad \text{See equation 12}$$

$$= 1 - \left((1-\rho(x,z))^{-\theta} + \sum_{i=1}^{j+1} c_i\right)^{-\frac{1}{\theta}}, \qquad \text{See equation 17}$$

which completes the proof.

Proof (Proof of Proposition 6). Taking the limit of Eq. 12 as n tends to infinity yields Eq. 13.

References

1. Alsmeyer, G.: Linear fractional Galton–Watson processes in random environment and perpetuities. Stochast. Qual. Control **36**(2), 111–127 (2021)
2. Daniely, A., Frostig, R., Singer, Y.: Toward deeper understanding of neural networks: the power of initialization and a dual view on expressivity. In: Advances in Neural Information Processing Systems, vol. 29. Curran Associates, Inc. (2016)

3. Fischer, H.: A history of the Central Limit Theorem: From Classical to Modern Probability Theory. Springer (2011). https://doi.org/10.1007/978-0-387-87857-7
4. Gardner, J.R., Pleiss, G., Bindel, D., Weinberger, K.Q., Wilson, A.G.: GPyTorch: Blackbox matrix-matrix Gaussian process inference with GPU acceleration. In: Advances in Neural Information Processing Systems (2018)
5. Grimmett, G., Stirzaker, D.: Probability and Random Processes. Oxford university press, 4th edn. (2020)
6. Guinness, J., Fuentes, M.: Isotropic covariance functions on spheres: some properties and modeling considerations. J. Multivar. Anal. **143**, 143–152 (2016)
7. Harris, T.E.: The Theory of Branching Processes. Springer-Verlag (1963). https://doi.org/10.2307/3614869
8. Kersting, G., Vatutin, V.: Discrete time branching processes in random environment. Wiley (2017)
9. Kozlov, M.V.: On the asymptotic behavior of the probability of non-extinction for critical branching processes in a random environment. Theory Prob. Appl. **21**(4), 791–804 (1977)
10. Lei, H., Akhtar, N., Mian, A.: Octree guided CNN with spherical kernels for 3D point clouds. In: 2019 IEEE/CVF Conference on Computer Vision and Pattern Recognition (CVPR), pp. 9623–9632 (2019)
11. Liang, T., Tran-Bach, H.: Mehler's formula, branching process, and compositional kernels of deep neural networks. J. Am. Stat. Assoc. **117**, 1324–1337 (2021)
12. Marques, R., Bouville, C., Bouatouch, K.: Gaussian process for radiance functions on the $\mathbb{S}^2$ sphere. Comput. Graphics Forum **41**(6), 67–81 (2022)
13. Milios, D., Camoriano, R., Michiardi, P., Rosasco, L., Filippone, M.: Dirichlet-based Gaussian processes for large-scale calibrated classification. In: Proceedings of the 32nd International Conference on Neural Information Processing Systems, pp. 6008–6018. NIPS 2018, Curran Associates Inc. (2018)
14. Pennington, J., Yu, F.X.X., Kumar, S.: Spherical random features for polynomial kernels. In: Advances in Neural Information Processing Systems, vol. 28. Curran Associates, Inc. (2015)
15. Rasmussen, C.E., Williams, C.K.: Gaussian Processes for Machine Learning. MIT Press (2005)
16. Sagitov, S., Lindo, A.: A special family of galton-watson processes with explosions. In: del Puerto, I.M., et al. (eds.) Branching Processes and Their Applications. LNS, vol. 219, pp. 237–254. Springer, Cham (2016). https://doi.org/10.1007/978-3-319-31641-3_14
17. Seal, H.L.: The historical development of the use of generating functions in probability theory. Mitt Verein Schweiz Versich Math. **49**, 209–228 (1949)
18. Smola, A., Óvári, Z., Williamson, R.C.: Regularization with dot-product kernels. In: Advances in Neural Information Processing Systems, vol. 13. MIT Press (2001)
19. Thomas, M.: A generalization of Poisson's binomial limit for use in ecology. Biometrika **36**(1/2), 18–25 (1949)
20. Wilson, A.G., Hu, Z., Salakhutdinov, R., Xing, E.P.: Deep kernel learning. In: Proceedings of the 19th International Conference on Artificial Intelligence and Statistics. Proceedings of Machine Learning Research, vol. 51, pp. 370–378. PMLR (2016)
21. Zhao, C., Song, J.S.: Exact heat kernel on a hypersphere and its applications in kernel SVM. Front. Appl. Math. Stat. **4**, 1 (2018)

Tailored Finite Point Operator Networks for Interface Problems

Ye Li[1][(✉)], Ting Du[2], and Zhongyi Huang[2]

[1] College of Computer Science and Technology, Nanjing University of Aeronautics and Astronautics, Nanjing 211106, China
yeli20@nuaa.edu.cn
[2] Department of Mathematical Sciences, Tsinghua University, Beijing 100084, China
dt20@mails.tsinghua.edu.cn, zhongyih@tsinghua.edu.cn

Abstract. Interface problems pose significant challenges due to the discontinuity of their solutions, particularly when they involve singular perturbations or high-contrast coefficients, resulting in intricate singularities that complicate resolution. The increasing adoption of deep learning techniques for solving partial differential equations has spurred our exploration of these methods for addressing interface problems. In this study, we introduce Tailored Finite Point Operator Networks (TFPONets) as a novel approach for tackling parameterized interface problems. Leveraging DeepONets and integrating the Tailored Finite Point method (TFPM), TFPONets offer enhanced accuracy in reconstructing solutions without the need for intricate equation manipulation. Experimental analyses conducted in both one- and two-dimensional scenarios reveal that, in comparison to existing methods such as DeepONet and IONet, TFPONets demonstrate superior learning and generalization capabilities even with limited locations.

Keywords: Interface problems · Tailored Finite Point method · DeepONets

1 Introduction

Elliptic and parabolic problems featuring discontinuous coefficients or sources are prevalent across diverse fields, including fluid mechanics [2,21], materials science [16,22], electromagnetics [5], and biomimetics [8]. Typically, while solutions to interface problems exhibit smoothness within homogenous material or fluid subdomains, the global solution regularity can be significantly compromised, manifesting singularities at interfaces [1,7,9]. The variable nature of singularities and the complex geometries at interfaces render traditional finite element or finite difference methods less effective for achieving precision.

The integration of deep learning into computational science has initiated a transformative approach to solving partial differential equations (PDEs). Innovations such as physics-informed neural networks [19], the Deep Ritz Method [25], and the Deep Galerkin Method [20] leverage neural networks to navigate these challenges. Moreover, operator learning strategies, including the Fourier Neural Operator (FNO) [13] and Deep Operator Networks (DeepONets) [17], are emerging as powerful tools capable of approximating mappings between infinite-dimensional spaces. Addressing interface problems-where solutions demonstrate piecewise continuity and are subject to jump conditions at interfaces-poses a challenge for vanilla neural networks methods, often leading to poor performance near the interfaces.

In the manuscript, we introduce Tailored Finite Point Operator Networks (TFPONets), a novel integration of the Tailored Finite Point method (TFPM) [7] with DeepONets, designed to resolve parameterized interface problems with high precision. TFPONets are adept at handling not only general interface problems but also singularly perturbed and high-contrast interface problems, as our experimental results demonstrate robust performance even in these complex scenarios. While our model utilizes DeepONets, its architecture is versatile, allowing for the substitution of other operator networks or neural operators, such as the Fourier Neural Operator (FNO). The key contributions of our work are as follows:

- We introduce TFPONets, a method that combines TFPM for accurate solution reconstruction with the learning capabilities of DeepONets, eliminating the need for complex numerical PDEs discretization and calculation. By incorporating the prior knowledge of basis functions into the neural network design, our method accelerates the learning process without imposing additional training complexity.
- TFPONets preserve the meshless nature inherent in DeepONets, allowing for the utilization of either randomly sampled or fixed grid points without constraint. Our models can be trained on coarse grids while delivering accurate predictions for functions beyond the training set and finer locations, exhibiting remarkable generalization capability.
- We demonstrate the efficacy of our model across one-dimensional and two-dimensional interface problems, including those with singular perturbations and high-contrast coefficients. Our findings indicate that TFPONets, even with limited locations in the training dataset, can resolve complex interface problems, effectively capturing boundary and interior layers.

2   Preliminaries

2.1   Interface Problems

In this manuscript, we develop learning methods to solve an elliptic interface problem specified by the following equation:

$$-\nabla \cdot (a\nabla u) + bu = f, \text{ in } \Omega/\Gamma, \ [u]_\Gamma = g_D, \ [a\nabla u \cdot \boldsymbol{n}]_\Gamma = g_N, \ u|_{\partial\Omega} = h, \quad (1)$$

where Ω is the domain comprising N subdomains $\{\Omega_i, i = 1, \ldots, N\}$ separated by the interface Γ. Here, $a(\boldsymbol{x}) > 0$ and $b(\boldsymbol{x}) \geq 0$ are piecewise smooth functions.

The terms $[\cdot]\,\Gamma$ and $[a\nabla u \cdot n]\,\Gamma$ represent the jump across the interface, i.e., for a point $x0 \in \Gamma$ between Ω_i and Ω_{i+1},

$$[u]\,(x_0) = \lim_{x \in \Omega_{i+1}, x \to x_0} u(x) - \lim_{x \in \Omega_i, x \to x_0} u(x),$$

$$[a\nabla u \cdot n]\,(x_0) = \lim_{x \in \Omega_{i+1}, x \to x_0} a(x)\nabla u(x) \cdot n - \lim_{x \in \Omega_i, x \to x_0} a(x)\nabla u(x) \cdot n.$$

We also address two more challenging types of interface problems:

- **singularly perturbed interface problem.** When the coefficient function $a(x)$ in the equation is a small positive constant, the resulting solutions display rapid changes in the form of boundary, interior, and corner layers.
- **high-constrast interface problem.** These arise when the coefficient function $a(x)$, while smooth within each subdomain, exhibits significant variations between different subdomains. An example is the modeling of elastic materials with thin layers.

2.2   Current Approaches

Numerical Solver. Numerical methods like the immersed boundary (IB) method [18], immersed interface method (IIM) [10], ghost fluid method (GFM) [15], and tailored finite point method (TFPM) [7] have been developed to address these problems, each improving the accuracy of capturing interface discontinuities and complex conditions. Nevertheless, general numerical methods necessitate meticulous region discretization and equation formulation.

Neural Network Solver. In the realm of neural networks, piecewise networks have been employed to approximate solutions by assigning a network to each subdomain and using jump conditions as penalty terms in the loss function [3,4]. Despite their effectiveness, these methods can be complicated by the need to balance various loss terms during optimization. Domain decomposition methods, such as the Schwarz method [14], have been adapted with neural networks to ease these challenges [11,12]. Additionally, approaches like the discontinuity capturing shallow neural network (DCSNN) [6] and deep Nitsche-type method [23] have been proposed to manage high-contrast discontinuous coefficients and complex boundary conditions. However, these methods are confined to solving a single equation.

Neural Operator. For parametric PDE problems involving repeated evaluations across similar inputs, operator learning techniques like Fourier Neural Operators (FNO) and Deep Operator Networks (DeepONets) have shown promise. However, research in solving parametric interface problems is limited, with some studies like IONet [24] using DeepONets for interface problems across subdomains. Challenges remain in tackling problems with more pronounced singularities, like singularly perturbed and high-contrast interface problems.

3   Method

3.1   Tailored Finite Point Method

Theoretical findings from [7] indicate that Tailored Finite Point Method (TFPM) not only resolves general interface problems with high precision but also exhibits uniform convergence with respect to the perturbation parameter. This section outlines the Tailored Finite Point method for solving interface problems, as proposed by [7], covering both one-dimensional and two-dimensional scenarios.

For the one-dimensional case with domain $\Omega = (\alpha, \beta)$, the transformation $y(x) = \int_\alpha^x 1/a(\xi)d\xi$ enables the reformulation of Eq. (1). In the two-dimensional case, where the domain is $\Omega = (\alpha_0, \beta_0) \times (\alpha_1, \beta_1)$, a corresponding transformation $\boldsymbol{y} = (y_1, y_2) = \left(\int_{\alpha_0}^{x_1} 1/a(\xi_1, x_2)d\xi_1, \int_{\alpha_1}^{x_2} 1/a(x_1, \xi_2)d\xi_2 \right)$ is applied to achieve a similar reformulation. The transformed equation takes the form:

$$- \Delta u(\boldsymbol{y}) + c(\boldsymbol{y})u(\boldsymbol{y}) = F(\boldsymbol{y}), \tag{2}$$

where $c(\boldsymbol{y}) = a(\boldsymbol{x}(\boldsymbol{y}))b(\boldsymbol{x}(\boldsymbol{y}))$ represents a transformed coefficient and $F(\boldsymbol{y}) = a(\boldsymbol{x}(\boldsymbol{y}))f(\boldsymbol{x}(\boldsymbol{y}))$ is the transformed source term. Based on this reformulated equation, we will introduce the Tailored Finite Point method (TFPM).

One-Dimensional Case. In a one-dimensional setting with a partition $\{y_j\}$, the solution to Eq. (2) within each subinterval (y_{j-1}, y_j) can be approximated by:

$$u_h(y) = \alpha_j A_1^j(y) + \beta_j A_2^j(y) + \int_{y_{j-1}}^{y_j} F(s)G_j(y(x), s)ds, \tag{3}$$

where G_j is the Green's function associated with the subinterval. The functions $A_1(y) = \{A_1^j(y)\}$ and $A_2(y) = \{A_2^j(y)\}$ serve as a pair of local bases, reflecting the distinct characteristics of the various subintervals. They are piecewise defined for each subinterval as follows:

$$A_1^j(y), \ A_2^j(y) = \begin{cases} 1, \ y, & a_j = b_j = 0, \\ \exp(y\sqrt{b_j}), \ \exp(-y\sqrt{b_j}), & a_j = 0, \ b_j \neq 0, \\ \mathrm{Ai}(c_h(y)a_j^{-2/3}), \ \mathrm{Bi}(c_h(y)a_j^{-2/3}), & a_j \neq 0, \ b_j \neq 0, \end{cases} \tag{4}$$

where $c_h(y)$ is a piecewise linear approximation of the function $c(y)$, characterized by a slope a_j and an intercept b_j over the j-th subinterval. Ai($\cdot$) and Bi($\cdot$) denote the Airy functions of the first and second kind, respectively. The coefficients α_j and β_j are determined by a linear system that ensures continuity or enforces jump conditions at the interfaces between subintervals.

Two-Dimensional Case. In two dimensions, we subdivide the domain Ω into a set of subdomains $\{\Omega^j\}$. By selecting a reference point $\boldsymbol{y}^j$ within each Ω^j, the solution to Eq. (2) over Ω^j can be approximated by [7] :

$$u_h(\boldsymbol{y}) = u_j(r) + \sum_{n=0}^{+\infty}(a_n^j I_n(\mu_j r) \cos n\theta + b_n^j I_n(\mu_j r) \sin n\theta), \tag{5}$$

where $u_j(r) = F_j \sum_{n=0}^{+\infty} \frac{(\mu_j)^{2n-2} r^{2n}}{4(n!)^2}$. Here, μ_j is a constant approximation of the function $\sqrt{c(\boldsymbol{y})}$ and F_j approximates $F(\boldsymbol{y})$ within the j-th subdomain. The coordinates (r, θ) represent the polar coordinates centered at $\boldsymbol{y}^j$, and I_n denotes the modified Bessel function of the first kind of order n.

Finally, we truncate u_h to include only the leading terms of the series expansion, leaving several unknowns a_n^j and b_n^j. These parameters are determined by ensuring u and $\partial u/\partial n$ are continuous at Ω^j's boundary.

3.2 Deep Operator Networks

Deep Operator Networks (DeepONets) are designed to learn nonlinear mappings between infinite-dimensional function spaces, particularly adept at solving parametric partial differential equations (PDEs) by establishing relationships between two such spaces. DeepONet operates by processing two inputs: a function f and a location x. It consists of two sub-networks: the branch network handles the function f at predetermined points $\{x_i\}$, while the trunk network handles the location x. The final output is the dot product of the outputs from these sub-networks, as shown in the DeepONet block in Fig. 1.

Due to the inherent discontinuity of solutions to interface problems, which exhibit significant variations across different sub-regions, direct application of DeepONet often leads to inaccuracies at the interface. While IONet [24] employs multiple DeepONets to handle solution at different sub-regions, it struggles with more singular problems, such as those involving singular perturbations and high-contrast coefficients. To address these challenges, we will propose an enhanced model capable of accurately solving interface problems, including the aforementioned singular cases.

3.3 Improved Network Structure

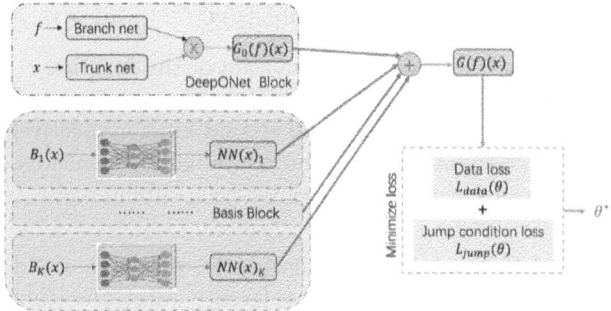

Fig. 1. TFPONet integrates a DeepONet with multiple neural networks that receive inputs of the form $(f, x, B_1(x), B_2(x), \dots)$. Specifically, DeepONet is fed with the tuple (f, x), while the i-th auxiliary neural network within TFPONet is tasked with processing the local basis $B_i(x)$.

In Sect. 3.1, we establish that the solution to interface problems can be effectively approximated by a reconstructed solution u_h (see Eq. (3) and (5)), which utilizes a series expansion over a finite or infinite set of local basis functions. This approach not only yields high-fidelity approximations but also maintains parameter consistency for singular perturbation problems without necessitating a dense discretization. Consequently, an accurate approximation of u_h directly translates to a precise representation of the interface problem's solution, with the added benefit of a simpler computational framework. This realization underpins the development of the Tailored Finite Point Operator Network (TFPONet), depicted in Fig. 1, which augments the vanilla Deep Operator Network (DeepONet).

The TFPONet framework integrates a DeepONet block with multiple basis blocks, each underpinned by a shallow neural network that processes a distinct basis function evaluated at point x. The resulting model is encapsulated by the following formulation:

$$
\text{TFPONet}(f)(x) = \text{DeepONet}(f)(x) + \sum_{i=1}^{K} \text{NN}_i(B_i(x)), \tag{6}
$$

where $\text{DeepONet}(f)(x)$ represents the output from the DeepONet block, and NN_i ($i = 1, 2, \ldots, K$) denotes the shallow neural networks corresponding to the "$NN(x)_i$" blocks in Fig. 1.

As illustrated in Fig. 1, TFPONet comprises a DeepONet block and a series of basis blocks. Each basis block processes a basis function at location x, contributing to the approximation of the corresponding term in the reconstructed solution as described by Eq. (3) or (5). In the one-dimensional case, these terms are $\alpha_j A_1^j(y(x))$ and $\beta_j A_2^j(y(x))$ from Eq. (3), while in the two-dimensional case, these terms are $(a_n^j \cos n\theta + b_n^j \sin n\theta)I_n(\mu_j r)$ from Eq. (5). The DeepONet block is designed to receive the function f and location x, enabling the approximation of non-basis terms, exemplified by $\int_{y_{j-1}}^{y_j} F(s)G_j(y(x), s)ds$ in the context of Eq. (3), or $u_j(r)$ in Eq. (5). This block further serves as an error corrector. Approximation inaccuracies in the basis blocks, stemming from both their intrinsic limitations and the finite number of blocks implementable within TFPONet, necessitate this error correction mechanism. The integration of the DeepONet block harnesses its advanced approximation capabilities to mitigate these errors effectively.

Constructing the local basis functions for basis blocks is straightforward. In the one-dimensional scenario, either a single or a pair of basis blocks suffice, corresponding to inputs $A_1(y(x))$ or $A_1(y(x)), A_2(y(x))$, respectively, as delineated in Eq. (4). For the two-dimensional case, we employ the 0-th to $(K-1)$-th order modified Bessel functions of first kind as inputs for K basis blocks. Each basis block is tasked with approximating each local basis term. Consequently, employing a greater number of basis blocks generally enhances the solution's accuracy. However, this also leads to an expanded set of neural networks, which, while potentially improving training accuracy and simplifying the training process, necessitates consideration of the inherent trade-off due to the resultant increase in parameter count.

3.4   Optimization Problem and Loss Function

For interface problems where the solution exhibits a non-zero jump, indicating discontinuity at the interface, the solution varies significantly across sub-regions. To preserve this discontinuity, a suitable approach is the deployment of distinct, smaller TFPONet for each subdomain, with each network dedicated to its respective region. This composite model is henceforth referred to as TFPONets.

If we partition the specified region Ω into N sub-regions using the interface Γ, such that $\Omega = \overset{N}{\underset{i=1}{\cup}} \Omega_i$, we address the underlying problem with a model where

$$\mathcal{N}(f)(x) = \mathcal{N}_i(f)(x), \text{ if } x \in \Omega_i, \tag{7}$$

with $\mathcal{N}_i$ being the TFPONet for sub-region Ω_i. The overall approach involves training all subproblems simultaneously, with each subproblem optimizing the parameters $\boldsymbol{\theta}_i$ of $\mathcal{N}_i$ within its region Ω_i. This is achieved by utilizing a total loss function that encompasses both individual subproblem losses and the interface loss.

The model's parameters across TFPONets are $\boldsymbol{\theta} = \overset{N}{\underset{i=1}{\cup}} \boldsymbol{\theta}_i$, optimized by minimizing the loss function

$$\mathcal{L}(\boldsymbol{\theta}) = \mathcal{L}_{data}(\boldsymbol{\theta}) + \gamma \cdot \mathcal{L}_{jump}(\boldsymbol{\theta}). \tag{8}$$

Here, $\mathcal{L}_{data}$ is the data fit term, computed as

$$\mathcal{L}_{data}(\boldsymbol{\theta}) = \frac{1}{MJ} \sum_{m=1}^{M} \sum_{j=1}^{J} |u_m(x_j) - \mathcal{N}_{\boldsymbol{\theta}}(F_m)(x_j)|^2, \tag{9}$$

where the ground truth, u_m, is obtained through high-precision numerical methods, specifically using TFPM on a uniform mesh. The samples $F_1, \ldots, F_M$ are independently drawn from Gaussian random fields, and the points $x_1, \ldots, x_J$ are positioned on a uniform grid within the region. $\mathcal{L}_{jump}$ is a soft constraint enforcing jump conditions at interfaces, with γ as the penalizing parameter, as specified by

$$\begin{aligned} \mathcal{L}_{jump}(\boldsymbol{\theta}) = \frac{1}{MJ_0} &\left(\sum_{m=1}^{M} \sum_{i=1}^{J_0} |[\mathcal{N}_{\boldsymbol{\theta}}(F_m)](x_i^{\gamma}) - g_D(x_i^{\gamma})|^2 \right. \\ &\left. + \sum_{m=1}^{M} \sum_{i=1}^{J_0} |\left[a\nabla\mathcal{N}_{\boldsymbol{\theta}}(F_m) \cdot \boldsymbol{n} \right](x_i^{\gamma}) - g_N(x_i^{\gamma})|^2 \right), \end{aligned} \tag{10}$$

where $[\cdot](x_i^{\gamma})$ denotes the jump across the interface at x_i^{γ}, similar to the definition in Sect. 2.1.

4   Experiments

Experiment Details. In this section, we explore the application of TFPONets to various interface problems encompassing both one-dimensional and two-dimensional scenarios, including those with singular perturbations and high-contrast coefficients. Additionally, we conduct a comparative analysis of our experimental results with those obtained using IONet [24], which is an improvement based on the vanilla DeepONet. IONet employs a DeepONet in each sub-region to handle the solution's singularity at the interface. If the solution is continuous across the entire region, this method degenerates to the vanilla Deep-ONet. The training dataset is composed of numerous triplets (f, x, u), with f samples being independently drawn from Gaussian random fields to serve as the input function. Here, x denotes a location, chosen without any specific constraints, and for subsequent experiments, uniform grid points are utilized as these points. The solution u of the governing equations at each location x for a given input function f is derived using TFPM. When indicating that the training dataset comprises $n_1 \times n_2$ triplets (f, x, u), we mean that the input data includes n_1 random samples f and n_2 locations x, resulting in a corresponding output dataset with $n_1 \times n_2$.

TFPONets Settings. In all experiments, each TFPONet within the respective sub-region utilizes two basis blocks. Specifically, for one-dimensional scenarios, $A_1(y(x))$ and $A_2(y(x))$-as defined by Eq. (4)-serve as the inputs for these blocks, while for two-dimensional scenarios, the inputs are the zeroth and first-order modified Bessel functions of the first kind. In the one-dimensional scenario, both the branch net, trunk net, and basis block within the DeepONet block utilize a fully connected neural network with a single hidden layer. Additionally, in the two-dimensional setting, they all employ neural networks with two hidden layers. The activation functions across all layers are ReLU.

One-Dimensional Settings. Our initial experiments address the following one-dimensional problem:

$$- (au')' + bu = f, \text{ in } \Omega/\Gamma, \; [u]_\Gamma = g_D, \; [u']_\Gamma = g_N, \; u|_{\partial\Omega} = 0. \qquad (11)$$

where $\Omega = (0, 1)$ and the interface, denoted by $\Gamma = \{0.5\}$, consists of a single point. Unless otherwise indicated, our goal is to infer the mapping $f \mapsto u$ using multiple sets of input-output data pairs.

Example 1. We explore a straightforward scenario wherein both the solution to the problem and its derivative demonstrate continuity at the interface Γ, that is, $g_D = g_N = 0$. The coefficient $b(x)$ and the source term $f(x)$ in Eq. (11) are piecewise functions given by:

$$b(x) = \begin{cases} 2x + 1, & x \in [0, 0.5), \\ 2(1 - x) + 1, & x \in [0.5, 1], \end{cases} \quad f(x) = \begin{cases} f_1(x), & x \in [0, 0.5), \\ f_2(x), & x \in [0.5, 1], \end{cases}$$

where $f_1(x)$ and $f_2(x)$ are functions drawn from two specified distribution spaces. The discontinuity in these functions or their derivatives makes the solution appear singular at the interface, even though the solution u and its derivatives are continuous. Our objective is to deduce the mapping from f to u.

We assess the efficacy of our proposed TFPONets on problems characterized by singular perturbations and high-contrast coefficients. Given the continuity of the solution across the interface, a single TFPONet suffices for the entire domain. This approach will be benchmarked against the vanilla DeepONet.

In the singular perturbation case, setting the coefficient $a(x)$ in Eq. (11) to 0.0001 induces rapid solution transitions within interior and boundary layers, as illustrated in Fig. 2a). For high-contrast problems, we define $a(x) = 1$ for $x \in [0, 0.5)$ and $a(x) = 0.0001$ for $x \in [0.5, 1]$, resulting in solutions with distinct singularities at $x = 0.5$ and $x = 1$, shown in Fig. 3a).

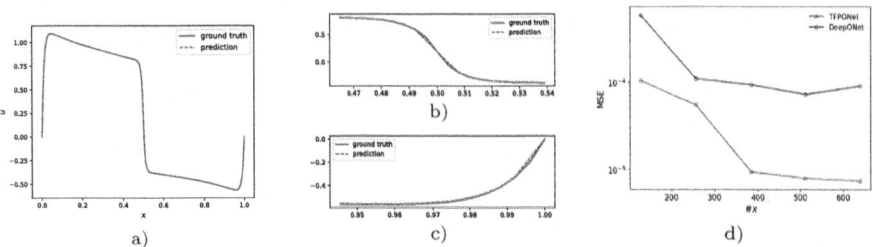

Fig. 2. a): Ground truth and TFPONet's prediction for Eq. (11) with $a(x) = 0.0001$ on a random f and 257 locations. b), c): Zoomed-in view. d): Test error (MSE) vs. Training resolution ($\#x$) for TFPOnet and DeepONet.

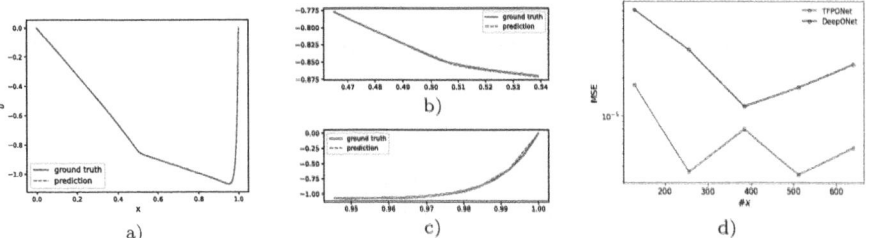

Fig. 3. a): Ground truth and TFPONet's prediction for Eq. (11), where $a(x) = 1$ for $x \in [0, 0.5)$ and $a(x) = 0.0001$ for $x \in [0.5, 1]$, across 257 locations using a random f. b), c): Zoomed-in view. d): Test error (MSE) vs. Training resolution ($\#x$) for TFPOnet and DeepONet.

We train the TFPONet using datasets comprising 1000×129 triples (f, x, u) for both problems. To assess performance, we test the trained model at 257 locations using a random f sample that is independent from the training data. Subplots a), b), and c) in Figs. 2 and 3 compare the predictions with the ground truth, highlighting the model's accuracy in critical regions. Then we train both TFPONet and DeepONet across varying position counts-129, 257, 385, 513, and 641-and assessed them on 100 distinct f samples at 1001 positions. Using mean square error (MSE) as the evaluation metric, we charted test errors relative to the number of training positions ($\#x$) in Figs. 2d) and 3d). The results consistently demonstrate TFPONet's superior learning and generalization over vanilla DeepONet for these interface problems.

Example 2. We next examine an interface problem as defined by Eq. (11), with the solution exhibiting a discontinuity at the interface $\Gamma = \{0.5\}$, specifically where $g_D = g_N = 1$. The coefficients are given by $a(x) = 1$ and a piecewise function for $b(x)$:

$$b(x) = 5000 \text{ for } x \in [0, 0.5) \text{ and } b(x) = 100(4 + 32x) \text{ for } x \in (0.5, 1].$$

The source term f is continuous, sampled from a Gaussian random field.

To train the TFPONets, which comprise two TFPONets each dedicated to one of the subdomains $[0, 0.5)$ and $(0.5, 1]$, we utilized a dataset consisting of 1000×128 triplets (f, x, u), with 64 locations uniformly distributed within each subdomain. The trained network is then evaluated using a random f sample at 256 locations. The results, displayed in subplot a) of Fig. 4, illustrate the prediction versus the ground truth.

We proceed to train both TFPONets and IONet across varying position counts: 128, 256, 384, 512, and 640. Subsequently, we evaluate their performance on 100 distinct f samples at 1001 positions. Subplot b) of Fig. 4 illustrates the testing errors relative to the training resolution. This analysis showcases the robust generalization capabilities of our model, coupled with its remarkable accuracy in handling general interface problems.

Two-Dimensional Settings. We consider the following equation:

$$- a\Delta u + bu = 0, \ \boldsymbol{x} \in \Omega/\Gamma, \ [u]_\Gamma = 1, \ [\nabla u \cdot \boldsymbol{n}]_\Gamma = 0, \ u|_{\Gamma_1} = 0, \ u|_{\Gamma_2 \cup \Gamma_3} = f. \tag{12}$$

Here $\Omega = (-1, 1) \times (-1, 1)$, $a(\boldsymbol{x}) = 0.001$ and $b(\boldsymbol{x})$ and $f(x_1)$ are piecewise-defined functions:

$$b(\boldsymbol{x}) = \begin{cases} (1 - x_1^2)^2, & x_1 < 0, \\ 0.001, & x_1 > 0, \end{cases} \quad f(x_1) = \begin{cases} f_1(x_1), & x_1 < 0, \\ f_2(x_1), & x_1 > 0. \end{cases}$$

The equation is governed by jump conditions on Γ and boundary conditions on Γ_1, Γ_2, and Γ_3, where:

$$\Gamma = \{\boldsymbol{x} \in \mathbb{R}^2 | x_1 = 0, -1 \leq x_2 \leq 1\}, \ \Gamma_1 = \{\boldsymbol{x} \in \mathbb{R}^2 | x_1 = \pm 1, -1 \leq x_2 \leq 1\},$$
$$\Gamma_2 = \{\boldsymbol{x} \in \mathbb{R}^2 | -1 \leq x_1 < 0, x_2 = \pm 1\}, \ \Gamma_3 = \{\boldsymbol{x} \in \mathbb{R}^2 | 0 < x_1 \leq 1, x_2 = \pm 1\}.$$

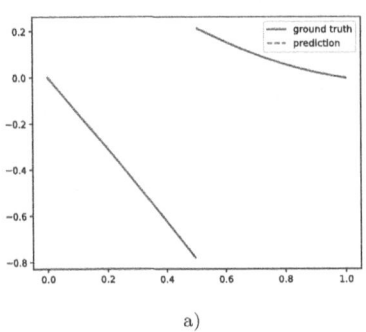
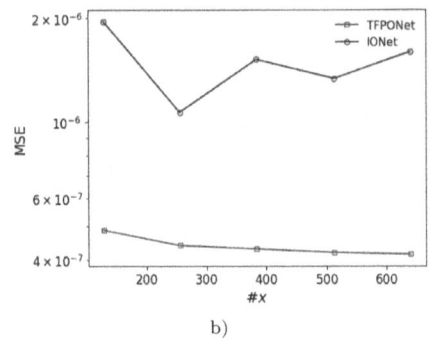

a)

b)

Fig. 4. a): Ground truth and TFPONet's prediction for Example 4. b): Test error (MSE) vs. Training resolution ($\#x$) for TFPONets and IONet.

Example 3. We focus on Eq. (12) as our research subject, aiming to learn the mapping $f \mapsto u$, where f_1 and f_2 are independently drawn from two given distribution spaces. We randomly select 500 f samples, along with the grid points on a 65×65 uniform grid (excluding points on $x_1 = 0$) as locations. The center point in each generated cell serves as the reference point. Additionally, we employ IONet, utilizing two DeepONets for each subdomain, trained on the same set of f samples and locations. The prediction capabilities of these models for a random f at identical locations are illustrated in Fig. 5. Experiments demonstrate that our model excels in capturing the inner layer of solutions to two-dimensional problems compared to IONet.

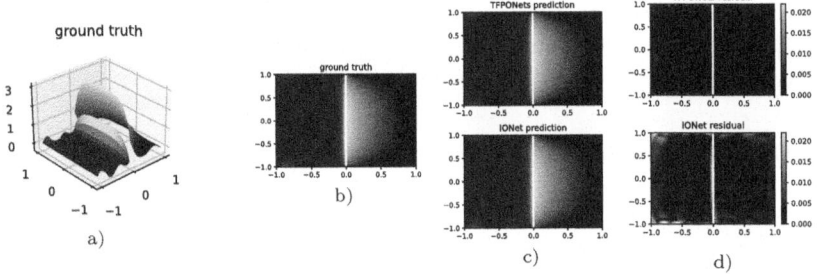

a)

b)

c)

d)

Fig. 5. Performance of TFPONets and IONet on Eq. (12). a), b): Ground truth. c): Predictions of TFPONets (above) and IONet (below). d): Absolute Point-wise error of TFPONets (above) and IONet (below).

5   Conclusion

In this work, we introduce Tailored Finite Point Operator Networks (TFPONets), a novel integration of the tailored finite point method (TFPM) with Deep Operator Networks (DeepONets), designed to tackle parameterized interface problems. The motivation behind TFPONets stems from the strengths of TFPM, which avoids the need for specialized spatial discretization and complex equation manipulation, using local basis expansion to approximate solutions to interface problems. TFPM has demonstrated high accuracy for standard interface problems and maintains consistency under parameter perturbations in singular perturbation scenarios. By combining TFPM's robustness with DeepONets' powerful learning capabilities, TFPONets effectively address one-dimensional and two-dimensional interface problems, including those with singular perturbations and high-contrast coefficients. Our analysis shows that TFPONets surpass the performance of vanilla DeepONets and IONet in terms of prediction accuracy and generalization, even with spatially sparse training samples.

Despite our experiments being limited to one- and two-dimensional scenarios, the inherent adaptability of our model inspires confidence in its potential applicability to more complex, higher-dimensional problems.

Acknowledgements. Z.Y. Huang was partially supported by the NSFC Projects No. 12025104. Y. Li was partially supported by the NSFC Project No. 62106103.

References

1. Babuška, I.: The finite element method for elliptic equations with discontinuous coefficients. Computing **5**(3), 207–213 (1970)
2. Fadlun, E.A., Verzicco, R., Orlandi, P., Mohd-Yusof, J.: Combined immersed-boundary finite-difference methods for three-dimensional complex flow simulations. J. Comput. Phys. **161**(1), 35–60 (2000)
3. Guo, H., Yang, X.: Deep unfitted Nitsche method for elliptic interface problems. arXiv preprint arXiv:2107.05325 (2021)
4. He, C., Hu, X., Mu, L.: A mesh-free method using piecewise deep neural network for elliptic interface problems. J. Comput. Appl. Math. **412**, 114358 (2022)
5. Hesthaven, J.S.: High-order accurate methods in time-domain computational electromagnetics: a review. Advances in imaging and electron physics **127**, 59–123 (2003)
6. Hu, W.F., Lin, T.S., Lai, M.C.: A discontinuity capturing shallow neural network for elliptic interface problems. J. Comput. Phys. **469**, 111576 (2022)
7. Huang, Z.: Tailored finite point method for the interface problem. Netw. Heterog. Media **4**(1), 91–106 (2009)
8. Ji, N., Liu, T., Xu, J., Shen, L.Q., Lu, B.: A finite element solution of lateral periodic Poisson-Boltzmann model for membrane channel proteins. Int. J. Mol. Sci. **19**(3), 695 (2018)
9. Kellogg, R.: Singularities in interface problems. In: Numerical Solution of Partial Differential Equations–II, pp. 351–400. Elsevier (1971)
10. LeVeque, R.J., Li, Z.: The immersed interface method for elliptic equations with discontinuous coefficients and singular sources. SIAM J. Numer. Anal. **31**(4), 1019–1044 (1994)
11. Li, K., Tang, K., Wu, T., Liao, Q.: D3M: a deep domain decomposition method for partial differential equations. IEEE Access **8**, 5283–5294 (2019)

12. Li, W., Xiang, X., Xu, Y.: Deep domain decomposition method: Elliptic problems. In: Mathematical and Scientific Machine Learning, pp. 269–286. PMLR (2020)
13. Li, Z., et al.: Fourier neural operator for parametric partial differential equations. In: International Conference on Learning Representations (2020)
14. Lions, P.L., et al.: On the schwarz alternating method. I. In: First International Symposium on Domain Decomposition Methods for Partial Differential Equations, vol. 1, p. 42. Paris, France (1988)
15. Liu, X.D., Fedkiw, R.P., Kang, M.: A boundary condition capturing method for poisson's equation on irregular domains. J. Comput. Phys. **160**(1), 151–178 (2000)
16. Liu, Y., Sussman, M., Lian, Y., Yousuff Hussaini, M.: A moment-of-fluid method for diffusion equations on irregular domains in multi-material systems. J. Comput. Phys. **402**, 109017 (2020)
17. Lu, L., Jin, P., Pang, G., Zhang, Z., Karniadakis, G.E.: Learning nonlinear operators via DeepONet based on the universal approximation theorem of operators. Nat. Mach. Intell. **3**(3), 218–229 (2021)
18. Peskin, C.S.: Numerical analysis of blood flow in the heart. J. Comput. Phys. **25**(3), 220–252 (1977)
19. Raissi, M., Perdikaris, P., Karniadakis, G.E.: Physics-informed neural networks: a deep learning framework for solving forward and inverse problems involving nonlinear partial differential equations. J. Comput. Phys. **378**, 686–707 (2019)
20. Sirignano, J., Spiliopoulos, K.: DGM: a deep learning algorithm for solving partial differential equations. J. Comput. Phys. **375**, 1339–1364 (2018)
21. Sussman, M., Fatemi, E.: An efficient, interface-preserving level set redistancing algorithm and its application to interfacial incompressible fluid flow. SIAM J. Sci. Comput. **20**(4), 1165–1191 (1999)
22. Wang, L., Zheng, H., Lu, X., Shi, L.: A Petrov-Galerkin finite element interface method for interface problems with Bloch-periodic boundary conditions and its application in Phononic crystals. J. Comput. Phys. **393**, 117–138 (2019)
23. Wang, Z., Zhang, Z.: A mesh-free method for interface problems using the deep learning approach. J. Comput. Phys. **400**, 108963 (2020)
24. Wu, S., Zhu, A., Tang, Y., Lu, B.: Solving parametric elliptic interface problems via interfaced operator network. J. Comput. Phys. **514**, 113217 (2024)
25. E, W., Yu, B.: The deep Ritz method: a deep learning-based numerical algorithm for solving variational problems. Commun. Math. Stat. **6**(1), 1–12 (2018). https://doi.org/10.1007/s40304-018-0127-z

Novel Methods in Machine Learning

A Simple Task-Aware Contrastive Local Descriptor Selection Strategy for Few-Shot Learning Between Inter Class and Intra Class

Qian Qiao, Yu Xie, Shaoyao Huang, and Fanzhang Li[✉]

School of Computer Science and Technology, Soochow University,
Suzhou 215006, China
{qqiao,20224227022}@stu.suda.edu.cn, lfzh@suda.edu.cn

Abstract. Few-shot image classification aims to classify novel classes with few labeled samples. Recent research indicates that deep local descriptors have better representational capabilities. These studies recognize the impact of background noise on classification performance. They typically filter query descriptors using all local descriptors in the support classes or engage in bidirectional selection between local descriptors in support and query sets. However, they ignore the fact that background features may be useful for the classification performance of specific tasks. This paper proposes a novel task-aware contrastive local descriptor selection network (TCDSNet). First, we calculate the contrastive discriminative score for each local descriptor in the support class, and select discriminative local descriptors to form a support descriptor subset. Finally, we leverage support descriptor subsets to adaptively select discriminative query descriptors for specific tasks. Extensive experiments demonstrate that our method outperforms state-of-the-art methods on both general and fine-grained datasets.

Keywords: few-shot learning · task-aware · local descriptor · image classification

1 Introduction

The purpose of few-shot learning is to enable models to adapt quickly to new tasks with only a small number of training samples in scenarios where data is scarce. Generally, these methods can be divided into three groups: optimization-based methods [1,5,11], metric-based methods [10,23,24], and data augmentation-based methods [2,6,9,21,29].

This work is based on a few-shot learning method using local descriptors, falling within the realm of metric learning. Features based on local descriptors exhibit superior representational capabilities compared to image-level features. In previous works, [14] proposed DN4, which directly utilizes all query descriptors. It selects k support descriptors directly for each query local descriptor

© The Author(s), under exclusive license to Springer Nature Switzerland AG 2024
M. Wand et al. (Eds.): ICANN 2024, LNCS 15016, pp. 75–88, 2024.
https://doi.org/10.1007/978-3-031-72332-2_6

through k-nearest neighbors (k-NN) and approximates the relationship between query samples and support classes using cosine similarity distances. Based on DN4, [16] introduced DMN4, which believes that not all query descriptors are task-relevant and contain significant background noise. DMN4 establishes mutual nearest neighbor (MNN) relationships, explicitly selecting query descriptors most relevant to each task, thereby avoiding the impact of background noise on classification performance. Similarly, based on DN4, [4,30] proposed ATL-Net and TADNet, respectively. Both methods measure the relationships between each query local descriptor and all support classes, adaptively selecting discriminative query descriptors for classification. [19] introduced TALDS-Net, recognizing background noise in query descriptors. It first adaptively selects optimal descriptor subsets composed of support class local descriptors and then adaptively chooses query descriptors for classification from the optimal descriptor subset. However, all these methods aim to eliminate background noise to prevent its influence on the feature representation of local descriptors, either by filtering query descriptors through all support class local descriptors or by bidirectionally selecting between support class local descriptors and query descriptors. Their goal is to remove background noise. We observe that from a human cognitive perspective, for example, considering an image of a dog and an image of a dolphin, not only do the target features differ significantly, but the background features also exhibit substantial distinctions (*e.g.*, dolphin backgrounds are unlikely to be grassy, whereas dog backgrounds might include grass). In such cases, background features can contribute to classification. Conversely, for two images both belonging to the dolphin category, the differences in background features might not be as pronounced, and in this scenario, background features can be considered as noise. For instance, when dealing with unfamiliar images, if the background is an ocean, it can help narrow down the classification to objects commonly found only in the ocean. This aids in identifying the target category among familiar images. Thus, background features within the same category might positively impact classification performance. Furthermore, background features between different categories might also contribute to enhancing classification performance. Determining discriminative local descriptors for methods based on local descriptors is a challenging task. Moreover, it is essential to judiciously retain or discard background noise in the process.

In response to this challenge, a straightforward solution is to select local descriptors in the support class to form a support descriptor subset, and then use the support descriptor subset to select query descriptors. Experimental results have also demonstrated the effectiveness of this simple method.

The above-described method is our proposed Task-Aware Contrastive Discriminative Local Descriptor Selection Network (TCDSNet). Specifically, we first select local descriptors from the support class. For each support descriptor, we calculate the sum of its similarities with the remaining support descriptors in the same category as the intra-class similarity score. Next, we compute the sum of its similarities with support descriptors from other categories as the inter-class similarity score. A high intra-class similarity score indicates that the support descriptor has strong representational capabilities for that class, while a low inter-class similarity score suggests that the support descriptor has high discriminative power across other classes. We calculate the discriminative score by dividing the intra-class similarity score by the inter-class similarity score, which we term the contrastive discriminative score. Then we select K support descriptors in descending order of their discriminative scores. Finally, we utilize the selected support descriptors to choose query descriptors. For selecting discriminative query descriptors, we employ a simple learnable module to adaptively predict a threshold. Using the learned threshold and a score map, we select the most discriminative descriptors for final classification. This approach enhances the model's classification and generalization capabilities.

In summary, our main contributions are three folds:

– We propose a novel method that calculates discriminative scores ($\mathcal{CDS}$) for local descriptors in the support class. This enhances the model's adaptability to different tasks and strengthens the performance of local descriptors in few-shot learning tasks.
– We propose a novel Task-Aware Contrastive Discriminative Local Descriptor Selection Network (TCDSNet) that not only selects a subset of support descriptors based on discriminative scores but also incorporates a learnable module for adaptively choosing the discriminative query descriptors.
– Extensive experimental results demonstrate that TCDSNet outperforms state-of-the-art methods on multiple general and fine-grained datasets.

2 Method

Figure 1 shows an overview of the proposed method.

2.1 Problem Definition

In this paper, we follow the same setting as previous methods [4,14,19,30]. Given a support set S, a query set Q, and an auxiliary set A, where the label space of the auxiliary set A is disjoint from S and is used to learn transferable knowledge. The support set S contains C classes, each with K labeled samples, while the samples in the query set Q are unlabeled and share the same label space as S. We are given a support set consisting of n classes, each with k samples, and a query image, and the task is to classify the query image into one of the n support classes. This constitutes the n-way k-shot few-shot classification problem. Under this setting, we introduce a meta-training mechanism [25] called the episodic training mechanism. We randomly sample from the auxiliary set A to construct an n-way k-shot task. Each task consists of a support set A_S and a query set A_Q. During the training phase, we construct tens of thousands of tasks to learn transferable knowledge.

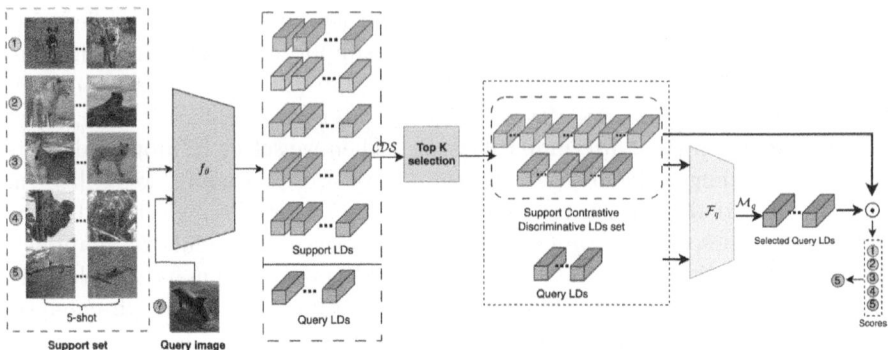

Fig. 1. The overall architecture of the proposed method under a 5-way 5-shot setting. The model primarily consists of three components: a feature extraction module f_θ for extracting features, a model for selecting K discriminative LDs, and $\mathcal{F}_q$ for adaptively selecting query LDs.

2.2 Image Representation Based on Local Descriptors

We obtain a three-dimensional feature representation $f_\theta(X) \in \mathbb{R}^{h \times w \times d}$ for the image X through the embedding module $f_\theta(\cdot)$, which is considered as a set of d-dimensional local descriptors (LDs):

$$f_\theta(X) = [l_1, l_2, \cdots, l_m] \in \mathbb{R}^{m \times d} \tag{1}$$

where l_i denotes the i-th deep local descriptor (LD). Similar to other descriptor-based Methods [4,14,16,19,30], we consider it as a set of m d-dimensional descriptors, and $m = h \times w$.

In each episode, each support class has k images. We denote the descriptor set from category c as $\mathcal{L}_c^S$, where there are n classes in total, and represent the descriptor representation for each query image as l^q. When using shallower embedding modules (*e.g.*, Conv-4), each support category is represented in its original form. When using deeper embedding modules (*e.g.*, ResNet-12), each support category is represented by the empirical mean of its support descriptors.

2.3 Contrastive Discriminative Scores for Support Local Descriptors Selection

As mentioned above, X_S represents an image in the support class, fed into the embedding module f_θ to obtain local descriptors $\mathcal{L}^S = f_\theta(X_S) \in \mathbb{R}^{m \times d}$, where $m = h \times w$. Here, l^s denotes a supporting local descriptor in $\mathcal{L}_S$, $\hat{\mathcal{L}}^S$ represents the set of remaining support descriptors in $\mathcal{L}_S$ excluding the current l^s, and $\bar{\mathcal{L}}^S$ represents the set of local descriptors from the remaining support classes. Thus, we obtain m d-dimensional local descriptors (LDs) for an image in the support class. Under the n-way k-shot setting, there are a total of nkm d-dimensional support LDs. Previous methods [19] only considered the average

Table 1. The classification accuracies on the miniImageNet and tieredImageNet datasets in the 5-way 1-shot and 5-shot settings using Conv-4 and ResNet-12 as backbones with 95% confidence interval. All the results of comparative methods are from the exiting literature ('-' not reported). The methods below "hline" are LDs-based methods.

Method	Conv-4				ResNet-12			
	miniImageNet		tieredImageNet		miniImageNet		tieredImageNet	
MatchingNet [25]	43.56±0.84	55.31±0.73	–	–	63.08±0.20	75.99±0.15	68.50±0.92	80.60±0.71
ProtoNet [23]	51.20±0.26	68.94±0.78	53.45±0.15	72.32±0.57	62.33±0.12	80.88±0.41	68.40±0.14	84.06±0.26
RelationNet [24]	50.44±0.82	65.32±0.70	54.48±0.93	71.31±0.78	60.97	75.12	64.71	78.41
FRN [28]	54.87	71.56	55.54	74.68	66.45±0.19	82.83±0.13	72.06±0.22	86.89±0.14
Meta-OLE [27]	56.82±0.84	73.87±0.67	58.82±0.88	75.85±0.87	67.04±0.72	82.23±0.67	68.82±0.71	85.51±0.59
Approximate GAP [12]	53.52±0.88	70.75±0.67	57.47±0.99	71.66±0.76	–	–	–	–
GAP [12]	54.86±0.85	71.55±0.61	58.56±0.93	72.82 ±0.77	–	–	–	–
DeepEMD [31]	51.72±0.20	65.10±0.39	51.22±0.14	65.81±0.68	65.91±0.82	82.41±0.56	71.16±0.87	86.03±0.58
DN4 [14]	51.24±0.74	71.02±0.64	52.89±0.23	73.36±0.73	65.35	81.10	69.60	83.41
DMN4 [16]	55.77	74.22	56.99	74.13	66.58	83.52	72.10	85.72
ATL-Net [4]	54.30±0.76	73.22±0.63	–	–	–	–	–	–
TADNet [30]	56.14±0.20	74.68±0.15	57.88±0.21	75.98±0.17	67.26±0.20	84.23±0.13	71.29± 0.22	86.46±0.15
TCDSNet(ours)	57.14±0.22	75.89±0.35	58.67±0.61	76.06±0.33	68.53±0.19	85.12±0.42	72.43±0.72	87.35±0.55

similarity between each LD and the remaining LDs within the same class as the discriminative score. However, our goal is not only to maintain discriminative relationships within the same class but also across other classes. For each l^s, we calculate its average similarity with all other LDs within the same support class, referred to as intra-class similarity, and then calculate its average similarity with LDs from the remaining support classes, referred to as inter-class similarity. We seek support LDs with high intra-class similarity and low inter-class similarity. High intra-class similarity indicates strong representational capabilities of the support LD for its corresponding class, while low inter-class similarity signifies poorer representational capabilities of the support LD for other classes. Support LDs exhibiting these characteristics suggest discriminative capabilities, potentially incorporating discriminative background features to enhance classification results. Therefore, the calculations for intra-class and inter-class similarities are as follows:

$$\text{SIM}_{intra} = \frac{1}{m-1} \sum_{\hat{l}^s \in \hat{\mathcal{L}}^S} \cos(l^s, \hat{l}^s),$$

$$\text{SIM}_{inter} = \frac{1}{(n-1)m} \sum_{\bar{l}^s \in \bar{\mathcal{L}}^S} \cos(l^s, \bar{l}^s) \tag{2}$$

where $\hat{\mathcal{L}}^S$ represents the set of remaining support descriptors in $\mathcal{L}_S$ excluding the current l^s (in the case of 1-shot, it corresponds to the remaining local descriptors of the current image). $\bar{\mathcal{L}}^S$ denotes the set of local descriptors from the remaining support classes, SIM_{intra} denotes the intra-class similarity score, and SIM_{inter} denotes the inter-class similarity score. Furthermore, we normalize these two

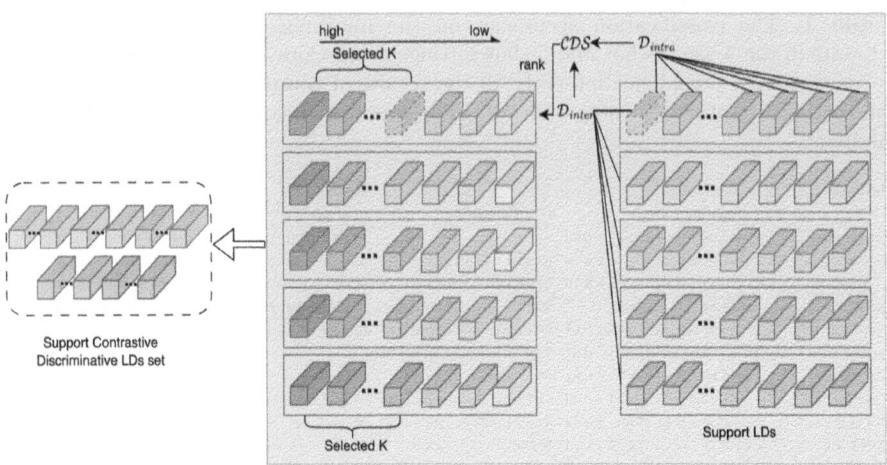

Fig. 2. Selecting K LDs to form a discriminative support descriptor set is achieved by computing the CDS for each LD in each support class.

similarity scores and subsequently calculate their discriminative scores:

$$\begin{aligned}\mathcal{D}_{intra} &= \text{softmax}(\text{SIM}_{intra}) \\ \mathcal{D}_{inter} &= \text{softmax}(\text{SIM}_{inter})\end{aligned} \quad (3)$$

where $\mathcal{D}_{intra}$ denotes the discriminative score of the local descriptor l^s within its own class, and $\mathcal{D}_{inter}$ represents its discriminative score across classes.

Based on the above results, we can calculate the two discriminative scores $\mathcal{D}_{intra}$ and $\mathcal{D}_{inter}$ for each support descriptor using a comparative approach. Subsequently, an optimized Contrastive Discriminative Score (CDS) can be computed:

$$CDS = \sigma(\frac{\mathcal{D}_{intra}}{\mathcal{D}_{inter}}) \quad (4)$$

We can observe that CDS aligns well with our initial idea, indicating that the current local descriptor exhibits high similarity with other local descriptors within the same class and low similarity with local descriptors from other classes, where σ is a sigmoid function. Furthermore, in Fig. 2, based on the descending order of CDS, we select the top K support descriptors with the contrastive discriminative scores for each class, forming a discriminative support descriptor set:

$$\mathcal{L}_c^{CDS} = \text{Top K}_{l_i}(CDS) \quad (5)$$

The value of K will be discussed in Ablation Studies 3.4. We will form a set $\mathcal{L}_{CDS}$ with the support descriptors selected after screening.

2.4 Query Local Descriptors Selection

Given a query image X_q embedded as $\mathcal{L}^Q = f_\theta(X_q) \in \mathbb{R}^{m \times d}$. l_i^q denotes a query descriptor in $\mathcal{L}^Q$. Previous works [19,30] employed k-NN to select k support descriptors from each support class. However, we have observed that, after computing the discriminative support descriptor set $\mathcal{L}^{\mathcal{CDS}}$, it is not necessary to use k-NN for selecting k support LDs from $\mathcal{L}^{\mathcal{CDS}}$. We directly compute the sum of similarities between each query descriptor l_i^q and the discriminative support descriptor set $\mathcal{L}_c^{\mathcal{CDS}}$ for each support class c:

$$\text{SIM}_c^{l_i^q} = \sum_{l_c \in \mathcal{L}_c^{\mathcal{CDS}}} \cos(l_i^q, l_c) \tag{6}$$

where $c \in \{1, 2, \ldots, k\}$ denotes a support class, and l_c is one discriminative support LD from the discriminative support descriptor set of support class c. Similarly, the discriminative score for each query descriptor l_i^q is calculated as:

$$\mathcal{D}^{l_i^q} = \max_c \left(\frac{\text{SIM}_c^{l_i^q}}{\sum_{c=1}^n \text{SIM}_c^{l_i^q}} \right) \tag{7}$$

Previous works [4,14,16,19,30] have employed methods that directly select query descriptors by using a fixed threshold $\mathcal{V}$ and the top-k query descriptors with the highest similarity. However, both of these methods suffer from poor generalization, as they may overlook some discriminative LDs. Thus, inspired by [4,8,19,30], we employ a network $\mathcal{F}_q$ consisting of two fully connected layers as an MLP to adaptively predict the threshold $\mathcal{V}^{l_i^q}$ for each query descriptor. Finally, we utilize the predicted threshold $\mathcal{V}$ to learn a query descriptor weights map $\mathcal{M}_q$. We feed the discriminative support descriptor set $\mathcal{L}_c^{\mathcal{CDS}}$ and the query descriptor l_i^q into $\mathcal{F}_q$, ultimately predicting the threshold $\mathcal{V}^{l_i^q}$:

$$\mathcal{V}^{l_i^q} = \sigma(\mathcal{F}_q(l_i^q, \mathcal{L}_c^{\mathcal{CDS}})) \tag{8}$$

where $i \in \{1, 2, \ldots, m\}$ denotes a query LD, and $c \in \{1, 2, \ldots, n\}$ denotes a support class. The final calculation for the query descriptor weights map $\mathcal{M}_q$ is as follows:

$$\mathcal{M}_q = 1/(1 + \exp^{-\lambda(\mathcal{D}^{l_i^q} - \mathcal{V}^{l_i^q})}) \tag{9}$$

where, when λ is sufficiently large and $\mathcal{D}^{l_i^q} > \mathcal{V}^{l_i^q}$, the values of $\mathcal{M}_q$ approximates 1. Conversely, the values of $\mathcal{M}_q$ approximates 0.

Therefore, we can utilize $\mathcal{M}_q$ to select query LDs. The calculation for the similarity scores between each query image X_q and each support class c is as follows:

$$\text{Score}(X_q, c) = \sum_{l_i^q \in \mathcal{L}_c^Q} \mathcal{V}^{l_i^q} \mathcal{M}_q \tag{10}$$

The cross-entropy loss is used to meta-train the network:

$$p_\phi \left(y = c \mid X_q\right) = \frac{\exp\left(\text{score}(X_q, c)\right)}{\sum_{c'=1}^{n} \exp\left(\text{score}(X_q, c')\right)} \qquad (11)$$

$$\mathcal{J}(\phi) = -\frac{1}{|A_Q|} \sum_{X_q \in A_Q} \sum_{c=1}^{n} y \log p_\phi \left(y = c \mid X_q\right) \qquad (12)$$

3   Experiments

In this section, we validate the effectiveness of our proposed method on several few-shot benchmark datasets and compare it with other state-of-the-art LDs-based methods. Additionally, we compare our method with small-sample methods using different parameter settings. Furthermore, we conduct ablation experiments to further analyze and validate the effectiveness of our proposed method.

3.1   Datasets

miniImageNet [25] is a subset of ImageNet [3]. It is divided into a training set with 64 classes, a validation set with 16 classes, and a test set with 20 classes. Each class consists of 600 image samples, each of size 84×84 pixels.

tieredImageNet [20] is another subset of ImageNet. It comprises 608 classes, with each class containing 1281 images. These 608 classes are divided into 351 for training, 97 for validation, and 160 for testing.

CUB-200 [26] is a fine-grained dataset that consists of 11788 bird images, encompassing 200 different bird species. We partition it into 100 classes for training, 50 classes for validation, and 50 classes for testing. For fine-grained datasets, we resized the images in them to the same size as miniImageNet, which is 84×84 pixels.

3.2   Implementation Details

Model Architecture. We use Conv-4 and ResNet-12 as feature extraction networks f_θ, similar to previous work [4,14,19,30]. Conv-4 consists of 4 convolutional blocks, each containing a convolutional layer, batch normalization layer, and Leaky ReLU layer. ResNet-12 is composed of 4 residual blocks, with each block consisting of 3 convolutional layers with 3×3 kernels, 3 batch normalization layers, 3 Leaky ReLU layers, and a 2×2 max-pooling layer. Conv-4 and ResNet-12 generate feature maps of size $19 \times 19 \times 64$ and $5 \times 5 \times 640$ for 84×84 images, respectively. These feature maps are then mapped through a transformation layer f_ϕ, which consists of a 1×1 convolutional layer, a batch normalization layer, and a LeakyReLU layer. Finally, $\mathcal{F}_q$ is implemented with two fully connected layers.

Table 2. The classification accuracies on the CUB-200 dataset in the 5-way 1-shot and 5-shot settings using Conv-4 and ResNet-12 as backbones, The confidence intervals of our method are all below 0.20.

Method	Conv-4		ResNet-12	
	1-shot	5-shot	1-shot	5-shot
ProtoNet [23]	63.73	81.50	66.09	82.50
DSN [22]	66.01	85.41	80.80	91.19
FRN [28]	73.48	88.43	83.16	92.59
Meta-OLE [27]	71.32	86.11	–	–
Approximate GAP [12]	43.77	62.92	–	–
GAP [12]	44.74	64.88	–	–
DeepEMD [31]	–	–	77.14	88.98
DN4 [14]	73.42	90.38	–	–
DMN4 [16]	78.36	92.16	–	–
TADNet [30]	82.47	93.36	87.62	94.80
TCDSNet(ours)	**82.73**	**95.04**	**88.71**	**95.82**

Table 3. Ablation study on miniImageNet and CUB-200 datasets for the influence of Top K in support LDs selection.

Conv-4			ResNet-12		
K	miniImagenet	CUB-200	**K**	minImagenet	CUB-200
1%	74.94	90.11	3%	83.92	89.21
2%	**75.89**	90.23	5%	**85.12**	89.25
5%	74.23	92.37	10%	84.39	92.33
10%	72.02	**95.04**	25%	84.41	**95.82**
30%	71.11	94.57	30%	83.83	94.29

Training and Evaluation Details. During the meta-training phase, we followed the settings in [16, 19, 30]. For Conv-4, we set the learning rate to $1e-3$ and decay 0.1 every 10 epoch, training for a total of 30 epochs using the Adam optimizer. For ResNet-12, we pre-trained it first and then conducted meta-training for 40 epochs using momentum SGD with an initial learning rate of $5e-4$ and decay 0.1 every 10 epochs. During the test, as in [16, 19, 30], we randomly constructed 10000 episodes from the test set to calculate the classification accuracy. This process was repeated five times, and we reported the average accuracy along with a 95% confidence interval.

3.3   Comparisons with State-of-the-Art Methods

We choose 7 generic few-shot learning state-of-the-art baselines [12, 23–25, 27, 28], as well as 5 SOTA baselines based on LDs [4, 14, 16, 30, 31]. For fine-grained datasets, we also selected 9 state-of-the-art baselines [12, 14, 16, 22, 23, 27, 28, 30, 31].

Results on MiniImageNet Dataset. As shown in Table 1, the performance of our method in the 5-way 1-shot and 5-shot settings exceeds that of all current LDs-based methods [4, 14, 16, 30, 31]. Compared to the baseline DN4, our method exhibits significant improvement. In the 5-way 1-shot and 5-shot settings, using Conv-4 as the backbone, it achieved improvements of 5.90% and 4.87%, respectively. Compared to the state-of-the-art (SOTA), our method also improved by 1% and 1.21%, respectively. When using ResNet-12 as the backbone, improvements of 3.18% and 4.02% were achieved, surpassing SOTA by 1.27% and 0.89%, respectively.

Results on TieredImageNet Dataset. As shown in Table 1, our method outperforms the current state-of-the-art LDs-based methods as well. In the 5-way 1-shot and 5-shot settings, when using Conv-4 as the backbone, our method improves by 0.79% and 0.08%, respectively, compared to the state-of-the-art method based on LDs. When using ResNet-12 as the backbone, our method improves by 1.14% and 0.89%, respectively, compared to the state-of-the-art method based on LDs.

Results on Fine-Grained CUB-200 Dataset. As shown in Table 2, our method also achieves state-of-the-art performance on fine-grained datasets. In the 5-way 1-shot and 5-shot settings, when using Conv-4 as the backbone, our method improves by 0.26% and 1.68%, respectively, compared to the state-of-the-art method based on LDs. When using ResNet-12 as the backbone, our method improves by 1.02% and 1.19%, respectively, compared to the state-of-the-art method based on LDs.

3.4   Ablation Studies

Influence of Top K in Support LDs Selection. In Subsect. 2.2, we selected K (K as a percentage) LDs based on $\mathcal{CDS}$ for each support class to form a discriminative LDs set. As shown in Table 3, we conducted experiments on the miniImagenet and CUB-200 datasets under the 5-way 5-shot setting. When using Conv-4 as the backbone, we set K to $1\%, 2\%, 5\%, 10\%, 30\%$ respectively. When using ResNet-12 as the backbone, we set K to $3\%, 5\%, 10\%, 25\%, 30\%$ respectively. Through experiments, we found that when using Conv-4 as the backbone, the performance is best when $K = 2\%$ and $K = 10\%$ on both datasets. When using ResNet-12 as the backbone, the performance is best when $K = 2\%$ and $K = 25\%$ on both datasets. The experimental results indicate that compared to general datasets, fine-grained datasets require more discriminative LDs. Similarly, under the 5-way 1-shot setting, the best performance is achieved with Conv-4 as the backbone when $K = 5\%$ and $K = 10\%$, and with ResNet-12 as the backbone, the best performance is achieved when $K = 5\%$ and $K = 10\%$.

Table 4. The classification accuracies on the miniImageNet dataset in the 5-way 1-shot and 5-shot settings for backbones with different parameters.

Method	Backbone	≈ Params	miniImageNet	
CTM [13]	ResNet-18	11.7 M	64.12 ± 0.82	80.51 ± 0.13
Neg-Cosine [15]	ResNet-18	11.7 M	62.33 ± 0.82	80.94 ± 0.59
UniSiam+dist [17]	ResNet-18	11.7 M	64.10 0.36	82.26 ± 0.25
Meta-OLE [12]	WRN-28-10	36.5 M	75.22 ± 0.30	86.12 ± 0.28
MetaQDA [32]	WRN-28-10	36.5 M	67.83 ± 0.64	84.28 ± 0.69
OM [18]	WRN-28-10	36.5 M	66.78 ± 0.30	85.29 ± 0.41
FewTURE [7]	ViT-Small	22 M	68.02 ± 0.88	84.51 ± 0.53
FewTURE [7]	Swin-Tiny	29 M	**72.40 ± 0.78**	**86.38 ± 0.49**
TCDSNet(ours)	ResNet-12	**12.4 M**	68.53 ± 0.19	85.12 ± 0.42

Comparison with Methods using Backbones with Different Parameters. As shown in Table 4, we selected three baselines [13,15,17] using ResNet-18 as the backbone, three baselines [12,18,32] using WRN-28-10 as the backbone, and baselines [7] using ViT-Small and Swin-Tiny as the backbone. These methods are not LDs-based baselines. Compared to the baselines using ResNet-18 as the backbone, our method outperforms the best-performing method by 4.41% and 2.86% in the 1-shot and 5-shot settings, respectively. When compared to baselines using WRN-28-10 as the backbone [12,18,32], our method achieves a 0.70% improvement in the 1-shot setting and is only 0.17% lower than OM [18] in the 5-shot setting, despite WRN-28-10 having three times the parameters of ResNet-12. Compared to FewTURE [7] with ViT-Small as the backbone, our method achieves improvements of 0.51% and 0.61%, and is only 1.26% lower than FewTURE with Swin-Tiny as the backbone in the 5-shot setting. However, Swin-Tiny has 2.3 times the parameters of ResNet-12. Additionally, FewTURE's ViT-Small and Swin-Tiny were trained using 4 and 8 Nvidia A100 40GB GPUs, respectively, making their GPU requirements relatively high.

4   Conclusion

We propose a novel Task-Aware Contrastive Discriminative Local Descriptor Selection Network (TCDSNet), which utilizes a novel contrastive discriminative measure to filter discriminative local descriptors from the support class. Subsequently, it further selects discriminative query local descriptors from the filtered discriminative support descriptors, ensuring the selection of task-relevant query local descriptors. Extensive experiments validate the superiority and effectiveness of our proposed method. We anticipate that TCDSNet provides a new perspective for research in few-shot learning based on local descriptors.

Acknowledgment. This work was supported in part by the National Key R&D Program of China (2018YFA0701700; 2018YFA0701701) and by the National Natural Science Foundation of China under Grant No.61672364, No.62176172 and No.62002253.

References

1. Antoniou, A., Edwards, H., Storkey, A.: How to train your MAML (2018). arXiv preprint arXiv:1810.09502
2. Antoniou, A., Storkey, A., Edwards, H.: Data augmentation generative adversarial networks (2017). arXiv preprint arXiv:1711.04340
3. Deng, J., Dong, W., Socher, R., Li, L.J., Li, K., Fei-Fei, L.: ImageNet: a large-scale hierarchical image database. In: 2009 IEEE Conference on Computer Vision and Pattern Recognition, pp. 248–255. IEEE (2009)
4. Dong, C., Li, W., Huo, J., Gu, Z., Gao, Y.: Learning task-aware local representations for few-shot learning. In: Proceedings of the Twenty-Ninth International Conference on International Joint Conferences on Artificial Intelligence, pp. 716–722 (2021)
5. Finn, C., Abbeel, P., Levine, S.: Model-agnostic meta-learning for fast adaptation of deep networks. In: International Conference on Machine Learning, pp. 1126–1135. PMLR (2017)
6. He, F., Li, G., Zhang, M., Yan, L., Si, L., Li, F.: FreeStyle: Free lunch for text-guided style transfer using diffusion models (2024)
7. Hiller, M., Ma, R., Harandi, M., Drummond, T.: Rethinking generalization in few-shot classification. Adv. Neural. Inf. Process. Syst. **35**, 3582–3595 (2022)
8. Huang, S., Cao, Z., Qin, L., Gao, J., Zhang, J.: Contrastive learning with high-quality and low-quality augmented data for query-focused summarization. In: ICASSP 2024-2024 IEEE International Conference on Acoustics, Speech and Signal Processing (ICASSP), pp. 11536–11540. IEEE (2024)
9. Huang, S., Qin, L., Cao, Z.: Diffusion language model with query-document relevance for query-focused summarization. In: Findings of the Association for Computational Linguistics: EMNLP 2023, pp. 11020–11030 (2023)
10. Jiang, M., Li, F.: Lie group continual meta learning algorithm. Appl. Intell. **52**(10), 10965–10978 (2022)
11. Jiang, M., Li, F., Liu, L.: Continual meta-learning algorithm. Appl. Intell. **52**(4), 1–16 (2022)
12. Kang, S., Hwang, D., Eo, M., Kim, T., Rhee, W.: Meta-learning with a geometry-adaptive preconditioner. In: Proceedings of the IEEE/CVF Conference on Computer Vision and Pattern Recognition, pp. 16080–16090 (2023)
13. Li, H., Eigen, D., Dodge, S., Zeiler, M., Wang, X.: Finding task-relevant features for few-shot learning by category traversal. In: Proceedings of the IEEE/CVF Conference on Computer Vision and Pattern Recognition, pp. 1–10 (2019)

14. Li, W., Wang, L., Xu, J., Huo, J., Gao, Y., Luo, J.: Revisiting local descriptor based image-to-class measure for few-shot learning. In: Proceedings of the IEEE/CVF Conference on Computer Vision and Pattern Recognition, pp. 7260–7268 (2019)
15. Liu, B., et al.: Negative margin matters: understanding margin in few-shot classification. In: ECCV (2020)
16. Liu, Y., Zheng, T., Song, J., Cai, D., He, X.: DMN4: few-shot learning via discriminative mutual nearest neighbor neural network. In: Proceedings of the AAAI Conference on Artificial Intelligence. vol. 36, pp. 1828–1836 (2022)
17. Lu, Y., Wen, L., Liu, J., Liu, Y., Tian, X.: Self-supervision can be a good few-shot learner. In: Avidan, S., Brostow, G., Cissé, M., Farinella, G.M., Hassner, T. (eds.) Computer Vision - ECCV 2022. ECCV 2022. LNCS, vol. 13679. Springer, Cham (2022). https://doi.org/10.1007/978-3-031-19800-7_43
18. Qi, G., Yu, H., Lu, Z., Li, S.: Transductive few-shot classification on the oblique manifold. In: Proceedings of the IEEE/CVF International Conference on Computer Vision, pp. 8412–8422 (2021)
19. Qiao, Q., Xie, Y., Zeng, Z., Li, F.: TALDS-Net: task-aware adaptive local descriptors selection for few-shot image classification. In: ICASSP 2024-2024 IEEE International Conference on Acoustics, Speech and Signal Processing (ICASSP), pp. 3750–3754. IEEE (2024)
20. Ren, M., et al.: Meta-learning for semi-supervised few-shot classification (2018). arXiv preprint arXiv:1803.00676
21. Schwartz, E., et al.: Delta-encoder: an effective sample synthesis method for few-shot object recognition. In: Advances in Neural Information Processing Systems, vol. 31 (2018)
22. Simon, C., Koniusz, P., Nock, R., Harandi, M.: Adaptive subspaces for few-shot learning. In: Proceedings of the IEEE/CVF Conference on Computer Vision and Pattern Recognition, pp. 4136–4145 (2020)
23. Snell, J., Swersky, K., Zemel, R.: Prototypical networks for few-shot learning. In: Advances in Neural Information Processing Systems, vol. 30 (2017)
24. Sung, F., Yang, Y., Zhang, L., Xiang, T., Torr, P.H., Hospedales, T.M.: Learning to compare: relation network for few-shot learning. In: Proceedings of the IEEE Conference on Computer Vision and Pattern Recognition, pp. 1199–1208 (2018)
25. Vinyals, O., Blundell, C., Lillicrap, T., Wierstra, D., et al.: Matching networks for one shot learning. In: Advances in Neural Information Processing Systems, vol. 29 (2016)
26. Wah, C., Branson, S., Welinder, P., Perona, P., Belongie, S.: The caltech-UCSD birds-200-2011 dataset (2011)
27. Wang, Z., Lu, Y., Qiu, Q.: Meta-OLE: meta-learned orthogonal low-rank embedding. In: Proceedings of the IEEE/CVF Winter Conference on Applications of Computer Vision, pp. 5305–5314 (2023)
28. Wertheimer, D., Tang, L., Hariharan, B.: Few-shot classification with feature map reconstruction networks. In: Proceedings of the IEEE/CVF Conference on Computer Vision and Pattern Recognition, pp. 8012–8021 (2021)
29. Xian, Y., Sharma, S., Schiele, B., Akata, Z.: F-VAEGAN-D2: a feature generating framework for any-shot learning. In: Proceedings of the IEEE/CVF Conference on Computer Vision and Pattern Recognition, pp. 10275–10284 (2019)
30. Yan, L., Li, F., Zheng, X., Zhang, L.: Few-shot learning via task-aware discriminant local descriptors network. In: Proceedings of the 32nd ACM International Conference on Information and Knowledge Management, pp. 2887–2894 (2023)

31. Zhang, C., Cai, Y., Lin, G., Shen, C.: DeepEMD: few-shot image classification with differentiable earth mover's distance and structured classifiers. In: Proceedings of the IEEE/CVF Conference on Computer Vision and Pattern Recognition, pp. 12203–12213 (2020)
32. Zhang, X., Meng, D., Gouk, H., Hospedales, T.M.: Shallow bayesian meta learning for real-world few-shot recognition. In: Proceedings of the IEEE/CVF International Conference on Computer Vision, pp. 651–660 (2021)

Adaptive Compression of the Latent Space in Variational Autoencoders

Gabriela Sejnova(✉)[iD], Michal Vavrecka[iD], and Karla Stepanova[iD]

Czech Institute of Informatics, Robotics and Cybernetics, Czech Technical University in Prague, 16000 Prague, Czech Republic
`gabriela.sejnova@cvut.cz`

Abstract. Variational Autoencoders (VAEs) are powerful generative models that have been widely used in various fields, including image and text generation. However, one of the known challenges in using VAEs is the model's sensitivity to its hyperparameters, such as the latent space size. This paper presents a simple extension of VAEs for automatically determining the optimal latent space size during the training process by gradually decreasing the latent size through neuron removal and observing the model performance. The proposed method is compared to the traditional and computationally costly hyperparameter grid search and is shown to be significantly faster while still achieving the best optimal dimensionality on four image datasets. Furthermore, we show that the final performance as well as the speed of our method is comparable to training on the optimal latent size from scratch, and might thus serve as a convenient substitute for example in low-resource scenarios.

Keywords: Variational Autoencoders · Hyperparameter tuning · Representation learning

1 Introduction

Variational Autoencoders (VAEs) [10] are generative models that learn a probabilistic mapping between the data and the latent space, which allows for the generation of new, unseen data samples that are similar to the training data. Although VAEs have proven to be highly effective in this task, they are known for their sensitivity toward individual hyperparameters [8] and particularly the size (dimensionality) of the latent space. The selection of optimal latent dimensionality in VAEs is a crucial step in achieving the desired trade-off between reconstruction quality and clustering capability. Several works have studied and tried to mitigate this problem [7,9,15], however, in most current models, the optimal latent dimensionality is still chosen based on the specific task and data at hand.

Traditionally, determining the optimal latent space size has been done through a process of trial and error or by manually tuning the hyperparameters through grid search and observing the model's performance. However, this process can be time-consuming (as there are several hyperparameters to tune in

© The Author(s), under exclusive license to Springer Nature Switzerland AG 2024
M. Wand et al. (Eds.): ICANN 2024, LNCS 15016, pp. 89–101, 2024.
https://doi.org/10.1007/978-3-031-72332-2_7

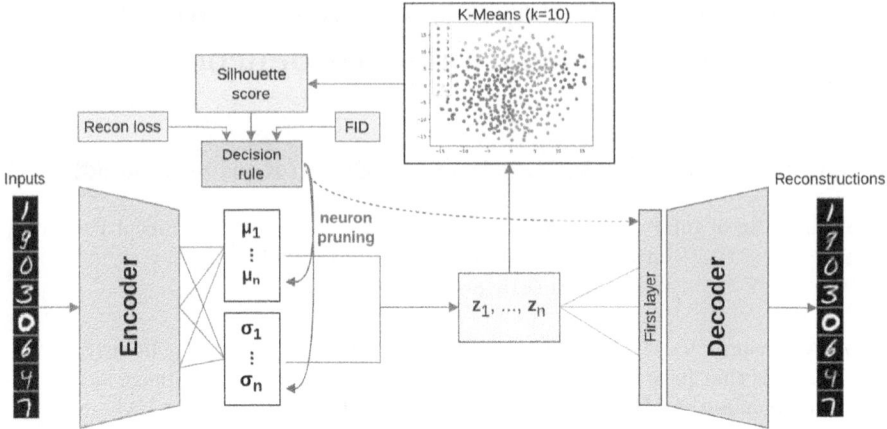

Fig. 1. Overview of our training procedure with shrinking latent space size. After each epoch, run k-means clustering on encoded samples from the validation dataset. We then calculate the Silhouette score based on the detected clusters. The Decision rule observes the approximate slope of the Silhouette score, FID and the reconstruction loss over epochs. We prune 1–5 neurons after every 5 epochs from the encoder's μ and σ layers and the decoder's first layer until the slopes meet our criteria (see Sect. 3.2), then the latent space size is fixed.

parallel) and unreliable, making VAEs less practical for real-world applications or scenarios with limited computational resources [1,16].

In this paper, we propose ALD-VAE (Adaptive Latent Dimensionality VAE) - a novel, automated heuristic for determining the optimal latent space size in VAEs. Our approach involves gradually decreasing the latent space size by removing neurons during the training process and observing the reconstruction loss, Frchet Inception Distance (FID) and cluster quality of the latent samples using the Silhouette score.

We show that the ALD-VAE approximates the optimal number of dimensions and thus avoids the need for manual tuning, plus reaches the same accuracy as if it was trained on the optimal latent dimensionality from scratch. We demonstrate the effectiveness of our approach on four image datasets ranging from toy-level (SPRITES, MNIST) to real-world noisy data (EuroSAT). With our proposed method, VAEs can be made more practical for real-world applications and the need for manual tuning can be eliminated. The corresponding code and more detailed results can be found in the GitHub repository[1].

2   Related Work

Hyperparameter tuning in VAEs has become a common subject of research in recent years as it is usually a necessary step for obtaining optimal results on the

---

[1] https://github.com/gabinsane/ald-vae.

chosen dataset [12], but the tuning process can also influence the results of comparison between various VAE architectures [8]. Several studies provided a comparison of VAE performance under various latent space dimensionalities explored through grid search. For example, Ahmed et al. [1] examine the optimal latent space size for electroencephalographic (EEG) signals by comparing 25 different latent sizes, Way et al. [16] compare different latent space dimensionalities for gene expression data.

Considering the high computational costs of training multiple models with different hyperparameters for comparison, other works focused on hyperparameter tuning on the fly during a single training. Mondal [13] proposed a method to automatically estimate the optimal latent dimensionality in adversarial autoencoders by introducing an additional trainable mask layer that hides the least informative dimensions. However, this method relies on a discriminator network which is not present in VAEs. A different approach is the two-stage VAE [3], where first a model with larger than optimal latent dimensionality is trained and then another VAE is finetuned during the second stage to learn the correct probability measure. Although close to our work, this approach does not focus on the clustering capability in VAEs.

In this paper, we focus on an adaptive decrease of the latent space size in VAEs during a single training which would be applicable to any VAE or its adaptation without requiring extra computational costs. We gradually find the optimal dimensionality based on the observation of the model's reconstructing and clustering capabilities in order to maximize the VAE's performance.

3   Adaptively Compressing Latent Space

As proposed by Kingma [10], VAEs are trained end-to-end in an unsupervised manner: first, the encoder maps the input data into a distribution over the latent space, typically by using multiple layers of neural networks. Second, samples are drawn from the distribution and the decoder maps them back to the original data space. The objective of the VAE is to maximize the evidence lower bound (ELBO), which is equivalent to maximizing the likelihood of the data and regularizing the approximate posterior to be close to the prior. The ELBO loss term is denoted as follows:

$$\mathcal{L}_{ELBO} = \mathbb{E}_{q(z|x)} \log p(x|z) - KL(q(z|x)||p(z)) \tag{1}$$

where $\mathbb{E}_{q(z|x)} \log p(x|z)$ is the expected log-likelihood of the data under the approximated posterior distribution, $KL(q(z|x)||p(z))$ is the Kullback-Leibler divergence between the approximated posterior distribution and the prior distribution over the latent variables. The size of the latent space is equal to the dimensionality of the multivariate distribution learned by the encoder. The parameters of the distribution (a vector of means, μ, and a vector of variances, σ) are usually provided by the two fully connected output layers of the encoder network, each of them having the dimensionality n_z. Since the output size of the layer is conventionally decided at model initialization, the latent space remains fixed during training.

In this section, we first formulate the problem of latent space compression and the task we are focusing on. Next, we define our proposed ALD-VAE training algorithm for automatic estimation of the optimal latent dimensionality.

3.1   Problem Formulation

Let $\mathcal{D}$ be a dataset consisting of datapoints $\mathbf{x}$ lying in an n-dimensional manifold $\mathcal{X}$ embedded in $\mathbb{R}^d$. Variational autoencoders [10] then establish a probabilistic relationship between $\mathcal{X}$ and a learned low-dimensional latent space $\mathcal{Z}$. It is our goal to find the optimal latent dimensionality n_z within $\mathcal{Z}$ that matches the intrinsic dimensionality n of the data. In real-world datasets, the intrinsic dimensionality n has been shown to be much smaller than the dimensionality of $\mathbb{R}^d$, i.e. $n \ll d$ [14], and using $n_z \gg n$ might thus lead to redundant dimensions and noisy outputs. One of the metrics that can show the qualitative performance of the model is Frchet Inception Distance (FID) [6] and, as demonstrated by Dai [2], too low as well as too high n_z is non-optimal (see the typical u-shape of FID for different latent dimensionality e.g., in Fig. 2).

One way to find the optimal n_z would be thus by finding the minimum of FID. However, in many cases, it is also desirable to cluster the data in the latent space based on the mutual similarity between the data, e.g., for image datasets with multiple classes. Quality of the clusters in the latent space can be evaluated using Silhouette score $S = \frac{1}{N} * \sum_{i=1}^{N} s(x_i)$ calculated over N samples encoded by the encoder network. For each sample x_i, $s(x_i)$ can be computed as:

$$s(x_i) = \frac{b_i - a_i}{max(b_i, a_i)}, \tag{2}$$

where b_i is the inter-cluster distance defined as the average distance to the closest cluster C of datapoint x_i (except for the one it is part of):

$$b_i = \min_{k}, k \neq i \frac{1}{|C_k|} \sum_{x_j \in C_k} d(x_i, x_j), \text{where } x_i \in C_i,$$

$|C_k|$ is the number of points belonging to cluster C_k and a_i is the intra-cluster distance defined as the average distance to all other points within the same cluster C:

$$a_i = \frac{1}{|C_i| - 1} \sum_{x_j \in C_i, i \neq j} d(x_i, x_j), \text{where } x_i \in C_i.$$

Note that for the k-means analysis, we need to provide the k expected number of clusters. For this, we can either use our prior knowledge about the dataset (e.g., MNIST has 10-digit classes), or we can first perform the k-means analysis over the training dataset with various k values to calculate Silhouette scores and choose the optimal k based on the highest score.

As shown in Fig. 2, the Silhouette score also changes with n_z, but can be optimal in different dimensionality than FID. Our objective is thus to find such latent dimensionality n_z, that minimizes the FID score and reconstruction loss $\mathcal{L}_r$ (which is a part of the VAE objective loss function, see Eq. 1), while maximizing the Silhouette score S:

$$n_z = \underset{n_z \in \mathbb{Z}}{\operatorname{argmin}} FID(n_z) \wedge \underset{n_z \in \mathbb{Z}}{\operatorname{argmin}} \mathcal{L}_r(n_z) \wedge \underset{n_z \in \mathbb{Z}}{\operatorname{argmax}} S(n_z). \tag{3}$$

However, as the Silhouette score (reflecting the clustering capability) increases together with the reconstruction loss and FID (reconstruction capability, see Fig. 2), it is not possible to find an optimal solution for n_z, that would satisfy the constraints in Eq. 3.

In Sect. 3.2, we propose a heuristic solution that balances all three metrics simultaneously during training.

3.2   Proposed Algorithm

We strive to minimize the latent space dimensionality n_z during the VAE training procedure to a value that is optimal in terms of the model's reconstruction accuracy and latent space regularization/clustering capability, i.e. the Silhouette score (see Eq. 3). In this section, we first describe the process of downsizing n_z during training without the need of retraining from scratch, then we summarize the stopping mechanisms used to determine whether the adapted dimensionality is an optimal solution or not. For a high-level overview of the whole procedure, see Fig. 1 or the pseudocode in Algorithm 1.

Latent Dimensionality Reduction. As mentioned in Sect. 3, the latent space size n_z is given by the output feature size of the two fully-connected layers $\boldsymbol{\mu}$ (learning the means of the multivariate distribution) and $\boldsymbol{\sigma}$ (learning the variances) which process the encoder's output.

The latent dimensionality reduction works as follows (see also Algorithm 1):

1. Initialization phase (choose a random, high-enough initial dimensionality n_z)
2. Train the VAE model for p epochs.
3. Stopping criterion check (see Eq. 6). If the stopping criterion is not met, proceed to 4., otherwise fix the current n_z and finish training.
4. Dimensionality reduction (remove n neurons from each of the two last encoder layers of the VAE model), then proceed to 2.

In each reduction process (step 4), we replace the last two fully connected layers of the trained VAE encoder with new ones, which have the original output feature dimension reduced by n. We then copy the learned weights and biases from the old layer into the newly initialized layer. As the latent vectors z have a reduced dimensionality after downsizing the encoder layers, we perform the same operation on the first decoder network layer. In the current scenario, we always prune the last neuron (neurons) from both the encoder's $\boldsymbol{\mu}$ and $\boldsymbol{\sigma}$ layers and the first decoder layer. In the future, an additional metric might be used for the selection of the least useful neuron, e.g., the dimension-specific KL divergence.

We perform the latent space compression during the training reactive to the Silhouette score and FID values. In the beginning, we remove $n = 5$ neurons (corresponds to the parameter *latent_decrease* in Alg. 1) after each $p = 5$ epochs (corresponds to the parameter patience p in Alg. 1), until the slope of the Silhouette score over the last 10 epochs becomes positive (see Eq. 7 for the slope calculation). Afterwards, we gradually decrease the pruning rate by setting n to 1, thereby pruning only one neuron after every p = 5 epochs. This is done iteratively until the stopping mechanism decides that the current dimensionality is optimal and n_z remains fixed. Please note that although we choose the initial n and p empirically, the algorithm was tested on various numbers n ($n \in 2 - 10$) and p ($p \in 2 - 20$) with similar results.

The total training time required for ALD-VAE is influenced by both the initial value n chosen by the user and the optimal dimensionality. A larger difference between these values results in a longer convergence time for the algorithm. However, in our experiments using various datasets, we observed that ALD-VAE typically identifies the optimal dimensionality in the approximately same number of epochs as needed by a standard VAE with a fixed dimensionality to converge (for our datasets approx. 200 epochs). Consequently, compared to performing a grid search over k different values of the latent space size, our approach yields an optimal solution approximately k-times faster.

Algorithm 1 Adaptively compressing latent space

1: **Input:** data x_{train}, x_{val}, number of data classes $k\_classes$
2: **Parameters:** init latent dims n_z, $latent\_decrease$, patience p, floating window w, thresholds for reconstruction and clustering metrics τ_r, τ_c
3: s_scores, recon_losses, fid_g, fid_r ← empty list
4: $compressing \leftarrow 1$
5: **for** $epoch$ in num_epochs **do**
6:     train_loop(x_{train})
7:     val_loop(x_{val})
8:     labels = KMeans.fit($z\_samples$, k=$k\_classes$)
9:     s_score = silhouette_score($z\_samples$, labels=$labels$)
10:     $recon\_loss$ = calculate_nll(model, data=x_{val})
11:     FID_g, FID_r = calculate_fid(model, data=x_{val})
12:     append s_score to s_scores
13:     append $recon\_loss$ to recon_losses
14:     append FID_g to fid_g, FID_r to fid_r
15:     **if** $epoch$ mod p = 0 **then**
16:         e1 = polyfit_slope(recon_losses[-w:], deg=1)
17:         e2 = polyfit_slope(s_scores[-w:], deg=1)
18:         e3 = polyfit_slope(fid_r[-w:], deg=1)
19:         e4 = polyfit_slope(fid_g[-w:], deg=1)
20:         **if** e1 > τ_r and e2 > τ_c and e3 > τ_r and e4 > τ_c **then**
21:             // stop with optimal latent dimensionality n_z
22:             $compressing \leftarrow 0$
23:         **else if** e2 >0 and $latent\_decrease$ ¿ 1 **then**
24:             $latent\_decrease \leftarrow 1$
25:         **end if**
26:     **end if**
27:     **if** $compressing$ = 1 **then**
28:         $n_z = n_z$ - $latent\_decrease$
29:         model.update_latent_dim(new_dim=n_z)
30:     **end if**
31: **end for**

Stopping Mechanism. As demonstrated in Fig. 2, the Silhouette score, as well as FID and the reconstruction loss react to the latent space size throughout the training. We seek such latent space dimensionality n_z, under which we get the maximal Silhouette score (S) and minimal reconstruction loss ($-\mathbb{E}_{q(z|x)} \log p(x|z)$) as well as FID ($FID$). Since the Silhouette score has an optimum in different n_z than reconstruction loss and FID, it is not possible to satisfy all conditions at the same time, we thus seek a balance among them. Here we first describe the general approach of when and how we adjust the dimensionality n_z, then we specify the decision rule for each of the three strategies.

As stated in the previous subsection, we apply the decision mechanism for possible neuron pruning after each $p = 5$ training epochs. As the Silhouette score and FID are increasing, we need to make a decision on whether to further adjust the latent space or not. Since we do not know the absolute optimal values of these metrics, we use the trends of the curves to estimate whether reducing the dimensionality will be beneficial. At each evaluation epoch e, we calculate the slope of the Silhouette score, FID and reconstruction losses using a sliding window of 20 epochs. For each window, we calculate the first-degree polynomial using the least squares polynomial fit to obtain the slope of the metric curve given as $y = mx + b$, where m is the slope of the curve and b is the y-intercept. The formula for the slope m is given by:

$$m = \frac{N \sum_{i=1}^{N} x_i y_i - \sum_{i=1}^{N} x_i \sum_{i=1}^{N} y_i}{N \sum_{i=1}^{N} x_i^2 - \left(\sum_{i=1}^{N} x_i\right)^2} \tag{4}$$

and the formula for the intercept b is given by:

$$b = \frac{\sum_{i=1}^{N} y_i - m \sum_{i=1}^{N} x_i}{N}, \tag{5}$$

where N is the number of data points, and x_i and y_i are the individual data points.

We use the slopes of the Silhouette score, reconstruction loss, FID for reconstructed images and FID for generated images to determine the stopping epoch e after which we no longer reduce the latent dimensionality. The stopping epoch e_s is thus defined as follows:

$$e_s = \min\{e|(S'(e) > \tau_c \wedge (\mathcal{L}'_r(e) > \tau_r) \wedge (FID'_r(e) > \tau_r) \wedge (FID'_g(e) > \tau_c)\} \tag{6}$$

where $\tau_{c,r}$ are chosen thresholds (in the default scenario, we choose $\tau_{c,r} = 0$) and $(S', \mathcal{L}'_r, FID'_r, FID'_g)$ are the first-degree polynomial scores p' (for the given metric curve) obtained through minimization of the squared error:

$$p' = \text{argmin}_p \sum_{j=0}^{k} |p(x_j) - y_j|^2, \text{where } p \in \{S, \mathcal{L}_r, FID_r, FID_g\}, \tag{7}$$

where k is the size of the sliding window (we choose $k=20$), x is the x-coordinate of each sample point, y is the y-value of the given sample point for the given metric (Silhouette score, Reconstruction loss, and FID for reconstructed or generated images). In other words, we seek the first epoch for which the slopes of the observed metrics calculated over the last $k=20$ epochs are all positive. This means that FID and Reconstruction loss have already reached the optimal peak value and are starting to increase (i.e., deteriorate), while the Silhouette score is improving. This scenario provides a balanced compromise between the reconstruction quality and clustering capability. However, it is also possible to bias the model towards one of those metrics by changing the thresholds τ of the stopping mechanism for individual metrics. There are thus three options the user can choose from:

1. **Balancing the reconstruction and clustering quality** (labelled as ALD_b approach): We stop reducing D when the slopes of all observed metrics become positive (i.e., $\tau_{c,r} = 0$)
2. **Prioritizing the reconstruction quality** (ALD_r): D remains constant immediately after the slopes of the reconstruction quality metrics stop decreasing (improving). We choose $\tau_{c,r} = -0.05$.
3. **Prioritizing the clustering capability** (ALD_c): D is fixed after the slope of the Silhouette score reaches a minimal threshold. We choose $\tau_r = 0$ and $\tau_c = 0.05$.

The user can select which of the schemes to follow based on the specific task and dataset. We selected the threshold values for both metrics empirically based on the experiments performed over the 4 datasets. For a different overview of our method, see Algorithm 1.

4   Experiments

To show the robustness and efficiency of our ALD-VAE approach, we compare its performance to grid search for 4 image datasets of different complexity (SPRITES, MNIST, FashionMNIST, EuroSAT) with known labels that enable evaluation of the clustering capability. For each dataset, we evaluate the model's performance on generative and reconstructive FID, Reconstruction loss and Silhouette score. As a baseline, we first

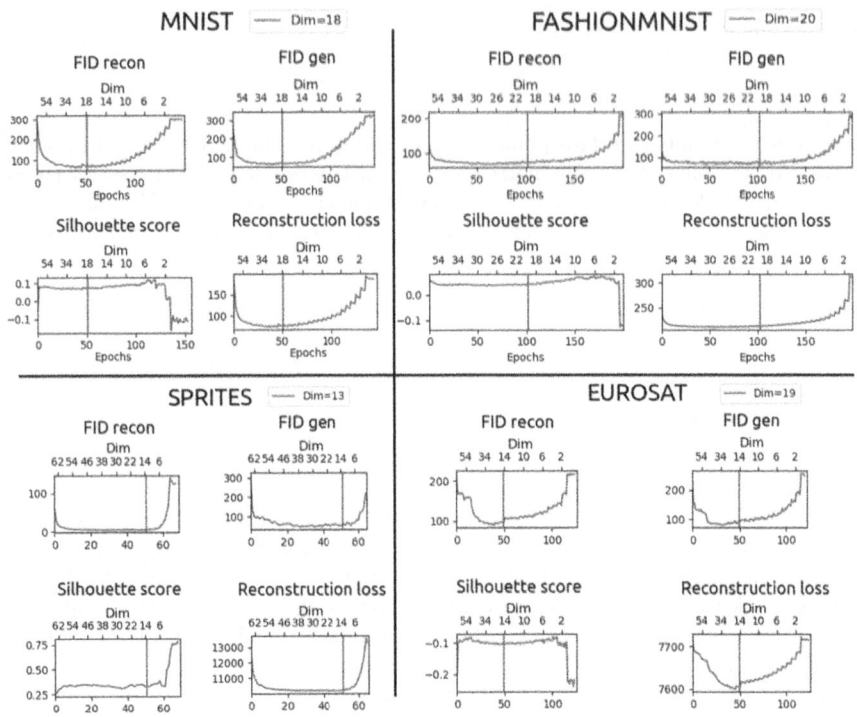

Fig. 2. The values of the four observed metrics (FID for reconstructions, FID for generations, Silhouette score and Reconstruction loss) during training when the latent space is gradually reduced until Dim=2. The upper axis shows the dimensionality at the given epoch, the vertical red line shows when would our ALD algorithm stop the compression. We show the results for four datasets. (Color figure online)

train a new VAE model for different (fixed) latent dimensionalities (grid search). Next, we train the proposed ALD-VAE model (using the three balancing schemes proposed in Sect. 3.2) by gradually compressing latent space. We train all models in all experiments over 5 different seeds. The number of epochs as well as the used encoder and decoder networks are mentioned for each dataset. All encoder networks also include the two additional fully connected layers to produce the μ and σ vectors.

4.1   Datasets Description and VAE Models Specification

The **SPRITES dataset** [11] contains colour images of animated characters (sprites). Each input is a sequence of 8 images showing random game characters performing one of 9 actions. For the encoder and decoder networks, we used a 3D convolutional layer adapted from VideoGPT [18]. We train all models for 300 epochs, the fixed models with latent sizes 12, 16, 24 and 32 and ALD-VAE with initial latent sizes $n_z = 100, 80, 64$.

MNIST [4] is a widely used dataset comprising grayscale images with written digits. For the encoder and decoder networks, we use 2 fully connected layers with 400 hidden dimensions. We train for 200 epochs with latent sizes 4, 8, 12, 16, 24 and 32 for

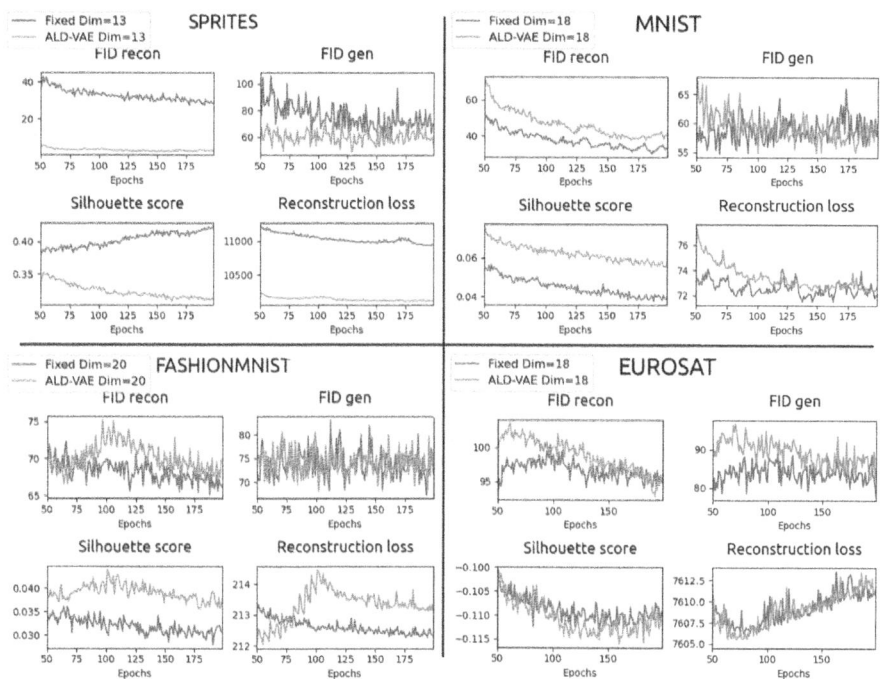

Fig. 3. Comparison of the four metrics when using a fixed latent dimensionality for the whole training (blue curves) or our ALD-VAE latent space compression that converged to the same final dimensionality (orange curves). We show FID for reconstructed and generated images (lower is better), Silhouette score (higher is better) and Reconstruction loss (lower is better). (Color figure online)

the fixed dimensionality baseline scenarios and initial latent sizes $n_z = 100, 80, 64$ for our adaptive ALD-VAE scenario.

FashionMNIST [17] contains 28×28 grayscale images of fashion articles from 10 different categories. For this dataset, we use the same encoder and decoder networks as for the MNIST dataset, i.e. 2 fully-connected layers with 400 hidden dimensions. We trained the models for 200 epochs and we compared the latent sizes 8, 16, 24, 32 and 64 in the fixed dimensionality experiments and the initial latent sizes $n_z = 100, 80, 64$ for ALD-VAE.

EuroSAT [5] contains 64×64 RGB satellite images showing pieces of land from 10 categories (e.g. *forest, river, industrial* etc.). For the encoder and decoder networks, we use $4 \times 2D$ convolutional layers with 256 hidden dimensions and SiLU activation functions. Compared to the other datasets, EuroSAT has a higher level of detail and variability in the images and is thus more difficult to reconstruct and even cluster. We trained the models again for 200 epochs and compared the latent sizes 16, 24, 32 and 64 in the fixed dimensionality experiments. The initial latent sizes for ALD-VAE were $n_z = 100, 80, 64$.

Table 1. Final reconstruction FID (FID_r, lower is better) and Silhouette scores (S, higher is better) for the four datasets. We compare standard VAE models trained with fixed dimensionalities (16, 24 or 32) with ALD-VAE in the balanced scenario (ALD_b) or biased either towards reconstruction quality (ALD_r) or clustering capability (ALD_c). The final dimensionality for ALD-VAE for each dataset is shown in parentheses. The reported values are means calculated over 5 seeds.

	SPRITES		MNIST		FashionMNIST		EuroSAT	
Dim	FID_r	S	FID_r	S	FID_r	S	FID_r	S
16	28.3	0.38	45.2	0.06	72.3	0.03	103.5	−0.10
24	33.1	0.34	43.5	0.04	65.2	0.02	84.2	−0.11
32	35.6	0.28	45.6	0.03	71.4	0.01	88.2	−0.10
ALD_r	12.6 (15)	0.28 (15)	38.5 (22)	0.03 (22)	66.2 (24)	0.02 (24)	85.2 (25)	−0.11 (25)
ALD_b	14.3 (13)	0.33 (13)	44.2 (18)	0.06 (18)	67.6 (20)	0.04 (20)	96.4 (18)	−0.11 (18)
ALD_c	17.4 (12)	0.44 (12)	58.2 (15)	0.09 (15)	74.6 (17)	0.05 (17)	124.5 (13)	−0.09 (13)

5   Results

ALD-VAE Performance on Different Datasets. First, we show in Fig. 2 the quality of the latent space dimensionality n_z as detected online by the proposed ALD-VAE algorithm. To show the quality of the online detection, we calculated the corresponding scores (Silhouette score, FID, reconstruction losses) for the model as if it would decrease n_z until $n_z = 2$, ignoring the stopping criterion, and visualise the point, where the ALD-VAE actually decided to stop the reduction of the latent space dimensionality. As can be seen in Fig. 2, the ALD-VAE finds a balance between the four observed metrics. As expected, the algorithm stops for all the datasets at the minimum (or directly after) of FID and Reconstruction loss, right after the Silhouette score started increasing.

Comparison to Training a VAE with Fixed Optimal Dimensionality. Next, we compared the ALD-VAE algorithm with another model trained with a fixed dimensionality that is the same as the final n_z found by ALD-VAE. The results are shown in Fig. 3. On average, it took 50 epochs until the model converged to the optimal dimensionality except for FashionMNIST, which took 100 epochs to converge. As seen in the plots, the training curves either converge to very similar values (e.g. for the

Table 2. Table with the overall mean training times (min) for each of the four datasets. We compare the parameter grid search (here we sum the total training time of 4 models with the following latent sizes: 8, 16, 24, 32) with the ALD-VAE approach. We show the standard deviations in parentheses.

Time [min]	MNIST	FashionMNIST	SPRITES	EuroSAT
grid search	3 224 (26.3)	3 985 (31.6)	8 896 (46.2)	7 287 (27.2)
ALD-VAE	**822 (15.3)**	**1016 (18.2)**	**2 267 (31.7)**	**1 827 (26.5)**

EuroSAT dataset), or there is a trade-off between some of the metrics - e.g., better Silhouette score for ALD-VAE in FashionMNIST, but worse final reconstruction loss. However, the differences are in all cases marginal and the overall outcome is almost identical when training with ALD-VAE compared to the optimal dimensionality fixed from the beginning. Moreover, we show the total training times for ALD-VAE compared to the grid search in Table 2. We can see that the ALD-VAE method is approximately $4\times$ faster than hyperparameter grid search over 4 latent dimensionalities.

Moreover, in Table 1 we show the final reconstruction FID values and Silhouette scores for VAEs with fixed dimensionalities as well as for ALD-VAE which was either in the balanced scenario (ALD_b, default option) or biased either towards reconstruction quality (ALD_r) or the clustering capability (ALD_c). Note that while ALD_r and ALD_c converge to different optimal dimensionality than ALD_b (values shown in parentheses), they produce better results in terms of reconstruction FID or Silhouette score, respectively (FID_r represents here the reconstruction quality and S represents the clustering capability).

In conclusion, we did not find any drawback to using the significantly faster ALD-VAE compared to hyperparameter grid search over multiple dimensionalities. Moreover, we demonstrated the three options for balancing reconstruction or clustering capabilities with ALD-VAE to satisfy the user's individual needs.

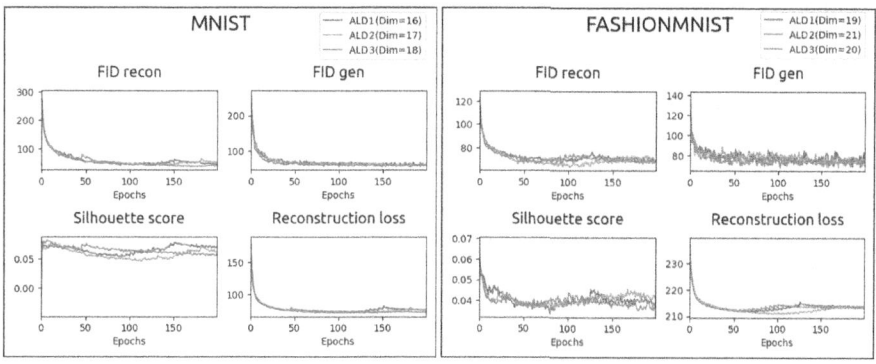

Fig. 4. Training our ALD-VAE with different initial conditions. $ALD1$ was trained with initial latent dimensionality $n_z = 100$, $ALD2$ was trained with initial $n_z = 80$ and $ALD3$ was trained with initial $n_z=64$. We show the results for the MNIST (left) and FashionMNIST (right) datasets. The Dim values in brackets for each ALD-VAE show the final converged n_z.

Influence of Initialization Values. Moreover, to see how the initial n_z influences the final optimal dimensionality, we trained three models for each of the four datasets with different n_z values of 100, 80 and 64. In Fig. 4, you can see results for MNIST and FashionMNIST. Both figures show the final optimal n_z for each model. The results for SPRITES and EuroSAT had a similar distribution of the final optimal n_z values. The found optimal dimensionalities differed only by 1–2 dimensions across the models and the four observed metrics had very similar final values. Note that we did not have to increase the number of training epochs for larger n_z as the adaptive neuron pruning can speed up the initial latent compression based on the decreasing Silhouette score.

6   Conclusion

We have introduced a new method, ALD-VAE, for automatically determining the optimal size of the latent space in Variational Autoencoders during a single training. Compared to parameter grid search exploring K latent dimensions, our rule-based approach is able to converge to the optimal dimensionality K-times faster.

Our method gradually reduces the size of the latent space during the training process and evaluates the quality of the reconstruction and clustering using the FID score, Reconstruction loss and Silhouette score. We present three methods to reach the optimal latent dimensionality: one with respect to minimizing the size of the latent space that still produces accurate reconstructions, one for finding the optimal size for clustering and one for balancing the reconstruction and clustering quality. We compare our method to the results obtained using a random hyperparameter search on four different image datasets and show that our proposed method can approximate the optimal size of the latent space and eliminate the need for manual tuning. Furthermore, ALD-VAE reaches comparable accuracies as the baselines trained with the same latent size from scratch. As ALD-VAE requires only simple adjustments to the original architecture and is easy to implement, it might be a convincing alternative to the classical models that require manual tuning of the latent space dimensionality.

Acknowledgement. This work was supported by the Czech Science Foundation (GA ČR) grants no. 21-31000S and 22-04080L.

References

1. Ahmed, T., Longo, L.: Examining the size of the latent space of convolutional variational autoencoders trained with spectral topographic maps of EEG frequency bands. IEEE Access **10**, 107575–107586 (2022)
2. Dai, B., Wipf, D.: Diagnosing and enhancing VAE models (2019). arXiv preprint arXiv:1903.05789
3. Dai, B., Wipf, D.P.: Diagnosing and enhancing VAE models. In: 7th International Conference on Learning Representations, ICLR 2019, New Orleans, LA, USA, May 6-9 (2019)
4. Deng, L.: The MNIST database of handwritten digit images for machine learning research. IEEE Signal Process. Mag. **29**(6), 141–142 (2012)
5. Helber, P., Bischke, B., Dengel, A., Borth, D.: EuroSAT: a novel dataset and deep learning benchmark for land use and land cover classification. IEEE J. Selected Topics Appl. Earth Observations Remote Sensing **12**(7), 2217–2226 (2019)
6. Heusel, M., Ramsauer, H., Unterthiner, T., Nessler, B., Hochreiter, S.: Gans trained by a two time-scale update rule converge to a local Nash equilibrium. Adv. Neural Inf. Proces. Syst. **30** (2017)
7. Higgins, I., et al.: beta-vae: Learning basic visual concepts with a constrained variational framework (2016)
8. Hu, Q., Greene, C.S.: Parameter tuning is a key part of dimensionality reduction via deep variational autoencoders for single cell RNA transcriptomics. In: BIO-COMPUTING 2019: Proceedings of the Pacific Symposium, pp. 362–373. World Scientific (2018)
9. Kim, H., Mnih, A.: Disentangling by factorising. In: International Conference on Machine Learning, pp. 2649–2658. PMLR (2018)

10. Kingma, D.P., Welling, M.: Auto-encoding variational bayes. In: 2nd International Conference on Learning Representations (ICLR) (2014)
11. Li, Y., Mandt, S.: Disentangled sequential autoencoder. In: International Conference on Machine Learning (2018)
12. Locatello, F., et al.: Challenging common assumptions in the unsupervised learning of disentangled representations. In: International Conference on Machine Learning, pp. 4114–4124. PMLR (2019)
13. Mondal, A., Pal Chowdhury, S., Jayendran, A., Asnani, H., Singla, P., A P, P.: MaskAAE: latent space optimization for adversarial auto-encoders. In: Proceedings of the 36th Conference on Uncertainty in Artificial Intelligence (UAI), vol. 124, pp. 689–698. PMLR (2020)
14. Narayanan, H., Mitter, S.: Sample complexity of testing the manifold hypothesis. In: Advances in Neural Information Processing Systems, vol. 23 (2010)
15. Shao, H., et al.: Controlvae: controllable variational autoencoder. In: International Conference on Machine Learning, pp. 8655–8664. PMLR (2020)
16. Way, G.P., Zietz, M., Rubinetti, V., Himmelstein, D.S., Greene, C.S.: Compressing gene expression data using multiple latent space dimensionalities learns complementary biological representations. Genome Biol. **21**(1), 1–27 (2020)
17. Xiao, H., Rasul, K., Vollgraf, R.: Fashion-MNIST: a novel image dataset for benchmarking machine learning algorithms (2017)
18. Yan, W., Zhang, Y., Abbeel, P., Srinivas, A.: VideoGPT: Video generation using VQ-VAE and transformers (2021). arXiv preprint arXiv:2104.10157

Asymmetric Isomap for Dimensionality Reduction and Data Visualization

Dominik Olszewski$^{(\boxtimes)}$

Faculty of Electrical Engineering, Warsaw University of Technology, Warsaw, Poland
`dominik.olszewski@pw.edu.pl`

Abstract. We propose an asymmetric version of the Isomap dimensionality reduction and data visualization approach. Our improvement uses the information on asymmetric input data relationships, and in this way, it determines the input dissimilarities more accurately than original Isomap. We introduce as well the asymmetric coefficients discovering and expressing the asymmetric properties of the input data. These coefficients asymmetrize geodesic distances in Isomap making this method asymmetric. The experiments on two real datasets confirm the effectiveness of our proposal.

Keywords: Isomap · asymmetric Isomap · asymmetric data analysis · dimensionality reduction · data visualization

1 Introduction

Dimensionality reduction is a branch of artificial intelligence and machine learning gaining constantly increasing interest and attention in recent years. The general goal of dimensionality reduction is to design a linear or nonlinear projection from an original input high-dimensional data space onto a resulting output low-dimensional data space in a way preserving similarities between the data points in the input data space [3,4]. A data transformation and processing formulated and described in this way can be, naturally, recognized as well as a specific form of data compression, and this observation clearly points out one of the important gains and advantages of reducing the number of dimensions in data. Consequently, dimensionality reduction can be regarded as a data compression based on a criterion of data similarities maintaining and restoration.

However, beside the compression context, there is another, probably most significant and intuitive justification of decreasing the number of data dimensions, and that is data visualization. Speaking more specifically, in case of carrying out dimensionality reduction with the output number of dimensions set to 1, 2, or 3, one obtains a constructed output data space eligible for graphical presentation. It means that the input high-dimensional data has been transformed in a way making possible to observe, analyze, and interpret it purely visually. As a result, one gains a novel kind of insight into the data enabling to notice various phenomena, effects, and relationships easily and straightforwardly, especially

M. Wand et al. (Eds.): ICANN 2024, LNCS 15016, pp. 102–115, 2024.
https://doi.org/10.1007/978-3-031-72332-2_8

taking into consideration a difficulty of such an analysis in case of the original input multidimensional data [3,4].

Within a wide and various range of existing dimensionality reduction solutions, there is a category, which employs a precise, well-justified, and well-formulated criterion of constructing the output data space. Namely, one strives to build a low-dimensional differential manifold embedded in the linear input Euclidean high-dimensional data space. The output space is obtained after projection from this manifold. Among these methods and approaches, a pioneer and fundamental one is Isomap [17]. The technique approximates the mentioned low-dimensional manifold by constructing a k-Nearest Neighbors (k-NN) graph representing the original input data. Based on the graph, Isomap computes the geodesic distances matrix, which is an improvement with respect to the traditional Euclidean distances matrix. Isomap with its geodesic distances extracts the inner geometry of the input data, and speaking informally, but intuitively, Isomap estimates the "shape" of the input data. This kind of enhancement and extension to traditional dimensionality reduction in a manner of MultiDimensional Scaling (MDS) enriches significantly the information about the relationships within the input data based solely on the standard Euclidean distances. And this method and way of thinking have been a point of departure for our original proposal presented in this paper.

1.1 Our Proposal

We propose an asymmetric version of the Isomap algorithm. Our version is a generalization of the conventional Isomap technique, and it is extended by a mechanism making it capable to handle asymmetric relationships within the input dataset. The mentioned asymmetric relationships arise as a consequence of hierarchical input data structure. The hierarchy in data stems, in turn, from a difference in generality degree between the data points as it is explained in more detail in Sect. 3. In this way, beside the advantage of representing the data geometry, our form of Isomap computes the dissimilarities between the data points more accurately than its basic counterpart.

Recapitulating, we propose the following contributions:

- asymmetric coefficients reflecting the generality degree of the data points calculated on the basis of the strength of the k-NN graph representing the input dataset,
- asymmetric geodesic dissimilarities obtained using the introduced asymmetric coefficients being a generalization of the standard Isomap's geodesic distances,
- the final asymmetric improved version of the Isomap method,
- experimental study confirming the effectiveness and usefulness of our approach, as well as sustaining the theoretical claims in our work and theoretical justification of the introduced method.

2   Related Work

Generally, the dimensionality reduction methods can be divided into 4 groups.

The first one refers to linear-algebra-based approaches. The solutions belonging here are based on linear algebraic transformation leading from the input high-dimensional data space to an output low-dimensional space. The transformation itself relies on projection matrix composed, e.g., from the appropriately chosen eigenvectors. The examples of the particular algorithms: Principal Component Analysis (PCA) and its variants.

In the second category, one can find artificial-neural-networks-based techniques. A major position on this group belongs to the Self-Organizing Map (SOM) by T. Kohonen [7]. It is a 2-dimensional regular grid of neurons, which constitutes a screen for data visualization. Recent publications in this field include [12,13].

As the third class, we can give a set of geometric approaches to dimensionality reduction. This time, the manifold learning is achieved using the k-NN graph representing the input data. The graph is a discrete approximation of the low-dimensional manifold embedded in the linear input Euclidean high-dimensional data space. The examples of such methods are the following: Isomap [17] and Laplacian Eigenmaps [1].

Finally, the fourth and last group of dimensionality reduction techniques gathers the Neighborhood Preserving Projections (NPPs). An idea behind this methodology is to analyze the data locally, i.e., the neighborhood of each data point, and to maintain the neighborhoods in the output data space. Consequently, the locations of the data points in the output space are chosen so as to restore the neighborhoods of the corresponding data points in the input space. The examples of the methods include: Locally Linear Embedding (LLE) [16], Stochastic Neighbor Embedding (SNE) [6], t-SNE [8], Neighbor Retrieval Visualizer (NeRV) [11,14,18], and Uniform Manifold Approximation and Projection (UMAP) [10].

In our research, we focus especially on the third geometry-based group as it contains the Isomap algorithm, which is the basis for our improvement development.

Isomap gains constant attention of the researchers, and the recent proposals in the literature regarding this solution include [5,15,19].

In the work [15], the authors introduce the A*-FastIsomap method for overcoming the shortest distance path and high computational complexity problems. This Isomap's extension is based on the A*Search algorithm with the Double Buckets algorithm.

The article [19] provides an effective inverse design method for three-dimensional aerodynamic configuration based on Isomap in manifold learning. In this approach, the aerodynamic shape and parameterized control volume are regarded as the spatial nonlinear coupling surface, which is then projected onto a low-dimensional manifold via Isomap to produce the manifold structure.

On the other hand, the paper [5] employs Isomap in an adaptive sampling method for reduced-order models, which are widely used in physics-based modeling and simulation.

3   Asymmetry in Data Analysis

Asymmetry in data arises, whenever the data presents a hierarchical structure, i.e., the entities within a dataset are characterized by a different degree of generality. This claim has been stated and argued in the work [9]. A helpful example illustrating that phenomenon refers to a textual dataset consisting of words belonging to classes of different level of hierarchy. Assuming that this hypothetical dataset contains mathematical vocabulary and the words "equation" and "Poisson" among others, then it may be easy to notice and understand that the word "equation" is more general than "Poisson" and these both words, therefore, reside on a different level in the hierarchical decomposition of the dataset. Consequently, any measure of dissimilarity should return greater value in the direction from "equation" to "Poisson" and lower value in the opposite direction. This is an asymmetric relationship. Similarly, one may recall a well-known example given by A. Tversky that son resembles father, but the relationship could not be presented in the opposite order. In this scenario, the asymmetry appeared as a consequence of different prominence between the roles of father and son, which has the same effect as difference in generality degree.

Taking into consideration that asymmetric data requires proper handling, the data analysis methods should be equipped with mechanisms and abilities to adapt and adjust to this kind of data and specific phenomenon affecting the relationships within the data. This goal can be achieved by providing a numerical representation of generality degree of data entities. Then, using a given measure correctly reflecting the asymmetric data associations, one may formulate asymmetric versions of dissimilarities, and finally, asymmetric forms of entire data analysis methods. This kind of methods' extensions always lead to generalizing their area of effectiveness and expanding their operating range and possibilities. In our work, we express the generality level of data points in the input data space using the asymmetric coefficients introduced in this paper and described in Sect. 5.2. Our asymmetric coefficients are based on the strength of the k-NN graph, which is a crucial part of Isomap.

4   Traditional Isomap

The standard Isomap is build on the basis of the MDS classical dimensionality reduction technique. Isomap replaces Euclidean distances matrix, which is a conventional input to MDS, with geodesic distances matrix. A geodesic distance measures the dissimilarity between two input data points as the shortest path in the k-NN graph representing the input dataset, being a discrete approximation of the low-dimensional manifold embedded in the linear input Euclidean high-dimensional data space. In this way, Isomap uses the information on the geometric structure of the input data (informally, the "shape" of the input data), and consequently, geodesic distances are obeying that geometry and are reflecting the genuine dissimilarity between data points more accurately than Euclidean distances. A clear and convincing illustration of the superiority of geodesic distances over Euclidean ones is an example using the well-known Swiss roll synthetic dataset presented in [17], where the geometry assumes a form of coils, which greatly affects the dissimilarity relationships between the data points, and the resulting difference between the Euclidean and geodesic distances is major.

The Isomap algorithm consists of the three steps presented in Algorithm 1.

Algorithm 1. Isomap dimensionality reduction procedure

Require: High-dimensional input dataset X.
Ensure: Output dataset Y of reduced number of dimensions.
 1: Construct the k-NN graph representing the input dataset.
 2: Compute the geodesic distances matrix as the shortest paths in the constructed k-NN graph, e.g., using the Dijkstra or Bellman—Ford algorithm.
 3: Perform standard MDS on the geodesic distances matrix.

---

5   Asymmetric Isomap

We propose an improvement of the traditional Isomap capable to handle the asymmetric relationships between the data points in the input data space. Our method grasps the asymmetry degree using the asymmetric coefficients defined in Sect. 5.2. Since the asymmetry itself is caused by hierarchical data structure and differences in generality degree of data points, our asymmetric coefficients aim to reflect and express those generality degrees. Afterwards, they are used to asymmetrize geodesic distances, and consequently, obtain asymmetric Isomap.

5.1   Theoretical Justification of the Proposed Method

In order to present a convincing argumentation of using our particular Isomap's extension, first of all, we justify that the asymmetric coefficients indeed convey the information about the asymmetric relationships between the input data points. The goal is to express numerically the generality degree of the data points. In order to achieve it, we calculate the average Euclidean distance from a given point to its k nearest neighbors, where k is the parameter of Isomap's k-NN graph. This value shows the data density in the neighborhood of this point. The density provides the information on the frequency of occurrences of that point and the ones highly similar to it. And this information refers to the generality level of the considered point.

As a result of this reasoning, one needs to compute the average distance from each point to its neighborhood. Considering that the Isomap method commences from constructing the k-NN graph of the input data, our task comes down to computing the graph's strength, defined as the vector, in which each entry corresponds to a data point, and the value of this entry is the sum of weights of edges attached to the considered point. The strength of the graph is then almost a vector of asymmetric coefficients. It remains only to divide the entries of the strength vector by the respective numbers of edges attached to the point corresponding to this entry (number of the point's neighbors), and we obtain the exact asymmetric coefficients vector.

Now, when the formulation of the asymmetric coefficients has been explained, the next thing is to apply them to asymmetrize the geodesic distances and obtain asymmetric geodesic distances matrix A_G. The asymmetric geodesic distances must reflect the difference in generality level (position in hierarchy) of the data points given as their arguments. Hence, by applying the ratio of the asymmetric coefficients as it is shown in Eq. (3), they will return greater values in the direction from more general point to more specific one and lower values in the opposite direction.

The last operation remaining is to execute the conventional MDS on A_G.

5.2   Asymmetric Coefficients

The asymmetric coefficients are intended to grasp, express, and convey the information on asymmetric associations within the input data. According to justification and explanation delivered in Sect. 5.1, they are defined in the following way.

$$a_i = \frac{1}{k} \sum_{j=1}^{k} d_{\text{Euc}}\left(x_i, x_j\right),$$ (1)

where x_i, $i = 1, \ldots, N$ is an input data point, N is the total number of the data points in both – input and output datasets, x_j, $i = 1, \ldots, k$ are the data points in the neighborhood of the point x_i, k is the number of nearest neighbors of the data point x_i, and d_{Euc} is the Euclidean distance.

A vector of asymmetric coefficients for the entire input dataset is $a = \text{strength}(G)/k$, where G is the k-NN graph representing the input dataset.

5.3   Procedure of Our Approach

The asymmetric coefficients are incorporated in the geodesic distances formulae, which are the basis for the asymmetric geodesic distances matrix composition.

The detailed algorithmic description of our approach is presented as Algorithm 2.

Algorithm 2. The algorithm of our Isomap's enhancement

Require: High-dimensional input dataset X.

Ensure: Output dataset Y of reduced number of dimensions.

1: Construct the k-NN graph representing the input dataset.

2: Calculate the asymmetric coefficients according to Eq. (1).

3: Compute the asymmetric geodesic distances matrix A_G according to the following formulae:

$$A_G = \{d_G^{\text{Asymm}}(x_i, x_j)\}, \tag{2}$$

$$d_G^{\text{Asymm}}(x_i, x_j) = \frac{a_j}{a_i} d_G(x_i, x_j), \tag{3}$$

where A_G is the asymmetric geodesic distances matrix, $d_G^{\text{Asymm}}(\cdot, \cdot)$ is the asymmetric geodesic distance, $d_G(\cdot, \cdot)$ is the standard geodesic distance, x_i, $i = 1, \ldots, N$ are the input data points, a_i, $i = 1, \ldots, N$ are the asymmetric coefficients, and N is the total number of the data points in both – input and output datasets.

4: Apply traditional MDS on the asymmetric geodesic distances matrix by minimizing the following objective function:

$$E = \|\tau(A_G), \tau(D_Y)\|_{L^2}, \tag{4}$$

$$D_Y = \{d_{\text{Euc}}(y_i, y_j)\}, \tag{5}$$

where y_i, $i = 1, \ldots, N$ are the output data points, N is the total number of the data points in both – input and output datasets, $\|X\|_{L^2} = \sqrt{\sum_{i,j} X_{i,j}^2}$ is the L^2 matrix norm, and τ is the operator converting distances to inner-products, which uniquely characterize the geometry of the data in a form that supports efficient optimization. The global minimum of Eq. (4) is achieved by setting the coordinates y_i to the top d eigenvectors of the matrix $\tau(A_G)$ as it is stated in the paper [17] for the case of the classical Isomap's objective function minimization.

---

6 Experiments

In our experimental research, we have evaluated the effectiveness of the proposed asymmetric version of the Isomap algorithm on the basis of a comparison to the five reference dimensionality reduction and data visualization methods. In order to obtain a numerical measurable assessment criterion, we have continued the dimensionality reduction by data clustering. Consequently, the clustering has been conducted in the 2-dimensional visualization space. The clustering results could be evaluated better than the visualization itself, and therefore, in order to evaluate the clustering outcomes, we have utilized two assessment criteria, i.e., the accuracy rate q (external index) measuring the number of points correctly assigned to clusters and the uncertainty degree U_d (internal index) measuring the number of points in the overlapping areas between clusters.

Two clustering techniques have been used: the well-known k-means algorithm and the Density-Based Spatial Clustering for Applications with Noise

(DBSCAN). They represent two different clustering concepts categories – partitioning clustering and density-based clustering, respectively.

Taking into consideration that the methods for dimensionality reduction and clustering have certain randomness in their behavior, all the experiments have been run 50 times, and the values of the accuracy rates and the uncertainty degrees in Tables 1 and 2 are arithmetic averages from these repetitions.

6.1 Reference Methods

Five reference dimensionality reduction methods have been selected, namely, PCA, LLE, t-SNE, UMAP, and standard Isomap. This selection of data visualization approaches is intended to represent wide spectrum of possible solutions, and it represents members of different dimensionality reduction categories given and described in Sect. 2, i.e., linear-algebra-based solutions (PCA), geometric approaches (Isomap), and NPP group (LLE, t-SNE, and UMAP). We use an example of a linear method (PCA) and nonlinear ones (all the remaining methods). However, most importantly, we have included in our experimental study the conventional Isomap technique, because it is a basis for our enhancement's development, and by comparing its results against the ones returned by our version of Isomap, we can essentially verify, whether we have indeed improved the traditional Isomap approach.

6.2 Evaluation Criteria

We have evaluated our experimental results on the basis of the *a posteriori* output data clustering, which has been assessed using the two criteria: the accuracy rate and the uncertainty degree.

1. **Accuracy rate.** It determines the number of correctly assigned data points divided by the total number of points. It is an external clustering evaluation index, because it uses the ground truth knowledge.
   The accuracy rate is defined as follows:

$$q = \frac{N_c}{N},\tag{6}$$

   where N_c is the number of correctly assigned data points, and N is the total number of data points in the whole dataset.
   The accuracy rate q belongs to the range $\langle 0, 1\rangle$. Bigger values are preferred.
2. **Uncertainty degree.** It measures the number of overlapping data points divided by the total number of data points in a dataset. The data points in the overlapping area are the ones, for which the ratio of the Euclidean distances between them and any two clusters centroids belongs to the range $\langle 0.99, 1.01\rangle$. This criterion is an internal clustering index, because it does not require the ground truth, and uses only the data scattering information.
   The uncertainty degree is defined in the following way:

$$U_d = \frac{N_o}{N},\tag{7}$$

where N_o is the number of overlapping data points in the dataset, and N is the total number of data points in the dataset.

The uncertainty degree belongs to the range $\langle 0, \ 1 \rangle$. Lower values are expected.

6.3   Gas Sensors Array Drift Dataset

The first of our analyzed datasets is an excerpt from the collection [2]. It is a time series dataset, and it contains 3000 measurements from 16 chemical sensors exposed to 6 different gases at various concentration levels. Therefore, the number of instances in this dataset is 3000. Each measurement vector contains the 8 features extracted from each particular sensor, resulting in a 128-dimensional feature vector (8 features × 16 sensors). The dataset comprises recordings from 6 distinct pure gaseous substances, namely, Ammonia, Acetaldehyde, Acetone, Ethylene, Ethanol, and Toluene. Our visualization and clustering aimed to form clusters representing those gaseous substances. Therefore, the target number of clusters was 6.

The results are presented in Fig. 1 and in Table 1. We include selected most interesting plots of the graphical presentation of our results for a reason of text length limit.

The colors on the plots in Fig. 1 correspond to different clusters formed in the 2-dimensional data visualization space constructed as a result of dimensionality reduction. In case of the DBSCAN clustering algorithm, the number of the generated clusters might appear to be different than the desirable 6 clusters. This happens, because the DBSCAN clustering technique does not assume the target number of clusters as an input parameter, but it automatically determines it itself as a part of the density-based clustering procedure. Therefore, the number of colors on the plots depicting the DBSCAN results may be different than 6, which is, naturally, erroneous and unwanted behavior and result and has been treated in such a way in our evaluation of the experimental results.

According to the results of this part of our empirical research, the proposed Isomap's extension & DBSCAN appeared to outperform all the other investigated methods taking into consideration the accuracy rate, however, in case of the uncertainty degree, two of the benchmark techniques returned lower (better) results, i.e., PCA & DBSCAN and UMAP & DBSCAN. This observation shows that from the perspective of separation margin maximization, our approach was surpassed in a moderate extent by these two methods. Nevertheless, the uncertainty degrees value of 0.1107 corresponding to our approach may be recognized as satisfactory. Furthermore, the accuracy rate is a primary and most important evaluation criterion, and this one points out clearly the superiority of our Isomap's improvement over the other investigated algorithms, and consequently, the correctness of our theoretical assertions.

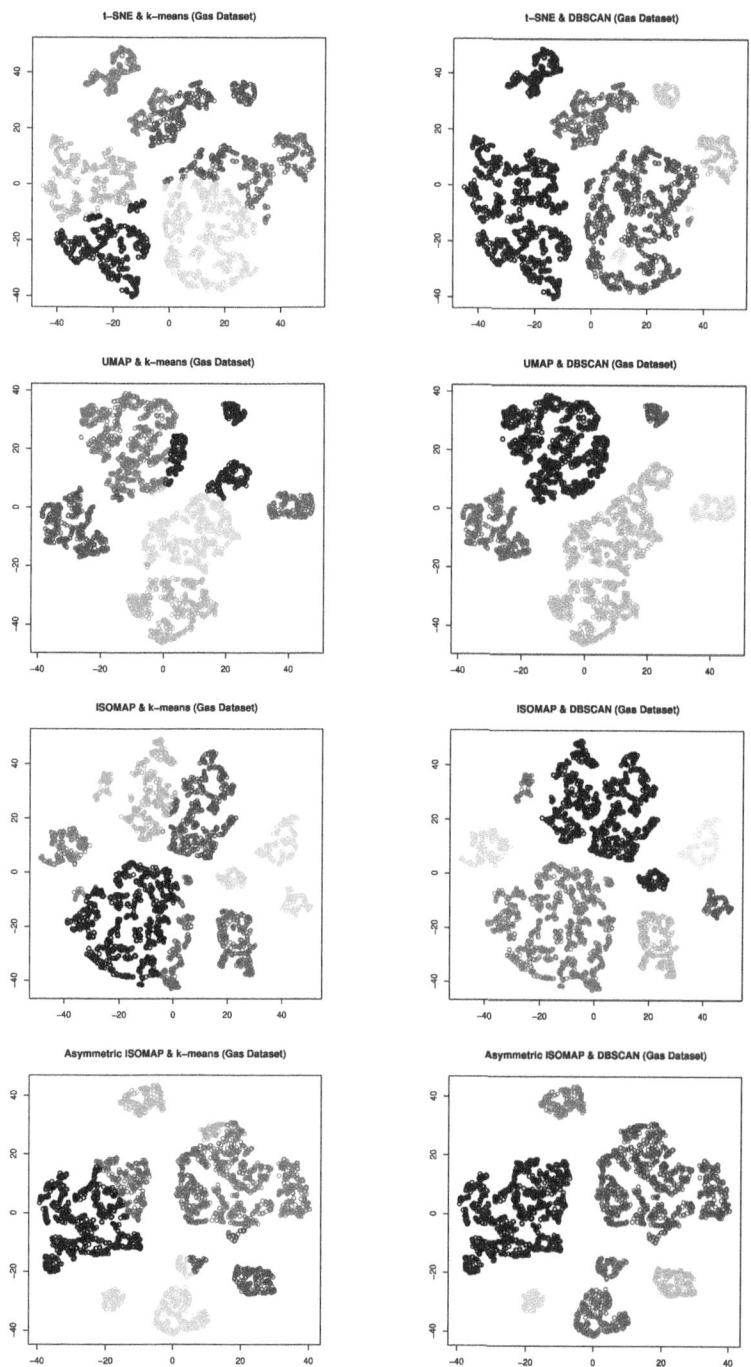

Fig. 1. Results of gases visualization and clustering.

Table 1. Accuracy rates and uncertainty degrees for gases visualization and clustering

	q	U_d
PCA & k-means	2111 / 3000 = 0.7037	733 / 3000 = 0.2443
PCA & DBSCAN	2400 / 3000 = 0.8000	154 / 3000 = 0.0513
LLE & k-means	2150 / 3000 = 0.7167	1678 / 3000 = 0.5593
LLE & DBSCAN	2571 / 3000 = 0.8570	513 / 3000 = 0.1710
t-SNE & k-means	2200 / 3000 = 0.7333	1239 / 3000 = 0.4130
t-SNE & DBSCAN	2953 / 3000 = 0.9843	1910 / 3000 = 0.6367
UMAP & k-means	2354 / 3000 = 0.7847	1005 / 3000 = 0.3350
UMAP & DBSCAN	2782 / 3000 = 0.9273	269 / 3000 = 0.0897
Isomap & k-means	2320 / 3000 = 0.7733	383 / 3000 = 0.1277
Isomap & DBSCAN	2891 / 3000 = 0.9637	751 / 3000 = 0.2503
Asymmetric Isomap & k-means	2379 / 3000 = 0.7930	551 / 3000 = 0.1837
Asymmetric Isomap & DBSCAN	**3000 / 3000 = 1.0000**	**332 / 3000 = 0.1107**

6.4 DeliciousMIL Dataset

In the second part of our experiments, we used a textual dataset consisting of 1944 text documents extracted from DeliciousT140 dataset from [2]. The text corpus was numerically represented using the bag-of-words textual model. The resulting number of features is 8519. From the dataset, we extracted text parts and chose 5 tags in order to form 5 clusters corresponding to these tags. The tags were the following: programming, internet, grammar, style, and language.

The results are presented in Fig. 2 and in Table 2. We include selected most interesting plots of the graphical presentation of our results for a reason of text length limit.

Similarly as in the case of the gases dataset, the colors on the plots in Fig. 2 correspond to different clusters generated in the 2-dimensional data visualization space constructed as a result of dimensionality reduction. The DBSCAN clustering method determines the number of clusters automatically as a result of the density-based clustering procedure, and therefore, the number of clusters formed by this algorithm might be different than 5, which we have recognized as incorrect outcome during our experimental assessment.

Similarly as in the case of the results from the previous part of our experiments, the introduced approach (cooperating with k-means) appeared to be superior over the benchmark algorithms with respect to the accuracy rate. This time, there was only one case of better uncertainty degree provided by a reference method, and that is a combination of standard Isomap & k-means. Therefore, again, the value of the uncertainty degree of our approach can be regarded as sufficiently low, even though there was one other method producing lower.

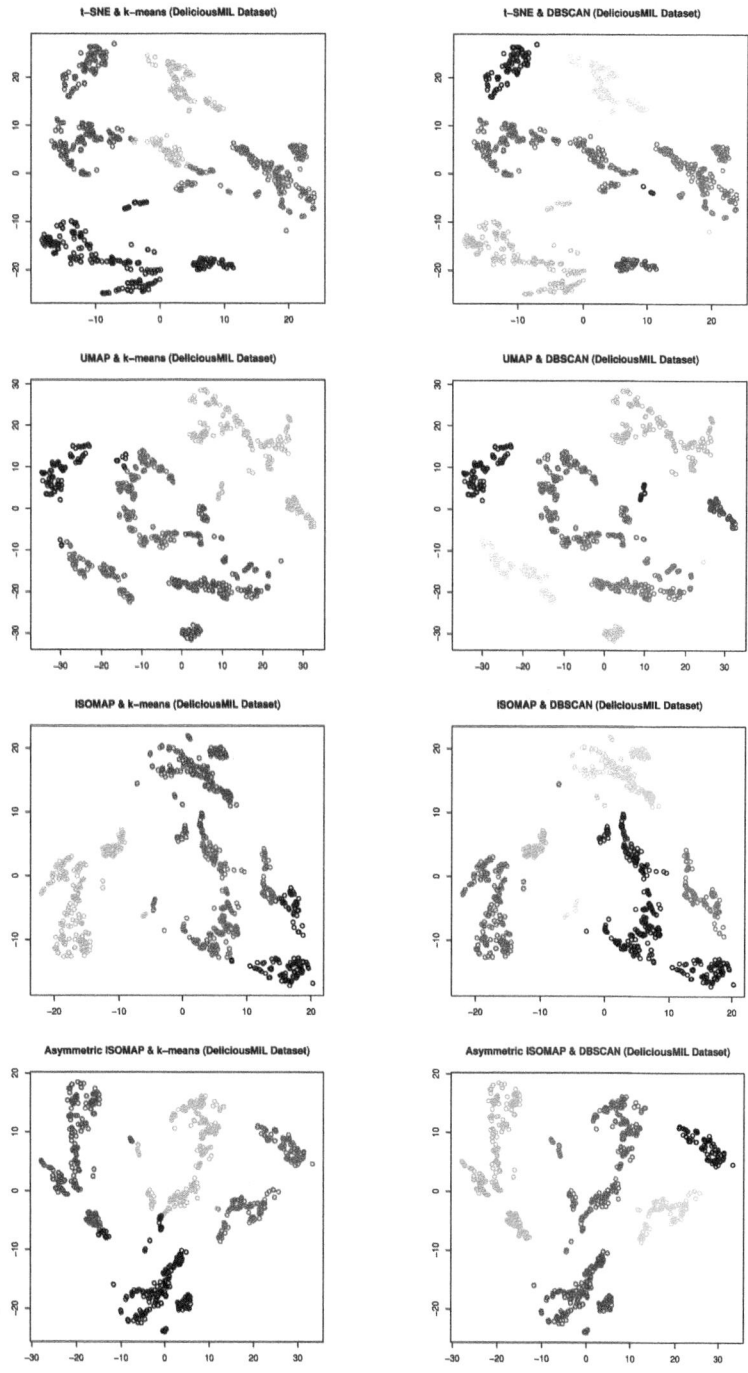

Fig. 2. Results of documents visualization and clustering.

Table 2. Accuracy rates and uncertainty degrees for text documents visualization and clustering

	q	U_d
PCA & k-means	1703 / 1944 = 0.8760	147 / 1944 = 0.0756
PCA & DBSCAN	1630 / 1944 = 0.8385	491 / 1944 = 0.2526
LLE & k-means	1876 / 1944 = 0.9650	157 / 1944 = 0.0808
LLE & DBSCAN	1545 / 1944 = 0.7948	686 / 1944 = 0.3529
t-SNE & k-means	1860 / 1944 = 0.9568	183 / 1944 = 0.0941
t-SNE & DBSCAN	1569 / 1944 = 0.8071	1095 / 1944 = 0.5633
UMAP & k-means	1782 / 1944 = 0.9167	119 / 1944 = 0.0612
UMAP & DBSCAN	1507 / 1944 = 0.7752	691 / 1944 = 0.3555
Isomap & k-means	1844 / 1944 = 0.9486	52 / 1944 = 0.0267
Isomap & DBSCAN	1651 / 1944 = 0.8493	1051 / 1944 = 0.5406
Asymmetric Isomap & k-means	**1893 / 1944 = 0.9738**	**114 / 1944 = 0.0586**
Asymmetric Isomap & DBSCAN	1705 / 1944 = 0.8771	273 / 1944 = 0.1404

7   Conclusions

We proposed an asymmetric version of the classical Isomap method. As we justified theoretically, our extension is a generalization of Isomap, and it reflects the dissimilarities in the data more accurately, in particular, it handles correctly asymmetric data.

In the experimental verification, we confirmed these theoretical claims and the justification of our Isomap's improvement. Moreover, our experiments on two real datasets showed superiority of asymmetric Isomap over the five reference methods cooperating with two clustering algorithms.

In the future study, one may consider: asymmetrizing other dimensionality reduction methods, using other ways to grasp the asymmetric relationships in data than the asymmetric coefficients, or finally, discovering the data geometry by different means than the k-NN graph.

References

1. Belkin, M., Niyogi, P.: Laplacian eigenmaps and spectral techniques for embedding and clustering. Adv. Neural Inf. Process. Syst. **14**, 586–691. MIT Press (2001)
2. Dua, D., Graff, C.: UCI machine learning repository (2019). http://archive.ics.uci.edu/ml
3. Garzon, M., Yang, C., Venugopal, D., Kumar, N., Jana, K., Deng, L.: Dimensionality Reduction in Data Science. Springer International Publishing (2022). https://doi.org/10.1007/978-3-031-05371-9
4. Ghojogh, B., Crowley, M., Karray, F., Ghodsi, A.: Elements of Dimensionality Reduction and Manifold Learning. Springer International Publishing (2023). https://doi.org/10.1007/978-3-031-10602-6

5. Halder, R., Fidkowski, K.J., Maki, K.J.: An adaptive sampling algorithm for reduced-order models using Isomap. Int. J. Numer. Meth. Eng. **125**(8), 7427 (2024)
6. Hinton, G., Roweis, S.T.: Stochastic neighbor embedding. Adv. Neural. Inf. Process. Syst. **14**, 833–840 (2002)
7. Kohonen, T.: Self-Organizing Maps. Springer (2001). Third Edition https://doi.org/10.1007/978-3-642-56927-2
8. van der Maaten, L., Hinton, G.E.: Visualizing data using t-SNE. J. Mach. Learn. Res. **9**, 2579–2605 (2008)
9. Martín-Merino, M., Muñoz, A.: Visualizing asymmetric proximities with SOM and MDS models. Neurocomputing **63**, 171–192 (2005)
10. McInnes, L., Healy, J., Melville, J.: UMAP: Uniform Manifold Approximation and Projection for Dimension Reduction (2020). https://doi.org/10.48550/arXiv.1802.03426
11. Olszewski, D.: A clustering-based adaptive Neighborhood Retrieval Visualizer. Neural Netw. **140**, 247–260 (2021)
12. Olszewski, D.: Clustering-based adaptive self-organizing map. In: Artificial Intelligence and Soft Computing (ICAISC 2021), Lecture Notes in Computer Science, vol. 12854, pp. 182–192 (2021)
13. Olszewski, D.: A data-scattering-preserving adaptive self-organizing map. Eng. Appl. Artif. Intell. **105**, 104420 (2021)
14. Olszewski, D.: An asymmetric topology-preserving neighborhood retrieval visualizer. Expert Syst. Appl. **225**, 120175 (2023)
15. Rehman, T.U., Yousaf, M., Jing, L.: A*-FastIsomap: an improved performance of classical Isomap based on A*Search algorithm. Neural Process. Lett. **55**, 12719–12736 (2023)
16. Roweis, S.T., Saul, L.K.: Nonlinear dimensionality reduction by locally linear embedding. Science **290**(5500), 2323–2326 (2000)
17. Tenenbaum, J.B., de Silva, V., Langford, J.C.: A global geometric framework for nonlinear dimensionality reduction. Science **290**, 2319–2323 (2000)
18. Venna, J., Peltonen, J., Nybo, K., Aidos, H., Kaski, S.: information retrieval perspective to nonlinear dimensionality reduction for data visualization. J. Mach. Learn. Res. **11**, 451–490 (2010)
19. Xiao, L., Chao, S., Zhu, Z., Yang, L.H., Sheng, L.Q., Tao, T.: Three-dimensional aerodynamic shape inverse design based on ISOMAP. Aerosp. Sci. Technol. **139**, 108409 (2023)

CALICO: Confident Active Learning with Integrated Calibration

Lorenzo S. Querol[1(✉)], Hajime Nagahara[1,2], and Hideaki Hayashi[1]

[1] Osaka University, Suita, Japan
lorenzoquerol@is.ids.osaka-u.ac.jp,
{nagahara,hayashi}@ids.osaka-u.ac.jp
[2] Premium Research Institute for Human Metaverse Medicine (WPI-PRIMe),
Suita, Japan

Abstract. The growing use of deep learning in safety-critical applications, such as medical imaging, has raised concerns about limited labeled data, where this demand is amplified as model complexity increases, posing hurdles for domain experts to annotate data. In response to this, active learning (AL) is used to efficiently train models with limited annotation costs. In the context of deep neural networks (DNNs), AL often uses confidence or probability outputs as a score for selecting the most informative samples. However, modern DNNs exhibit unreliable confidence outputs, making calibration essential. We propose an AL framework that self-calibrates the confidence used for sample selection during the training process, referred to as Confident Active Learning with Integrated CalibratiOn (CALICO). CALICO incorporates the joint training of a classifier and an energy-based model, instead of the standard softmax-based classifier. This approach allows for simultaneous estimation of the input data distribution and the class probabilities during training, improving calibration without needing an additional labeled dataset. Experimental results showcase improved classification performance compared to a softmax-based classifier with fewer labeled samples. Furthermore, the calibration stability of the model is observed to depend on the prior class distribution of the data.

Keywords: Active learning · confidence calibration · energy-based models · medical imaging

1 Introduction

The growing complexity of modern deep neural networks (DNNs) poses a challenge by demanding a substantial increase in labeled data needed to achieve state-of-the-art performance [8,27]. In real-world applications, obtaining labeled data is a logistically expensive process. This challenge is particularly profound in medical imaging, where images are complex and difficult to interpret, requiring the need for domain experts with clinical experience. This implies that a longer turnaround time is needed to finalize ground-truth annotations [3,27]. Given this

© The Author(s), under exclusive license to Springer Nature Switzerland AG 2024
M. Wand et al. (Eds.): ICANN 2024, LNCS 15016, pp. 116–130, 2024.
https://doi.org/10.1007/978-3-031-72332-2_9

time-consuming process, the ratio between labeled and unlabeled data samples becomes aggravated. To tackle this issue, methods have been devised to optimize data efficiency in training models.

Active learning (AL), or human-in-the-loop learning [27], is one of the existing methods that aims to reduce the need for extensive labeled data. The intuition behind AL is to involve human knowledge in the learning process by iteratively selecting samples that are considered most informative by some heuristic function [20]. A subset of these samples is then given to a domain expert for annotation, thereby maximizing the performance of the model while concurrently minimizing the annotation costs.

In AL, the confidence outputs of a DNN are commonly used to select the most informative samples [24]. In classification using DNNs, confidence is typically defined by the maximum value of the class posterior probabilities (i.e., $\max_c p(c \mid x)$ for class c given an input sample x) calculated by the softmax function of the final layer. A lower confidence level indicates greater uncertainty in the model's prediction, making it more beneficial to label such samples. Consequently, these low-confidence samples are prioritized as the most informative for selection.

However, the straightforward method of using a softmax-based classifier was revealed to produce uncalibrated outputs [9,10,18]. The problem arises from the characteristic of the cross-entropy loss function used in DNN training, where the loss decreases as the model's confidence approaches one. This happens when the posterior probability for a particular class approaches one while diminishing to zero for other classes. As a result, the classifier may erroneously exhibit high confidence even for input samples that are difficult to classify. This phenomenon is commonly referred to as the over-confidence issue. In the context of an AL paradigm, uncalibrated confidence outputs could impede reliable decision-making during the selection of informative samples. Consequently, this may lead to poor performance of the learned model on unseen data.

Hence, a concept called *confidence calibration* in neural networks becomes a crucial aspect of developing modern intelligent systems. Confidence calibration is defined as the reflection of a model's accuracy with its predictive confidence [10, 16]. For instance, in scenarios where a classifier provides 100 predictions, each with 95% confidence, it is statistically expected that 95 of those predictions should be correct. In general, the calibration of neural networks is performed in a post-hoc manner, which calibrates the confidence output of the model after training. A common method for this is temperature scaling, which modifies the softmax function in the final layer by incorporating a temperature parameter, thereby calibrating the confidence output. However, post-hoc methods typically require a separate labeled dataset, inefficient in the context of AL as it consumes the already limited labeled data.

To calibrate the confidence output during the AL loop without relying on validation samples, our key idea involves leveraging the distribution of unlabeled training data. This is achieved through the simultaneous learning of a classifier and generative model. Figure 1 outlines the advantages of this joint learning approach. Training a classifier in isolation often leads to inaccurately high posterior

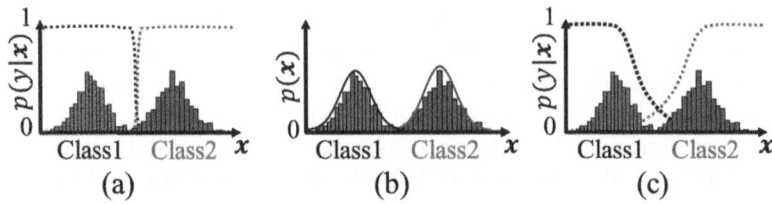

Fig. 1. Joint learning of a classifier and generative model and its advantage. (a) Training solely a classifier can lead to inaccurately high posterior probability near the class decision boundary. (b) Estimating the input distribution using a generative model allows us to account for the frequency of data occurrences. (c) Joint learning of the classifier and generative model helps to calibrate the confidence scores.

probability near the decision boundary. However, by concurrently estimating the input data distribution with a generative model, the classifier is trained to account for the frequency of data occurrences. This approach naturally lowers the posterior probabilities for ambiguous data points near the decision boundary, leading to an inherent self-calibration of confidence levels.

We propose an AL framework designed to self-calibrate confidence during the training process. The method, called CALICO (Confident Active Learning with Integrated CalibratiOn), involves the joint training of a neural network-based classifier with an energy-based model (EBM) [14] in a semi-supervised manner. This joint training enhances the model's understanding of the input data distribution, thereby calibrating the confidence outputs. The key idea is to use these calibrated confidence outputs as input for a query strategy, specifically using the least confidence strategy. This approach enhances decision reliability in selecting samples for annotation, with the overall goal of minimizing model miscalibration and improving accuracy with a minimal number of samples.

The contributions of this paper are as follows:

1. We propose an AL framework termed CALICO, which is designed to self-calibrate confidence during the training process. Our method involves joint training of a classifier and EBM to achieve calibration of confidence without separate validation samples, utilizing the calibrated confidence outputs to select the most informative samples.
2. We demonstrate that the self-calibration approach, which involves simultaneous learning of a classifier and a generative model, is effective for AL in terms of improving accuracy and decreasing calibration error using fewer labeled data than straightforward baseline methods.
3. We revealed the potential of class distribution balancing to enhance CALICO's performance on datasets with class imbalance.

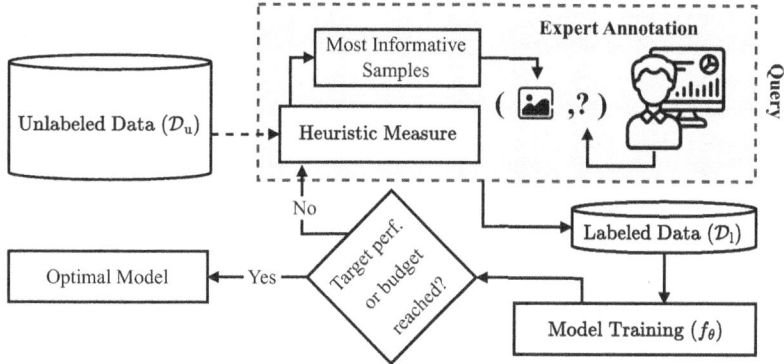

Fig. 2. An overview of the typical active learning (AL) cycle.

2 Background and Related Works

2.1 Active Learning

According to [20], allowing a learnable algorithm to select the data from which it learns can lead to more effective learning. In other words, AL seeks to maximize neural network performance with a smaller selection of data. As shown in Fig. 2, a typical AL scenario involves an *active learner* (model) continuously seeking new samples from a large pool of unlabeled data and inquiring the *oracle* (domain expert) for ground-truth annotations. The goal is to achieve a specific target performance, such as accuracy, while minimizing labeling costs. AL has been widely utilized in traditional machine learning tasks [6,13,20], but with the emerging use of DL achieving superior results in various tasks [8,24] comes with increasing dependence on the amount of labeled data gathered. Acquiring labeled data in task-specific or real-world problems is laborious and time-consuming, thus methods such as AL have become more relevant with its use in combination with DL, hence commonly referred to as deep AL [33].

Sample acquisition can be categorized into three main frameworks: membership query synthesis, stream-based sampling, and pool-based sampling. In real-world scenarios, it is common to acquire a large batch of unlabeled data at once, prompting the use of pool-based sampling [20]. The pool-based sampling framework assumes that there is a limited amount of labeled data and a more extensive pool of unlabeled data. To guide the active learner in selecting which data points to request labels for, samples are selected greedily via a *query strategy* [24]. The labels for the query are obtained by inquiring the oracle, and the data pools are consequently updated. The model is then retrained iteratively until a certain metric reaches a target value or the unlabeled data pool is exhausted. Various common query strategies, such as maximum entropy, margin, least confidence, and mean standard deviation-based approaches, have been established in AL. Moreover, the rise of frameworks like generative adversarial networks and Bayesian deep learning [7,23] have also contributed to the development of enhanced query strategies in this field.

Despite extensive research on query strategies in AL, extending these methods to DL remains challenging, especially due to model uncertainty and inadequate labeled data [33]. The often utilized softmax response for deriving the class probability in DNNs exhibits overconfidence [25], posing possible difficulties in evaluating unlabeled data. This observation has attracted research attention towards deep Bayesian AL [7,8,23]. Furthermore, recent advances in uncertainty quantification have received interest for its use in AL paradigms [1,12]. However, these strategies frequently result in increased computation time for model training and inference, and may require altering the model architecture itself. To solve the dilemma of limited labeled data in deep learning, an approach involves reducing the issue by merging semi-supervised learning with AL [21]. This strategy uses unlabeled data to gather more information about the data distribution and improves the model's overall performance.

2.2   Energy-Based Models

An EBM is a type of generative model that directly models a negative log-probability, which is also known as the energy function. The model's probability density function is derived by normalizing the energy function, expressed as $p_\theta(x) = \exp(-E_\theta(x))/Z_\theta$. In this formulation, E_θ is the energy function parameterized by θ, and Z_θ is the normalizing constant, computed as $Z_\theta = \int_x \exp(-E_\theta(x))dx$. A key challenge in working with EBMs is that this integral for the normalizing term is typically intractable. To estimate it, advanced sampling techniques are often employed, such as using Markov chain Monte Carlo (MCMC) and stochastic gradient Langevin dynamics (SGLD) [26].

EBMs are utilized in a vast range of applications such as image generation [5], texture generation [28], and text generation [2,4]. EBMs have also applied in the context of dropout and pruning within NNs [19]. Additionally, EBMs can be utilized for the complex problem of continuous inverse optimal control [29]. There have been several approaches to simultaneously training an EBM and a classifier [11,15,31,32], revealing the effectiveness of EBMs for semi-supervised learning, outlier detection, and confidence calibration.

2.3   Confidence Calibration

Formally, considering a dataset with input x and outcome $y \in \{1, \ldots, K\}$, a neural network is considered *perfectly calibrated* if:

$$p(Y = y \mid \hat{c} = c) = c, \quad \forall c \in [0, 1] \tag{1}$$

Here, $\hat{c}$ represents the probability of a predicted label Y, and y is the ground-truth label. Model calibration is frequently depicted using reliability diagrams [10], where the expected accuracy is plotted as a function of confidence. Perfect

---

Algorithm 1 CALICO: Confident Active Learning with Integrated Calibration

---

1: **Input:** Labeled dataset $\mathcal{D}_l$, Unlabeled pool $\mathcal{D}_u$, Query Size Q, Number of queries N_q, Query strategy α^{LC}, Oracle g
2: **Initialize:** Model f_θ
3: **while** $1 \leq i \leq N_q$ **do**
4:     **Train:** Train f_θ over $\mathcal{D}_l \cup \mathcal{D}_u$
5:     $\mathcal{D}_q \leftarrow \alpha^{LC}(\mathcal{D}_u, Q, f_\theta)$
6:     $\mathcal{D}_l \leftarrow \mathcal{D}_l \cup g(\mathcal{D}_q)$
7:     $\mathcal{D}_u \leftarrow \mathcal{D}_u \setminus \mathcal{D}_q$
8:     **Evaluate:** Compute performance of f_θ on a valid. set
9: **end while**
10: **Output:** Trained model f_θ

---

calibration is plotted similarly as an identity function, and any deviation from the diagonal signifies miscalibration.

Literature reveals that the utilization of modern neural networks with increasing complexity often results in poor calibration [10,16]. Therefore, various methods have been proposed to calibrate confidence. The calibration methods can be classified into two categories: post-hoc calibration and train-time calibration. Post-hoc calibration is one performed after training. It has been a primal approach, and many methods have been proposed [10]. Among such methods, temperature scaling [10] is considered simple and effective. It uses a softmax function with temperature instead of the usual softmax in the final layer of the NN and tunes the temperature parameter to optimize the negative likelihood. In contrast, train-time calibration is performed during training, which involves simultaneous learning with generative models [9,11] and the use of soft labels [22]. The main difference between post-hoc and train-time methods is whether the validation data is used to calibrate the confidence.

3   CALICO: Confident Active Learning with Integrated Calibration

The proposed CALICO achieves efficient learning by utilizing calibrated confidence when selecting samples to be labeled. Confidence calibration is performed during the AL process by simultaneously estimating the input data distribution with an EBM and the class posterior probabilities with a classifier.

3.1   Algorithm of CALICO

The details of CALICO are described in Algorithm 1. Suppose that we have a limited amount of labeled data $\mathcal{D}_l = \{(\boldsymbol{x}_i, y_i)\}_{i=1}^M$, and a more extensive pool of unlabeled data $\mathcal{D}_u = \{\boldsymbol{x}_i\}_{i=1}^N$, where $M < N$, and $y_i \in \{1, \ldots, K\}$ is the class label of input $\boldsymbol{x}_i$. The model f_θ is trained over both $\mathcal{D}_l$ and $\mathcal{D}_u$. The use of unlabeled data is enabled by simultaneous learning with an EBM, and this point differs from traditional AL processes. The query $\mathcal{D}_q$ is a subset of $\mathcal{D}_u$ and is selected based on a predefined query size Q and query strategy α^{LC}, which is detailed in the next subsection. The labels for $\mathcal{D}_q$ are obtained by inquiring with the oracle g, and $\mathcal{D}_q$ with the obtained labels is denoted as $g(\mathcal{D}_q)$. Subsequently, both data pools are updated accordingly where $\mathcal{D}_l \leftarrow \mathcal{D}_l \cup g(\mathcal{D}_q)$, and $\mathcal{D}_u \leftarrow \mathcal{D}_u \setminus \mathcal{D}_q$. The model f_θ is then retrained iteratively until a certain metric reaches a target value or the unlabeled data pool $\mathcal{D}_u$ is exhausted.

3.2   Query Strategy

We utilize a least confidence strategy [24]. The least confidence query strategy is designed to acquire samples with the smallest probability among the maximum activations. Given an unlabeled data pool $\mathcal{D}_u$, trained model f_θ, and the query size Q, the strategy α^{LC} is defined as follows:

1. Compute the posterior probability $p_\theta(y \mid x)$ for all $x \in \mathcal{D}_u$ based on (2).
2. Obtain the corresponding confidence $\max_y(p_\theta(y|x))$.
3. Sort the samples in $\mathcal{D}_u$ in ascending order with respect to the confidence.
4. The strategy α^{LC} returns the top Q samples.

Based on this strategy, a set of samples is selected from the unlabeled data pool to query the oracle for annotation.

3.3   Joint Learning of a Classifier and an Energy-Based Model

Among various methods proposed for joint learning of a classifier and an EBM [9, 11,31,32], we construct our model f_θ with reference to the structure of JEM [9]. The model consists of a single neural network with a multi-head output for classification and EBM. Given an input $x \in \mathbb{R}^D$, the model first outputs a real-valued K-dimensional vector, i.e., $f_\theta : \mathbb{R}^D \to \mathbb{R}^K$. The vector is then converted into the class posterior probability in the classification head and the probability density of input data in the EBM head. In the classification head, the posterior probability of class $y \in \{1, \ldots, K\}$ is calculated through the standard softmax transfer function, as defined below:

$$p_\theta(y \mid x) = \frac{\exp(f_\theta(x)[y])}{\sum_{y'=1}^{K} \exp(f_\theta(x)[y'])}, \tag{2}$$

where $f_\theta(x)[y]$ indicates the logit corresponding to y-th class. In the EBM head, the probability density $p(x)$ is computed as follows:

$$p_\theta(x) = \frac{\sum_{y=1}^{K} \exp(f_\theta(x)[y])}{Z(\theta)}, \tag{3}$$

where $Z(\theta) = \int_x \sum_{y=1}^{K} \exp(f_\theta(x)[y]) \mathrm{d}x$ is the normalizing constant, otherwise known as the partition function.

In the training of the hybrid model, we minimize the following loss function over the union of labeled and unlabeled datasets $\{(x_n, y_n)\}_{i=1}^{M} \cup \{x_i\}_{i=M+1}^{M+N}$.

$$\mathcal{L} = -\sum_{i=1}^{M} \log p(y_i \mid x_i) - \sum_{i=1}^{M+N} \log p(x_i), \qquad (4)$$

where the first term on the right-hand side corresponds to cross-entropy and is optimized via conventional stochastic gradient descent. The second term is the negative log-likelihood for the EBM optimization whose gradient can be computed using stochastic gradient Langevin dynamics (SGLD).

As a distinction from the original JEM training algorithm, we incorporate techniques from recent EBM studies to stabilize and accelerate training. First, we adopt the informative initialization [31], which uses samples from a Gaussian mixture distribution estimated from the training dataset instead of random noise samples for initializing the SGLD chain. Second, we employ Proximal-YOPO-SGLD [31], which freezes the gradient with respect to the second and subsequent layers during the sample updates. Third, we exclude data augmentation from the maximum likelihood estimation pipeline to alleviate the adverse effects of data augmentation on image generation quality [32].

4   Experiments

4.1   Experimental Conditions

To verify the validity of CALICO, we conducted experiments using medical image datasets. We used five benchmark medical imaging datasets found in the MedMNIST collection [30], namely Blood, Derma, OrganS, OrganC, and Pneumonia. These medical imaging datasets consist of preprocessed 28×28 two-dimensional images, accompanied by their corresponding class labels. The classification tasks within these datasets range from binary to multi-class, serving as benchmarks for foundational models in the medical imaging domain.

The model closely followed the setup of [31]. However, we changed the ReLU activation function used by the Wide-ResNet architecture to a Swish activation function for the added stability observed by [5]. As the computational training time of JEM++ is also relatively long, we limited the number of queries for all datasets (more details in Appendix).

We evaluated the results based on classification accuracy and calibration errors. Miscalibration can be condensed into a convenient scalar metric, often quantified as the expected calibration error (ECE) [17]. This metric discretizes probability intervals into a fixed number of bins and the calibration error is calculated as the difference between the fraction of correct predictions (accuracy), and the mean of the probabilities in the bin (confidence). ECE computes a weighted average of this error across bins:

$$\text{ECE} = \sum_{m=1}^{M_{\text{bin}}} \frac{|\mathcal{B}_m|}{N_{\text{data}}} \mid \text{acc}(\mathcal{B}_m) - \text{conf}(\mathcal{B}_m) \mid \qquad (5)$$

Table 1. Performance Comparison of Test Accuracy (↑) and ECE (↓) Values

Dataset	Criterion (%)	Baseline	Active	CALICO
Blood	Best ACC / ECE	95.82 / 0.95	95.47 / 1.73	**96.43 / 0.54**
	Final ACC / ECE	-	95.44 / 1.87	**96.00 / 0.56**
Derma	Best ACC / ECE	74.02 / 3.50	74.87 / **1.82**	**76.37** / 1.89
	Final ACC / ECE	-	74.56 / **1.82**	**75.66** / 1.89
OrganS	Best ACC / ECE	78.54 / **6.75**	78.55 / 10.04	**79.90** / 7.00
	Final ACC / ECE	-	78.55 / 10.04	**79.90 / 7.00**
OrganC	Best ACC / ECE	89.54 / 5.31	89.79 / 4.17	**90.34 / 3.03**
	Final ACC / ECE	-	88.58 / 5.76	**90.26 / 3.03**
Pneumonia	Best ACC / ECE	87.66 / 12.33	**89.26 / 9.73**	89.26 / 9.82
	Final ACC / ECE	-	**87.66** / 12.33	86.22 / **11.88**

where $\mathcal{B}_m$ represents the subset of samples whose predicted confidence levels lie within the interval $I_m = (\frac{m-1}{M_{\text{bin}}}, \frac{m}{M_{\text{bin}}}]$, N_{data} is the total number of data points, and $\text{acc}(\mathcal{B}_m)$ and $\text{conf}(\mathcal{B}_m)$ denote the accuracy and confidence of $\mathcal{B}_m$, respectively. For this study, we utilize ECE as the primary metric for measuring calibration, as well as reliability diagrams for visualization.

We establish the baseline reference by maximizing only the $\log p(y|x)$ objective through the utilization of the entire dataset, with a softmax-based classifier (named **Baseline**), involving the evaluation of calibration error using the probability outputs from the softmax activation function and calculating the ECE. We also used a softmax-based classifier paired with an AL framework using the least confidence query strategy (named **Active**) as an additional baseline reference.

4.2   Performance Comparison

Table 1 shows that CALICO consistently outperformed the baseline accuracy across all evaluated datasets. In parallel with the accuracy improvements, CALICO also showed reduced ECEs, indicating its effectiveness in minimizing miscalibration. We observed that using a softmax-based classifier within an AL paradigm, as opposed to straightforward training methods, resulted in better calibration. However, CALICO demonstrated a more substantial increase in performance when compared to the baseline, suggesting its effectiveness in improving the classifier's performance in an AL paradigm. While CALICO generally achieved lower ECE values across most datasets, the use of a softmax-based classifier in an AL paradigm demonstrated comparable efficacy to CALICO in some cases, such as Derma and OrganS.

4.3   Confidence Calibration

The final reliability diagrams are depicted in Fig. 3. A perfectly calibrated model would align with the red-diagonal dashed line, effectively creating an identity

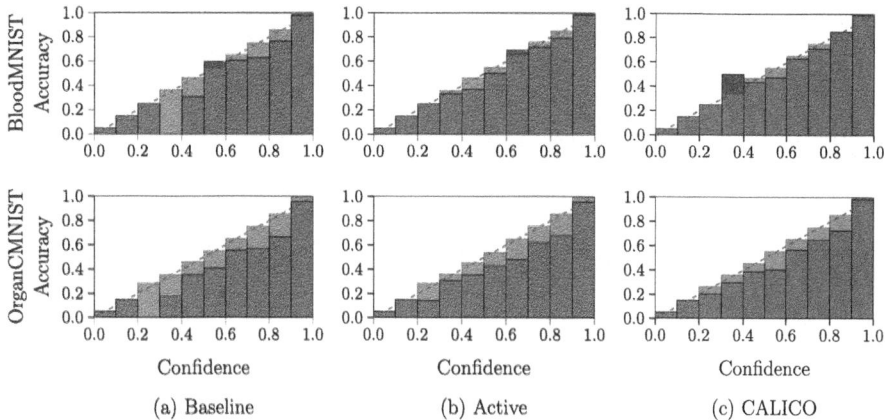

Fig. 3. The final reliability diagrams for the baseline, active, and CALICO. In comparison, CALICO demonstrated a substantial improvement in calibration across confidence intervals when compared to the other evaluated methods that used softmax-based classifiers.

function between the y-axis (accuracy) and the x-axis (confidence intervals). Any deviation above or below this diagonal indicates underconfidence or overconfidence, respectively, representing miscalibration. It is prominent that there is a marginal difference in the improvement of overconfident intervals between using softmax-based classifiers and CALICO, emphasizing the effectiveness of calibrated confidence outputs for iterative training in an AL paradigm.

4.4 Learning Curve Analysis

Learning curves play a crucial role in assessing how well a model performs in an AL paradigm, particularly by observing a specific metric with respect to the number of labeled samples. In this study, the ECE is of certain importance. Similar to the commonly used accuracy, the goal is to determine whether miscalibration can be minimized with significantly fewer labeled samples through the utilization of JEMs. As illustrated in Fig. 4, the overall calibration trend using CALICO is observed to be lower than that of the softmax-based classifiers. Additionally, it is noteworthy that a lower or comparable ECE can be achieved with fewer samples in comparison to the baseline reference.

4.5 Performance Comparison of Test Accuracy and ECE Values with an Equal Class Distribution

We explored the impact of class distribution balancing on CALICO's performance(Fig. 5). While many confidence calibration studies emphasize calibration error within balanced class distributions, achieving such balance is not guaranteed in AL due to sampling methods' greedy nature. Calibration learning curves

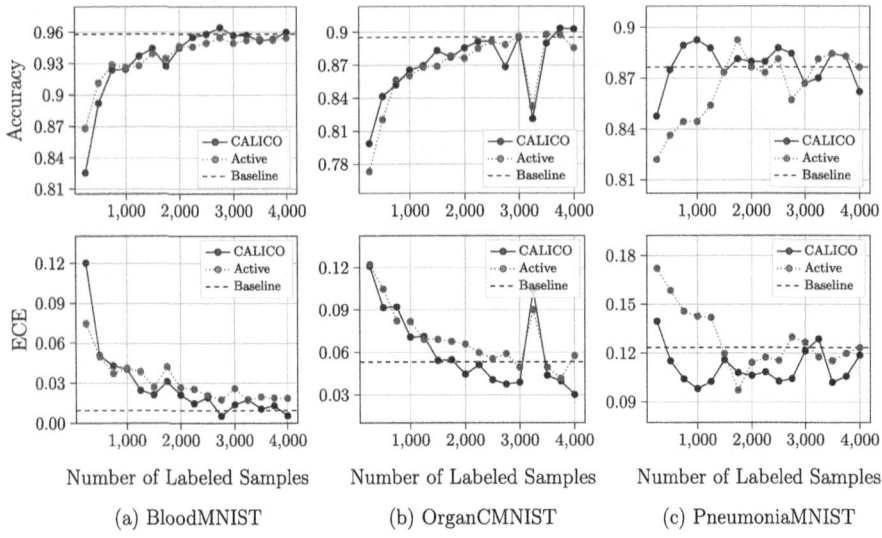

Fig. 4. The test accuracy (top) and ECE (bottom) values plotted against the number of labeled samples per AL iteration. It can be observed that a lower or comparable ECE value can be achieved with a lesser number of labeled samples.

(Fig. 4) exhibited instability, possibly due to overconfidence from uncertainty-based sampling. Additionally, inherent dataset class imbalances can lead to querying uninformative samples, creating a mode collapse problem and miscalibration. To simulate literature setups, CALICO's performance was evaluated with an equal class distribution (named **Equal**), enforcing limits based on the dataset's least represented class. Varying labels per class allowed for sufficient iterations to analyze learning curves. Experimental setup details are provided in Table 3 of Appendix.

We observed instances where having an equal class distribution yielded better calibration or more stable learning curves across the evaluated datasets, such as the results on the PneumoniaMNIST dataset. However, Table 2 also highlights instances where equal class distribution did not result in better calibration compared to the original CALICO. One possible explanation for this disparity is the nature of the datasets; PneumoniaMNIST is an imbalanced binary dataset, while others are relatively balanced multi-class datasets, and balancing the class ratio in sample selection facilitated effective learning of information from the minority class. These findings imply the potential of class distribution balancing to enhance CALICO's performance on imbalanced datasets.

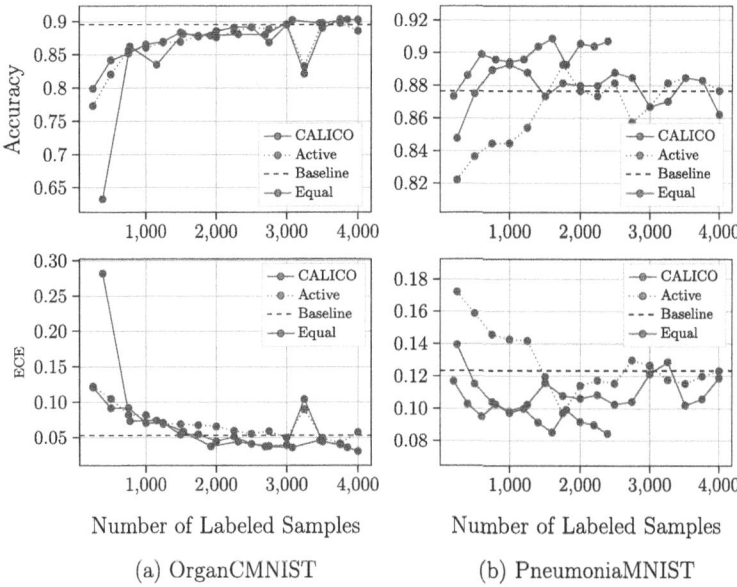

Number of Labeled Samples   Number of Labeled Samples

(a) OrganCMNIST   (b) PneumoniaMNIST

Fig. 5. The test accuracy (top) and ECE (bottom) values of CALICO with an equal class distribution. Note that the *Equal* method was designed to strictly enforce an equal distribution by setting a limit based on the dataset's class label with the fewest samples. Additionally, the number of labels per class varied per dataset to allow for an adequate number of iterations to properly analyze the learning curve.

5   Conclusion

We proposed an AL method called CALICO, which aimed to use the calibrated confidence outputs as the input for a query strategy in an AL paradigm. CALICO incorporates the joint training of a classifier and en EBM, allowing self-calibration of confidence used for sample selection in AL. Experimental results demonstrated that CALICO outperformed the baseline accuracy and achieved a lower ECE with less labeled data, compared to a softmax-based classifier.

One of the limitations of this study is scalability because training an EBM on high-resolution images requires considerable hyperparameter tuning, and CALICO was only evaluated on small-resolution images. Future research could explore the applications of CALICO with other AL methods, harnessing the power of Bayesian approaches for better uncertainty quantification. Additionally, further evaluation of larger datasets in other domains can broaden the scope and applicability of this research domain.

Acknowledgments. This work was supported by JSPS KAKENHI Grant Number JP24K03010 and the World Premier International Research Center Initiative (WPI), MEXT, Japan.

Table 2. Min. and Final Test Accuracy (↑) and ECE (↓) Values in Comparison to an Equal Class Distribution

Dataset	Criterion (%)	Baseline	CALICO	Equal
Blood	Best ACC / ECE	95.82 / 0.95	**96.43 / 0.54**	96.25 / 0.89
	Final ACC / ECE	-	96.00 / **0.56**	**96.05** / 0.89
OrganS	Best ACC / ECE	78.54 / 6.75	**79.90 / 7.00**	78.98 / 10.49
	Final ACC / ECE	-	**79.90 / 7.00**	78.07 / 10.49
OrganC	Best ACC / ECE	89.54 / 5.31	**90.34 / 3.03**	90.25 / 3.56
	Final ACC / ECE	-	90.26 / **3.03**	**90.32** / 3.57
Pneumonia	Best ACC / ECE	87.66 / 12.33	89.26 / 9.82	**90.87 / 8.46**
	Final ACC / ECE	-	86.22 / 11.88	**90.71 / 8.46**

Table 3. Experimental Setup for Equal Class Distribution

Dataset	Lowest Count (Class)	Limit	Labels per Class
Blood	849 (lymphocyte)	4,000	50
OrganS	614 (femur-right)	3,850	35
OrganC	600 (heart)	3,850	35
Pneumonia	1,214 (normal)	2,400	100

Appendix

Experimental Setup. To ensure consistency across all experiments, the computational runtime constraints necessitated limiting each dataset to 4000 labeled samples, with a query size of 250 for each iteration, resulting in a total of 16 iterations. In the ablation study that focused on an equal class distribution, the experimental setup was detailed in Table 3. The number of labeled samples per class for each iteration was determined by the class with the fewest samples to create enough iterations for analysis. The only exception to the ablation study was the DermaMNIST dataset, where the class with the lowest number of samples was 89. It was decided not to include this dataset in the ablation study.

Hyperparameter Settings. All datasets, except for PneumoniaMNIST, adapted the default hyperparameters from the original literature of JEM++ [31]. This included using an SGD optimizer with a learning rate of 0.1. However, for PneumoniaMNIST, an Adam optimizer with a learning rate of 0.0001 was utilized, as it exhibited a more stable calibration performance.

References

1. Abdar, M., et al.: A review of uncertainty quantification in deep learning: techniques, applications and challenges. Inf. Fusion **76**, 243–297 (2021)
2. Bakhtin, A., Deng, Y., Gross, S., Ott, M., Ranzato, M., Szlam, A.: Residual energy-based models for text. J. Mach. Learn. Res. **22**(40), 1–41 (2021)
3. Budd, S., Robinson, E.C., Kainz, B.: A survey on active learning and human-in-the-loop deep learning for medical image analysis. Med. Image Anal. **71**, 102062 (2021)
4. Deng, Y., Bakhtin, A., Ott, M., Szlam, A., Ranzato, M.: Residual energy-based models for text generation. In: Proceedings of the International Conference on Learning Representations (2020)
5. Du, Y., Mordatch, I.: Implicit generation and modeling with energy based models. In: Proceedings of the Annual Conference on Neural Information Processing Systems. vol. 32 (2019)
6. Fu, Y., Zhu, X., Li, B.: A survey on instance selection for active learning. Knowl. Inf. Syst. **35**, 249–283 (2013)
7. Gal, Y., Ghahramani, Z.: Bayesian convolutional neural networks with Bernoulli approximate variational inference. arXiv preprint arXiv:1506.02158 (2015)
8. Gal, Y., Islam, R., Ghahramani, Z.: Deep Bayesian active learning with image data. In: Proceedings of the International Conference on Machine Learning, pp. 1183–1192. PMLR (2017)
9. Grathwohl, W., Wang, K.C., Jacobsen, J.H., Duvenaud, D., Norouzi, M., Swersky, K.: Your classifier is secretly an energy based model and you should treat it like one. In: Proceedings of the International Conference on Learning Representations (2020)
10. Guo, C., Pleiss, G., Sun, Y., Weinberger, K.Q.: On calibration of modern neural networks. In: In Proceedings of the International Conference on Machine Learning, pp. 1321–1330 (2017)
11. Hayashi, H.: A hybrid of generative and discriminative models based on the Gaussian-coupled softmax layer. IEEE Trans. Neural Netw. Learn. Syst. (Early Access) (2024)
12. Hein, A., et al.: A comparison of uncertainty quantification methods for active learning in image classification. In: Proceedings of the International Joint Conference on Neural Networks, pp. 1–8 (2022)
13. Kumar, P., Gupta, A.: Active learning query strategies for classification, regression, and clustering: a survey. J. Comput. Sci. Technol. **35**, 913–945 (2020)
14. LeCun, Y., Chopra, S., Hadsell, R., Ranzato, M., Huang, F.: A tutorial on energy-based learning. Predicting Structured Data **1** (2006)
15. Liu, X., Staudt, D., Lin, C.T., Zach, C.: Effortless training of joint energy-based models with sliced score matching. In: Proceedings of the International Conference on Pattern Recognition, pp. 2643–2649 (2022)
16. Minderer, M., et al.: Revisiting the calibration of modern neural networks. In: Proceedings of the Advances in Neural Information Processing Systems. vol. 34, pp. 15682–15694 (2021)
17. Naeini, M.P., Cooper, G., Hauskrecht, M.: Obtaining well calibrated probabilities using Bayesian binning. In: Proceedings of the AAAI Conference on Artificial Intelligence. vol. 29 (2015)
18. Ren, P., et al.: A survey of deep active learning. ACM Comput. Surv. **54**, 1–40 (2021)

19. Salehinejad, H., Valaee, S.: EDropout: energy-based dropout and pruning of deep neural networks. IEEE Trans. Neural Netw. Learn. Syst. **33**(10), 5279–5292 (2022)
20. Settles, B.: Active learning literature survey. Comput. Sci. Tech. Rep. 1648, University of Wisconsin–Madison (2009)
21. Siméoni, O., Budnik, M., Avrithis, Y., Gravier, G.: Rethinking deep active learning: Using unlabeled data at model training. In: Proceedings of the International Conference on Pattern Recognition, pp. 1220–1227 (2021)
22. Thulasidasan, S., Chennupati, G., Bilmes, J.A., Bhattacharya, T., Michalak, S.: On mixup training: improved calibration and predictive uncertainty for deep neural networks. Proc. Annu. Conf. Neural Inf. Proc. Syst. **32** (2019)
23. Tran, T., Do, T.T., Reid, I., Carneiro, G.: Bayesian generative active deep learning. In: Proceedings of the International Conference on Machine Learning, pp. 6295–6304 (2019)
24. Wang, D., Shang, Y.: A new active labeling method for deep learning. In: Proceedings of the International Joint Conference on Neural Networks, pp. 112–119. IEEE (2014)
25. Wang, K., Zhang, D., Li, Y., Zhang, R., Lin, L.: Cost-effective active learning for deep image classification. IEEE Trans. Circuits Syst. Video Technol. **27**(12), 2591–2600 (2016)
26. Welling, M., Teh, Y.W.: Bayesian learning via stochastic gradient Langevin dynamics. In: Proceedings of the International Conference on Machine Learning, pp. 681–688 (2011)
27. Wu, X., Xiao, L., Sun, Y., Zhang, J., Ma, T., He, L.: A survey of human-in-the-loop for machine learning. Futur. Gener. Comput. Syst. **135**, 364–381 (2022)
28. Xie, J., Lu, Y., Gao, R., Wu, Y.N.: Cooperative learning of energy-based model and latent variable model via MCMC teaching. In: Proceedings of the Annual AAAI Conference on Artificial Intelligence. vol. 32 (2018)
29. Xu, Y., Xie, J., Zhao, T., Baker, C., Zhao, Y., Wu, Y.N.: Energy-based continuous inverse optimal control. IEEE Trans. Neural Netw. Learn. Syst. 1–15 (2022)
30. Yang, J., et al.: MedMNIST v2-A large-scale lightweight benchmark for 2D and 3D biomedical image classification. Sci. Data **10**(1), 41 (2023)
31. Yang, X., Ji, S.: JEM++: improved techniques for training JEM. In: Proceedings of the IEEE/CVF International Conference on Computer Vision, pp. 6494–6503 (2021)
32. Yang, X., Su, Q., Ji, S.: Towards bridging the performance gaps of joint energy-based models. In: Proceedings of the IEEE/CVF Conference on Computer Vision and Pattern Recognition, pp. 15732–15741 (2023)
33. Zhang, Z., Strubell, E., Hovy, E.: A survey of active learning for natural language processing. In: Proceedings of the Conference on Empirical Methods in Natural Language Processing, pp. 6166–6190. Association for Computational Linguistics (2022)

Improved Multi-hop Reasoning Through Sampling and Aggregating

Mengyu Luo$^{(\boxtimes)}$, Jianxia Chen, Qi Yan, Gaohang Jiang, Shi Dong, Liang Xiao, and Zhongwei Huang

School of Computer Science, Hubei University of Technology, Wuhan 430068, China
`102211172@hbut.edu.cn`

Abstract. Multi-hop reasoning over is helpful in knowledge graphs to unearth intricate associations and implicit information among entities through multi-step reasoning jumps. However, existing approaches still face the challenges of noise and sparsity. This is due to the fact that this issue it is difficult to identify head and tail entities along long and complex paths. To address this issue, we propose a novel multi-hop reasoning model based on Dual Sampling strategies and aggregation of Multi-Relational Types of entities and relationships, named DSMRT in short. In particular, the dual sampling strategy to address noise and sparsity of training data entities, including the forward traversal and the backward verification. Afterward, we distinguish the semantic types of different entities and relationships by constructing their type perception representations. Extensive experiments demonstrate that the proposed DSMRT model can adeptly oversee the sampling process, ensuring both balance and representativeness of the data. Additionally, it successfully mitigates challenges like noise and information gaps through the judicious application of type information.

Keywords: Knowledge Graph · Multi-Hop Reasoning · Dual Sampling · Type Aggregation

1 Introduction

A Knowledge Graph (KG) is mainly a knowledge-base graph structure constructed entities and their relationships [1]. As one of the most important branches of KGs, knowledge graph reasoning (KGR) aims to reveal the hidden relationships between entities. This has achieved remarkable success in many research fields such as recommendation system [2], question answering [3], and commonsense reasoning [4]. Current research efforts of KGR largely focus on dealing with two kinds of tasks including single-hop reasoning(SHR) and multi-hop reasoning (MHR). In particular, SHR utilizes the direct relationships to predict whether an edge connecting by a pair of nodes exists in the graph [5]. As shown in Fig. 1(a), if to know *"Is the capital of the United States is Washington or not?"*, the nodes of the *"United States"* and *"Washington"* are identified, then whether there is a *"capital"* relationship between them is judged. It is

M. Wand et al. (Eds.): ICANN 2024, LNCS 15016, pp. 131–146, 2024.
https://doi.org/10.1007/978-3-031-72332-2_10

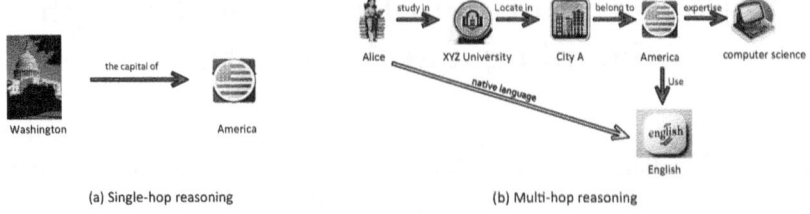

Fig. 1. Two types of tasks for knowledge reasoning

obvious that the SHR process is relatively simple. However, it is challenging to address more complex issues. This is because KGs often involve complex relationships between multi-steps and multi-entities in the real world. To this end, MHR techniques have been proposed to address this limitation. For example, in Fig. 1(b), to answer the question *"What is Alice's native language?"*. In the initial analysis, we scrutinize the entity Alice, discerning her enrollment at *"XYZ University"*. Positioned in *"City A"*, *"XYZ University"* is situated within *"the United States"*, where *"English"* serves as the official language. Consequently, through a multi-hop reasoning approach, we derive the inference that English constitutes Alice's native language. The entire process constitutes a straightforward MHR. Formally, MHR typically refers to acquiring new information by traversing paths across multiple entities and relationships [6].

Traditional MHR methods aim to trace all intermediate entities during inference, leading to a rapid increase in computational burden. For example, when addressing a simple query as depicted in Fig. 1(b), multiple steps must be traversed to locate the desired entity, potentially resulting in an exponential growth of intermediate entities and increased computational costs. To mitigate this challenge, query embedding (QE) techniques have been proposed for MHRs. QE-based approaches seek to convert queries or questions into low-dimensional vector representations. In MHR tasks, QE operates within an embedded space, enabling rapid answer discovery without tracing all intermediate entities [7]. Notably, QE-based models focus on inferring the underlying logic of MHR tasks [8–10]. For instance, the GQE [8] model handles incoming conjunction and existential logic in queries, the Q2B [9] model deals with incoming conjunction, existential, and disjunction logic, while the BETAE [10] model addresses negative logic in MHRs.

The scarcity of training entities in the knowledge graph poses a significant challenge. Researchers have explored various sampling techniques [15–17] to extract relevant information from alternative entities. Typically, in constructing query structures, an entity is randomly chosen as an anchor point, with subsequent relationships selected to fill other segments. However, a potential drawback of this approach is that the selected entity may not align with the expected type, resulting in entities inconsistent with our assumptions. This requires model resampling, leading to increased computational costs. Sampling negative entities presents similar complexities, with common methods randomly selecting entities unrelated to the query entity. However, since a single query may have multiple correct answer entities, there is a risk of mistakenly classifying certain correct entities as negative. As a result, several sampling techniques struggle with challenges such as heightened computational expenses and inherent sampling bias.

Moreover, existing MHR approaches in KGs attempt to construct a specific multiple semantic parsing and retrieving modules to handle complex questions [12]. Most of them encounter unexpected type mismatches [13]. The primary reason is that entity types embed all the attributes of the entity during the reasoning process. However, only certain attributes of this entity may be relevant, while others remain irrelevant. This can lead to noise and other unrelated information, ultimately diminishing accuracy [14].

In this context, we introduce a novel Multi-Hop Query Embedding (MHQE) model, named DSMRT, based on a Dual Sampling strategy and Multi-Relational Types of entities and relationships. To overcome the limitations of existing sampling methods, we propose a new sampling approach in the DSMRT model to obtain high-quality training data. During sampling, ensuring that query structure instantiation yields representative positive samples mitigates challenges posed by sample sparsity from random sampling. Negative sample acquisition employs bidirectional rejection sampling to prevent misclassification of non-negative entities as negative instances, minimizing model confusion. Additionally, we construct type perception representations for entities and relationships in the KG, enhancing KG content by distinguishing semantic types. Extensive experiments demonstrate that the DSMRT model offers more precise sampling control, ensuring balanced and representative datasets, and significantly improving overall execution efficiency. Our main contributions are summarized as follows:

- Propose a novel sampling strategy including a forward traversal and a backward verification, which not only more accurately control the sampling process but also greatly improve the execution efficiency.
- Propose the knowledge representations of entity types and relation types of the KG, distinguishing their semantic features to enrich the KG contents.
- Experimental results demonstrate that the DSMRT model has much better performance on generalization and deductive reasoning than the state-of-the-art baselines.

2   Related Work

2.1   Type-Aware Embedding of Multi-Hop Reasonings

Type information in knowledge graphs is utilized in various NLP tasks, encompassing entity type [14,22] and relation type information [23]. Entity types, among these, offer less noise and furnish broader semantics for each entity. For example, TKRL [24] utilizes hierarchical entity type information to encode entities into diverse representations. However, sole reliance on entity information inadequately captures entity knowledge representation. TransT [25] effectively integrates structural information and entity types to capture inherent entity and relationship distributions. Nonetheless, this method necessitates multiple embedded representations, consuming significant memory space. ConnectE [21] explores entity type inference by combining local entity type information with global KG triplet knowledge. Additionally, BLP [26] learns entity vector representations from descriptions, supplementing entity type information with text and emphasizing text-KG correlations. However, these type-based models require explicit type supervision to function on KGs. Furthermore, JOIE [27] integrates instance-view and ontology-view graphs for embedding, yet ontology information might be overly broad or noisy to provide precise entity types.

Additionally, current entity type models overlook the integration of various entity types in KGs as a topic entity for different relations. For instance, P. Jain et al. [28] explore the compatibility between entity type and relation embeddings for link prediction. AutoETER [23] proposes a KG embedding framework to autonomously learn entity type representations and utilizes relational-aware projection to learn potential entity type embeddings. CET [14] identifies missing entity types in KGR and infers entity types using entity context information. However, these approaches face challenges in directly answering first-order logical (FOL) queries due to uncertainties arising from multi-hop reasoning and intermediate entities. To address this limitation and guided by insights from entity type information, we introduce a perceptual representation approach grounded in both entity and relationship types. This method aims to enhance the abstract representation of elements in the knowledge graph, intricately modeling entities and relationships through comprehensive semantic integration. As a result, we expect improved effectiveness in handling uncertainty stemming from noise and information gaps during the multi-hop reasoning process.

2.2 Sampling Methods of Multi-Hop Reasonings

When selecting samples for knowledge reasoning, it's crucial to consider both positive and negative samples to form a balanced training set, facilitating effective learning of task-critical features by the model. Traditional sampling methods, like random sampling [32], negative sampling [33], or heuristic-based sampling [34], may introduce bias and noise [18]. Early methods like uniform random sampling [20] randomly select entities and relationships from the knowledge graph to construct positive and negative samples. However, they fail to fully explore multi-hop paths for multi-hop reasoning, hindering the model's capability to handle complex tasks. Adversarial negative sampling [29] addresses negative sample quality but requires significant computational resources for multi-hop reasoning with multiple relationships, potentially limiting scalability. Subgraph sampling [35] provides more contextual information but may lead to information loss or excessive inclusion, challenging the model's ability to capture complex reasoning patterns

Given these limitations, we propose a novel sampling method that integrates forward traversal and backward verification strategies to obtain high-quality positive and negative samples. Forward traversal captures relevant information effectively, while backward verification ensures negative sample quality.

3 Preliminaries

We begin by first formally defining the FOL queries, and briefly introducing a query calculation plan based on KGs and muti-relational types graphs based on MHRs.

3.1 First-Order Logical Queries

First-order logical queries involve four fundamental logical operations: conjunction ($\wedge$), disjunction ($\vee$), negation ($\neg$), and existential quantification ($\exists$). Query q comprises a set of non-variable anchor entities $\mathcal{V}q \subseteq \mathcal{V}$ with quantized constraint variables $V_1, \cdots, V_k$ and target variables $\mathcal{V}_?$. The disjunctive normal form with the query is defined in (1):

$$q[V_?] = V_? . \exists V_1, \ldots, V_k : c_1 \vee c_2 \vee \ldots \vee c_n \qquad (1)$$

In which:

- e is a positive or negative atomic formula.
- c is one or more e.
- $c_i = e_{i1}$ is the conjunction from e_{i1} through e_{im}.
- e_{ij} is a formula selected from $r(v_a, V), \neg r(v_a, V), r(V', V), \neg r(V', V)$.
- $v_a \in \mathcal{V}_q, V \in \{V_?, V_1, \ldots, V_k\}, V' \in \{V_1, \ldots, V_k\}, V \neq V', r \in \mathcal{R}$.

Consequently, the method of finding the answer entity of the query q is similar to finding the answer set of v, in which $q[v] = True$.

3.2 Query Calculation Plan Based on KGs

In this paper, KGs are represented in the standard format of RDF. We define $\mathcal{G}$ as a tuple $(\mathcal{E}, \mathcal{R}, \mathcal{C}, \mathcal{T})$, in which $\mathcal{E}$ and $\mathcal{C}$ are a set of entities and their types, $\mathcal{R}$ is a set of relation types, and $\mathcal{T}$ is a set of triples. The triples in the $\mathcal{T}$ are relational assertions (h, r, t), $h \in \mathcal{E}$ and $t \in \mathcal{E}$ is a head and a tail entity, and $r \in \mathcal{R}$ is the edge between h and t. Also, the triplet can be represented as an entity type assertion $(e, type, c)$, in which e and c represent an entity and its type, and $type$ represents a relationship instance.

The query computation plan (QCP) is comprised of nodes from V_q and $\{V_1, \cdots, V_k, V_?\}$, in which each node corresponds to a set of entities of the KG, and the edges in the QCP represent logical/relational transformations in this set, including projection, intersection, union, and complement/inverse.

- Projection: computing neighboring entities $\cup_{v \in S} A_r(v), A_r(\nu) \equiv \{\nu' \in V : (\nu, r, \nu') \in \mathcal{E}\}$ if there is a set of entities $S \subseteq V$ and relation types $r \in R$.
- Intersection: computing entities' intersection $\cap_{i=1}^{n} S_i$ if there is a set of entities $S \subseteq V$.
- Complement/Negation: computing entities' complement $\bar{S} \equiv V \backslash S$ if there is a set of entities $S \subseteq V$.
- Union: Given a set of entity sets $S_1, S_2, \cdots, S_n$, this operator obtains $\cup_{i=1}^{n} S_i$.

3.3 Muti-Relational Types Graphs Based on MHRs

Given a KG and a set of queries (e_h, r_q, e_t) where e_t is an unknown entity. Multi-hop reasoning is the inference of the target entity through a multi-step search path. This task not only predicts the target tail entity but also predicts the path to the target entity $\{(e_h, r_1, e_1), (e_1, r_2, e_2), \cdots, (e_{T-1}, r_T, e_t)\}$, T is the last step in the inference path.

In order to construct the relation type graph, we give a definition of the relation type graph $\mathcal{G}_{type}$, for $\mathcal{G} = (\mathcal{E}, \mathcal{R}, \mathcal{C}, \mathcal{T})$. In the type graph $\mathcal{G}_{type}$, we define the set of head entity types, the same as the format of the relation assertion $\mathrm{tp}_t(\mathrm{hd}(t)) = \{c \mid (\mathrm{hd}(t), type, c) \in \mathcal{T}\}$ and $\mathrm{tp}_r^{\mathrm{hd}}(\mathcal{G}) = \underset{t \in \mathcal{T}_r}{\cap} \mathrm{tp}_t(\mathrm{hd}(t))$. Moreover, we define the set of tail entity types of the relation assertion format as $\mathrm{tp}_r^{\mathrm{tl}}(\mathcal{G}) = \underset{t \in \mathcal{T}_r}{\cap} \mathrm{tp}_t(\mathrm{tl}(t))$. Therefore, a type graph can be represented as $\mathcal{G}_{type} = (V, E, T)$, in which $V = \underset{r \in \mathcal{R}}{\cup} \mathrm{tp}_r^{\mathrm{hd}}(\mathcal{G}) \cup \mathrm{tp}_r^{\mathrm{tl}}(\mathcal{G}), E \in \mathcal{R}, T = (v, r, v\prime)$ and $\mathrm{v} = \mathrm{tp}_t(\mathrm{hd}(t)), v\prime = \mathrm{tp}_t(\mathrm{tl}(t))$.

4   The DSMRT Model

As shown in Fig. 2, the DSMRT model is composed of three components, including dual sampling strategy, entity type and relationship type aggregation, and type feature fusion. The specific contents of these modules are described as follows in detail.

4.1   Dual Sampling Strategy

We propose a dual sampling strategy to improve the calculation efficacy of the MHR effectively. We utilize a multiple threading including forward traversal and backward validation to attain high-quality samples. In the training sample acquisition process for Multi-Hop Reasoning (MHR), it is common to initiate by instantiating a query structure to generate queries q. Subsequently, positive A_q^G and negative entities N_q^G are obtained by traversing the entire KG.

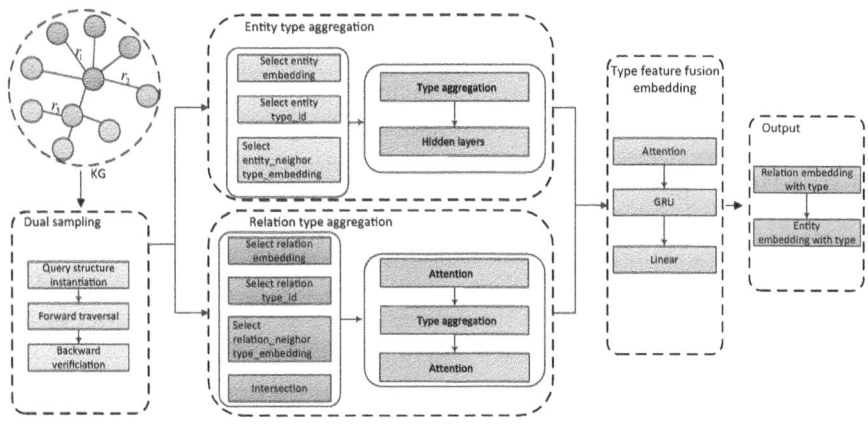

Fig. 2. The overall of DSMRT model

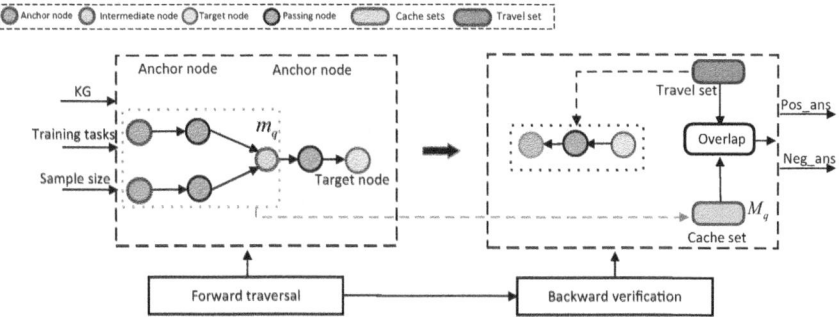

Fig. 3. The overall of Dual Sampling

Forward Traversal Knowledge Graphs. We propose a more efficient way to instantiate the query structure, which is in the opposite direction. As shown in the left head side of Fig. 3. Specifically, the identification of the root node (Anchor node), representing the answer node (Target node), initiates the process, followed by a systematic backward traversal towards the leaf nodes, denoting the anchor nodes. After instantiating the query through forward traversal, a triplet (q, V_q, a_q) can be obtained as a positive sample. This procedural aspect aligns with a depth-first traversal (DFS) paradigm, strategically exploring the query structure from the root to the leaf nodes. The method's noteworthy advantage lies in its consistent capacity to instantiate a given query. This is attributed to the meticulous orchestration of the entire reverse process, ensuring a comprehensive alignment between each node and edge within the query structure and their corresponding entities and relationships.

Backward Verification Strategy. We propose a bidirectional rejection sampling method. Since negative samples do not need to be entirely included during model training, we adopt a rejection sampling approach to identify a subset of negative entities. In other words, for a randomly selected entity V, we only need to check $V \in A_q^G$ without exhaustively enumerating each positive entity. The key in the entire process is to find an intermediate node m_q, which is a set of nodes capable of separating all paths between the root node and the leaf node, with each path containing only one node from the set m_q. The entire sampling process is as follows: Forward Caching: Initially, traverse from the leaf node to the intermediate node and store the entities obtained during the traversal in a caching set. Subsequently, perform backward validation: Traverse from the root node to the intermediate node, store the obtained entities in a traversal set and check for overlap between the nodes in the traversal set and the caching set. If there is an overlap, the root node for the path coming from this traversal is considered a positive entity; otherwise, it is a negative entity. The detailed execution process is illustrated in the right head side of Fig. 3.

4.2   Entity and Relation Type Aggregation

In MHRs, there may be long paths in which the anchor entity and the target variable are projected through multiple relationships, resulting in loose correlation between them. Moreover, an exponential cost in the search space can be triggered by the continuous projections, to enhance calculation complexity greatly. Within KGs, the salience of relationship types alongside entity classifications is evident. Remedying this inherent lack of relevance stands to notably enhance the comprehensive portrayal of diverse relational categories. Since existing KGs have not specific relational-type representations, we construct a new relational-type graph based on the original graph. Aggregating the same type of information on this graph, we can obtain the type embedding of each particular relationship.

Entity Type Aggregation. To enrich the contents of the KGs, we construct the type perception representations of entities and relationships, distinguishing the semantic types of different entities and relationships. Aggregating the same entity types aids the inference of entity identities or attributes.

To capture entity types, we employ a multi-layer GNN. The hidden layer state $\mathcal{H}_s^i \in \mathbb{R}^{d \times n}$ representing the i_{th} iteration. In which d is the vector size, and n is the number of types for the entity. The specifically, shown in (2):

$$\mathcal{H}_s^{i+1} = g * (W_i \prime \mathcal{H}_s^{\ i} + b_i \prime) + (1 - g) * \mathcal{H}_s^{\ i} \tag{2}$$

$$g = \sigma(W_i \mathcal{H}_s^i + b_i) \tag{3}$$

In which,

- σ is the sigmoid function.
- $W \in \mathbb{R}^{d \times d}$ and $b \in \mathbb{R}^{d \times 1}$ are learnable matrices.

Followings K iterations, $\mathcal{H}_s^K \in \mathbb{R}^{d \times n}$ is obtained finally. Through a linear transformation, we obtain $\widetilde{\mathcal{H}_s^K} \in \mathbb{R}^{d \times 1}$, $\widetilde{\mathcal{H}_s^K}$ is then concatenated with the initial entity $\mathcal{H}_s^0$ to enhance the representation, finally get the ultimate entity representation $\mathcal{H}_e$. As indicated in (4) (5), where [·] denotes the concatenation function:

$$\widetilde{\mathcal{H}_s^K} = W \mathcal{H}_s^K + b \tag{4}$$

$$\mathcal{H}^e = W'[\widetilde{\mathcal{H}_s^K}, \mathcal{H}_s^0] + b' \tag{5}$$

Relationship Type Aggregation. Within a graph of relationship types, a relationship might encompass multiple types, denoted as $\text{tp}_r{}^i(\mathcal{G}_{type})$, where, $1 < i < |\text{tp}_r(\mathcal{G}_{type})|$, but not all relevant to the query q. Therefore, an oversimplified concatenation of all relationship types would be problematic. Instead, we utilize an attention neural network to handle the types of relationships, denoted as $\mathcal{H}_s{}^i$, as depicted in (6).

$$\mathcal{H}^s = \sum_i a_i \odot \mathcal{H}_s^i \tag{6}$$

$$a_i = \frac{exp(\mathbf{MLP}(\mathcal{H}_r^i))}{\sum_j exp(\mathbf{MLP}(\mathcal{H}_r^j))} \tag{7}$$

4.3   Type Feature Fusion Embedding

For any query, it may be related to multiple types. Yet, not all of these types are relevant. Our focus lies in understanding relationships pertinent to the query and capturing the alignment between the query and the KGs. To precisely capture this constraint, we employ a bidirectional attention mechanism on the previously obtained $\mathcal{H}^e, \mathcal{H}^r$ and $\mathcal{H}^s$, integrating them pairwise such as entity-relation, entity-type, and relationship-type. The computations of them are expressed in (8) (9):

$$\mathcal{G}^{er} = Relu(W_1 \begin{bmatrix} \mathcal{H}^e \ominus \mathcal{H}^r \\ \mathcal{H}^e \otimes \mathcal{H}^r \end{bmatrix} + b_1) \tag{8}$$

$$\mathcal{G}^{re} = Relu(W_2 \begin{bmatrix} \mathcal{H}^r \ominus \mathcal{H}^e \\ \mathcal{H}^r \otimes \mathcal{H}^e \end{bmatrix} + b_2) \tag{9}$$

In which, $\{W_1, W_2\} \in \mathbb{R}^{2d \times 2d}$ and $\{b_1, b_2\} \in \mathbb{R}^{2d \times 1}$ are learnable parameters. $\mathcal{G}^{er}$, $\mathcal{G}^{re}$ are bidirectional integrations of entities and relationships, capturing interactions between entities and relationships. Similarly, the calculations for entity relation and entity type are as shown in (10) (11):

$$\mathcal{G}^{rs} = Relu(W_3 \begin{bmatrix} \mathcal{H}^r \oplus \mathcal{H}^s \\ \mathcal{H}^r \otimes \mathcal{H}^s \end{bmatrix} + b_3) \tag{10}$$

$$\mathcal{G}^{es} = Relu(W_4 \begin{bmatrix} \mathcal{H}^e \oplus \mathcal{H}^s \\ \mathcal{H}^e \otimes \mathcal{H}^s \end{bmatrix} + b_4) \tag{11}$$

Then, we fuse the outcomes generated by employing a gating mechanism, by the bidirectional attention to obtain the entity fusion $\widetilde{\mathcal{G}^e}$ and relation fusion $\widetilde{\mathcal{G}^r}$. Taking $\widetilde{\mathcal{G}^e}$ as an example, the calculation of the final representation of entities is illustrated in (12) (13):

$$g = \sigma(W_5\mathcal{G}^{er} + W_6\mathcal{G}^{es} + b_5 + b_6) \tag{12}$$

$$\widetilde{\mathcal{G}^e} = g * \mathcal{G}^{er} + (1 - g) * \mathcal{G}^{es} \tag{13}$$

Similarly, the calculation of $\widetilde{\mathcal{G}^r}$ is depicted in (15):

$$g = \sigma(W_7\mathcal{G}^{rs} + W_8\mathcal{G}^{re} + b_7 + b_8) \tag{14}$$

$$\widetilde{\mathcal{G}^r} = g * \mathcal{G}^{rs} + (1 - g) * \mathcal{G}^{re} \tag{15}$$

After that, the fused feature vectors need to be transformed back to the original vector size through a linear layer. The transformation formula is illustrated in (16) (17):

$$\overline{\mathcal{H}^e} = W_9\widetilde{\mathcal{G}^e} + b_9 \tag{16}$$

$$\overline{\mathcal{H}^r} = W_{10}\widetilde{\mathcal{G}^r} + b_{10} \tag{17}$$

In which, $\{W_3, \cdots, W_{10}\} \in \mathbb{R}^{2d \times 2d}$ and $\{b_3, \cdots, b_{10}\} \in \mathbb{R}^{2d \times 1}$ are learnable parameters. Whereas g represents a reset gate. $\widetilde{\mathcal{G}^e}, \widetilde{\mathcal{G}^r} \in \mathbb{R}^{2d \times 1}$ is the final entity representation.

5 Case Study

We use a NELL-995 example to show our model's effectiveness in predicting if Adam will attend a music festival. Our model considers Adam's entity and relation types, aggregating them for classification. For instance, Adam's types include *"pop music enthusiast"*, *"music festival participant"*, and *"basketball game winner"*. However, as our query focuses on *"music festivals"*, only relevant types are considered, making *"basketball game winner"* irrelevant. Ultimately, we predict Adam's attendance by analyzing his correlation with music festivals.

6 Experiments

Our main objective in the experiments is to integrate our model into different QE models and assess its generalization and reasoning capability across various models.

6.1 Datasets

To evaluate generalization and reasoning capabilities, we conducted experiments using three datasets: FB15k [20], FB15k-237 [19] and NELL (NELL-995) [30].

Furthermore, each of the aforementioned knowledge graph datasets is accompanied by two distinct types of query sub-datasets: Query2Box (Q2B) [9], which encompasses 9 query structures devoid of negation, and BETAE [10], which consists of 9 query structures without negation and 5 query structures incorporating negation.

Table 1. Hits@3 on Q2B datasets (Generalization)

Databases	Methods	1p	2p	3p	2i	3i	pi	ip	2u	up	AVG	Improved(%)
FB15k	GQE	71.1	36.1	25.9	47.5	**60.8**	30.0	15.8	45.2	28.1	40.1	
	GQE+Ours	**78.6**	**43.1**	**35.7**	**51.7**	60.8	**40.7**	**23.7**	**61.7**	**32.9**	**47.6**	+18.70
	Q2B	82.1	43.0	31.7	62.7	**74.5**	44.4	22.4	66.7	**33.5**	51.2	
	Q2B+Ours	**85.0**	**46.8**	**39.2**	**65.4**	74.5	**45.3**	**24.1**	**71.7**	29.5	**53.5**	+4.50
	BETAE	75.0	45.1	40.0	62.2	**75.4**	47.1	22.3	58.9	29.6	50.6	
	BETAE+Ours	**79.5**	**51.2**	**45.7**	**62.7**	73.4	**47.9**	**23.4**	**64.5**	29.8	**53.1**	+5.0
	LOGICE	80.5	50.5	45.8	62.2	72.5	47.5	**28.3**	63.9	36.8	54.2	
	LOGICE+Ours	**83.0**	**51.5**	**46.1**	**63.2**	**73.3**	**48.0**	28.3	**67.0**	**36.9**	**55.2**	+1.85
FB15k-237	GQE	41.3	21.3	15.3	26.8	38.1	16.8	8.7	17.4	15.8	22.3	
	GQE+Ours	**47.7**	**25.9**	**22.3**	**33.3**	**44.0**	**24.1**	**14.4**	**26.8**	**19.9**	**28.7**	+28.70
	Q2B	**47**	**24.9**	19.1	33.2	46.1	21.4	10.9	25.9	**18.8**	27.5	
	Q2B+Ours	46.1	24.3	**20.1**	**34.4**	**46.3**	**23.3**	**13.0**	**26.5**	18.5	**28.0**	+1.81
	BETAE	42.6	25.4	21.6	30.2	43.3	20.7	9.2	24.2	**18.3**	26.2	
	BETAE+Ours	**45.5**	**27.1**	**23.6**	**33.4**	**46.7**	**23.7**	**11.4**	**27.9**	18.3	**28.6**	+9.16
	LOGICE	45.6	27.8	24.1	34.7	46.5	23.5	12.0	27.1	20.8	29.1	
	LOGICE+Ours	**46.5**	**28.5**	**24.3**	**35.1**	**47.5**	**24.0**	**12.7**	**27.9**	**20.9**	**29.7**	+2.06
NELL	GQE	42.7	21.6	19.3	27.0	37.4	17.7	10.3	21.7	13.4	23.5	
	GQE+Ours	**44.7**	**23.2**	**20.7**	**27.8**	**38.5**	**18.1**	**10.5**	**23.3**	**14.4**	**24.6**	+4.68
	Q2B	56.2	27.1	24.1	34.7	49.2	21.8	12.7	37.5	16.5	31.1	
	Q2B+Ours	**58.5**	**30.3**	**26.2**	**37.0**	**50.2**	**22.9**	**13.1**	**39.7**	**18.2**	**32.9**	+5.79
	BETAE	58.4	29.1	30.7	35.2	48.4	**22.5**	10.5	44.5	21.2	33.4	
	BETAE+Ours	**58.6**	**30.7**	**31.0**	**36.3**	**50.3**	22.5	**11.7**	43.5	**21.3**	**34.0**	+1.80
	LOGICE	**64.3**	35.9	**36.0**	**41.1**	54.8	26.8	**14.7**	51.0	27.6	39.1	
	LOGICE+Ours	64.3	**36.3**	35.7	41.0	**55.3**	**26.9**	14.6	**51.2**	**27.7**	**39.2**	+0.30

6.2 Baselines

We will conduct comparisons between our proposed DRMTR model and four foundational models, all tailored for addressing complex queries: Q2B [9], BETAE [10], GQE [8], and LOGICE [11].

6.3   Experimental Results and Analysis

The result of the performance comparison of models in terms of generalization and deductive reasoning capabilities are shown in Table. 1 and Table. 2.

Generalization: Our evaluation of the model's generalizability aims at validating its performance on unknown data. Table. 1 displays our model's generalization across three datasets compared to other models. The results demonstrate that the DSMRT model exhibits improvements in Hit@3 results as high as 18.7%, 28.7%, and 5.79% on three datasets, respectively. In particular, the results of handling 2p and 3p queries are much better than those of 1p queries, showcasing its applicability in dealing with longer-chain queries.

Deductive Reasoning: The objective is to verify the hypothesis that the model can deduce new information solely based on known facts and logical rules, showcasing its deductive reasoning capabilities. The experimental results in Table. 2

Table 2. Hits@3 on Q2B datasets (Deductive)

Databases	Methods	1p	2p	3p	2i	3i	pi	ip	2u	up	AVG	Improved(%)
FB15k	GQE	73.8	40.5	32.1	49.8	64.7	36.1	18.9	47.2	30.4	44.5	
	GQE+Ours	**77.8**	**46.9**	**43.0**	**64.6**	**73.8**	**44.5**	**20.3**	**50.5**	**32.0**	**50.3**	**+13.0**
	Q2B	68.0	39.4	32.7	48.5	65.3	32.9	16.2	61.4	28.9	43.7	
	Q2B+Ours	**86.8**	**48.1**	**47.3**	**68.7**	**82.6**	**50.8**	**27.8**	**76.3**	**38.4**	**58.5**	**+33.87**
	BETAE	83.2	57.3	51.0	71.1	**81.4**	56.9	32.7	70.4	41.0	60.6	
	BETAE+Ours	**84.2**	**57.9**	**51.4**	**67.7**	80.4	**57.6**	**35.8**	**71.5**	**41.9**	**60.9**	**+1.49**
	LOGICE	88.4	64.0	57.9	70.8	80.6	59.0	41.0	76.6	51.0	65.5	
	LOGICE+Ours	**89.1**	**64.9**	**58.5**	**71.3**	**82.0**	**60.5**	**42.1**	**78.4**	**51.3**	**66.4**	**+1.37**
FB15k-237	GQE	56.4	30.1	24.5	35.9	51.2	25.1	13.0	25.8	**22.0**	31.6	
	GQE+Ours	**58.0**	**31.0**	**25.2**	**36.9**	**53.0**	**26.9**	**13.7**	**26.4**	**22.0**	**32.5**	**+2.84**
	Q2B	58.5	34.3	28.1	44.7	62.1	23.9	11.7	40.5	**22.0**	36.2	
	Q2B+Ours	**81.5**	**40.9**	**28.7**	**58.2**	**66.3**	**37.7**	**17.2**	**67.9**	29.2	**47.5**	**+31.21**
	BETAE	77.9	52.6	44.5	59.2	67.8	42.2	23.5	63.7	35.1	51.8	
	BETAE+Ours	**79.4**	**57.6**	**48.3**	**61.8**	**70.1**	**45.6**	**28.3**	**65.6**	**39.1**	**55.1**	**+6.37**
	LOGICE	81.5	54.2	46.0	58.1	67.1	44.0	28.5	66.6	40.8	54.1	
	LOGICE+Ours	**82.1**	**56.7**	**47.8**	**60.2**	66.1	**45.9**	**31.7**	**68.8**	**43.4**	**55.8**	**+3.14**
NELL	GQE	72.8	58.0	55.2	45.9	57.3	34.2	24.8	59.0	40.7	49.8	
	GQE+Ours	**73.8**	**58.7**	**56.6**	**47.9**	**58.6**	**35.1**	**26.1**	**60.0**	**42.3**	**51.0**	**+2.41**
	Q2B	83.9	57.7	47.8	49.9	66.3	29.6	19.9	73.7	31.0	51.1	
	Q2B+Ours	**86.3**	**70.4**	**66.0**	**72.2**	**86.1**	**65.2**	**49.5**	**93.5**	**82.8**	**74.6**	**+45.99**
	BETAE	94.3	88.2	76.2	84.0	90.2	68.8	46.6	92.5	81.4	80.2	
	BETAE+Ours	**95.3**	**89.0**	**77.5**	**84.1**	89.9	**69.2**	**49.1**	**92.7**	**82.3**	**81.0**	**+1.00**
	LOGICE	96.2	90.7	**84.1**	**84.1**	**89.5**	**76.0**	**65.2**	94.7	**87.1**	**85.3**	
	LOGICE+Ours	**96.4**	**91.0**	84.0	83.1	89.3	72.6	64.1	**95.2**	87.0	84.7	-0.70

demonstrate that the proposed DSMRT model accuracy of Hit@3 results has been improved by 33.87%,31.21% and 45.99% at most on three datasets respectively. However, we find an interesting fact that the LOGICE [11] in deductive reasoning is a little better than our model due to its logic circuits.

The above analysis highlights a significant performance enhancement of our model when handling type information across multiple knowledge graphs. The integration of type information enables the model to more accurately infer and comprehend relationships between entities, thereby enhancing the precision and robustness of knowledge graph reasoning. This indicates the potential and effectiveness of our model in advancing knowledge graph inference.

7   Ablation Studies

Conduct ablation experiments on the sampler, Entity Type Aggregation (ETA), and Relation Type Aggregation (RTA) individually, and compare the average MRR results, as delineated in Table. 3. Based on three benchmark datasets, it is obvious that the average accuracy of the DTSMRT model has been great by improved. In particular, RTA exerts the most pronounced influence on the performance, since the average accuracy can be improved by 16.5% to 20.9%. Meanwhile, ETA also yields a marked enhancement of the accuracy ranging from 4.6% to 8.0%. Moreover, the strategic utilization of the sampling improves the accuracy ranging from 0.2% to 7.2% for the DTSMRT model.

Table 3. Ablation Experiments

FB15k	Sampler	ETA	RTA	Avg	Improved(%)
	✓	✓	✗	42.1	**+20.9**
	✗	✓	✓	50.8	**+0.2**
	✓	✗	✓	48.8	**+5.1**
	✓	✓	✓	50.9	
FB15k-237	Sample	ETA	RTA	Avg	Improved(%)
	✓	✓	✗	23.5	**+20.9**
	✗	✓	✓	26.3	**+7.2**
	✓	✗	✓	26.1	**+8.0**
	✓	✓	✓	28.2	
NELL	Sample	ETA	RTA	Avg	Improved(%)
	✓	✓	✗	29.1	**+16.5**
	✗	✓	✓	33.4	**+1.5**
	✓	✗	✓	32.4	**+4.6**
	✓	✓	✓	33.9	

Therefore, the impact of the above three modules offers substantial precision and accuracy performance in MHR tasks, encompassing semantic comprehension and relationship inference.

8   Conclusion

In this paper, we propose a novel DSMRT model integrating entity and relationship type features in KGs to address loosely connected head-tail entity relations in long-chain reasoning. We also introduce a new method for sampling positive and negative instances, improving the quality of negative samples and reducing noise and sparsity in KGs. Experimental results show that our DSMRT model outperforms four baselines on three benchmark datasets, enhancing generalization and reasoning capabilities

In future work, we aim to incorporate a conditional encoder-decoder for entity and relationship type identification. This will hopefully, integrate type representations into the encoding or decoding architecture for improving entity or relationship type selection. Beyond evaluating the model's generalization and deductive reasoning capabilities as presented in this study, we will further investigate the model's inductive reasoning capabilities. This could help validate the model's capability to infer general patterns from limited data and make reasonable inferences and predictions in unfamiliar domains.

Acknowledgment. This work is supported by the National Natural Science Foundation of China (No.623061062), and Natural Science Foundation of Hubei Province under Grant 2023AFB377.

References

1. Chen, Y., Chen, Y., et al.: An overview of knowledge graph reasoning: key technologies and applications. J. Sens. Actuator Netw. **11**(4), 78 (2022)
2. Wang, H., Zhang, F., Xie, X., Guo, M.: DKN: deep knowledge-aware network for news recommendation. In: Proceedings of the 2018 World Wide Web Conference, pp. 1835–1844 (2018)
3. Saxena, A., Tripathi, A., Talukdar, P.: Improving multi-hop question answering over knowledge graphs using knowledge base embeddings. In: Proceedings of the 58th annual meeting of the association for computational linguistics, pp. 4498–4507 (2020)
4. Lin, B.Y., Chen, X., Chen, J., Ren, X.: Kagnet: knowledge-aware graph networks for commonsense reasoning, arXiv preprint arXiv:1909.02151 (2019)
5. Yang, S., et al.: Inductive link prediction with interactive structure learning on attributed graph. In: Machine Learning and Knowledge Discovery in Databases. Research Track: European Conference, ECML PKDD 2021, Bilbao, Spain, September 13–17, 2021, Proceedings, Part II 21, Springer, pp. 383–398 (2021). https://doi.org/10.1007/978-3-030-86520-7_24
6. Lin, X., Socher, R., Xiong, C.: Multi-hop knowledge graph reasoning with reward shaping. arxiv 2018, arXiv preprint arXiv:1808.10568

7. Long, X., Zhuang, L., Aodi, L., Wang, S., Li, H.: Neural-based mixture probabilistic query embedding for answering FOL queries on knowledge graphs. In: Proceedings of the 2022 Conference on Empirical Methods in Natural Language Processing, pp. 3001–3013 (2022)
8. Hamilton, W., Bajaj, P., Zitnik, M., Jurafsky, D., Leskovec, J.: Embedding logical queries on knowledge graphs. Adv. Neural Inf. Proc. Syst. **31** (2018)
9. Ren, H., Hu, W., Leskovec, J.: Query2box: Reasoning over knowledge graphs in vector space using box embeddings. arXiv preprint arXiv:2002.05969 (2020)
10. Ren, H., Leskovec, J.: Beta embeddings for multi-hop logical reasoning in knowledge graphs. Adv. Neural. Inf. Process. Syst. **33**, 19716–19726 (2020)
11. Luus, F., et al.: Logic embeddings for complex query answering. arXiv preprint arXiv:2103.00418 (2021)
12. Lan, Y., He, G., Jiang, J., Jiang, J., Zhao, W.X., Wen, J.R.: A survey on complex knowledge base question answering: Methods, challenges and solutions. arXiv preprint arXiv:2105.11644 (2021)
13. Jin, W., Zhao, B., Yu, H., Tao, X., Yin, R., Liu, G.: Improving embedded knowledge graph multi-hop question answering by introducing relational chain reasoning. Data Min. Knowl. Disc. **37**(1), 255–288 (2023)
14. Pan, W., Wei, W., Mao, X.-L.: Context-aware entity typing in knowledge graphs. arXiv preprint arXiv:2109.07990 (2021)
15. Ying, R., He, R., Chen, K., Eksombatchai, P., Hamilton, W.L., Leskovec, J.: Graph convolutional neural networks for web-scale recommender systems. In: Proceedings of the 24th ACM SIGKDD International Conference on Knowledge Discovery & Data Mining, pp. 974–983 (2018)
16. Zou, D., Hu, Z., Wang, Y., Jiang, S., Sun, Y., Gu, Q.: Layer-dependent importance sampling for training deep and large graph convolutional networks. Adv. Neural Inf. Proc. Syst. **32** (2019)
17. Wang, R., et al.: Rete: retrieval-enhanced temporal event forecasting on unified query product evolutionary graph. In: Proceedings of the ACM Web Conference 2022, pp. 462–472 (2022)
18. Tao, D., Li, X., Maybank, S.J.: Negative samples analysis in relevance feedback. IEEE Trans. Knowl. Data Eng. **19**(4), 568–580 (2007)
19. Zhu, Z., Xu, S., Tang, J., Qu, M.: Graphvite: a high-performance CPU-GPU hybrid system for node embedding. In: The World Wide Web Conference, pp. 2494–2504 (2019)
20. Bordes, A., Usunier, N., Garcia-Duran, A., Weston, J., Yakhnenko, O.: Translating embeddings for modeling multi-relational data. Adv. Neural Inf. Proc. Syst. **26** (2013)
21. Zhao, Y., Zhang, A., Xie, R., Liu, K., Wang, X.: Connecting embeddings for knowledge graph entity typing. arXiv preprint arXiv:2007.10873 (2020)
22. Zhou, B., Khashabi, D., Tsai, C.-T., Roth, D.: Zero-shot open entity typing as type-compatible grounding. arXiv preprint arXiv:1907.03228 (2019)
23. Niu, G., Li, B., Zhang, Y., Pu, S., Li, J.: Autoeter: Automated entity type representation for knowledge graph embedding. arXiv preprint arXiv:2009.12030 (2020)
24. Xie, R., et al.: Representation learning of knowledge graphs with hierarchical types. In: IJCAI, Vol. 2016, pp. 2965–2971 (2016)
25. Ma, S., Ding, J., Jia, W., Wang, K., Guo, M.: Transt: type-based multiple embedding representations for knowledge graph completion. In: Machine Learning and Knowledge Discovery in Databases: European Conference, ECML PKDD (2017)
26. Daza, D., Cochez, M., Groth, P.: Inductive entity representations from text via link prediction. In: Proceedings of the Web Conference 2021, pp. 798–808 (2021)

27. Kumar, S., Zhang, X., Leskovec, J.: Predicting dynamic embedding trajectory in temporal interaction networks. In: Proceedings of the 25th ACM SIGKDD International Conference On Knowledge Discovery & Data Mining, pp. 1269–1278 (2019)

28. Jain, P., et al.: Type-sensitive knowledge base inference without explicit type supervision. In: Proceedings of the 56th Annual Meeting of the Association for Computational Linguistics (Vol. 2 Short Papers), pp. 75–80 (2018)

29. Wang, Z., Zhang, J., Feng, J., Chen, Z.: Knowledge graph embedding by translating on hyperplanes. In: Proceedings of the AAAI Conference on Artificial Intelligence. Vol. 28 (2014)

30. Xiong, W., Hoang, T., Wang, W.Y.: Deeppath: a reinforcement learning method for knowledge graph reasoning. arXiv preprint arXiv:1707.06690 (2017)

31. Carlson, A., Betteridge, J., Kisiel, B., Settles, B., Hruschka, E., Mitchell, T.: Toward an architecture for never-ending language learning. In: Proceedings of the AAAI Conference on Artificial Intelligence, Vol. 24, pp. 1306–1313 (2010)

32. Yaniv, Z.: Random sample consensus (ransac) algorithm, a generic implementation, Imaging (2010)

33. Mikolov, T., Sutskever, I., Chen, K., Corrado, G.S., Dean, J.: Distributed representations of words and phrases and their compositionality. Adv. Neural Inf. Proc. Syst. **26** (2013)

34. Li, B., Hou, Y., Che, W.: Data augmentation approaches in natural language processing: a survey. Ai Open **3**, 71–90 (2022)

35. Lao, N., Cohen, W.W.: Relational retrieval using a combination of path-constrained random walks. Mach. Learn. **81**, 53–67 (2010)

Learning Solutions of Stochastic Optimization Problems with Bayesian Neural Networks

Alan A. Lahoud$^{(\boxtimes)}$, Erik Schaffernicht, and Johannes A. Stork

Center for Applied Autonomous Sensor Systems (AASS), Örebro University,
Örebro, Sweden
`alan.lahoud@oru.se`

Abstract. Mathematical solvers use parametrized Optimization Problems (OPs) as inputs to yield optimal decisions. In many real-world settings, some of these parameters are unknown or uncertain. Recent research focuses on predicting the value of these unknown parameters using available contextual features, aiming to decrease decision *regret* by adopting end-to-end learning approaches. However, these approaches disregard prediction uncertainty and therefore make the mathematical solver susceptible to provide erroneous decisions in case of low-confidence predictions. We propose a novel framework that models prediction uncertainty with Bayesian Neural Networks (BNNs) and propagates this uncertainty into the mathematical solver with a Stochastic Programming technique. The differentiable nature of BNNs and differentiable mathematical solvers allow for two different learning approaches: In the *Decoupled* learning approach, we update the BNN weights to increase the quality of the predictions' distribution of the OP parameters, while in the *Combined* learning approach, we update the weights aiming to directly minimize the expected OP's cost function in a stochastic end-to-end fashion. We do an extensive evaluation using synthetic data with various noise properties and a real dataset, showing that decisions *regret* are generally lower (better) with both proposed methods. The code is available at https://github.com/AlanLahoud/BNNSOP.

Keywords: Neural Networks · Uncertainty · Constrained Optimization

1 Introduction

A mathematical solver uses optimization techniques to find solutions for Optimization Problems (OPs), defined by a cost function and a feasible set, aiming to reach global optima. This process minimizes the cost function by finding the best decisions within the feasible set. In real-world scenarios, some parameters of the OP might be unknown during decision-making and must be estimated using available information. Examples include optimizing trading decisions (e.g., minimizing risk) based on unknown market fluctuations; and optimizing energy

M. Wand et al. (Eds.): ICANN 2024, LNCS 15016, pp. 147–162, 2024.
https://doi.org/10.1007/978-3-031-72332-2_11

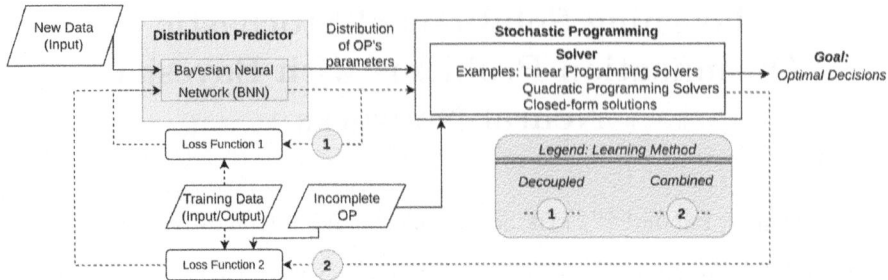

Fig. 1. We propose the addition of a BNN predictor block before solving data-driven OPs with a mathematical solver in a stochastic fashion. The solid lines illustrate the decision inference given new input data, the dashed indicate the learning process, where the BNN weights are learned based on prediction quality (*Decoupled* learning, 1) or decision quality (*Combined* learning, 2).

scheduling considering estimated energy demand. This paper focuses on modeling the uncertainty of these unknown parameters (e.g., market fluctuations; energy demand) of the OP using input-output training data, where the target variables represent the unknown OP parameters.

From a Machine Learning perspective, one could predict those unknown parameters in a supervised fashion using contextual features. Predicting these parameters completes the OP (albeit with estimated parameters), allowing mathematical solvers to seek optimal decisions. However, inaccuracies in predictions due to noise or lack of data can affect decision quality in ways that may vary significantly depending on the OP type [13,17]. Some recent methods align the loss function with the OP's cost function to avoid this problem [10,24,28]. However, they remain deterministic and thus ignore model and data uncertainty.

In contrast, Stochastic Programming approaches provide a more robust solution by considering different sets of possible OP parameters, modeling them as distributions. Mathematical solvers then find optimal decisions that account for this distribution [6]. In this paper, we explore the treatment of data-driven OPs as Stochastic OPs, introducing a novel approach to model the distribution of their unknown parameters focusing on decision quality. To achieve this, we leverage Bayesian Neural Networks' (BNNs) architecture due to its significant advantages in modeling complex and non-linear data relationships using gradient descent approaches.

The main motivation of our work is to enhance decision quality by connecting probabilistic models to data-dependent OPs within a task-oriented learning framework. More specifically, our contributions include:

– Our first method, the *Decoupled* Learning, integrates existing learning techniques within BNNs to predict the actual distribution of the unknown parameters of conditional Stochastic OPs. Then, it propagates the predicted distribution to provide decisions that minimizes the expected final task cost in a Stochastic Programming fashion (See Fig. 1, *Decoupled*).

– Our second method, the *Combined* Learning, introduces an end-to-end gradient computation. It innovates the way of learning BNNs by implementing a task-oriented method, where the goal is not to fit the data distribution, but to learn a distribution that directly minimizes the OP cost in a Stochastic Programming fashion (See Fig. 1, *Combined*).
– Through experimental results, we show that our *Decoupled Learning* method generally outperforms other uncertainty estimation methods, while our *Combined Learning* method surpasses previous end-to-end learning approaches. We further delineate the specific scenarios where either the *Combined* or the *Decoupled* approach holds an advantage.

2   Background

In this section, we review the notation on data-driven OPs and BNNs since we use them as our probabilistic model.

2.1   Problem Formulation

OPs under Uncertainty. In OPs under uncertainty, the goal is to identify the optimal decisions $z^* \in S \subset \mathbb{R}^{d_z}$, which depend on an unknown parameter $y \in \mathbb{R}^{d_y}$. This is expressed as $z^*(y) = \arg\min_z f(z, y)$ subject to $z \in S$, where f is a cost function (task loss) and S is the feasible decision set. Given y's uncertainty, a common approach is to approximate it with a parametric distribution $\hat{y}$ [4,6,14], leading to the approximation $z^*(y) \approx z^*(\hat{y})$ where $z^*(\hat{y}) := \arg\min_z \mathbb{E}_{\hat{y}}[f(z, \hat{y})]$ subject to $z \in S$. Here, $\hat{y}$ is a predicted distribution. This equation captures many real-world problems related to decision-making under uncertainty [26], where practitioners need to make decisions before having the actual observation of the outcome and aim to minimize the cost f in the long term on average. With specific combinations of OPs (i.e., f and S) and simple parametric distributions for $\hat{y}$, the argmin can be reduced in a way that standard mathematical solvers (e.g., linear programming (LP) solvers, quadratic programming (QP) solvers) are able to provide optimal decisions [6]. By varying the form of f and S, it is also possible to represent a broad set of known problems such as conditional values at risk and chance-constrained problems [4]. In general, we are interested in OP formulations where propagating uncertainty from predictions to the OP is effective. This is valid when the equality $\arg\min_z \mathbb{E}_{\hat{y}}[f(z, \hat{y})] = \arg\min_z f(z, \mathbb{E}[\hat{y}])$ does not hold, otherwise solving the expectation in the prediction step would lead to the same result as solving the expectation in the decision step. Appendix A provides details regarding this limitation.

Dataset and Problem Statement. Let us denote the input-output training data (let superscript t refer to training samples) as $\mathcal{D}^t = \{(x_i{}^t, y_i{}^t)\}_{i=1}^{N^t}$. This paper focus on learning a probabilistic model h_ω that outputs predicted distributions for the unknown parameters, i.e., $h_\omega := P_\omega(y \mid x)$. In our first proposed method

(*Decoupled*), we explore the case where data-driven decisions are achieved by trying to approximate the model output to the actual unknown parameters distributions $P(y \mid x)$, while in our second proposed method (*Combined*), we consider a direct minimization to the expected loss of the downstream task, which although can lead to considerable differences between $P(y \mid x)$ and $P_\omega(y \mid x)$, the final goal can still be achieved. In both cases, data-driven decisions $z^*(x, h_\omega)$ are provided as

$$z^*(y) \approx z^*(x, h_\omega) := \arg\min_z \mathbb{E}_{\hat{y} \sim h_\omega(x)}[f(z, \hat{y})]$$
$$\text{subject to } z \in S \tag{1}$$

where $x \in \mathbb{R}^{d_x}$ is a feature vector conditioning the unknown parameters y and is available at decision-making time. We restrict the problem definition to scenarios where decisions do not impact the actual observations of y, a common assumption also reflected in prior works [9,10].

2.2 Probabilistic Model as Bayesian Neural Networks

We have selected BNNs to represent the probabilistic model h_ω due to the following key characteristics. First, BNNs are adept at modeling uncertainty, both epistemic and aleatoric [16,19]. Second, the flexibility of BNNs allows them to capture complex relationships between inputs and outputs, making them suitable to be applied into a wide range of datasets. Finally, BNNs are compatible with gradient descent optimization methods, an attribute that is particularly valuable in our proposed *Combined* method due to its end-to-end learning manner.

In short, BNNs are Neural Networks that contain stochastic components [7,18]. In Variational Inference, BNNs' weights are treated as a distribution Q parametrized by θ (e.g., Gaussian class), and the aim is to optimize the evidence lower bound (ELBO). The ELBO optimization is often rewritten as

$$\theta^* = \arg\min_\theta \mathbb{E}_{\omega \sim Q_\theta(\omega)}[C_\omega - \sum_{i=1}^{N^t} \log P(y_i{}^t \mid \omega, x_i{}^t)] \tag{2}$$

where $C_\omega = \log Q_\theta(\omega) - \log P(\omega)$ works as a regularization term between the weights' distribution and their provided prior distribution $P(\omega)$. Also, assuming a Gaussian likelihood and $\hat{y}$ continuous, and following [19], the negative log-likelihood term of the above equation is proportional to a data loss that captures epistemic and aleatoric uncertainties, and can be written as $\frac{1}{N^t}\sum_{i=1}^{N^t} \exp(-h_\omega^\sigma(x_i{}^t))(y_i{}^t - h_\omega^\mu(x_i{}^t))^2 + h_\omega^\sigma(x_i{}^t)$, where both $h_\omega^\mu(x_i{}^t)$ and $h_\omega^\sigma(x_i{}^t)$ are BNN outputs representing a stochastic mean and a stochastic variance of the predicted distribution. In practice, M^t weights combinations are sampled to approximate Eq. (2) [7], and backpropagation is used to compute gradients with the help of the reparametrization trick [20]. Once θ is trained, predictions are sampled from the BNN as $h_\omega(x_i) = h_\omega^\mu(x_i) + \epsilon \circ \sqrt{\exp(h_\omega^\sigma(x_i))}$ where $\omega \sim Q_\theta(\omega)$, $\epsilon \in \mathbb{R}^{d_y}$ is a sample from the multivariate normal distribution and $\circ$ denotes element-wise multiplication.

3   Methods

This section presents two methods to learn a BNN h_ω. The *Decoupled* Learning focus on approximating the Stochastic OPs' parameters to their actual distribution, while the *Combined* Learning learns a distribution for the Stochastic OPs' parameters by minimizing the OP cost function directly. In both cases, new decisions are made by propagating learned Stochastic OPs' parameters to solve Eq. (1) for a new set of input data.

3.1   Decoupled Learning with BNN

If a trained BNN h_ω fits the actual data accurately, then $z^*(h_\omega, x) = z^*(y)$, indicating that the model leads to optimal data-driven decisions. This insight serves as a motivation for our *Decoupled* framework. This approach initially leverages common BNN learning techniques to approximate the data distribution of the OP parameters directly from the training data $\mathcal{D}^t$. Then, it integrates prediction samples into a Stochastic Programming block.

Inferring decisions from a trained BNN is not straightforward because Eq. (1) needs to be solved. More specifically, the expected value operator of the downstream task cost function in this equation makes it to be non-trivial. Therefore, we approximate the expectation by sampling M predictions from the learned model, denoted as $\hat{y}^{(j)} \sim h_\omega(x)$, and then we propagate those predictions into the argmin operator of a single and complete Stochastic OP with no unknown parameters as follows:

$$z^*(x, h_\omega, M) = \arg\min_z \frac{1}{M} \sum_{j=1}^{M} f(z, \hat{y}^{(j)})$$

(3)

subject to $z \in S$.

In order to understand better the scalability of this complete Stochastic OP, the cost function f can be rewritten as $f_d(z_d) + \frac{1}{M} \sum_{j=1}^{M} f_s(u^{(j)})$ (as done in [12,27]). Here, f_d represents a deterministic objective function with $z_d \in \mathbb{R}^{d_{z_d}}$ as decision variables, and f_s is a stochastic objective function with $u \in \mathbb{R}^{d_u}$ as auxiliary decision variables that depends on prediction samples. The transformation into a Stochastic OP makes the number of decision variables to increase from d_z to $d_{z_d} + M * d_u$. This transformation and the relationship between the values of d_z, d_{z_d} and d_u depends on the OP's specific structure, which we detail for our experimental problems in Appendices B and C. It is noteworthy that M works as a hyperparameter. A higher value of M provides a more representative estimate of the unknown OP parameters' distribution, resulting in a better decision result. However, it also increases the size of decision variables in the Stochastic OP, making it more time consuming to solve. Careful choice of M is therefore essential to find a balance between solution accuracy and computational efficiency.

3.2   Combined Learning with BNN

Leveraging the availability of the OP information during training, i.e., f and S are given, decisions can be computed and evaluated during training time. This context drives us to refine the BNN training process, transforming it into a *Combined* learning-optimization differentiable block in order to directly enhance the decision quality of the downstream task. To achieve this, we leverage the same BNN structure as the previous method, but we modify the loss function to minimize the OP cost f subject to S. The forward process of the BNN involves computing M^t output samples from the training input as $h_\omega(\boldsymbol{x}^t) = h_\omega^\mu(\boldsymbol{x}^t) + \epsilon \circ \sqrt{\exp(h_\omega^\sigma(\boldsymbol{x}^t))}$. Then, the argmin value $\boldsymbol{z}^*(\boldsymbol{x}^t, h_\omega, M^t)$ (decision) is calculated following Eq. (3) and evaluated within the OP cost function f. Additionally, we introduce a regularization term C_ω to the end-to-end loss to address overfitting and convergence difficulties. The combined loss function is then expressed as

$$\theta^* = \arg\min_\theta \left(\mathbb{E}_\omega[C_\omega] + \frac{K}{N^t} \sum_{i=1}^{N^t} f\big(\boldsymbol{z}^*(h_\omega, \boldsymbol{x_i}^t, M^t), \boldsymbol{y_i}^t\big)\right) \tag{4}$$

where $\mathbb{E}_\omega[C_\omega]$ represents the same Regularization as the previous method sampling $\omega \sim Q_\theta(\omega)$, and K is a hyperparameter that adjusts the trade-off between the end-to-end loss and the regularization term. Note that $\boldsymbol{z}^*(h_\omega, \boldsymbol{x_i}^t, M^t)$ represents a Stochastic OP solution with $d_{z_d} + M^t * d_u$ decision variables. While this approach does increase the training time since the Stochastic OP needs to be solved during the training process, our hypothesis is that the sampling size M^t does not need to be high. This is because the final task solution is not dependent on the accurate reconstruction of the actual data distribution, but on a learned latent distribution.

Gradient descent is used during the training process, and computing the gradients of the right-hand side of Eq. (4) with respect to θ requires to solve the challenge of computing the chain $\frac{\partial f}{\partial z^*} \frac{\partial z^*}{\partial h_\omega} \frac{\partial h_\omega}{\partial \theta}$. Specifically, the partial $\frac{\partial z^*}{\partial h_\omega}$ is computed through the Stochastic Programming block, i.e., Eq. 3, and the argmin differentiation can be complicated because the gradients have to be computed through an OP solver. To overcome this, we leverage specialized methods [1,2] to perform KKT differentiation.

The learning process yields BNN parameters θ^*, and then we solve Eq. (3) during decision inference given a new input $\boldsymbol{x}$.

4   Evaluation

In decision theory, the quality of a decision is often evaluated using the regret metric [5,6,8]. The average regret, R, for a dataset $\mathcal{D} = \{(\boldsymbol{x_i}, \boldsymbol{y_i})\}_{i=1}^N$ with a trained BNN model h_ω is given by $R = \frac{1}{N} \sum_{i=1}^N f(\boldsymbol{z}^*(\boldsymbol{x_i}, h_\omega), \boldsymbol{y_i}) - f(\boldsymbol{z}^*(\boldsymbol{y_i}), \boldsymbol{y_i})$. However, this metric may not accurately reflect model performance on noisy data. To address this, we also calculate a free-aleatoric version of the regret (FR), or the expected regret [23], defined when the data distribution $P(\boldsymbol{y} \mid \boldsymbol{x})$ is known, crucial for a proof of concept in synthetic problems:

$$FR = \frac{1}{N} \sum_{i=1}^{N} f(\boldsymbol{z}^*(\boldsymbol{x}_i, h_\omega), \boldsymbol{y}_i) - f(\boldsymbol{z}^*(\boldsymbol{y}_i^{dist}), \boldsymbol{y}_i),$$

where $\boldsymbol{z}^*(\boldsymbol{y}_i^{dist})$ is the argmin over the actual conditional distribution, computed as

$$\boldsymbol{z}^*(\boldsymbol{y}_i^{dist}) := \arg\min_{\boldsymbol{z}} \mathbb{E}_{\boldsymbol{y}_i^{dist} \sim P(\boldsymbol{y}_i | \boldsymbol{x}_i)}[f(\boldsymbol{z}, \boldsymbol{y}_i^{dist})] \quad \text{subject to } \boldsymbol{z} \in S.$$

An effective model has $FR << R$, indicating the remaining regret is due to data noise. These metrics are leveraged to evaluate methods on synthetic datasets through Monte Carlo simulations in our experiments, while in real-datasets we compute only R.

5 Experiments

In this section, we design three data-driven OPs for evaluation, detail the chosen baselines, and present the experimental results. Further implementation details can be found in Appendix D.

5.1 Classical Newsvendor Problem

The classical newsvendor (NV) problem is defined as finding the optimal order quantity z^* that minimizes the cost function $z^*(y) = \arg\min_z c_s(y - z)_+ + c_e(z - y)_+$, subject to $z \geq 0$, where $(u)_+ = \max(u, 0)$ for demand $y \in \mathbb{R}$ and c_s, c_e denote shortage and excess unit costs respectively. The goal is to estimate demand y to minimize costs; ideally, $z^*(y) = y$ leads to zero cost. When demand follows a distribution, the optimal z^* aligns with the quantile $\frac{c_s}{c_s + c_e}$ of this distribution, offering a closed-form solution for minimizing expected costs [3].

Data and OP Parameters. We generate datasets of (x, y) pairs with non-linear relationships: 1800 for training, 1200 for validation, and 1200 for testing, where $x, y \in \mathbb{R}$. The data is used in two Newsvendor (NV) experiments, NV1 and NV2, introducing input-dependent Gaussian noise and Multimodal Gaussian noise, respectively, to simulate heteroscedastic uncertainty. Additionally, varying densities in the input space are used to simulate epistemic uncertainty. We set $c_s = 100$ and $c_e = 900$ to emphasize the cost imbalance. Appendix A.2 discusses that equalizing c_s and c_e negates the advantage of uncertainty propagation.

5.2 Quadratic Programming Newsvendor

We now consider a constrained and quadratic version of the Newsvendor problem (NVQP) with multiple items defined by the following equation:

$$\begin{aligned}
\boldsymbol{z}^*(\boldsymbol{y}) = \arg\min_{\boldsymbol{z}} \ & \boldsymbol{z}^\mathsf{T} \boldsymbol{Q} \boldsymbol{z} + \boldsymbol{c}^\mathsf{T} \boldsymbol{z} + (\boldsymbol{y} - \boldsymbol{z})_+^\mathsf{T} \boldsymbol{Q}_s (\boldsymbol{y} - \boldsymbol{z})_+ + \boldsymbol{c}_s^\mathsf{T} (\boldsymbol{y} - \boldsymbol{z})_+ \\
& + (\boldsymbol{z} - \boldsymbol{y})_+^\mathsf{T} \boldsymbol{Q}_e (\boldsymbol{z} - \boldsymbol{y})_+ + \boldsymbol{c}_e^\mathsf{T} (\boldsymbol{z} - \boldsymbol{y})_+ \quad \text{subject to } \boldsymbol{z} \succeq 0 \text{ and } \boldsymbol{p}^\mathsf{T} \boldsymbol{z} \leq B
\end{aligned}$$

(5)

where y is the unknown demand, $Q, Q_s, Q_e \in \mathbb{R}^{d_z \times d_z}$, and $c, c_s, c_e \in \mathbb{R}^{d_z}$ are quadratic and linear deterministic parameters in the cost function regarding fixed, shortage and excess costs of each item, and $p \in \mathbb{R}^{d_z}$ and $B \in \mathbb{R}$ are deterministic parameters in the inequality constraint regarding the unit price of items and the total budget. In this problem, $d_y = d_z$. With a few mathematical steps detailed in Appendix B, this problem is transformed into a standard QP formulation (i.e., $\arg\min_v \frac{1}{2}v^\mathsf{T} H v + k^\mathsf{T} v$ s.t. $Av \preceq b$).

Data and OP Parameters. We generate 4000/2000/2000 training/validation/test pair samples (x, y), where $x \in \mathbb{R}^4$ and $y \in \mathbb{R}^6$ (i.e., $d_x = 4$ and $d_y = 6$), with a nonlinear and noisy relationship between those variables, similar to the previous experiment. From each data sample, we seek to find $z^*(x) \in \mathbb{R}^6$ (i.e., $d_z = 6$). The noise of $y \mid x$ is generated by mixing different class of distributions across the outputs.

5.3   Portfolio Conditional Loss Minimization

Drawing from Conditional Value at Risk formulation [27], we address a Portfolio Optimization Problem (POP) aiming to minimize potential losses exceeding a threshold (here, zero) by resource allocation z across assets, given uncertain asset performance y. The optimization is framed as $z^*(y) = \arg\min_z(\max -y^\mathsf{T} z, 0)$, constrained by $z \succeq 0$ and a minimum expected return $p^\mathsf{T} z \geq R$, with p representing historical average returns. Appendix C explains transforming this into a LP problem, then approximating it as QP for compatibility with a quadratic solver, including a regularization term for solution refinement based on [28].

Data and OP Parameters. In this experiment, we use both synthetic (POP) and real datasets (POP2). As a synthetic dataset, we generate 1500/900/1500 training/validation/test pair samples (x_i, y_i), where $x_i \in \mathbb{R}^3$ and $y_i \in \mathbb{R}^{15}$ (i.e., $d_x = 3$ and $d_y = 15$), with a nonlinear relationship similar to the previous experiments. We also vary the number of the training dataset and the number of assets (d_y) for further analysis in the results section. The real dataset [15] includes daily data from 2010 to 2017 on major US stock indexes and features such as technical indicators, futures, commodity prices, global market indices, major US company prices, and treasury bill rates.

5.4   Methods and Baselines

Predictors. We implemented both methods BNN *Decoupled* and BNN *Combined* (we denote in this section as D-BNN and C-BNN, respectively) with a fully connected architecture. For the NV problem, we used three hidden layers with (128, 64, 64) neurons. For the NVQP and POP, we used three hidden layers with (512, 128, 128) neurons. We consider the respective standard neural networks ANN *Decoupled* and ANN *Combined* as baselines (we denote in this section as D-ANN and C-ANN, respectively), with the same number of hidden layers and neurons but without the uncertainty modeling. For the C-ANN baseline, our implementation is based on the [9] idea with a single output (instead of a fixed number of categories) for the NVQP, and based on the [28] (LP version with a quadratic additional term) for the POP experiment. We also implement Gaussian process (GP) as a decoupled baseline since they are commonly used for predictions' uncertainty. We considered the GP with radial basis function kernel with the white noise addition; this combination of kernels provided better results than other kernels and could model epistemic and homoscedastic aleatoric uncertainty.

OP Solvers. In the NV problem, z^* and its gradients are computed within a closed-form solution, so no specific mathematical solver is needed. In the NVQP and in the POP problems, z^* and the KKT differentiation were computed using the qpth library [2], which leverages the cvxpy QP solver.

5.5   Results

Main Results. Table 1 shows the results by running it five times varying the seed data generation and computing the average and standard deviation values. The table is divided into the presented experiments. It shows that the D-ANN method has the highest R and FR (worst result) for the experiments. Both R and FR decrease when modeling uncertainty with the GPs and BNN in a *Decoupled* fashion. Although the C-ANN is able to achieve reasonable results, we observed that it can sometimes converge to a bad local minima, resulting in a stagnation of the learning process, as it happened for the POP experiment. Finally, the C-BNN outperformed the other methods, but only with a small advantage compared to the D-BNN (in most cases). Indeed, we observed that the BNN sampling size that we use to approximate the expectation operations in training and inference (M^t and M) plays an important role in the results. Therefore, we investigate important differences between the two versions of the presented BNNs in the following analysis.

Table 1. Mean of R and FR (and std of FR between brackets) for all the experiments. Some results in this table were scaled to be represented as an integer.

Method	Exp: NV1 R FR	Exp: NV2 R FR	Exp: NVQP R FR	Exp: POP R FR	Exp: POP2 R
D - ANN	958 531 (70)	950 589 (67)	1456 419 (12)	2137 1938 (79)	1228 (283)
D - GP	617 191 (16)	583 222 (20)	1337 300 (15)	274 75 (13)	**852 (31)**
D - BNN	**460 33 (7)**	**421 60 (8)**	1252 214 (11)	**246 47 (15)**	944 (83)
C - ANN	461 33 (12)	410 49 (22)	1263 226 (30)	2138 1939 (84)	1022 (61)
C - BNN	457 30 (3)	**400 39 (11)**	**1242 204 (7)**	**245 46 (36)**	**721 (95)**

Table 2. The table shows the FR average (out of five runs) result values for the NVQP experiment for different sampling sizes (M^t, M).

Method/$(\boldsymbol{M^t}, \boldsymbol{M})$	(4, 8)	(8, 8)	(8, 16)	(16, 16)	(16, 32)	(16, 64)
FR (C-BNN)	259	234	221	217	209	204
FR (D-BNN)	354	331	262	259	231	214

Varying the Sampling Size of BNNs. To provide the main results for the NVQP experiment, the pair (M^t, M) was limited to $(16, 64)$ for both D-BNN and C-BNN. In the POP experiment, we have limited both BNNs to $(M^t, M) = (32, 64)$. Initially, the idea is that these values should be as large as possible in order to approximate the expectation operations of Eqs. 3 and 4, but increasing the number of samples can lead to solving an OP with more decision variables, as detailed in the methods. In Table 2, we show that, as expected, the values of FR decrease (better) by increasing both M^t and M for the experiments NV1 and NVQP. The same is valid for increasing only M while fixing M^t, as shown in Fig. 2a for the POP experiment. It is observed that the C-BNN requires less sampling to converge to small values of FR.

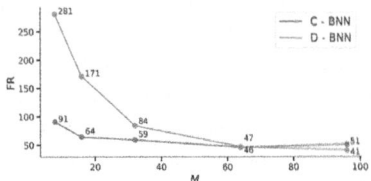 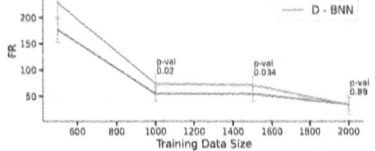

(a) Values of FR decrease by fixing $M^t = 32$ and increasing M.

(b) FR decreases by increasing the amount of training data available.

Fig. 2. Variation of FR with sampling size and training data.

Varying the Training Dataset Size. Fixing the sampling size to $(M^t = M = 32)$ and for $d_y = d_z = 10$, we have also analyzed how the quality of the decisions varies with the increase in training data availability in the POP experiment. Figure 2b shows that in a scenario with less training data, the C-BNN has significantly outperformed the D-BNN. The difference becomes insignificant as we increase the data size in the training set.

Interpreting Predicted OP Parameters. Building upon the simplicity of our NV1 experiment, Fig. 3 illustrates the contrasting behaviors of different methods in predicting OP parameters. D-ANN successfully predicts the average of the unknown parameters' distribution (upper-left graph) but fails in the downstream task due to data noise (Table 1). D-BNN captures the uncertainty of the OP parameters, improving final task performance by aligning the $\frac{c_s}{c_s + c_e}$ quantile of the predicted distribution with the actual one (bottom-left graph) if enough sampling size is chosen. Conversely, C-ANN predicts the $\frac{c_s}{c_s+c_e}$ quantile rather than the mean and performs well in the final task, but exhibits diminished performance in more complex OPs due to a lack of uncertainty modeling. Lastly, the C-BNN method (bottom-right graph) learns a distribution that minimizes the OP cost; even though the overall predicted distribution may not align closely with the actual one, the congruence of the $\frac{c_s}{c_s+c_e}$ quantiles illustrates its ability to focus on the most relevant aspect of the distribution. In both D-BNN and C-BNN graphs, the lower predicted quantile represents the $\frac{c_s}{c_s+c_e}$ quantile of the predicted distribution.

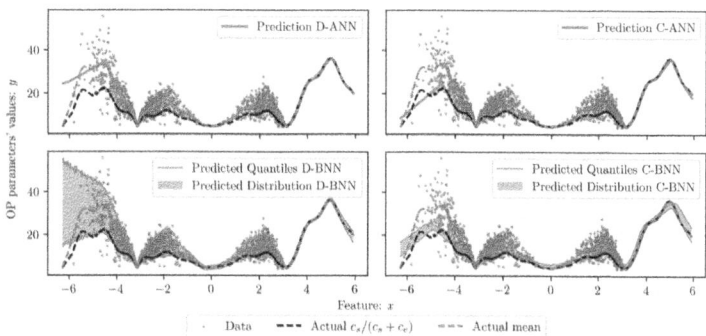

Fig. 3. For the NV1 OP, the C-ANN, D-BNN, and D-BNN methods achieve good decisions with different strategies for the OP parameters predictions.

6   Related Work

A common method of data-driven decision-making, known as "predict-then-optimize", is first to predict the unknown parameters of the OP and then using a solver to yield optimal decisions. This method has been criticized for propagating prediction errors to the optimization problem (OP) block [10]. This led to the development of "Smart predict, then optimize," which incorporates OP information into the learning process with a surrogate end-to-end loss function, though it focuses only on linear objectives and linear predictors.

Following the introduction of KKT differentiation [2], subsequent works have combined ANNs with QP solvers [9], increasing predictive complexity compared linear models. These methods laid the groundwork for LP with neural networks [28] and relaxation techniques for discrete OPs [11,22].

Like our approach, [9,21] aimed to minimize the expected objective function value stochastically. While [9] either lacked proper probabilistic modeling and manually discretized the target variables (OP parameters) before solving the OP or relied on analytical expectations, [21] is an *approximation* of a combined approach using energy-based models. Our method uniquely employs BNNs to model distributions within an task-based loss, enhancing decision quality through proper distribution modeling.

Other works have considered modeling predictions' distribution with BNNs [7,19,25] but without focusing on solving stochastic or constrained OPs. Our work differs by adapting BNNs within a predictor-optimizer framework to improve data-driven decision quality.

7 Conclusion

This paper presented a framework for solving uncertain optimization problems (OPs) using input-output training data by predicting unknown parameters probabilistically and applying a Stochastic Programming technique for near-optimal decisions. We used BNNs to treat predictions as distributions and presented two ways of learning their weights.

The proposed *Decoupled* BNN models aleatoric and epistemic uncertainty, leveraging Variational Inference techniques to predict the OP parameters' distribution. It revives decoupled approaches value by providing good decision results. Also, it can be used in OPs where differentiation over the argmin operator is impossible or time-consuming. The proposed *Combined* BNN learning approach, on the other hand, focused on directly minimizing the expected cost of the OP in an end-to-end fashion through a differentiable solver. Although its training process is more time consuming, we showed that it considerably outperforms state-of-the-art combined approaches in non-trivial OPs. It also outperforms the *Decoupled* BNN mainly in scenarios where there is a limitation of sampling size and training data size.

Acknowledgement. This work has been supported by the Industrial Graduate School Collaborative AI & Robotics funded by the Swedish Knowledge Foundation Dnr:20190128, and the Knut and Alice Wallenberg Foundation through Wallenberg AI, Autonomous Systems and Software Program (WASP).

Appendices

Appendix A. Limitations of Uncertainty Propagation

This paper focuses on minimizing $\arg\min_z \mathbb{E}_{\hat{y}} f(z, \hat{y})$. This problem simplifies to $\arg\min_z f(z, \mathbb{E}_{\hat{y}} \hat{y})$ when substituting the objective function's expected value with the expected value of predictions, but this simplification is only applicable in certain conditions. If these conditions are met, we recommend solving the argmin by directly calculating the expected value of predictions (Decoupled).

Appendix A.1. Linear Objective Functions with Respect to the Unknown Variable

If $f(z, y)$ is linear with respect to y, then $\mathbb{E}_{\hat{y}} f(z, \hat{y}) = f(z, \mathbb{E}_{\hat{y}} \hat{y})$. Applying the argmin with respect to z on both sides we have $\arg\min_z \mathbb{E}_{\hat{y}} f(z, \hat{y}) = \arg\min_z f(z, \mathbb{E}_{\hat{y}} \hat{y})$.

Appendix A.2. Balanced Newsvendor Problem

When $c_s = c_e$ in the NV problem, the optimal order quantity $\arg\min_z \mathbb{E}_{\hat{y}} f(z, \hat{y})$ corresponds to the median of $\hat{y}$'s distribution, given by the $\frac{c_s}{c_s+c_e} = 0.5$ quantile. If $\hat{y}$'s distribution is Gaussian, this median equals the mean, simplifying the argmin to the mean of $\hat{y}$. This observation extends to both Gaussian models and the NVQP, highlighting that propagating uncertainty becomes more beneficial with more imbalance between c_s and c_e.

Appendix B. Newsvendor Problem as Quadratic Programming

Following [9,12], we reformulate Eq. 8 by introducing new decision variables $z_s = y - z$ and $z_e = z - y$, with added constraints to align with the original problem's bounds. This leads to a QP formulation: $\arg\min_v \frac{1}{2}v^\mathsf{T}Hv + k^\mathsf{T}v$ subject to $Av \preceq b$, where $H = 2\text{diag}[Q, Q_s, Q_e]$, $v = [z, z_s, z_e]$, $k = [c, c_s, c_e]$, $A = [-I_{3d_z}, [-I_{d_z}, -I_{d_z}, 0], [I_{d_z}, 0, -I_{d_z}], [p, 0, 0]]^\mathsf{T}$, and $b = [0, 0, 0, -y, y, B]$. $z^*(y)$ is the primary variable of interest. Assuming H is positive-definite ensures convexity. The formulation's efficiency depends on the item count. It's initially suitable for single vector predictions y, but we propose a Stochastic Programming method for generalization to multiple predictions.

.1   Appendix B.1. Newsvendor Problem as Stochastic Quadratic Programming

When propagating the uncertainty of y in a Monte Carlo fashion with M samples, the formulation above becomes as $\arg\min_v \frac{1}{2}v^\mathsf{T}Hv + k^\mathsf{T}v$ s.t. $Av \preceq b$ where $H = 2\text{diag}([Q \quad \frac{Q_s}{M} \quad \cdots \quad \frac{Q_s}{M} \quad \frac{Q_e}{M} \quad \cdots \quad \frac{Q_e}{M}])$; $k = [c \quad \frac{c_s}{M} \quad \cdots \quad \frac{c_s}{M} \quad \frac{c_e}{M} \quad \cdots \quad \frac{c_e}{M}]$; $v = [z \quad z_s^{(1)} \quad \cdots \quad z_s^{(M)} \quad z_e^{(1)} \quad \cdots \quad z_e^{(M)}]$; $A = [-I_F, [-I_{B1}, -I_{BB}, 0_{BB}], [I_{B1}, 0_{BB}, -I_{BB}], [p, 0, 0]]^\mathsf{T}$; and $b = [0_F, -y^{(1)}, ..., -y^{(M)}, y^{(1)}, ..., y^{(M)}, B]$. Where $I_F = I_{d_z+2Md_z}$, $0_F = 0_{d_z+2Md_z}$ (1D vector), $I_{BB} = I_{Md_z}$, $0_{BB} = 0_{Md_z}$, $I_{B1} = I_{d_z}$ repeated for M rows. This is a generalization of the quadratic newsvendor experiment proposed in [Donti et al., 2017]. Note that $v \in \mathbb{R}^{d_z+2Md_z}$, $H \in \mathbb{R}^{d_z+2Md_z \times d_z+2Md_z}$, $k \in \mathbb{R}^{d_z+2Md_z}$, $A \in \mathbb{R}^{d_z+4Md_z+1 \times d_z+2Md_z}$ and $b \in \mathbb{R}^{d_z+4Md_z+1}$. Therefore, both the number of items and prediction sampling size play an important and approximately equal role on the time to solve each instance of the OP. In practice, the complexity of the QP problem depends on the decision variable dimension, which is $d_z + 2Md_z$.

Appendix C. Portfolio Risk Minimization as a LP

With the same strategy as in Appendix 7, we use the auxiliary variable $u = \max\{-y^\mathsf{T}z, 0\}$ to rewrite the POP formulation from the main text to

$$(z^*, u^*)(y) = \arg\min_{z,u} 0^\mathsf{T}z + u \quad \text{s.t.} \ -[z, u] \preceq 0, -y^\mathsf{T}z \preceq u, -p^\mathsf{T}z \leq -R. \tag{6}$$

Note that the zero constant in the objective function is only to reinforce that z is also part of the set of decision variables.

Appendix C.1. Portfolio Risk Minimization as a Stochastic LP

By giving a set of samples $y^{(j)}$ as input, as suggested in [Rockafellar et al., 2000], the equation above can be rewritten in a stochastic programming fashion as

$$(z^*(y), u^*(y)) = \arg\min_{z,u} 0^\mathsf{T} z + \frac{1}{M} \sum_{j=1}^{M} u^{(j)} \tag{7}$$

$$\text{s.t. } -z \preceq 0, -u^{(j)} \preceq 0, -y^{\mathsf{T}(j)} z - u^{(j)} \preceq 0 \quad \forall j \in 1..M, \quad -p^\mathsf{T} z \leq -R.$$

For implementation purpose, we followed [28] by adding a quadratic small term to LP in order to fit the OP into the Amos & Kolter QP solver.

Appendix D. Implementation Details

NNs were implemented with Pytorch and Adam optimizer, with learning rates of 0.0015 for NV, 0.002 for NVQP, and 0.001 for POP experiments. The *Decoupled* Bayesian Neural Network (BNN) had a learning rate of 0.0007, whereas the *Combined* BNN's rate ranged between 0.0004 and 0.0007. An exponential scheduler was used to adjust the learning rate by a factor of 0.99. Hyperparameter K balanced data loss and regularization, selected without optimization. Training occurred on Nvidia RTX 2080 GPUs, with models evaluated on the validation set before testing report. Gaussian Process baselines were implemented with Scikit-learn and radial basis function kernel, optimized the length scale and white noise. For multi-output tasks (NVQP and POP), separate Gaussian processes for each output proved more effective.

References

1. Agrawal, A., Amos, B., Barratt, S., Boyd, S., Diamond, S., Kolter, J.Z.: Differentiable Convex Optimization Layers, vol. 32. Curran Associates Inc. (2019)
2. Amos, B., Kolter, J.Z.: Optnet: differentiable optimization as a layer in neural networks. In: International Conference on Machine Learning, pp. 136–145. PMLR (2017)
3. Ban, G.Y., Rudin, C.: The big data newsvendor: practical insights from machine learning. Oper. Res. **67**(1), 90–108 (2019)
4. Bayraksan, G., Love, D.K.: Data-driven stochastic programming using phi-divergences. In: The operations research revolution, pp. 1–19. INFORMS (2015)
5. Bell, D.E.: Regret in decision making under uncertainty. Oper. Res. **30**(5), 961–981 (1982)

6. Birge, J.R., Louveaux, F.: Introduction to stochastic programming. Springer Science & Business Media (2011)
7. Blundell, C., Cornebise, J., Kavukcuoglu, K., Wierstra, D.: Weight uncertainty in neural network. In: International Conference on Machine Learning, pp. 1613–1622. PMLR (2015)
8. Demirović, E., et al.: An investigation into prediction + optimisation for the knapsack problem
9. Donti, P., Amos, B., Kolter, J.Z.: Task-based end-to-end model learning in stochastic optimization. Adv. Neural Inform. Process. Syst. **30** (2017)
10. Elmachtoub, A.N., Grigas, P.: Smart "predict, then optimize". Manage. Sci. **68**(1), 9–26 (2017)
11. Ferber, A., Wilder, B., Dilkina, B., Tambe, M.: Mipaal: Mixed integer program as a layer. In: Proceedings of the AAAI Conference on Artificial Intelligence, vol. 34, pp. 1504–1511 (2020)
12. Gah-Yi, B., Rudin, C.: The big data newsvendor: practical insights from machine learning. Oper. Res. **67**(1), 90–108 (2018)
13. Grimes, D., Ifrim, G., O'Sullivan, B., Simonis, H.: Analyzing the impact of electricity price forecasting on energy cost-aware scheduling. Sustainable Comput. Inform. Syst. **4**(4), 276–291 (2014), special Issue on Energy Aware Resource Management and Scheduling (EARMS)
14. Hannah, L.A.: Stochastic optimization. Inter. Encycl. Soc. Behav. Sci. **2**, 473–481 (2015)
15. Hoseinzade, E., Haratizadeh, S.: Cnnpred: Cnn-based stock market prediction using a diverse set of variables. Expert Syst. Appl. **129**, 273–285 (2019)
16. Hüllermeier, E., Waegeman, W.: Aleatoric and epistemic uncertainty in machine learning: an introduction to concepts and methods. Mach. Learn. **110**(3), 457–506 (2021)
17. Ifrim, G., O'Sullivan, B., Simonis, H.: Properties of energy-price forecasts for scheduling. In: Milano, M. (ed.) CP 2012. LNCS, pp. 957–972. Springer, Heidelberg (2012). https://doi.org/10.1007/978-3-642-33558-7_68
18. Jospin, L.V., Laga, H., Boussaid, F., Buntine, W., Bennamoun, M.: Hands-on bayesian neural networks-a tutorial for deep learning users. IEEE Comput. Intell. Mag. **17**(2), 29–48 (2022)
19. Kendall, A., Gal, Y.: What uncertainties do we need in bayesian deep learning for computer vision? Adv. Neural Inform. Process. Syst. **30** (2017)
20. Kingma, D.P., Welling, M.: Auto-encoding variational bayes. In: Bengio, Y., LeCun, Y. (eds.) 2nd International Conference on Learning Representations, ICLR 2014, Banff, AB, Canada, 14-16 April 2014, Conference Track Proceedings (2014)
21. Kong, L., Cui, J., Zhuang, Y., Feng, R., Prakash, B.A., Zhang, C.: End-to-end stochastic optimization with energy-based model. Adv. Neural. Inf. Process. Syst. **35**, 11341–11354 (2022)
22. Lahoud, A.A., Schaffernicht, E., Stork, J.A.: Datasp: A differential all-to-all shortest path algorithm for learning costs and predicting paths with context. arXiv preprint arXiv:2405.04923 (2024)
23. Li, X., Shou, B., Qin, Z.: An expected regret minimization portfolio selection model. Eur. J. Oper. Res. **218**(2), 484–492 (2012)
24. Mandi, J., Guns, T.: Interior point solving for lp-based prediction+ optimisation. Adv. Neural. Inf. Process. Syst. **33**, 7272–7282 (2020)
25. Pearce, T., Leibfried, F., Brintrup, A.: Uncertainty in neural networks: Approximately bayesian ensembling. In: International Conference on Artificial Intelligence and Statistics, pp. 234–244. PMLR (2020)

26. Powell, W.B.: A unified framework for stochastic optimization. Eur. J. Oper. Res. **275**(3), 795–821 (2019)
27. Rockafellar, R.T., Uryasev, S., et al.: Optimization of conditional value-at-risk. J. Risk **2**, 21–42 (2000)
28. Wilder, B., Dilkina, B., Tambe, M.: Melding the data-decisions pipeline: decision-focused learning for combinatorial optimization. In: Proceedings of the AAAI Conference on Artificial Intelligence, vol. 33, pp. 1658–1665 (2019)

Revealing Unintentional Information Leakage in Low-Dimensional Facial Portrait Representations

Kathleen Anderson[✉] and Thomas Martinetz

Institute for Neuro- and Bioinformatics, University of Lübeck, Lübeck, Germany
{k.anderson,thomas.martinetz}@uni-luebeck.de

Abstract. We evaluate the information that can unintentionally leak into the low dimensional output of a neural network, by reconstructing an input image from a 40- or 32-element feature vector that intends to only describe abstract attributes of a facial portrait. The reconstruction uses blackbox-access to the image encoder which generates the feature vector. Other than previous work, we leverage recent knowledge about image generation and facial similarity, implementing a method that outperforms the current state-of-the-art. Our strategy uses a pretrained StyleGAN and a new loss function that compares the perceptual similarity of portraits by mapping them into the latent space of a FaceNet embedding. Additionally, we present a new technique that fuses the output of an ensemble, to deliberately generate specific aspects of the recreated image.

Keywords: feature vector reconstruction · face recognition · privacy

Despite the increasing ubiquity of AI solutions, the relevance of privacy in the context of machine learning is often overlooked. One important question is of how much unwanted information can leak into an attribute feature vector. Neural networks are applied to extract, e.g., the stress level of drivers [15], the engagement of students [3,37], a medical diagnosis [13] and abstract avatars[1] from photos of users. They are even proposed for explicitly anonymizing private data [2,6,24,26–29,35,38,39]. By most of these examples, it is neglected that the information processing of neural networks is not well enough understood to be able to guarantee the absence of private input information in their output. Information that is supposed to be removed could still be a part of the feature vector, only in an altered way.

This paper aims to raise awareness for the fact that we are currently unable to prevent private input information from leaking into the output of a neural network. We reconstruct an input image from a feature vector which should only describe a few attributes of the image, by training a *decoder* (depicted in Fig. 1)

---

[1] One example is the *Bitmoji* created by the *Snapchat* app, see https://www.bitmoji.com/.

© The Author(s), under exclusive license to Springer Nature Switzerland AG 2024
M. Wand et al. (Eds.): ICANN 2024, LNCS 15016, pp. 163–177, 2024.
https://doi.org/10.1007/978-3-031-72332-2_12

which reverts from a feature vector back to the input that created it, focusing on recreating components of the input that should not be a part of the feature vector. The focus of this paper is set on reconstructing human portraits, with the main target on their identity.

To achieve such a reconstruction, several challenges need to be overcome by the decoder. One is the nature of the mapping from image to feature vector: the relation is not bijective, more than one input can lead to the same output. To counter this problem, we leverage a pretrained StyleGAN [20] image generator to introduce a strong bias towards a defined target distribution. Another difficulty arises from the fact that the exact target of our reconstruction can be hard to define. An optimal decoder would recreate every pixel of the input, which is a complex task, even given our limited search space. This is, however, not necessary: a reconstruction that only incorporates the important parts of the input can be just as harmful (for example a photo of a person with the same identity, taken from a different angle with a different facial expression or background). Hence, optimizing only the pixel-wise error is not a clever approach, and might even favor the reconstruction of the wrong person in front of the right background. Instead, we map both input and reconstructed images into different latent spaces and compare their embeddings.

The combination of various specialized loss functions improves the reconstruction, but also introduces a third challenge: how to blend the result of decoders trained for different goals? We found that combining the losses into a single cost function oftentimes keeps the model from converging - even though all are theoretically aiming for the same perfect solution (an exact reconstruction). A reason might be that the paths of the cost functions during learning are too dissimilar. A solution presented in this paper is to make use of a special property of the StyleGAN image generator: each layer of the image generator creates a certain aspect of the image, like the shape of the face or the pose of the depicted person. By feeding specialized inputs to the individual StyleGAN layers, we can systematically define each component of the generated image.

By explicitly leveraging the different layers of a pretrained StyleGAN, to join the output of an ensemble trained for differnt specialized loss functions, we are the first to generate truly recognizable reconstructions of input images from attribute vectors. We further demonstrate that a sophisticated training method can be more valuable than a large network architecture: the components of our decoder ensemble are nothing but shallow multi layer perceptrons.

1   Related Work

There are many angles from which machine learning models can be attacked. The topic of this paper falls into the category of data reconstruction at inference time. This line of research can be split according to three properties:

- the **knowledge of the attacker**, who is either given access to all model parameters (whitebox) or only allowed to pose queries to the model and receive an answer, with no information about how that answer was calculated (blackbox),

- the **goal of the attack**: either to infer knowledge about the data used to train a model or to reconstruct a specific input, given an output, with the input not having been part of the training set,
- and the **type of encoder**: an encoder trained to extract features from a given image of a person without aiming at being able to reveal the person's identity or a facial embedding network with features explicitly created to encode the identity of a depicted person.

In the following, we pay particular attention to the two different attack goals. The two goals are similar enough to motivate coinciding strategies but are confronted with different limitations. To the best of our knowledge, the recreation of training data as an attack goal has only been done for encoders that classify whole identities - one class corresponding to exactly one individual. The challenge has been tackled from a whitebox [4,14,32,41] and blackbox [7,17,19] access, sometimes making a similar use of image generators as our method.

The most popular targets for the second goal, a feature vector reconstruction attack, are face embeddings: high-level representations of a face or identity. Again, by design, face embedding networks are trained to condense as much identity information as possible into the embedding vector. [5], for example, even use the reconstruction capability as a quality measure for the embedding. Many blackbox attacks [8,9,12,22,31,34] expand upon this goal, proving that if an encoder network is trained to create a condensed version of an identity, this version can be decompressed to reveal most of the original input. By now, major companies[2] have started to recognize this fact and carry out measures to protect such embeddings.

Our research faces a more difficult challenge by reconstructing from feature vectors that do not explicitly aim to keep the person's identity. A typical face embedding is made up of 512 or more elements [30], the attribute vectors that we are reconstructing from contain 40 and 32 items, describing abstract attributes. A large part of the information from the input image is not required to solve the given task, and should therefore, in theory, have been discarded in early layers of the network - giving reason for the intuitive assumption that they are "safe". We are not the first to implement a reconstruction from such attribute vectors. In 2015, 2016, and 2018, respectively, [11,21,23] published results on the reconstruction of images from different feature representations, like HOG and SIFT descriptors, and the feature maps of a (very small) CNN. All assume full access to the training dataset, and put their focus on understanding network properties, rather than attacking it. [21] iteratively optimize an input image, using gradient descent through the target to minimize a combination of encoder loss and a custom *image prior* term. An image prior can be as simple as the norm of the mean-subtracted image, encouraging the image to stay within a target interval. Their method represents a very early stage of what evolved into

---

[2] For example Microsoft (https://learn.microsoft.com/en-us/windows-hardware/design/device-experiences/windows-hello-enhanced-sign-in-security in September 2023) and Apple (https://support.apple.com/en-us/102381 in September 2023).

the utilization of image generators: implement additional knowledge about the overall nature of our target images into the loss.

[11] train a decoder with an inverted dataset: for a set of input images of the same type as the target image set (e.g. facial images), the feature vectors that E returns are determined, and these returned feature vectors then serve as input to D. The images define the optimal output of D. The authors follow up with an improved loss function shortly after [10], based on the image- and feature map difference, supplemented by the output of a (very simplistic) discriminator, that is trained alongside the decoder to create more realistic images. The (at the time) innovative use of discriminators makes the images much less blurry.

The same inverted dataset strategy (with no discriminator loss term) achieves better results with an improved decoder architecture in 2019 [40]. In a similar method [43] make use of additional model explanations as a second input to their decoder to improve the accuracy of their reconstruction.

2   Threat Model

As depicted in Fig. 1, our attack aims at an encoder E that takes as input an image (in our case a portrait of a person) and returns an output f describing certain properties of the input image (e.g. an assessment of the emotion of a patient). The vector f is considered safe, and shared (for example with the intent of scientific evaluation), without (intentionally) linking it to a specific person.

An attacker gains access to f, has blackbox knowledge about E, and uses additional assumptions about the kind of input image (facial portraits) to find the identity of the depicted person. It would allow them to leak exceedingly private personal information about their victim.

Generally, our scenario involves three datasets: one used to train E, one used by the attacker to train D, and data used at inference time, that was neither seen by E nor D. We consider two possibilities for the access to those datasets: **(A)**, the target encoder is trained on images from a public dataset. Then this public dataset can also be used to train D, and the challenge is to reconstruct images that were never seen by either D or E; and **(B)**, the target is trained on a private dataset. The attacker has no access to this dataset and has to use a more or less good substitute, which makes the task much more challenging.

3   Proposed Method

Our method is built around the idea of training a small mapping network to translate a feature vector into a generator input, whereas the strength of the mapping network does not come from its architecture, but rather the training strategy. The exploration of different training methods for that mapping network indicates that *image-based training* (Sect. 3.2) is a good choice for our context. This training method can be expanded by not only comparing the images directly but also comparing an embedding that is explicitly extracting features that are

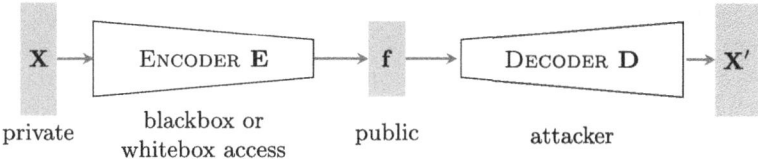

Fig. 1. The scenario of this paper: an encoder E creates an attribute vector f for an image, which is reconstructed by our decoder D.

relevant for our reconstruction goal (Sect. 3.3). To ultimately combine the results of separate mapping networks trained for specialized loss functions, we propose to make use of the style mixing abilities of a StyleGAN (Sect. 3.4).

3.1   Leveraging Image Generators to Limit the Search Space

The task of reconstructing an image from a low-dimensional feature vector is fundamentally ill-posed. [16] have shown that adding an imperceptibly small vector to an image can change the created feature vector significantly - two images that are visually the same create different outputs. Vice versa, a single feature vector could have been created by different images. For example, an image that seems like random noise to the human eye can lead to the same output as an actual image.

While we are unable to enforce a bijective relation between image and feature vector, the challenge can be significantly simplified by making sure that our reconstructed images are from a similar distribution as the images that the target encoder has been trained on. If we can define a set of constraints that all reconstructed images should adhere to (e.g. "facial portraits"), we can greatly limit the search space. To infuse that knowledge into the reconstruction pipeline, our approach is based on two steps: (1) train an image generator to create images that satisfy the known limitations and (2) freeze the generator and train a second network to search over the input space of our trained generator. Since a well-trained generator introduces a very strong bias towards a specific type of image, we are adding the same kind of bias to our reconstruction task.

3.2   PGE-Training

Our decoder is now made up of two components: G, a pretrained StyleGAN, and P, a pre-layer that maps a feature vector to a generator input. Figure 2 illustrates the relation between the three networks involved in the reconstruction. Note that we are only actively training P, both G and E remain frozen, and E can only be accessed as a blackbox.

The canonical way of training P would be *noise based training*: generate random noise vectors n drawn from a Gaussian distribution, feed n into G, then E, and determine the f that corresponds to n and minimize the squared distance between both. A significant advantage would be that we can generate an infinite number of samples, and iterate very quickly through large amounts of training data since the training only involves calculating the gradient for the very small network P. However, the distribution modeled by G will likely deviate from the specific distribution of X. In addition, even the very advanced StyleGAN2 sometimes generates images that are not realistic faces. Therefore, we use what we call *image based training*, which ensures that real face images are used and allows us to control the training data distribution:

1. Select a dataset that is as close as possible to the target dataset[3],
2. For each image X in the dataset, retrieve the matching feature vector f returned by E, to create a new dataset $f \to X$,
3. Train P on this dataset. To evaluate its output n, it needs to be mapped into the image space, using G, with the loss defined as $L(X, G(P(f)))$ for a distance measure L.

The strategy comes with the obvious disadvantage that even though G remains frozen, the gradient needs to be propagated through G, making the training slower. We are also once again limited to the number of samples of our training dataset. Additionally, P is no longer encouraged to create n according to Gaussian noise. This difficulty can be largely remedied by adding a term to the overall loss function which we call *distribution loss*:

$$L_{\text{dist}}(n) = \frac{1}{|n|} \sum_{n_i \in n} n_i + |1 - \sigma(n)|, \tag{1}$$

which at least encourages distributions of n with zero mean and standard deviation $\sigma = 1$.

The drawbacks of image-based training are nevertheless outweighed by its advantages: the training is bound to a specific dataset - if G generates unrealistic images, they are punished by the loss function that compares the output of G to images from our training set, discouraging any n that would make G create such an image while pushing for n that can reflect the full diversity of the training set. We are further directly comparing images instead of noise, allowing for the advanced loss function presented in Sect. 3.3.

Note, that we are not using the StyleGAN generator trained with the original method from [20], but a version that is architecturally the same, but trained slightly differently as proposed in [42]. The training leads to generators which are more robust to input that does not perfectly adhere to a Gaussian distribution.

3.3   Image Similarity Loss

The basic loss function for the image-based training (Sect. 3.2) simply compares the pixels of the original and the recreated image. For $n = P(f)$, obtained for the feature vector $f = E(X)$, the pixel-wise loss L_{pixel} is given by

$$L_{\text{pixel}}(n) = (X - G(n))^2 \tag{2}$$

---

[3] In our experiments, this is always the same dataset that G was trained on.

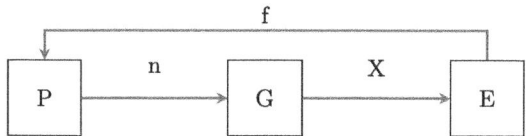

Fig. 2. The in- and outputs of the three components of the reconstruction task: the encoder E translates an image X into a feature vector f, the pre-generator P creates a generator input from the feature vector, and the generator G turns the input into an image. The combination of P and G makes up the decoder. As described in Sect. 3.2, our training method is based on keeping G frozen, but nonetheless comparing not n, but X to a defined target.

where G is the image of the pretrained image generator (see Fig. 2).

However, the goal of our attack is not to recreate every pixel but to identify the person seen in the image. Our training needs to focus on the information of the image that is relevant for a human to identify the depicted person, which is not easily defined. A reconstruction of a face that is taken from a different angle could still be identifiable. Contrarily, if a reconstruction gets the exact pixel values of 95% of the target image perfectly right, but fails for a few small, but significant areas, like the shape of the eyes, the depicted person could be unrecognizable. To train explicitly for our defined goal, we are instead comparing the embedding created by the FaceNet [30] network. Both target image and recreation are given to the embedding network, and the loss is given by the squared distance between them:

$$L_{\text{fn}}(n) = (F(X) - F(G(n)))^2. \tag{3}$$

If the embedding created for the target and reconstruction is similar, we can assume that the depicted people have similar identities. Other components of the image, like the pose or the background, are largely disregarded[4]. Less relevance is also given to the coloring of the image, including the color of hair and skin. Both depend on lighting conditions - the same person can appear to have a different hair color in a different image. Invariance to color is a desired property in the context of face recognition.

In contrast, for our reconstruction color of hair and skin is an important aspect. Thus, a good starting point for the FaceNet loss training is a P that has previously been trained to minimize the pixel-wise error - the pixel-wise error does not truly motivate similar facial features, but steers the overall image towards the right color composition.

---

[4] We found that even though FaceNet [30] is trained to ignore everything but the identity of the depicted person, images can end up recreating some parts of the pose or background even when trained with nothing but FaceNet embedding loss. This is in line with our main premise: additional information can leak into even very specialized networks.

In total, for a target image X and its corresponding output n from P, our loss is given by

$$L(n) = \lambda_p \cdot L_{\text{pixel}}(n) + \lambda_f \cdot L_{\text{fn}}(n) + \lambda_d \cdot L_{\text{dist}}(n). \qquad (4)$$

The distribution term is always weighted with a fixed $\lambda_d = 0.01$. Our training starts with pure pixel-wise loss ($\lambda_p = 1, \lambda_f = 0$) until the images no longer improve - we denote the resulting model P as L_{pixel}-model. Then the training is switched to facenet loss ($\lambda_p = 0, \lambda_f = 1$), continuing from the L_{pixel}-model, to compute the L_{fn}-model.

3.4 Using the Layers of a StyleGAN

A StyleGAN differs from other GANs by its *progressive* training method. Each network block is trained progressively for a different image resolution - starting at rough outlines in 4×4 images and ending with fine details created by the final block for the highest resolution. The strategy is meant to alleviate the drawback of convolutions and allow a convolutional neural network to model image-wide relations. By starting from tiny images, small enough to be covered by one convolutional kernel, global relations are (said to be) embedded into the lowest layers of the generator.

The training strategy is the reason for a very interesting property of a Style-GAN: each block defines different features of an image, features that can even be distinguished from a human perspective. In our examples, we are using a Style-GAN trained to create 256×256 images, whose architecture can be separated into seven blocks. Visible in the example in Fig. 3, its first block defines the pose of the depicted person, the second and third different facial features, and the fourth the background. Deeper layers are creating smaller details of the image.

Fig. 3. Stylemixing with a StyleGAN2, between an image A (leftmost image) and an image B (rightmost image). For each of the seven images between A and B, all generator inputs are the same as for image A, with the exception of a single vector that was taken from image B. For the i-th of those seven images, we replaced the i-th input vector with the B vector.

This allows to combine the L_{pixel}-model and the L_{fn}-model more robustly. We feed the output of P trained for L_{fn} to the second and third block, and the output of P trained for only L_{pixel} to the rest[5], deliberately assigning Ps optimized for different loss functions to different aspects of the image.

---

[5] Both L_{fn} and L_{pixel} are always used in combination with $0.01 \cdot L_{\text{dist}}$.

4   Experiments and Results

As the target for our attack, we train a ResNet18 [18] for 300 epochs, minimizing the mean squared distance between the model output and a target attribute vector. The output vectors are made up of 40 or less values, describing abstract properties of both face and image. There are only few publications with the same scenario as our attack, making the most direct competitor the method implemented by [40]. The result of their strategy for our setting is presented in the last row in Fig. 5 and 4. Code to recreate our results is available at https://github.com/ka-anderson/pge-reconstruction.

To compare our strategies quantitatively, we are assessing the squared distance (MSE) between images and feature vectors, in addition to the structural similarity index measure (SSIM) [36] between images and the similarity between face embedding extracted by VGG-Face [25] and OpenFace [1], face embedding networks similar to FaceNet that were not a part of our reconstruction training. Inspired by [33], we also calculate the distance scores between two distinct subsets of CelebA as a baseline, to find that if the images are from the same distribution, but otherwise completely different, the differences are at MSE 0.62, SSIM 0.06, VGG-Face 0.0005 and OpenFace 0.0037.

To the best of our knowledge, there is currently no perfect metric for the visual similarity of two portraits. A standard facial embedding disregards details, like the coloring of the image, while a measure created for the overall similarity of images weights all image areas equally. Additionally, while the statistics are useful to compare images from the same distribution, all are likely to be "fooled" by images from unexpected distributions, like the relatively blurry images from [40]. To compare our images to the previous state-of-the-art, we are instead evaluating the results of a designated user study (see Fig. 2): 19 participants are asked to determine the most likely original for a reconstruction, given a selection of five portraits that would have resulted in a similar attribute vector.

4.1   D and E Trained on the Same Dataset

In the scenario of an attack on an encoder that was trained on a public dataset, our E is trained to extract a 32-element attribute vector. The vector describes basic image and facial properties (like emotion, hair and image brightness), in addition to the position and angle of the depicted face[6]. Both E and D (i.e. the frozen image generator G and the pregenerator P) are trained on FFHQ. To evaluate the attack, we are feeding new images from CelebA to E and then applying our trained reconstruction networks to recreate the original image from the feature vector. Figure 4 presents a qualitative evaluation of nine randomly selected images.

Even though the FaceNet loss started from images that were approximately correct in terms of skin and hair color, the reconstruction is sometimes shifted

---

[6] The features are extracted from https://github.com/DCGM/ffhq-features-dataset.

towards a different coloring over time. The difference becomes especially evident from the fact that the FFHQ dataset exhibits some degree of diversity in terms of skin color, which CelebA is lacking entirely - while all nine random CelebA images are taken from subjects with light skin, some of the FaceNet reconstructions are different. The stylemixing fusion strategy helps to maintain the overall coloring scheme extracted by the pixel loss, combining them with the more detailed facial features extracted by the embedding loss. Looking at the recreation for the last random target image (last column) reveals that there is a small chance that our process fails entirely, maybe caused by an erroneous feature vector, maybe by an unusual combination of attributes.

The results by [40] are also recreating basic attributes, however, all reconstructed images are blurry and visually close to the same "average face". While their images usually represent attributes like age and pose correctly, additional information like the shape of the eyes, face and mouth, is barely distinguishable in their images. The images created by our predecessor do come with one advantage: because the images are not bound to the distribution of actual human faces, the scores for common image evaluation methods are quite high, indicating that they could be more useful when attempting to fool a facial recognition software (MSE 0.078, SSIM 0.403, VGG-Face 0.00015, OpenFace 0.0047). Seen in Fig. 2, both strategies achieve a maximum user recognition rate of 80%. In terms of average recognition rate and number of false answers ("subtract false answers"), our method outperforms the predecessor.

The metrics shown in Table 1 are in line with our qualitative evaluation: the pixel loss minimizes the difference between images pixels (visible in both mean squared distance and SSIM), while the FaceNet loss minimizes the loss between facial embeddings, though sacrificing pixel accuracy. The combination of both results in almost the same pixel distance as pure pixel based loss training, but with a lower embedding distance. Notable is that the stylemixing significantly decreases the distance between feature vectors - the images are most similar from the perspective of the encoder.

Interestingly, even though hair color is not included in the attribute vector, the reconstruction manages to distinguish "brown" and "not brown" hair with surprising reliability.

4.2   D and E Trained on Different Datasets

For the second setting, we evaluate our attack against an encoder that was trained to extract a 40-element binary attribute vector on CelebA, while our decoder is still trained on FFHQ. Note that the attribute vectors of this experiment include no information about the position or angle of the depicted face. The reconstruction is evaluated on a subset of CelebA that was never seen by E. Hence, D is not only trained on different images than E but even on images drawn from a different distribution.

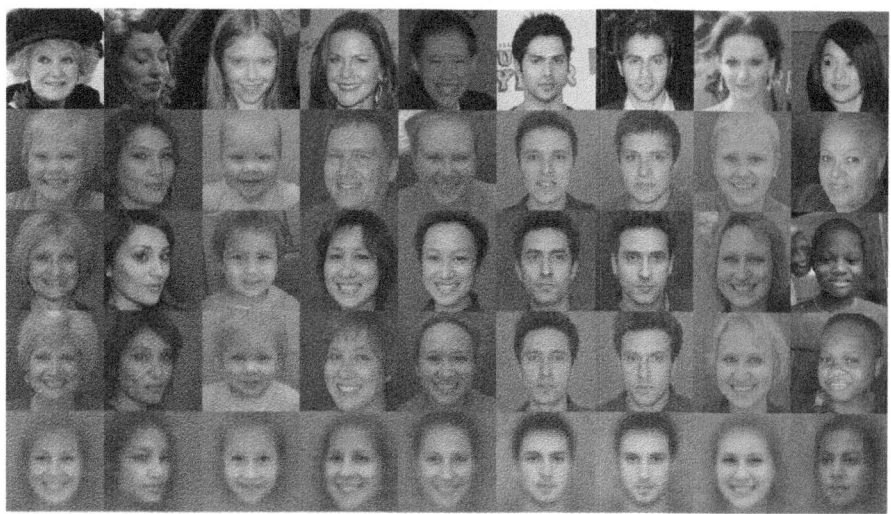

Fig. 4. Reconstructions for random images of the CelebA dataset, for a D and E both trained on FFHQ. From top to bottom, the rows present the target images, the result of the pixel-wise loss, the FaceNet embedding loss, and the fused version of the two. The final row displays the outcome of our predecessor, [40]. Details about training can be found in Sect. 4.1.

It is clear from both the quantitative evaluation of randomly selected example images (Fig. 5) and the calculated statistics (Table 1 and Fig. 2) that this task is more difficult. However, while the pixelwise loss is sometimes unable to create a similar image, the FaceNet loss usually helps the model project matching facial features onto the sometimes otherwise mismatched portraits, proving that this loss formulation is more stable to the challenge of diverging datasets.

The attribute vector used for this reconstruction setting does not include information about the pose of the depicted person. Interestingly, for the images taken from the side, the recreated person is photographed from a similar angle.

This information, e.g., is not extracted by [40]: all images are from a frontal perspective, and all are slight variations of an average human portrait, providing much less information about exact facial features than our reconstruction. But, like for the previous setting, scores like MSE and SSIM are higher - possibly because using the "average" for many pixels allows to minimize the loss, despite the absence of distinguishing features (MSE 0.075, SSIM 0.426, VGG-Face 0.00014, OpenFace 0.0050).

Fig. 5. Reconstructions for random images of the CelebA test set, for a D trained on FFHQ and E trained on the CelebA training set. From top to bottom, the rows present the target images, the result of the pixelwise loss, the FaceNet embedding loss, the fused version of the two and the results from [40]. Details about training can be found in Sect. 4.2.

Table 1. Quantitative evaluation of our reconstruction of CelebA test images for D trained on FFHQ, and E on FFHQ (top) and the CelebA training set (bottom).

		MSE		SSIM	VGG-Face	OpenFace
		X	f			
FFHQ	L_{pixel}	0.3535	7.8280	0.1200	0.0003272	0.004061
	L_{fn}	0.4017	4.5761	0.1247	0.0003367	0.003797
	mixed	0.3417	3.3914	0.1302	0.0003138	0.003723
CelebA	L_{pixel}	0.4387	0.2059	0.0934	0.0003947	0.003992
	L_{fn}	0.4399	0.1759	0.1016	0.0003631	0.003934
	mixed	0.4302	0.1744	0.0972	0.0003967	0.004053

Table 2. Results of a user study for the two scenarios (see Sect. 4.1 for the "FFHQ" 4.2 for "CelebA"), by our method and [40]. The score for an image is determined by the number of correct answers. In the second column, we are counting a correct answer as one, no answer as zero, and a false answer as minus one.

		% correct answers		% correct - % incorrect	
		best image	mean	best image	mean
FFHQ	ours	79	42	79	15
	Yang	79	29	73	-22
CelebA	ours	68	36	63	11
	Yang	53	39	16	5

5   Conclusion

This work introduced a novel reconstruction pipeline, operating with only a low-dimensional vector representing high-level attributes of a portrait and black-box access to the encoder network. The reconstruction surpassed the perceptual quality of previous work in this setting by leveraging recent advances in image generation and facial identification. A pivotal step involved merging the model trained on face embedding loss for mere individual identity with the model from pixel loss for finer image details. This integration was achieved by directing their outputs into distinct layers of the StyleGAN.

Our results show that already a low-dimensional attribute vector and only black-box access to the encoder network can allow unexpected conclusions about the original image - most notable the identity of the depicted person. In a user study we observed recognition rates up to 79%. But also additional details like the angle from which the portrait was taken can be extracted. Despite being intended to encode nothing but a small number of abstract features, the target network inadvertently captures more information from its input than just those features. It's crucial to recognize that a neural network, when employed conventionally, can convey much more information from a high-dimensional input to a low-dimensional output than expected. Consequently, the output of a neural network must be treated with the same level of care as the input.

References

1. Baltru?aitis, T., Robinson, P., Morency, L.P.: Openface: an open source facial behavior analysis toolkit. In: 2016 IEEE Winter Conference on Applications of Computer Vision (WACV) (2016)
2. Bertran, M., et al.: Adversarially learned representations for information obfuscation and inference. In: ICLR (2019)
3. Bosch, N., et al.: Detecting student emotions in computer-enabled classrooms. In: Proceedings of the Twenty-Fifth International Joint Conference on Artificial Intelligence (2016)
4. Chen, S., Kahla, M., Jia, R., Qi, G.J.: Knowledge-enriched distributional model inversion attacks. In: ICCV, pp. 16158–16167 (2020)
5. Deng, J., Guo, J., Xue, N., Zafeiriou, S.: Arcface: additive angular margin loss for deep face recognition. In: CVPR (2019)
6. Ding, X., Fang, H., Zhang, Z., Choo, K.K.R., Jin, H.: Privacy-preserving feature extraction via adversarial training. IEEE Trans. Knowl. Data Eng., 1967–1979 (2020)
7. Dionysiou, A., Vassiliades, V., Athanasopoulos, E.: Exploring model inversion attacks in the black-box setting. In: Proceedings on Privacy Enhancing Technologies, pp. 190–206 (2023)
8. Dong, X., Jin, Z., Guo, Z., Jin Teoh, A.B.: Towards generating high definition face images from deep templates. In: BIOSIG, pp. 1–11 (2021)
9. Dong, X., et al.: Reconstruct face from features based on genetic algorithm using gan generator as a distribution constraint. Comput. Sec. (2023)

10. Dosovitskiy, A., Brox, T.: Generating images with perceptual similarity metrics based on deep networks. In: Lee, D., Sugiyama, M., Luxburg, U., Guyon, I., Garnett, R. (eds.) Advances in Neural Information Processing Systems (2016)
11. Dosovitskiy, A., Brox, T.: Inverting visual representations with convolutional networks. In: CVPR (2016)
12. Duong, C.N., Truong, T.D., Luu, K., Quach, K.G., Bui, H., Roy, K.: Vec2face: Unveil human faces from their blackbox features in face recognition. In: CVPR, pp. 6132–6141 (2020)
13. Esteva, A., et al.: Dermatologist-level classification of skin cancer with deep neural networks. Nature **542**(7639) (2017)
14. Fredrikson, M., Jha, S., Ristenpart, T.: Model inversion attacks that exploit confidence information and basic countermeasures. In: ACM SIGSAC Conference on Computer and Communications Security (CCS), pp. 1322–1333 (2015)
15. Gao, H., Y?ce, A., Thiran, J.P.: Detecting emotional stress from facial expressions for driving safety. In: ICIP (2014)
16. Goodfellow, I., Shlens, J., Szegedy, C.: Explaining and harnessing adversarial examples. In: ICLR (2015)
17. Han, G., Choi, J., Lee, H., Kim, J.: Reinforcement learning-based black-box model inversion attacks. In: CVPR, pp. 20504–20513 (2023)
18. He, K., Zhang, X., Ren, S., Sun, J.: Deep residual learning for image recognition. In: CVPR, pp. 770–778 (2016)
19. Kahla, M., Chen, S., Just, H.A., Jia, R.: Label-only model inversion attacks via boundary repulsion. In: CVPR, pp. 15025–15033 (2022)
20. Karras, T., Laine, S., Aittala, M., Hellsten, J., Lehtinen, J., Aila, T.: Analyzing and improving the image quality of stylegan. In: CVPR, pp. 8110–8119 (2020)
21. Mahendran, A., Vedaldi, A.: Understanding deep image representations by inverting them. In: CVPR (2015)
22. Mai, G., Cao, K., Yuen, P.C., Jain, A.K.: On the reconstruction of face images from deep face templates. IEEE Trans. Pattern Anal. Mach. Intell., 1188–1202 (2019)
23. Nash, C., Kushman, N., Williams, C.K.I.: Inverting supervised representations with autoregressive neural density models. In: International Conference on Artificial Intelligence and Statistics (2018)
24. Osia, S.A., Taheri, A., Shamsabadi, A.S., Katevas, K., Haddadi, H., Rabiee, H.R.: Deep private-feature extraction. IEEE Trans. Knowl. Data Eng., 54–66 (2018)
25. Parkhi, O.M., Vedaldi, A., Zisserman, A.: Deep face recognition. In: British Machine Vision Conference (2015)
26. Pittaluga, F., Koppal, S., Chakrabarti, A.: Learning privacy preserving encodings through adversarial training. In: 2019 IEEE Winter Conference on Applications of Computer Vision (WACV), pp. 791–799. IEEE (2019)
27. Raval, N., Machanavajjhala, A., Cox, L.P.: Protecting visual secrets using adversarial nets. In: CVPR Workshops, pp. 1329–1332 (2017)
28. Ren, Z., Lee, Y.J., Ryoo, M.S.: Learning to anonymize faces for privacy preserving action detection. In: Ferrari, V., Hebert, M., Sminchisescu, C., Weiss, Y. (eds.) ECCV 2018. LNCS, vol. 11205, pp. 639–655. Springer, Cham (2018). https://doi.org/10.1007/978-3-030-01246-5_38
29. Roy, P.C., Boddeti, V.N.: Mitigating information leakage in image representations: A maximum entropy approach. In: CVPR, pp. 2586–2594 (2019)
30. Schroff, F., Kalenichenko, D., Philbin, J.: Facenet: a unified embedding for face recognition and clustering. In: CVPR, pp. 815–823 (2015)
31. Shahreza, H.O., Hahn, V.K., Marcel, S.: Face reconstruction from deep facial embeddings using a convolutional neural network. In: ICIP, pp. 1211–1215 (2022)

32. Simonyan, K., Vedaldi, A., Zisserman, A.: Deep inside convolutional networks: visualising image classification models and saliency maps. In: ICLR Workshops (2014)
33. Tinsley, P., Czajka, A., Flynn, P.: This face does not exist... but it might be yours! identity leakage in generative models, pp. 1319–1327
34. Vendrow, E., Vendrow, J.: Realistic face reconstruction from deep embeddings. In: NeurIPS 2021 Workshop Privacy in Machine Learning (2021)
35. Wang, J., Zhang, J., Bao, W., Zhu, X., Cao, B., Yu, P.S.: Not just privacy: Improving performance of private deep learning in mobile cloud. In: ACM SIGKDD International Conference on Knowledge Discovery & Data Mining, pp. 2407–2416 (2018)
36. Wang, Z., Bovik, A.C., Sheikh, H.R., Simoncelli, E.P.: Image quality assessment: From error visibility to structural similarity. Trans. Img. Proc. p. 600–612 (2004)
37. Whitehill, J., Serpell, Z., Lin, Y.C., Foster, A., Movellan, J.R.: The faces of engagement: automatic recognition of student engagementfrom facial expressions. IEEE Trans. Affective Comput. (2014)
38. Wu, Z., Wang, H., Wang, Z., Jin, H., Wang, Z.: Privacy-preserving deep action recognition: an adversarial learning framework and a new dataset. IEEE Trans. Pattern Anal. Mach. Intell., 2126–2139 (2020)
39. Wu, Z., Wang, Z., Wang, Z., Jin, H.: Towards privacy-preserving visual recognition via adversarial training: a pilot study. In: Ferrari, V., Hebert, M., Sminchisescu, C., Weiss, Y. (eds.) ECCV 2018. LNCS, vol. 11220, pp. 627–645. Springer, Cham (2018). https://doi.org/10.1007/978-3-030-01270-0_37
40. Yang, Z., Zhang, J., Chang, E.C., Liang, Z.: Neural network inversion in adversarial setting via background knowledge alignment. In: ACM SIGSAC Conference on Computer and Communications Security, pp. 225–240 (2019)
41. Zhang, Y., Jia, R., Pei, H., Wang, W., Li, B., Song, D.X.: The secret revealer: generative model-inversion attacks against deep neural networks. In: CVPR, pp. 250–258 (2019)
42. Zhao, S., Liu, Z., Lin, J., Zhu, J.Y., Han, S.: Differentiable augmentation for data-efficient gan training. Adv. Neural. Inf. Process. Syst. **33**, 7559–7570 (2020)
43. Zhao, X., Zhang, W., Xiao, X., Lim, B.Y.: Exploiting explanations for model inversion attacks. In: ICCV, pp. 662–672 (2021)

Safe Data Resampling Method Based on Counterfactuals Analysis

Diwen Liu[1], Xiaodong Yue[2,3(✉)], and Zhikang Xu[4]

[1] School of Computer Engineering and Science, Shanghai University, Shanghai, China
[2] Artificial Intelligence Institute, Shanghai University, Shanghai, China
[3] VLN Lab, NAVI MedTech Co., Ltd., Shanghai, China
`yswantfly@shu.edu.cn`
[4] School of Computer and Information Technology, Shanxi University, Shanxi, China

Abstract. The challenges of limited data availability and imbalance can significantly affect model performance, providing a strong motivation for developing robust data resampling strategies. However, existing resampling methods generally neglect the fact that different data samples and features have different importance, which can lead to irrelevant or incorrect resampled data. Counterfactual analysis aims to identify the minimum feature changes required to flip a model decision. Through this approach, it is possible to precisely measure the impact of each feature on the decision and evaluate the ease of flipping prediction of one single sample. Inspired by this, we propose two types of safeness evaluation metrics based on counterfactual instances to measure the safeness of features and samples, respectively. Then, we can achieve high quality data resampling by selecting safe features and samples, or by changing feature values within safe intervals. In addition, the proposed safeness evaluation metrics can be seamlessly integrated into existing data resampling methods to further enhance the performance. Experimental results show that our resampling method improves the data diversity while reducing the noise introduced by resampled data, thereby achieving safe resampling.

Keywords: Safe Resampling · Data Resampling · Counterfactuals · Safeness Metrics

1 Introduction

In numerous real-world application areas, such as finance and healthcare, the limited available data leads to the creation of small datasets with biased and unbalanced class distributions [2]. To tackle this challenge, data augmentation methods are promising strategies to overcome the limitations, potentially driving major breakthroughs in artificial intelligence.

In recent years, some effective data augmentation techniques have been introduced [6,16], such as resampling methods, GANs and counterfactuals-based methods. Resampling-based methods [5,11] typically generate new samples using simple interpolation strategies, resulting in a lack of diversity. In contrast, methods based on GANs [7] use adversarial learning to model the data distribution,

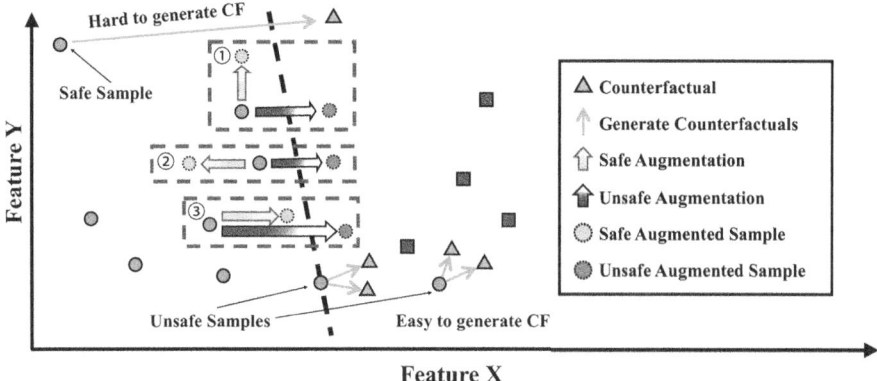

Fig. 1. An example of illustrating safe and unsafe augmentation.

which can significantly increase the diversity of generated samples. Moreover, counterfactual-based methods [9,14,15] directly change the feature values of a sample until the class label is flipped, and then the resulting counterfactual sample is used as the new augmented sample. However, these methods neglect the safeness of data augmentation, which can lead to the introduction of noisy data with potential impact on downstream tasks such as classification.

To address this issue, instead of directly augmenting new data by changing the feature values of samples, we use counterfactuals as a metric to measure the safeness or risk of data augmentation. As shown in Fig. 1, the safeness can be divided into sample and feature aspects. In terms of sample safeness, samples at classification boundaries or within other class distributions carry high risk and are unsafe for augmentation. In addition, the safeness of features can be measured from three granularities. Firstly, different features have different degrees of safeness. Changes in feature **X** (horizontal) may flip the class label, whereas any changes in feature **Y** (vertical) have minimal impact. Secondly, the direction of change in a single feature has different degrees of safeness. For positive samples (denoted as blue circles), a change to the left is safe, while a change to the right is unsafe. Thirdly, changing interval in the same direction has different degrees of safeness. Significant changes may introduce greater risk, while minor adjustments remain safe. It can be seen that the safeness of samples and features is closely related to the difficulty in generating counterfactuals. Therefore, utilizing counterfactual instances to derive safeness metrics is highly effective and intuitive.

In this paper, we propose a safe data resampling method based on counterfactuals which can increase data diversity while reducing the noise introduced by resampled data. Specifically, based on the generation of counterfactuals, we propose two types of safeness evaluation metrics to measure the safeness of features and samples, respectively. For safe samples and features, extensive modifications can be undertaken during data augmentation. In contrast, for others, minimum changes or even no changes should be made. In addition, these metrics are adaptable and can be seamless integrated into the existing data resampling methods to further enhance the performance. The contributions of this research are as follows:

– *Propose a safe resampling method based on the counterfactuals.* Inspired by the counterfactuals, the safeness of data resampling can be measured by the ease of flipping the class labels. Then, by selecting safe samples or modifying feature values within a relatively safe ranges, we can achieve safe resampling. In addition, the proposed method can be seamlessly integrated in the existing data resampling methods, allowing for generating safe and diverse data samples.
– *Propose two types of evaluation metrics for measuring the safeness of data resampling.* One metric is used to measure the safeness of feature from three different perspectives, including feature stability, feature sync-variation, and the range of feature changes. The other metric is used to evaluate the safeness of individual samples.

The rest of the paper is organized as follows. Section 2 provides a review of related work. Section 3 introduces the proposed method. Section 4 validates the effectiveness of the proposed method through comparative and ablation experiments. Section 5 gives the conclusion.

2   Related Works

2.1   Data Augmentation

Existing data augmentation methods can be roughly divided into three types, methods based on resampling, methods based on GANs and methods based on counterfactual analysis.

Resampling is a effective solution for addressing imbalanced datasets [5,11]. For example, SMOTE [3] is a popular resampling method for alleviating the imbalanced data distribution through interpolation between minority class samples. However, it may produce erroneous samples due to indiscriminate resampling. Based on this, Borderline-SMOTE [8] focuses on border samples between classes. SVM-SMOTE [19] uses SVM to identify and augment boundary samples. ADASYN [10] generates synthetic samples based on minority class density. However, these resampling methods may lack some diversity.

Methods based on GANs uses adversarial learning to model the distribution of data, and can significantly enhance the diversity of data augmentation, such as CTGAN [20] and WGAN [1]. However, these methods overly rely on the discriminator and may not able to accurately model the original data distribution, leading to the generation of samples with incorrect class labels, especially when the dataset contains noisy samples.

Similarly, methods based on counterfactuals directly add the generated counterfactual instances to the dataset as augmented samples to achieve data augmentation. [9,14,15]. However, these methods do not consider the underlying data structure, and the efficacy is closely to the capability of the counterfactual generator. This can lead to convergence to local optima and does not improve diversity of the data distribution.

Existing data augmentation methods generally neglect the safeness of the generated new data. This can lead to the introduction of unsafe and noisy data, thereby potentially affecting downstream tasks such as classification.

2.2 Counterfactuals Analysis

Counterfactual analysis, a model-agnostic interpretation approach as described by Verma et al. [4,17], focuses on how a model's output changes with modifications to its inputs. Wachter et al. [18] introduced an objective function to determine the minimum feature change required to flip the model prediction, forming the basis of counterfactual explanation [12]. The objective function can be formalized as follows:

$$L(x, c, y_c, \lambda) = \lambda \cdot (\hat{f}(c) - y_c)^2 + d(x, c), \tag{1}$$

$$c = \arg \min_c \max_\theta L(x, c, y_c, \lambda), \tag{2}$$

where $(\hat{f}(c) - y_c)^2$ is the distance between the model prediction of the counterfactual instance c and the predefined label y_c, and $d(x, c)$ is the distance between x and c. Additionally, the DiCE method [13] addresses the generation of multiple diverse counterfactuals by introducing a diversity term to the loss function. Additionally, it can also ensure that specific features remain unchanged while generating counterfactuals. Inspired by this, we adopt it as the counterfactual generator in this paper.

3 Safe Data Resampling Method Based on Counterfactuals Analysis

In this section, we introduce how to implement safe data resampling based on counterfactuals. For simplicity, the method we propose will be referred to as 'Safe-Resampling' in the following. As shown in Fig. 2, we evaluate the safeness of data augmentation from the perspective of samples and features, where feature-based safeness metrics includes three different perspectives.

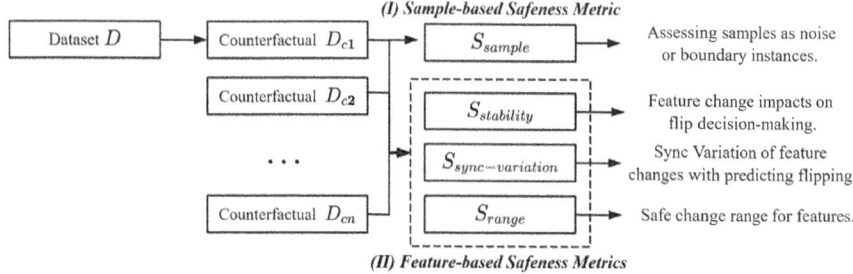

Fig. 2. Two types of safeness metric based on counterfactuals.

By combining the two kinds of safeness metrics, we can achieve safe data resampling. The overall framework is shown in Fig. 3. Initially, we employ the sample-based metric to filter the original dataset to remove noisy data. Next, we utilize feature-based metrics to perform data resampling. Then, the safeness values of each generated sample are recomputed, and only the samples that satisfy the safeness metric are retained, while the rest are regenerated.

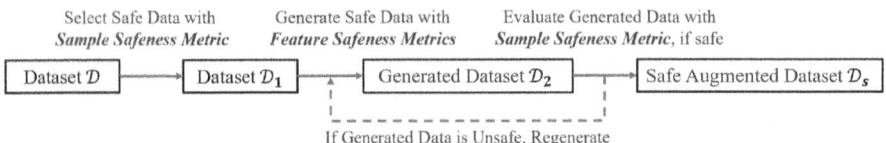

Fig. 3. Framework of safe data resampling based on counterfactuals.

3.1   Sample-Based Safeness Metric

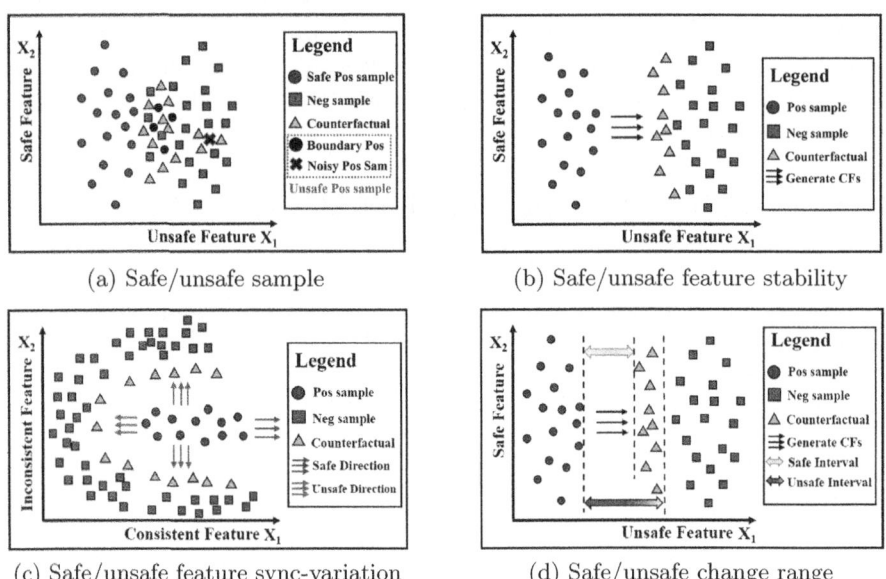

(a) Safe/unsafe sample

(b) Safe/unsafe feature stability

(c) Safe/unsafe feature sync-variation

(d) Safe/unsafe change range

Fig. 4. Safe/unsafe situation diagram.

Traditional data augmentation indiscriminately alters all samples in a dataset, which does not account for mislabeled or noisy samples. Augmenting these samples can introduce inaccurate and negative effects in classifiers training (As shown in Fig. 4(a)). Therefore, it is crucial to either minimize or entirely avoid augmenting samples that are unsafe or near classification boundaries.

Unsafe samples, typically overlapping with other categories, enable easier counterfactual generation. This capability in generating counterfactuals effortlessly can be a metric for sample safeness. In our approach, we define the sample safeness metric by taking into account both minimum number of feature variations required to flip the class label and the distance between the counterfactual instance and the original data point.

Definition 1. *For sample x_i and counterfactual c_i, nf_i is the number of features that can generate counterfactuals by changing slightly. S_{sample} can be defined as*

$$S_{sample_i} = \|\mathbf{c_i} - \mathbf{x_i}\|^2 \cdot nf_i \tag{3}$$

A high S_{sample_i} value implies that generating counterfactuals for $\mathbf{x_i}$ is challenging, which means that the sample is considered safe. After calculating S_{sample} for each sample, we set a parameter t to exclude samples under this threshold, preventing the production of inaccurate samples and realizing dataset cleaning.

3.2   Feature-Based Safeness Metrics

In this section, we introduce three feature safeness metrics, namely $S_{stability}$, $S_{sync-variation}$ and S_{range}. These metrics are determined by evaluating the frequency of feature change, the consistency of values during feature changes(sync-variation), and the degree of variation seen in features while generating counterfactuals. To demonstrate safe and unsafe scenarios, we present schematic 2D simulation diagrams depicted in Fig. 4(b)-(d).

Safeness Metric Based on Feature Stability. Counterfactual generation typically involves minimal changes in a few selected features. In unconstrained scenarios, certain features are frequently altered, suggesting a significant impact on model predictions. In addition, some counterfactuals generated only depend on a specific feature with small changes, representing that this feature has an independent impact on the model prediction. A feature is deemed unsafe if it meets either of two criteria: (a) it is often changed in counterfactuals or (b) it can independently alter the predictions with minimal adjustments. Therefore, our study explores the aforementioned scenarios through two distinct counterfactuals generation approaches, including one method that generates counterfactuals without any restrictions and another that focuses on change to a single feature.

Generate Counterfactuals without Restrictions. For a training dataset $\mathcal{X} = (\mathbf{x_1}, \mathbf{x_2}, \cdots, \mathbf{x_n})^T$, where $\mathbf{x_i}$ is a $1 \times \mathbf{m}$ vector. By generating $\mathbf{nc}$ counterfactuals for each data in $\mathcal{X}$, we obtain a counterfactual dataset $\mathcal{C} = (\mathbf{c_1}, \mathbf{c_2}, \cdots, \mathbf{c_n})$.

Dataset $\mathcal{X}$ can be rewritten as an $\mathbf{n} \times \mathbf{m}$ matrix $\mathbf{X}$. Assuming $nc = 1$, we obtain a matrix $\mathbf{C}$ with same size as $\mathbf{X}$. Subtracting matrix $\mathbf{X}$ from matrix $\mathbf{C}$ can derive the difference matrix $\mathbf{R}$

$$\mathbf{R} = \mathbf{C} - \mathbf{X} = (r_{ij}), \tag{4}$$

where r_{ij} refers to the difference between sample i and its corresponding counterfactual on the feature j. By counting the number of non-zero elements in the jth column, the change frequency of feature j can be obtained by

$$\ell_j = \sum_{i=0}^{n} \mathbb{1}(r_{ij} \neq 0). \tag{5}$$

Generate Counterfactuals with Single Feature Change. For the training dataset $\mathcal{X}$, we can generate counterfactuals by changing only the jth feature in a single generation while keeping the remaining features constant. Consequently, we obtain m counterfactual sets $(\mathcal{C}^1, \mathcal{C}^2, \cdots, \mathcal{C}^m)$, with each set corresponding to a different feature. Then, we count the number of counterfactual instances within each counterfactual set to derive a vector k for each feature

$$k_j = |\mathcal{C}^j|. \tag{6}$$

where $|\mathcal{C}^j|$ denotes the number of counterfactual instances in the set corresponding to the jth feature. Based on this, we provide the definition of feature stability.

Definition 2. *Feature stability measurement $S_{stability}$ can be calculated by ℓ_j and k_j as follows*

$$S_{stability\,j} = \left[\frac{H_j - \min(H) + \epsilon}{\max(H) - \min(H) + \epsilon} \right]^2, \tag{7}$$

where

$$H_j = 1 - \left[\alpha \cdot \frac{\ell_j}{\sum_{j=0}^{m} \ell_j} + (1 - \alpha) \cdot \frac{k_j}{\sum_{j=0}^{m} k_j} \right], \tag{8}$$

where α and ϵ are two hyperparameters. α is used to adjust the weighting of two components, and ϵ is used to reduce alteration probability for less stable features, thereby enhancing sample diversity. During the resampling process, $S_{stability}$ is used to modulate the probability of feature changes, ensuring the safeness of the resampled data.

Safeness Metric Based on Feature Sync-Variation. During counterfactual generation for some samples, we observe synchronized trends of 'increase' or 'decrease' in specific features, indicating that direction changes could substantially impact model predictions. To avoid mislabeling in augmented data, when the direction of feature change in counterfactuals is uniform, this feature is considered safe. Note that, even when the direction of feature changes varies in some counterfactuals, minor and consistent adjustments can still ensure augmentation safeness, as will be discussed in the following sections.

Note that not all samples in a dataset can generate counterfactuals successfully when altering only one feature. Rearrange the original dataset, which can generate counterfactual instances, into new datasets $(\mathcal{X}^1, \mathcal{X}^2, \cdots, \mathcal{X}^m)$, with corresponding counterfactual sets denoted as $(\mathcal{C}^1, \mathcal{C}^2, \cdots, \mathcal{C}^m)$. Due to the fact that the counterfactual sample matrix $\mathcal{C}$ and the original sample matrix $\mathcal{X}$ differ only in a single feature variation, we can simplify these matrices into two vector sets $(\mathbf{x}^1, \mathbf{x}^2, \cdots, \mathbf{x}^m)$ and $(\mathbf{c}^1, \mathbf{c}^2, \cdots, \mathbf{c}^m)$, representing original samples and their counterfactuals. Based on this, $S_{sync-variation}$ is derived from analyzing the overall trend of feature changes during the counterfactual generation.

Definition 3. *Suppose the number of generated counterfactuals is nc^j, $\mathbf{x}^j = (x_1^j, x_2^j, \cdots, x_{nc^j}^j)$ and $\mathbf{c}^j = (c_1^j, c_2^j, \cdots, c_{nc^j}^j)$, $S_{sync-variation}$ is defined as*

$$S_{sync-variation\,j} = \frac{|n_+^j - n_-^j|}{nc^j}, \tag{9}$$

where

$$n_+^j = \sum_{i=0} \mathbb{1}(c_i^j > x_i^j), \quad n_-^j = \sum_{i=0} \mathbb{1}(c_i^j < x_i^j). \tag{10}$$

Data Set			Counterfactual Set		
No.	x	y	No.	x	y
1	3.1	0	1	3.6 (+0.5)	1
2	3.5	0	2	3.8 (+0.3)	1
3	3.6	0	3	3.2 (-0.4)	1
4	3.2	0	4	3.5 (+0.3)	1
5	3.5	0	5	3.3 (-0.2)	1

$$n_+ = 3, \quad n_- = 2, \quad S_{sync-variation} = 0.2$$

Fig. 5. Illustration for the sync-variation metric.

An illustrative and accessible example of the Sync-Variation metric is shown in Fig. 5. $S_{sync-variation\,j} \in [0, 1]$ assesses the synchrony of changes in feature j during counterfactual generation. Values near 1 indicate consistent modification patterns, representing the feature is safe for augmentation, while values near 0 suggest unsafe and asynchronous changes, increasing the risk of prediction flipping. Careful modification of such features is crucial for maintaining data safeness. By comparing the values of n_+^j and n_-^j, we can determine the modification direction. When the value of n_+^j is greater, decreasing the feature preserves the label, whereas a larger value of n_-^j suggests the opposite. A larger value of $S_{sync-variation\,j}$ will allow for a greater range of feature modification when defining the safe range in the subsequent sections.

Considering that the stability score $S_{stability}$ of feature j proposed in definition 2, features with higher stability scores have a reduced impact on changing category predictions, we can enhance data diversity and feature flexibility by setting the consistency score $S_{sync-variation\,j}$ to 1 when $S_{stability\,j}$ exceeds the threshold t'

$$S_{sync-variation\,j} = 1, \quad if\ S_{stability\,j} > t'. \tag{11}$$

The primary application of $S_{sync-variation}$ is to guide the direction and range of feature changes during data resampling, thereby reducing mislabeling risks.

Safeness Metric Based on Feature Change Range. Except $s_{sync-variation}$, the discrepancy between original data and counterfactuals can also be viewed as a safeness measure. Inspired by this, we use standard deviation to measure the data variability and thereby setting boundaries for feature alterations. A higher standard deviation for a feature indicates greater variability, which suggests a reduction in safeness. Given the normalized data $\mathbf{r}^j = \mathbf{c}^j - \mathbf{x}^j = (r_1^j, r_2^j, \cdots, r_{nc^j}^j)$, the standard deviation is defined as

$$\sigma = \sqrt{\frac{\sum_{i=1}^{nc^j}(r_i^j - \bar{r}^j)^2}{nc^j}}. \tag{12}$$

This metric effectively quantifies the dispersion of changes in counterfactual generation. However, it might not be optimal for resampling since it treats increments and decrements equally across each example. To improve this metric, we incorporate $S_{sync-variation}$ to determine safe change ranges and appropriate sampling intervals for data augmentation.

The safe change range is defined by the values of $S_{sync-variation}$ and $S_{stability\,j}$. Then, the discussion can be segmented into three distinct cases: 1) $S_{sync-variation\,j} = 1$ and $S_{stability\,j} > t$; 2) $S_{sync-variation\,j} = 1$ and $S_{stability\,j} \le t$; 3) $S_{sync-variation\,j} < 1$. Based on this, the final safe change ranges are defined as follows.

Definition 4. *Give vector $\mathbf{r}^j = \mathbf{c}^j - \mathbf{x}^j$, where $\mathbf{r}_+^j$ encompasses all positive elements of $\mathbf{r}^j$, and $\mathbf{r}_-^j$ includes all negative elements, we have*

$$r_{max,+}^j = max(\mathbf{r}_+^j), \quad r_{min,+}^j = min(\mathbf{r}_+^j). \tag{13}$$

$$r_{max,-}^j = max(|\mathbf{r}_-^j|), \quad r_{min,-}^j = min(|\mathbf{r}_-^j|). \tag{14}$$

Based on this, we define S_{range} as follows

$$\begin{cases} [-r_{max,-}^j, \ r_{max,+}^j\,], & \boldsymbol{Ssync-v}_j=1 \ and\ \boldsymbol{Sstability}_j > t, \\ [\ r_{min,-}^j, \ r_{max,+}^j\,], & \boldsymbol{Ssync-v}_j=1 \ and\ n_+^j = 0, \\ [\ r_{max,-}^j, \ r_{min,+}^j\,], & \boldsymbol{Ssync-v}_j=1 \ and\ n_-^j = 0, \\ [\ r_{min,-}^j, \ r_{min,+}^j\,], & \boldsymbol{Ssync-v}_j < 1. \end{cases} \tag{15}$$

where $[\cdot, \cdot]$ denotes S_{range}. Defining S_{range} is the key for model robustness in safe resampling. Integrating it with $S_{sync-variation}$ and $S_{stability}$ enables safe and diverse data augmentation. For instance, when $S_{stability} > t$ and $|S_{sync-variation}| \approx 1$, we maximize the safe range in both directions to enhance diversity. If sync-variation changes are unidirectional ($n_+^j = 0$ or $n_-^j = 0$), we increase the safeness range in the counterfactual's opposite direction and minimize it in the same direction for greater safeness. Additionally, when $S_{sync-variation}$ is low, all modifications should be limited to the smallest range.

In summary, feature stability is defined based on the frequency of feature changes in counterfactuals and can be employed to determine the probability of feature modifications during resampling. Feature synchronicity-variation and the safe change range are delineated by the specific characteristics of feature modifications observed in the counterfactual generation process, which can guide the direction and intervals of feature adjustments in resampling.

3.3 Safe Resampling Algorithm

In this section, we introduce the safe resampling algorithm as detailed in Algorithm 1. The first step of the algorithm involves employing the sample safeness metric to filter out relatively safe data from the dataset designated for augmentation. It then guides the data resampling process in accordance with the feature safeness metrics. Finally, the generated data is reassessed using the sample safeness metric, and the data that exhibit higher safeness levels are retained.

Algorithm 1. Safe-Resampling.

1: **procedure** (X)
2: **Input:**$S_{stability}$; $S_{sync-variation}$; S_{range}; S_{sample}.
3: **Output:** Safe augmented dataset X_{aug}.
4:     Determine S_{range} by $S_{stability}$ and $S_{sync-variation}$ (defined above)
5:     **if** $S_{sample_i} <$ *threshold t* **then**
6:         Remove x_i from X.(* Clean the dataset *)
7:     **end if**
8:     **for** *Each feature attr* $\leftarrow 1$ *to numattrs* **do**
9:         Sampling a random number p in $[0, 1]$
10:        **if** *p greater than* $S_{stability}^{attr}$ **then**
11:            $x^{attr} = x^{attr}$ (* Unsafe feature .*)
12:        **else**
13:            Sampling a random value dif in S_{range}.
14:            $x^{attr} = x^{attr} + dif$
15:        **end if**
16:    **end for**(* Sample x augmentation completed.*)
17:    **return** Resampled dataset X_{aug}
18: **end procedure**

---

4 Experiments

4.1 Experimental Setup

To evaluate the effectiveness of our method, we conduct extensive experiments on various datasets from UCI and Kaggle (Abalone, Raison, Red and White, AI4I2020, Diabetes, Lending Club, Biodeg, Iris, Work Absenteeism, Mobilephone, Bupa, Concrete, Crx, and Students Achievement), including both balanced and unbalanced datasets. The value of the hyperparameters t and t' depend on the value of specific features. In the experiments, they have been set to 1 and 0.95 respectively. ϵ is set to 0.02, and α is set to 0.5. The determination of these hyperparameters is based on empirical evidence from multiple experimental trials.

We utilize the high-performing XGBoost algorithm as the primary classifier and the DiCE algorithm for counterfactual generation. Moreover, We compare the proposed method with five state-of-the-art (sota) data augmentation methods, SMOTE, Bd-SMOTE, ADANY, Svm-SMOTE and CTGAN, and employ *Accuracy*, *Precision*, *Recall* and *F1-score* as evaluation metrics. In the following context, we use 'Safe-Resam' to denote our proposed method.

4.2 Experimental Results

We first conduct comparison experiments with other data augmentation methods. To guarantee fairness in comparison, we adopted a 10-fold cross-validation method. The dataset was evenly split into 10 segments, with each cycle utilizing

Table 1. Classification Performance of Comparing Algorithms on 15 Datasets (mean±std. Deviation)

Compare	Accuracy ↑						
Algorithm	Base	SMOTE	Bd-SMOTE	ADANY	Svm-SMOTE	CTGAN	Safe-Resam
ABALONE	0.819	0.819±0.006	0.821 ± 0.004	0.826 ± 0.004	0.828 ± 0.003	0.816 ± 0.006	**0.831 ± 0.002**
RAISIN	0.867	0.860 ± 0.011	0.855 ± 0.009	0.868 ± 0.008	0.856 ± 0.004	0.861 ± 0.007	**0.877 ± 0.003**
RED	0.801	0.791 ± 0.009	0.809 ± 0.006	0.802 ± 0.010	0.782 ± 0.008	0.793 ± 0.011	**0.818 ± 0.006**
WHITE	0.947	0.947 ± 0.010	0.953 ± 0.008	0.948 ± 0.006	0.937 ± 0.009	0.945 ± 0.010	**0.958 ± 0.005**
AI4I2020	0.713	0.715 ± 0.004	0.678 ± 0.006	0.669 ± 0.009	0.693 ± 0.005	0.689 ± 0.005	**0.725 ± 0.003**
DIABETES	0.762	0.765 ± 0.013	0.730 ± 0.010	0.771 ± 0.009	0.717 ± 0.011	0.734 ± 0.011	**0.787 ± 0.005**
LENDING	0.895	0.892 ± 0.007	0.875 ± 0.006	0.849 ± 0.008	0.883 ± 0.004	0.893 ± 0.002	**0.902 ± 0.002**
BIODEG	0.826	0.823 ± 0.007	0.826 ± 0.006	0.821 ± 0.007	0.798 ± 0.009	0.821 ± 0.007	**0.834 ± 0.004**
IRIS	0.933	0.960 ± 0.006	0.933 ± 0.004	0.933 ± 0.003	0.906 ± 0.002	0.933 ± 0.007	**0.967 ± 0.001**
ABSENT	0.770	0.764 ± 0.007	0.757 ± 0.003	0.770 ± 0.004	0.763 ± 0.003	0.751 ± 0.002	**0.778 ± 0.001**
PHONE	0.887	0.895 ± 0.010	0.857 ± 0.014	0.879 ± 0.013	0.883 ± 0.009	0.891 ± 0.007	**0.924 ± 0.005**
BUPA	0.696	0.690 ± 0.035	0.681 ± 0.024	0.686 ± 0.013	0.667 ± 0.022	0.681 ± 0.017	**0.724 ± 0.007**
CONCRETE	0.922	0.927 ± 0.006	0.927 ± 0.004	0.931 ± 0.003	0.932 ± 0.002	0.927 ± 0.007	**0.938 ± 0.003**
CRX	0.824	0.827 ± 0.007	0.809 ± 0.013	0.845 ± 0.006	0.839 ± 0.009	0.809 ± 0.011	**0.877 ± 0.012**
STUDENTS	0.776	0.785 ± 0.004	0.772 ± 0.006	0.774 ± 0.005	0.767 ± 0.008	0.781 ± 0.004	**0.791 ± 0.002**

Table 1. continued

Metric	Precision ↑						
ABALONE	0.819	0.820 ± 0.006	0.820 ± 0.004	0.825 ± 0.004	0.828 ± 0.004	0.815 ± 0.006	**0.831 ± 0.002**
RAISIN	0.870	0.862 ± 0.011	0.856 ± 0.009	0.867 ± 0.008	0.856 ± 0.004	0.873 ± 0.007	**0.877 ± 0.003**
RED	0.803	0.794 ± 0.009	0.812 ± 0.006	0.800 ± 0.007	0.785 ± 0.010	0.801 ± 0.009	**0.821 ± 0.006**
WHITE	0.941	0.941 ± 0.010	0.942 ± 0.006	0.947 ± 0.007	0.926 ± 0.010	0.940 ± 0.008	**0.954 ± 0.005**
A14I2020	0.513	0.497 ± 0.014	0.478 ± 0.006	0.506 ± 0.008	0.512 ± 0.007	0.520 ± 0.005	**0.526 ± 0.008**
DIABETES	0.741	0.743 ± 0.015	0.735 ± 0.012	0.751 ± 0.008	0.701 ± 0.017	0.739 ± 0.011	**0.771 ± 0.006**
LENDING	0.860	0.852 ± 0.005	0.810 ± 0.006	0.765 ± 0.007	0.829 ± 0.006	0.870 ± 0.002	**0.887 ± 0.003**
BIODEG	0.809	0.804 ± 0.010	0.807 ± 0.009	0.803 ± 0.009	0.777 ± 0.011	0.800 ± 0.010	**0.816 ± 0.005**
IRIS	0.819	0.820 ± 0.006	0.819 ± 0.005	0.820 ± 0.003	0.809 ± 0.007	0.820 ± 0.007	**0.831 ± 0.001**
ABSENT	0.752	0.730 ± 0.008	0.737 ± 0.013	0.752 ± 0.006	0.745 ± 0.010	0.730 ± 0.012	**0.762 ± 0.008**
PHONE	0.887	0.885 ± 0.010	0.861 ± 0.017	0.881 ± 0.013	0.883 ± 0.011	0.883 ± 0.012	**0.921 ± 0.005**
BUPA	0.704	0.689 ± 0.035	0.688 ± 0.023	0.686 ± 0.019	0.672 ± 0.027	0.673 ± 0.031	**0.718 ± 0.007**
CONCRETE	0.927	0.926 ± 0.006	0.926 ± 0.007	0.938 ± 0.005	0.930 ± 0.006	0.928 ± 0.007	**0.940 ± 0.003**
CRX	0.823	0.841 ± 0.007	0.810 ± 0.013	0.844 ± 0.006	0.812 ± 0.009	0.812 ± 0.010	**0.880 ± 0.014**
STUDENTS	0.725	0.738 ± 0.006	0.719 ± 0.011	0.728 ± 0.008	0.718 ± 0.012	0.733 ± 0.005	**0.746 ± 0.003**
ABALONE	0.818	0.820 ± 0.006	0.820 ± 0.005	0.826 ± 0.004	0.827 ± 0.004	0.814 ± 0.006	**0.831 ± 0.002**
RAISIN	0.866	0.860 ± 0.010	0.855 ± 0.008	0.867 ± 0.008	0.855 ± 0.009	0.872 ± 0.007	**0.877 ± 0.003**
RED	0.799	0.790 ± 0.010	0.806 ± 0.006	0.801 ± 0.007	0.780 ± 0.010	0.795 ± 0.009	**0.817 ± 0.006**
WHITE	0.949	0.946 ± 0.009	0.947 ± 0.010	0.954 ± 0.008	0.943 ± 0.010	0.947 ± 0.010	**0.957 ± 0.006**
A14I2020	0.505	0.499 ± 0.005	0.496 ± 0.007	0.500 ± 0.006	0.501 ± 0.006	0.505 ± 0.005	**0.508 ± 0.003**
DIABETES	0.720	0.733 ± 0.016	0.714 ± 0.019	0.721 ± 0.017	0.704 ± 0.020	0.736 ± 0.013	**0.751 ± 0.007**
LENDING	0.783	0.782 ± 0.003	0.791 ± 0.002	0.795 ± 0.002	0.792 ± 0.002	0.791 ± 0.003	**0.796 ± 0.002**
BIODEG	0.798	0.800 ± 0.009	0.804 ± 0.007	0.795 ± 0.010	0.774 ± 0.012	0.804 ± 0.007	**0.809 ± 0.006**
IRIS	0.818	0.820 ± 0.006	0.818 ± 0.005	0.820 ± 0.006	0.810 ± 0.007	0.820 ± 0.007	**0.831 ± 0.001**
ABSENT	0.752	0.744 ± 0.008	0.753 ± 0.009	0.748 ± 0.006	0.747 ± 0.010	0.720 ± 0.012	**0.754 ± 0.010**
PHONE	0.887	0.885 ± 0.010	0.860 ± 0.016	0.880 ± 0.012	0.884 ± 0.011	0.892 ± 0.012	**0.922 ± 0.005**
BUPA	0.681	0.678 ± 0.032	0.673 ± 0.021	0.672 ± 0.020	0.651 ± 0.028	0.656 ± 0.031	**0.717 ± 0.016**
CONCRETE	0.909	0.915 ± 0.007	0.918 ± 0.007	0.926 ± 0.006	0.922 ± 0.008	0.918 ± 0.010	**0.928 ± 0.003**
CRX	0.822	0.832 ± 0.006	0.803 ± 0.013	0.837 ± 0.006	0.801 ± 0.016	0.802 ± 0.014	**0.874 ± 0.014**
STUDENTS	0.688	0.714 ± 0.006	0.689 ± 0.011	0.697 ± 0.009	0.690 ± 0.013	0.703 ± 0.006	**0.708 ± 0.002**
ABALONE	0.819	0.819 ± 0.006	0.821 ± 0.005	0.827 ± 0.004	0.829 ± 0.004	0.815 ± 0.006	**0.831 ± 0.002**
RAISIN	0.866	0.859 ± 0.010	0.855 ± 0.008	0.867 ± 0.008	0.855 ± 0.009	0.872 ± 0.007	**0.877 ± 0.003**
RED	0.799	0.790 ± 0.010	0.806 ± 0.006	0.801 ± 0.007	0.780 ± 0.010	0.795 ± 0.009	**0.818 ± 0.005**
WHITE	0.944	0.943 ± 0.011	0.942 ± 0.010	0.951 ± 0.008	0.936 ± 0.012	0.945 ± 0.009	**0.955 ± 0.006**
A14I2020	0.481	0.467 ± 0.007	0.480 ± 0.003	0.476 ± 0.004	0.472 ± 0.006	0.476 ± 0.005	**0.483 ± 0.004**
DIABETES	0.728	0.737 ± 0.016	0.713 ± 0.020	0.726 ± 0.016	0.698 ± 0.023	0.721 ± 0.013	**0.758 ± 0.006**
LENDING	0.814	0.810 ± 0.003	0.790 ± 0.003	0.768 ± 0.005	0.798 ± 0.002	0.794 ± 0.004	**0.820 ± 0.003**
BIODEG	0.803	0.802 ± 0.009	0.805 ± 0.007	0.799 ± 0.010	0.775 ± 0.012	0.802 ± 0.007	**0.812 ± 0.005**
IRIS	0.819	0.819 ± 0.006	0.819 ± 0.005	0.820 ± 0.006	0.812 ± 0.006	0.819 ± 0.007	**0.831 ± 0.002**
ABSENT	0.752	0.741 ± 0.007	0.736 ± 0.007	0.750 ± 0.006	0.746 ± 0.010	0.725 ± 0.010	**0.757 ± 0.010**
PHONE	0.887	0.883 ± 0.011	0.858 ± 0.013	0.879 ± 0.012	0.883 ± 0.010	0.891 ± 0.009	**0.924 ± 0.005**
BUPA	0.683	0.676 ± 0.032	0.673 ± 0.021	0.673 ± 0.020	0.652 ± 0.028	0.656 ± 0.031	**0.715 ± 0.016**
CONCRETE	0.917	0.920 ± 0.007	0.922 ± 0.007	0.928 ± 0.006	0.926 ± 0.008	0.922 ± 0.007	**0.933 ± 0.003**
CRX	0.822	0.837 ± 0.006	0.803 ± 0.013	0.841 ± 0.006	0.806 ± 0.015	0.807 ± 0.017	**0.874 ± 0.014**
STUDENTS	0.699	0.714 ± 0.006	0.706 ± 0.011	0.715 ± 0.009	0.708 ± 0.013	0.719 ± 0.006	**0.719 ± 0.002**

7 segments for training and 3 for testing. For each method, we use different augmentation ratios, ranging from 1 to 5 times the size of the training data, and report the average performance. The experimental results are shown in Table 1.

It can be observed that our method achieves sota performance across different datasets. Compared with the second best method, *accuracy, precision, recall* rate and *F1-score* are improved by 1.22%, 1.38%, 1.14%, 1.27%, respectively. Compared with the baseline method XGBoost, *accuracy, precision, recall* rate and *F1-score* are improved by 1.95%, 1.91%, 1.90%, 1.83%, respectively. These results well validate the effectiveness of our proposed data augmentation method based on counterfactuals.

In addition, our proposed method not only achieves superior performance in terms of *accuracy*, but also exhibits the smallest and most stable variance, which well verifies that our proposed augmentation method based on safe metrics possesses higher safeness and stability.

Furthermore, we vary the ratio of testing set to training set on eight datasets to further compare the augmentation performance of each comparing method, and set the ratio to 0.1, 0.2, 0.3, 0.4 and 0.5. We aim to simulate the situation of data scarcity in the real world by reducing the proportion of the training dataset. The experimental results are presented in Fig 6 and Table 2. It can be seen that, our proposed method achieves sota performance in term of *accuracy* across different data divide ratios. In contrast, the performance of other methods is not stable as the data ratio changes.

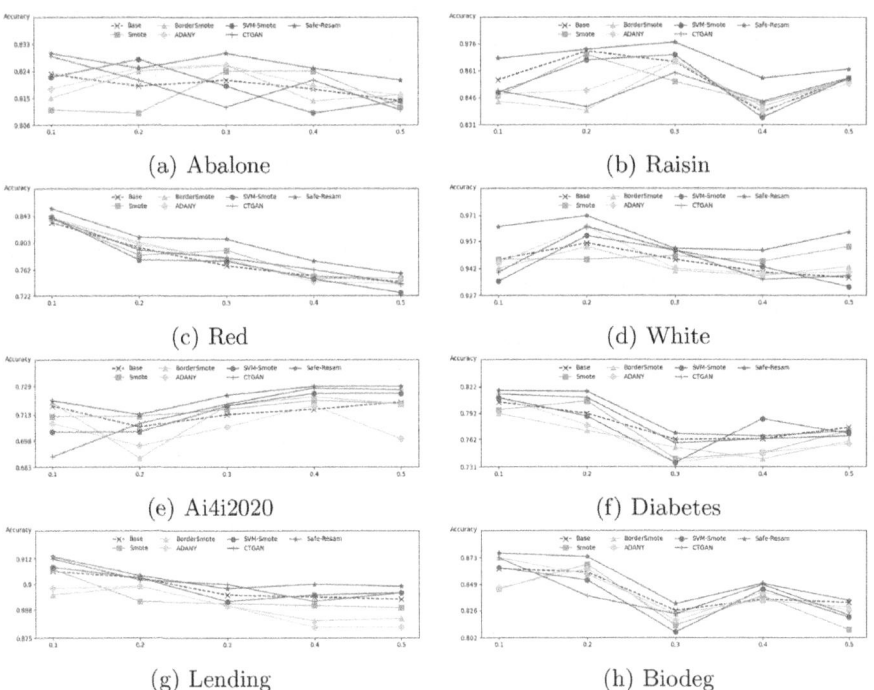

(a) Abalone   (b) Raisin

(c) Red   (d) White

(e) Ai4i2020   (f) Diabetes

(g) Lending   (h) Biodeg

Fig. 6. The accuracy of different methods on 8 dataset as the test set ratio varies.

Table 2. Comparison of 7 methods in term of accuracy on 8 dataset as the test set ratio varies. (mean±std. Deviation)

Datasets	Abalone	Raisin	Red	White	Ai4i2020	Diabetes	Lending Club	Biodeg
Algorithms				0.1 *test Accuracy* ↑				
Baseline	0.823	0.856	0.833	0.947	0.718	0.805	0.906	0.863
Smote	0.811±0.008	0.847±0.015	0.839±0.016	0.947±0.008	0.712±0.007	0.796±0.022	0.907±0.005	0.845±0.008
BorderSmote	0.815±0.009	0.844±0.012	0.840±0.012	0.942±0.010	0.720±0.012	0.792±0.010	0.895±0.010	0.873±0.008
ADANY	0.818±0.011	0.848±0.010	0.841±0.009	0.945±0.008	0.708±0.008	0.810±0.012	0.898±0.008	0.846±0.010
SVM-SMOTE	0.822±0.013	0.849±0.012	0.841±0.010	0.935±0.014	0.703±0.010	0.810±0.012	0.908±0.011	0.864±0.011
CTGAN	0.829±0.009	0.850±0.009	0.840±0.010	0.940±0.012	0.689±0.012	0.814±0.009	0.912±0.009	0.873±0.011
Safe-Resam	**0.830±0.006**	**0.868±0.004**	**0.854±0.006**	**0.965±0.001**	**0.721±0.003**	**0.818±0.010**	**0.913±0.002**	**0.877±0.005**
Algorithms				0.2 *test Accuracy* ↑				
Baseline	0.819	0.872	0.795	0.956	0.706	0.792	0.903	0.860
Smote	0.810±0.007	0.870±0.015	0.784±0.013	0.947±0.014	0.712±0.004	0.806±0.012	0.892±0.003	0.867±0.007
BorderSmote	0.824±0.009	0.839±0.014	0.803±0.009	0.954±0.012	0.688±0.012	0.772±0.016	0.899±0.011	0.858±0.012
ADANY	0.825±0.007	0.850±0.012	0.800±0.008	0.965±0.010	0.695±0.008	0.779±0.012	0.899±0.009	0.863±0.010
SVM-SMOTE	0.828±0.009	0.867±0.011	0.777±0.013	0.960±0.014	0.703±0.009	0.789±0.012	0.903±0.008	0.853±0.008
CTGAN	0.821±0.011	0.841±0.015	0.792±0.013	0.965±0.010	0.708±0.011	0.810±0.012	0.902±0.009	0.839±0.010
Safe-Resam	**0.825±0.003**	**0.873±0.002**	**0.811±0.003**	**0.971±0.004**	**0.713±0.002**	**0.817±0.006**	**0.904±0.002**	**0.874±0.006**
Algorithms				0.3 *test Accuracy* ↑				
Baseline	0.821	0.866	0.768	0.947	0.713	0.762	0.895	0.826
Smote	0.824±0.005	0.855±0.008	0.791±0.012	0.949±0.009	0.716±0.005	0.740±0.009	0.891±0.002	0.813±0.008
BorderSmote	0.826±0.006	0.866±0.014	0.776±0.009	0.941±0.008	0.718±0.012	0.753±0.012	0.890±0.008	0.823±0.010
ADANY	0.826±0.009	0.867±0.015	0.778±0.011	0.942±0.010	0.706±0.012	0.736±0.011	0.890±0.009	0.817±0.008
SVM-SMOTE	0.819±0.010	0.870±0.010	0.775±0.012	0.952±0.014	0.718±0.009	0.736±0.010	0.892±0.011	0.807±0.010
CTGAN	0.812±0.008	0.860±0.015	0.780±0.012	0.952±0.010	0.719±0.011	0.758±0.011	0.900±0.008	0.823±0.008
Safe-Resam	**0.830±0.002**	**0.877±0.003**	**0.808±0.005**	**0.953±0.004**	**0.724±0.002**	**0.769±0.007**	**0.898±0.001**	**0.832±0.005**
Algorithms				0.4 *test Accuracy* ↑				
Baseline	0.818	0.838	0.752	0.940	0.716	0.763	0.894	0.836
Smote	0.824±0.006	0.843±0.009	0.752±0.010	0.946±0.007	0.721±0.006	0.747±0.015	0.890±0.003	0.839±0.006
BorderSmote	0.814±0.008	0.841±0.015	0.748±0.009	0.938±0.014	0.723±0.008	0.740±0.013	0.883±0.008	0.835±0.014
ADANY	0.819±0.009	0.839±0.014	0.744±0.011	0.939±0.011	0.718±0.012	0.747±0.010	0.880±0.011	0.842±0.011
SVM-SMOTE	0.810±0.006	0.835±0.013	0.748±0.010	0.943±0.009	0.725±0.009	0.786±0.010	0.895±0.011	0.845±0.010
CTGAN	0.821±0.011	0.844±0.011	0.762±0.012	0.936±0.014	0.728±0.014	0.763±0.011	0.892±0.008	0.849±0.010
Safe-Resam	**0.825±0.002**	**0.857±0.005**	**0.775±0.004**	**0.952±0.004**	**0.729±0.001**	**0.766±0.006**	**0.900±0.001**	**0.850±0.005**
Algorithms				0.5 *test Accuracy* ↑				
Baseline	0.814	0.857	0.744	0.937	0.720	0.776	0.893	0.833
Smote	0.812±0.004	0.856±0.008	0.748±0.011	0.954±0.005	0.719±0.006	0.772±0.019	0.889±0.002	0.809±0.009
BorderSmote	0.816±0.011	0.856±0.015	0.749±0.011	0.943±0.014	0.719±0.006	0.760±0.013	0.884±0.010	0.829±0.013
ADANY	0.816±0.010	0.854±0.013	0.740±0.013	0.940±0.011	0.699±0.012	0.757±0.010	0.880±0.010	0.826±0.013
SVM-SMOTE	0.814±0.008	0.857±0.016	0.727±0.009	0.932±0.016	0.725±0.007	0.769±0.018	0.896±0.008	0.820±0.015
CTGAN	0.811±0.009	0.857±0.015	0.741±0.012	0.938±0.011	0.727±0.010	0.766±0.014	0.896±0.009	0.822±0.011
Safe-Resam	**0.821±0.001**	**0.862±0.003**	**0.756±0.003**	**0.962±0.004**	**0.729±0.001**	**0.772±0.004**	**0.899±0.001**	**0.835±0.004**

Moreover, we augment training data by using three different multipliers, 1, 2 and 5, respectively, and compare the performance in term of *accuray* of 7 methods on 8 different datasets. As shown in Table 3, as the augmentation ratio increases, our proposed augmentation method based on safeness metrics can remain stable, whereas other compared methods tend to generate numerous erroneous samples, thereby leading to further degradation in term of *accuracy*. There results validate effectiveness of the proposed method.

Table 3. Comparison of 6 methods in term of accuracy as the augmentation ratio varies. (mean±std. Deviation)

DATASETS	Abalone	Raisin	Red	White	Ai4i2020	Diabetes	Lending Club	Biodeg
BASELINE	0.819	0.867	0.795	0.947	0.706	0.772	0.895	0.826
ALGORITHMS				1x *Accuracy* ↑				
SMOTE	0.814±0.005	0.862±0.008	0.793±0.009	0.946±0.011	0.716±0.004	0.765±0.012	0.894±0.002	0.828±0.008
BORDERSMOTE	0.824±0.009	0.839±0.014	0.803±0.009	0.954±0.012	0.688±0.012	0.772±0.016	0.899±0.011	0.858±0.012
ADANY	0.825±0.007	0.850±0.012	0.800±0.008	0.965±0.010	0.695±0.008	0.779±0.012	0.899±0.009	0.863±0.010
SVM-SMOTE	0.828±0.009	0.867±0.011	0.777±0.013	0.960±0.014	0.703±0.009	0.789±0.012	0.903±0.008	0.853±0.008
CTGAN	0.821±0.011	0.841±0.015	0.792±0.013	0.965±0.010	0.708±0.011	0.810±0.012	0.902±0.009	0.839±0.010
SAFE-RESAM	**0.829±0.003**	**0.873±0.002**	**0.811±0.003**	**0.971±0.004**	**0.713±0.002**	**0.817±0.006**	**0.904±0.002**	**0.874±0.006**
ALGORITHMS				2x *Accuracy* ↑				
SMOTE	0.810±0.007	0.854±0.009	0.784±0.009	0.945±0.011	0.709±0.006	0.757±0.013	0.885±0.004	0.821±0.009
BORDERSMOTE	0.821±0.011	0.837±0.015	0.794±0.011	0.946±0.012	0.682±0.012	0.768±0.017	0.898±0.012	0.857±0.012
ADANY	0.816±0.009	0.844±0.013	0.796±0.010	0.959±0.012	0.694±0.010	0.769±0.012	0.895±0.011	0.853±0.010
SVM-SMOTE	0.823±0.011	0.864±0.013	0.776±0.015	0.956±0.015	0.700±0.011	0.782±0.014	0.897±0.011	0.844±0.010
CTGAN	0.820±0.013	0.835±0.018	0.787±0.015	0.958±0.012	0.699±0.013	0.790±0.013	0.900±0.011	0.839±0.012
SAFE-RESAM	**0.828±0.005**	**0.872±0.005**	**0.819±0.006**	**0.965±0.006**	**0.725±0.005**	**0.807±0.012**	**0.903±0.003**	**0.859±0.006**
ALGORITHMS				5x *Accuracy* ↑				
SMOTE	0.796±0.015	0.851±0.018	0.783±0.012	0.942±0.012	0.698±0.011	0.751±0.015	0.879±0.007	0.812±0.010
BORDERSMOTE	0.807±0.013	0.826±0.017	0.790±0.013	0.941±0.015	0.679±0.016	0.758±0.019	0.890±0.015	0.843±0.014
ADANY	0.809±0.011	0.835±0.014	0.793±0.013	0.950±0.013	0.692±0.014	0.757±0.014	0.892±0.012	0.833±0.011
SVM-SMOTE	0.810±0.013	0.859±0.015	0.774±0.016	0.945±0.020	0.686±0.010	0.776±0.017	0.896±0.013	0.830±0.015
CTGAN	0.808±0.014	0.832±0.017	0.773±0.017	0.947±0.013	0.692±0.013	0.797±0.014	0.895±0.011	0.826±0.014
SAFE-RESAM	**0.833±0.007**	**0.873±0.004**	**0.819±0.005**	**0.958±0.007**	**0.726±0.009**	**0.788±0.011**	**0.902±0.009**	**0.847±0.003**

5   Conclusion

In this study, we propose a safe data resampling method based on counterfactuals, where the safeness of generated new data is evaluated from both the sample and feature perspectives. We demonstrate that the application of safe resampling can effectively enhance the downstream machine learning tasks. In the future, we will extend the algorithm to more deep learning applications.

References

1. Arjovsky, M., Chintala, S., Bottou, L.: Wasserstein generative adversarial networks. In: International Conference on Machine Learning, pp. 214–223. PMLR (2017)
2. Chawla, N.V.: Data mining for imbalanced datasets: an overview. Data mining and knowledge discovery handbook, pp. 875–886 (2009)
3. Chawla, N.V., Bowyer, K.W., Hall, L.O., Kegelmeyer, W.P.: Smote: synthetic minority over-sampling technique. J. Artifi. Intell. Res. **16**, 321–357 (2002)
4. Del Ser, J., Barredo-Arrieta, A., Díaz-Rodríguez, N., Herrera, F., Saranti, A., Holzinger, A.: On generating trustworthy counterfactual explanations. Inf. Sci. **655**, 119898 (2024)
5. Elsobky, A.M., Keshk, A.E., Malhat, M.G.: A comparative study for different resampling techniques for imbalanced datasets. IJCI Inter. J. Comput. Inform. **10**(3), 147–156 (2023)

6. Good, P.I.: Resampling methods. Springer (2006)
7. Goodfellow, I., et al.: Generative adversarial nets. Adv. Neural Inform. Process. Syst. **27** (2014)
8. Han, H., Wang, W.-Y., Mao, B.-H.: Borderline-SMOTE: a new over-sampling method in imbalanced data sets learning. In: Huang, D.-S., Zhang, X.-P., Huang, G.-B. (eds.) ICIC 2005. LNCS, vol. 3644, pp. 878–887. Springer, Heidelberg (2005). https://doi.org/10.1007/11538059_91
9. Hasan, M.G.M.M., Talbert, D.A.: Counterfactual examples for data augmentation: a case study. In: The International FLAIRS Conference Proceedings, vol. 34 (2021)
10. He, H., Bai, Y., Garcia, E.A., Li, S.: Adasyn: adaptive synthetic sampling approach for imbalanced learning. In: 2008 IEEE International Joint Conference on Neural Networks (IEEE World Congress on Computational Intelligence), pp. 1322–1328. IEEE (2008)
11. James, G., Witten, D., Hastie, T., Tibshirani, R., Taylor, J.: Resampling methods. In: An Introduction to Statistical Learning: with Applications in Python, pp. 201–228. Springer (2023). https://doi.org/10.1007/978-3-031-38747-0_5
12. Laugel, T., Jeyasothy, A., Lesot, M.J., Marsala, C., Detyniecki, M.: Achieving diversity in counterfactual explanations: a review and discussion. In: Proceedings of the 2023 ACM Conference on Fairness, Accountability, and Transparency, pp. 1859–1869 (2023)
13. Mothilal, R.K., Sharma, A., Tan, C.: Explaining machine learning classifiers through diverse counterfactual explanations. In: Proceedings of the 2020 Conference on Fairness, Accountability, and Transparency, pp. 607–617 (2020)
14. Pitis, S., Creager, E., Garg, A.: Counterfactual data augmentation using locally factored dynamics. Adv. Neural. Inf. Process. Syst. **33**, 3976–3990 (2020)
15. Qiang, Y., Li, C., Brocanelli, M., Zhu, D.: Counterfactual interpolation augmentation (cia): a unified approach to enhance fairness and explainability of dnn. In: Proceedings of the Thirty-First International Joint Conference on Artificial Intelligence, IJCAI, pp. 732–739 (2022)
16. Shorten, C., Khoshgoftaar, T.M.: A survey on image data augmentation for deep learning. J. Big Data **6**(1), 1–48 (2019)
17. Verma, S., Boonsanong, V., Hoang, M., Hines, K.E., Dickerson, J.P., Shah, C.: Counterfactual explanations and algorithmic recourses for machine learning: A review. arXiv preprint arXiv:2010.10596 (2020)
18. Wachter, S., Mittelstadt, B., Russell, C.: Counterfactual explanations without opening the black box: automated decisions and the gdpr. Harv. JL Tech. **31**, 841 (2017)
19. Wang, Q., Luo, Z., Huang, J., Feng, Y., Liu, Z., et al.: A novel ensemble method for imbalanced data learning: bagging of extrapolation-smote svm. Comput. Intell. Neurosci. **2017** (2017)
20. Xu, L., Skoularidou, M., Cuesta-Infante, A., Veeramachaneni, K.: Modeling tabular data using conditional gan. Adv. Neural Inform. Process. Syst. **32** (2019)

Test-Time Augmentation for Traveling Salesperson Problem

Ryo Ishiyama$^{(\boxtimes)}$ ⓘ, Takahiro Shirakawa, Seiichi Uchida ⓘ,
and Shinnosuke Matsuo ⓘ

Kyushu University, Fukuoka, Japan
{ryo.ishiyama,shinnosuke.matsuo}@human.ait.kyushu-u.ac.jp

Abstract. We propose Test-Time Augmentation (TTA) as an effective
technique for addressing combinatorial optimization problems, including the Traveling Salesperson Problem. In general, deep learning models
possessing the property of invariance, where the output is uniquely determined regardless of the node indices, have been proposed to learn graph
structures efficiently. In contrast, we interpret the permutation of node
indices, which exchanges the elements of the distance matrix, as a TTA
scheme. The results demonstrate that our method is capable of obtaining shorter solutions than the latest models. Furthermore, we show that
the probability of finding a solution closer to an exact solution increases
depending on the augmentation size.

Keywords: Test-Time Augmentation (TTA) · Traveling Salesperson
Problem (TSP) · Transformer

1 Introduction

The traveling salesperson problem (TSP) is a well-known NP-hard combinatorial optimization problems [31]. TSP is a problem of finding the shortest-
length tour[1] (i.e., circuit) that visits each city once. More formally, as shown
in Fig. 1, the instance of TSP is a set $\{\boldsymbol{x}_1, \ldots, \boldsymbol{x}_i, \ldots, \boldsymbol{x}_N\}$, where $\boldsymbol{x}_i$ is the
coordinate of a city, and its solution is a tour represented as the city index
sequence, $\pi_1, \ldots, \pi_i, \ldots, \pi_N$, where $\pi_i \in [1, N]$. Traditional research focuses on
exact solvers [1,9] and approximation solvers based on algorithms such as min-
imum spanning tree, 2-opt [5,10,11,13,14,25]. In contrast, modern approximation solvers use machine-learning frameworks [2,4,6,12,16,18,19,22,29,30,34].
For TSP, it is difficult to use the standard fully-supervised machine-learning
framework because the ground-truth of TSP is hardly available due to its computational complexity. Therefore, deep reinforcement learning is often employed
because it does not need the ground-truth but only needs the evaluation of the
current solution (i.e., tour length) as a reward.

---

[1] This paper assumes "Euclidean" TSP unless otherwise mentioned.

M. Wand et al. (Eds.): ICANN 2024, LNCS 15016, pp. 194–208, 2024.
https://doi.org/10.1007/978-3-031-72332-2_14

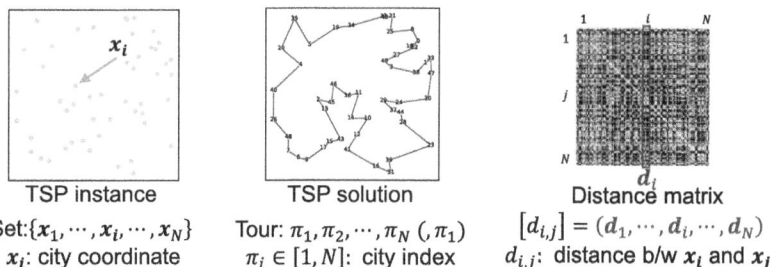

Fig. 1. An instance and its solution of the (Euclidean) traveling salesperson problem (TSP). Note that its distance matrix depends on the order of cities.

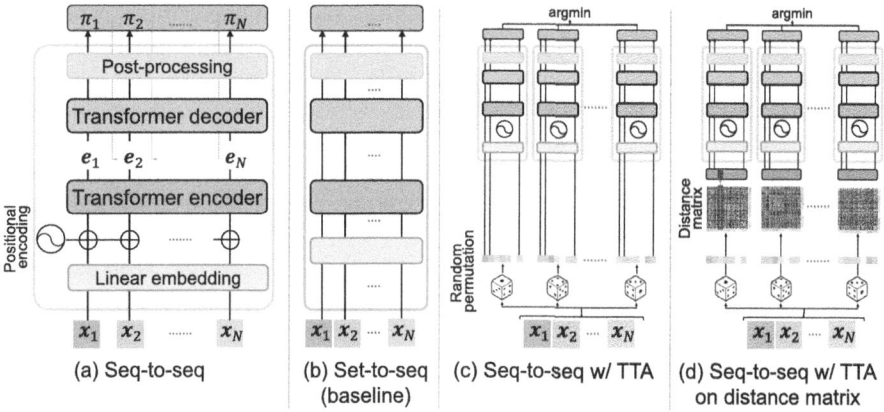

Fig. 2. Transformer encoder and decoder model. (a) The standard sequence-to-sequence model with positional encoding. (b) A set-to-sequence model for solving TSP [4,22], where positional encoding is removed from (a). (c) TTA with the M sequence-to-sequence model of (a). (d) The proposed model, where M distance matrices are used for TTA.

The latest and well-known machine learning-based solver is Bresson et al. [4]. Roughly speaking, their solver formulates TSP as a sequent-to-sequence conversion task (in fact, it is not exactly true as detailed soon) and solves the task using a Transformer-based solver. Figure 2(a) illustrates the standard sequence-to-sequence model. A sequence of city coordinates, $x_1, \ldots, x_N$, is fed to a Transformer encoder via linear embedding and positional encoding modules, and its output is fed to a Transformer decoder, which recursively outputs an approximated solution, i.e., the city index sequence, $\pi_1, \ldots, \pi_N$. The attention mechanism in Transformer can evaluate the mutual relationship between all city pairs and, therefore, is very favorable for TSP because the tour needs to be determined while looking at all cities.

Precisely speaking, Bresson et al. [4] does *not* treat TSP as a sequence-to-sequence task but a *set*-to-sequence task. In general applications of Transformers to sequence-to-sequence tasks (e.g., language translation), positional encoding is employed to differentiate the different sequences, such as $x_1, x_2, x_3, \ldots, x_N$ and $x_2, x_1, x_3, \ldots, x_N$. In contrast, Bresson et al. [4], as well as Kool et al. [22], do not employ positional encoding, as shown in Fig. 2(b). Without positional encoding, two input sequences $x_1, x_2, x_3, \ldots, x_N$ and $x_2, x_1, x_3, \ldots, x_N$ are treated as exactly the same inputs. Consequently, their solver can give the same solution regardless of the input order of cities. This property is so-called *permutation invariant* and seems very reasonable for TSP because the actual input of TSP is a set of cities, and the solution of TSP should not depend on the input order.

However, this paper goes against this traditional approach and experimentally proves that treating TSP as a sequence-to-sequence task is actually better by the introduction of *Test-Time Augmentation* (TTA). TTA is a data-augmentation technique. Different from standard data-augmentation techniques for the training phase, TTA is used in the test phase. Figure 2(c) illustrates TTA with the sequence-to-sequence model of (a). As noted above, the sequence-to-sequence model of Fig. 2(a) give different solutions with $x_1, x_2, x_3, \ldots, x_N$ and $x_2, x_1, x_3, \ldots, x_N$. This means if we generate M variants of a single TSP instance by a random permutation process, we have M different solutions from the M variants. Consequently, we will have an accurate solution by choosing the best solution among M solutions.

This paper proposes a Transformer-based model for solving TSP with the above idea about TTA. Figure 2(d) shows our model. In our model, we utilize the fact that a TSP instance can be represented as a $N \times N$ distance matrix of Fig. 1, whose (i, j)-th element is the distance $d_{i,j}$ between two cities, x_i and x_j. If we exchange the first and second cities (i.e., $x_1, x_2, x_3, \ldots, x_N$ to $x_2, x_1, x_3, \ldots, x_N$), the first and second rows and columns are flipped in the matrix. Similarly, by applying a random permutation to the city order, we can generate variants of the distance matrix. As shown in Fig. 1, the distance matrix is treated as a sequence of the column vector, $d_1, \ldots, d_N$, where $d_i \in \mathbb{R}^N$. Consequently, by TTA with M random permutations, we have M different column vector sequences, which become M different inputs to the model of Fig. 2(d) and give M different solutions. Since d_i comprises the distances from a certain city to all the N cities, it carries stable information (because $d_{i,j}$ is invariant to the geometric translation and rotation) than x_i. Moreover, since the random permutations affect not only columns but also rows of the distance matrix, our column vector representation realizes wider variations, which has a positive effect on TTA.

We experimentally show that our simple but effective model can achieve a quite low optimality gap of 0.01% for the public TSP dataset, TSP50, and 1.07% for TSP100. In addition, our result proves that the solution quality monotonically increases by the augmentation size M. This property is useful for users because they can set M by considering the trade-off between the solution quality and the computational complexity.

The main contributions of this paper are summarized as follows:

- We propose a machine learning-based model for solving TSP with an elaborated TTA. Specifically, our model first represents the TSP instance as a distance matrix and then generates its multiple variants by random permutations.
- Our model outperformed the latest methods on the public TSP instance datasets, TSP50 and TSP100 datasets.

2   Related Work

2.1   TSP Solver

Roughly speaking, TSP solvers can be classified into three types: exact solvers, (algorithmic) approximate solvers, and machine learning-based solvers. As an exact solver, we can find a classical Held-Karp algorithm [9], which is based on dynamic programming. This algorithm, however, becomes intractable with a large N (say, 40) due to the NP-hardness of TSP. In recent research, Concorde [1] has been used as an exact solver. Concorde is based on an algorithm that combines integer programming with cutting planes and branch-and-bound. Note that there are several exact solvers, such as [3], which focus only on Euclidean TSP; however, they also become intractable with a large N because TSP is still NP-hard even under the Euclidean metric condition.

Approximate solvers are computationally efficient at the expense of solution accuracy. The classical approximate solver by Christofides et al. [5] prepares the minimum spanning tree (with a linear time complexity) and then modifies it to have a tour. Its approximate solution is guaranteed that its tour length is less than 3/2 of the minimum tour length. Another classical approximate solver is 2-opt [13,25], where two edges are swapped if the tour becomes shorter. There are solvers which combine Christofides and 2-opt [10,11,26].

Machine learning-based solvers became alternatives to the above algorithm-based solvers because of the rapid development of deep neural networks. Vinyals et al. [34] proposed the Pointer Network, which is based on a recurrent neural network (RNN). Considering the difficulty of preparing the ground-truth for each training instance, which is NP-hard, Bello et al. [2] extended the Pointer Network to be trained in a reinforcement learning framework. Graph neural networks (GCN) are also common networks for solving TSP [16,22]. In these GCN-based models, each node of the input graph corresponds to a city, and each edge connects a pair of (roughly) neighboring cities. Then, GCN aggregates the information of the neighboring cities and finally determines the tour.

The majority of the recent machine learning-based TSP solvers use Transformers. An advantage of using the Transformer is its attention mechanism that evaluates the relationships between all N^2 pairs of N cities. Bresson et al. [4] proposed a method for predicting tours in the TSP based on the Transformer encoder and decoder model, which has been utilized in language translation. More precisely, as we saw in Fig. 2(b), their model uses the encoder-decoder model as a *set-to-sequence* converter by removing the positional encoding. Jung et al. [17] combines a convolutional neural network (CNN) with [4]. The CNN is responsible for extracting city features in the k-nearest neighborhood. Kwon et al. [23,24] also use Transformers to solve "asymmetric" TSPs by their bipartite graph representation.

There are two problem representation styles in the above machine learning-based TSP solvers. The first representation style is a set of city coordinates and is employed in most solvers [2,4,17,34]. This is a straightforward representation because the input of the original TSP is the set of city coordinates. The second style is a $N \times N$ distance matrix that comprises distances for all pairs of N cities. For GCN-based solver, where edges connecting two cities are crucial, employs this representation style. To the authors' knowledge, there is no past attempt that decomposes the distance matrix into a sequence of the N column vector $d_1, \ldots, d_N$.

2.2   Test-Time Augmentation (TTA)

TTA is a technique aimed at improving prediction accuracy by performing data augmentation during inference [20,21,27,28,32,35]. In general, TTA is a two-step procedure in the testing phase. First, from the input to be tested, M variants are generated by data augmentation techniques. Second, M model predictions for M variants are aggregated by, for example, majority voting or simple max/min operation to determine the final prediction. Augmentation and aggregation strategies depend on the characteristics of the target dataset and the model. Famous applications of TTA are image classification [20,27] and image segmentation [28]. Kimura et al. [21] have theoretically proved the effectiveness of TTA.

The effectiveness of TTA relies on the lack of invariance in prediction models. In other words, TTA has no effect if the target model is invariant to some specific changes (such as rotation for images and permutation for sequences), all M variants by the changes are identical for the model, and the M solutions are the same. One example of the invariant model is Fig. 2(b), which depicts an input order-invariant set-to-sequence model, such as Bresson et al. [4].

3   Test-Time Augmeantion (TTA) for the Traveling Salesperson Problem (TSP)

3.1   Overview

We propose the TTA method for TSP as shown in Fig. 2(d). As described in Sect. 1, the latest Transformer-based solvers [4,17] treat TSP as a *set*-to-sequence conversion problem as shown in Fig. 2(b). In other words, this solver is invariant to the permutation in the input city order and gives the same tour for different inputs, such as $x_1, x_2, x_3, \ldots, x_N$ and $x_2, x_1, x_3, \ldots, x_N$. In contrast, we treat TSP in a *sequence*-to-sequence model. Since we treat the input as a sequence, the model output will be sensitive to the city order of the input. Therefore, we can perform TTA with the model. Specifically, we first generate M variants of a single TSP instance by representing it as a distance matrix and applying random permutations of the input city order. Then, we feed them to the model and have M different solutions (i.e., tours). Finally, among M solutions, we chose the best solution with the minimum tour length.

Note that our model is based on the model architecture by Bresson et al. [4], while Jung et al. [17] is an improved version of [4]. This is because the model of [4] is simpler (as shown in Fig. 2(b)), and therefore, comparative studies will also become simpler. Moreover, another literature [37] shows Bresson et al. [4] outperforms [17] in a certain training condition.

3.2 TTA with Distance Matrix

We represent each TSP instance as a $N \times N$ distance matrix and use it as the input of the model. The (i,j)-th element of the matrix is the Euclidean distance $d_{i,j}$ between two cities x_i and x_j. As shown in Fig. 1, the distance matrix is treated as a sequence of the N column vectors, $d_1, \ldots, d_i, \ldots, d_N$, where $d_i \in \mathbb{R}^N$.

Why do we use d_i instead of x_i? In fact, if we use x_i, we can realize the model more simply, as shown in Fig. 2(c). There are several reasons to insist on using d_i. First, the distance $d_{i,j}$ in d_i is a representation invariant to spatial translation and rotation of the city coordinates $\{x_1, \ldots, x_N\}$, whereas the city coordinate x_i is variant to them. The solution of TSP should be invariant to these rigid transformations. Therefore, this invariant property by the distance representation is appropriate for treating TSP in a machine-learning framework. Second, d_i seems like a feature vector representing the positional relationships between the i-th city and all N cities. (In fact, a latest model [17] introduces a CNN to deal with such relationships among cities.)

The third and most important reason for using d_i instead of x_i is that d_i shows large variations by random permutation of city indices, and these variations are beneficial for TTA. The random perturbation affects not only columns but also rows of the distance matrix. Assume that we have a distance matrix D and its column vectors $d_1, \ldots, d_i, \ldots, d_N$, and we have another distance matrix D' by random index permutation and its column vectors $d'_1, \ldots, d'_i, \ldots, d'_N$. Since the permutation affects rows, no column vector in $\{d'_i\}$ is identical to d_i. In other words, our TTA by random permutation of city indices is not just a column permutation (like $d_1, d_2, d_3, \ldots, d_N \rightarrow d_2, d_1, d_3, \ldots, d_N$) but a more substantial permutation in both of the column and row directions.

3.3 Decoding to Determine the City Order

The decoder outputs the probability vectors $p_1, \ldots, p_i, \ldots, p_N$. Each vector $p_i = (p_{i,1}, \ldots, p_{i,j}, \ldots, p_{i,N}) \in \Delta^N$ represents the probability that each city becomes the next city. The decoder outputs the probability vectors $\{p_i\}$ in a recursive way from $i = 1$ to N. Specifically, p_i is given as the output by inputting p_{i-1} to the decoder after the post-processing (shown in Fig. 2) to mask the cities already visited. In this recursive process, we determine the city order in a greedy manner; we choose the city with the highest probability as the i-th city to visit, namely,

$$\pi_i = \arg\max_j \{p_{i,1}, \ldots, p_{i,j}, \ldots, p_{i,N}\}. \tag{1}$$

Through this procedure, our model determines the order of the cities, $\pi_1, \ldots, \pi_N$ from $i = 1$ to N in a greedy manner.

3.4 Model Architecture

Before feeding these N column vectors into the Transformer encoder, we apply traditional positional encoding (PE) with sinusoidal functions [33] to individual vectors among various PE strategies [7]. By PE, the Transformer encoder and decoder become input order-variant; namely, they can treat $x_1, x_2, x_3, \ldots, x_N$ and $x_2, x_1, x_3, \ldots, x_N$ as different inputs.

The Transformer encoder has six layers with multi-head attention to extract the relationship between all city pairs, and the Transformer decoder has two layers. The dimension of the latent variables (i.e., encoder outputs) is 512. Note that these model parameters are determined by following the latest and most famous model [4].

3.5 Optimization of the Model

Our model is trained not by supervised learning but by deep reinforcement learning, like the past attempts [4,17]. This is because it is difficult to obtain a ground-truth due to its computational complexity. In deep reinforcement learning, the city order $\pi_1, \ldots, \pi_N$ is learned by using the tour length as the reward. We use the traditional REINFORCE algorithm [36] to train our model.

4 Experiment Setup

4.1 Datasets and Comparative Models

For a fair comparison with the latest and most famous TSP solver by Bresson et al. [4], we follow their data preparation procedure. Each TSP instance is a set of $N = 50$ or 100 cities on a two-dimensional square plane $[0, 1]^2$. Therefore, each city coordinate x_i is represented by a two-dimensional vector. We consider Euclidean TSP, where Euclidean distance is used to evaluate the inter-city distance $d_{i,j} = \|x_i - x_j\|$ and therefore the tour length.

A training set for $N = 50$ was comprised of 100,000 instances, and each instance was comprised of 50 points randomly generated in $[0, 1]^2$. A training set for $N = 100$ was prepared in the same manner. Note that we do not know the ground-truth, i.e., the minimum length tour, of each instance. We, therefore, needed to use a reinforcement learning framework for training the model.

For testing, we used TSP50 and TSP100 [4] for $N = 50$ and 100, respectively, which are publicly available[2] Each of them contains 10,000 instances. For each test instance, a ground-truth is provided by using Concorde [1], which is a well-known algorithm-based TSP solver.

4.2 Implementation Details

We followed the past attempts [4,17] for setting various hyperparameters, such as the model architecture (the number of layers in the encoder and decoder, the vector dimensions, etc.), learning rate, optimizer, batch size, and the number of the training epochs.

We also follow the past attempts [4,17] for training our model in a deep reinforcement learning framework. Specifically, we used the REINFORCE algorithm [36]. Since reinforcement learning requires extensive computation time, we terminated the training process with 100 epochs by following [17].

For our model, we set its default augmented size M at 2500. We also conducted an experiment to see the effect of M; in the experiment, we changed M from 2 to 1024. As we will see later, the performance of our model is monotonically improved according to M.

4.3 Performance Metrics

By following the previous attenpts [4,16,17,22], the performance on the test data is evaluated by the optimality gap (%) and the average tour length:

$$\text{Optimality gap} = \frac{1}{K} \sum_{k=1}^{K} \left(\frac{\hat{l}_k^{\text{TSP}}}{l_k^{\text{TSP}}} - 1 \right), \tag{2}$$

$$\text{Average tour length} = \frac{1}{K} \sum_{k=1}^{K} \hat{l}_k^{\text{TSP}}, \tag{3}$$

where K is the number of instances, $\hat{l}_k^{\text{TSP}}$ is the predicted tour length for the k-th instance, and l_k^{TSP} is the shortest tour length provided by Concorde [1].

4.4 Comparative Models

We compared the following two latest solvers, Bresson et al. [4] and Jung et al. [17]. As noted before, both solvers are based on the Transformer encoder and decoder, and the latter introduces a CNN into the former framework. In other words, the latter is an improved version of the former. We employ them as comparative methods because they are the latest TSP solvers whose input is a set of city coordinates $\{\boldsymbol{x}_1, \ldots, \boldsymbol{x}_N\}$.

For a fair comparison with these latest solvers, which do not employ TTA, we employ the beam search in the decoding process. By the beam search, we can

---

[2] https://github.com/xbresson/TSP_Transformer.

obtain B tour candidates, where B is called "beam width". Each tour candidate has a different tour length, and we choose the one with the minimum tour length as our solution. The latest solvers [4,17] have already introduced this beam search to improve their solution. Since we set the default augmented size $M = 2500$, we also set the default beam size $B = 2500$. In the later experiment, we also evaluated the performance of these solvers with $B = 1$, which resulted in the greedy search for a minimum-length tour like ours.

Table 1. Average tour length (obj.) and optimality gap (Gap.). G and BS stand for the greedy search and the beam search in the decoding process, respectively.

Method	Decoding process	TSP50		TSP100	
		Obj.	Gap.	Obj.	Gap.
Concorde [1]	(Exact Solver)	5.690	0.00%	7.765	0.00%
Bresson et al. [4]	G	5.754	1.12%	8.005	3.09%
	BS	5.698	0.14%	7.862	1.25%
Jung et al. [17]	G	5.745	0.97%	7.985	2.83%
	BS	5.695	0.10%	7.851	1.11%
Our model	G	**5.690**	**0.01%**	**7.848**	**1.07%**
w/o Distance Matrix	G	5.731	0.72%	7.968	2.61%
w/o TTA	G	5.837	2.58%	8.283	6.68%
	BS	5.720	0.53%	8.039	3.53%

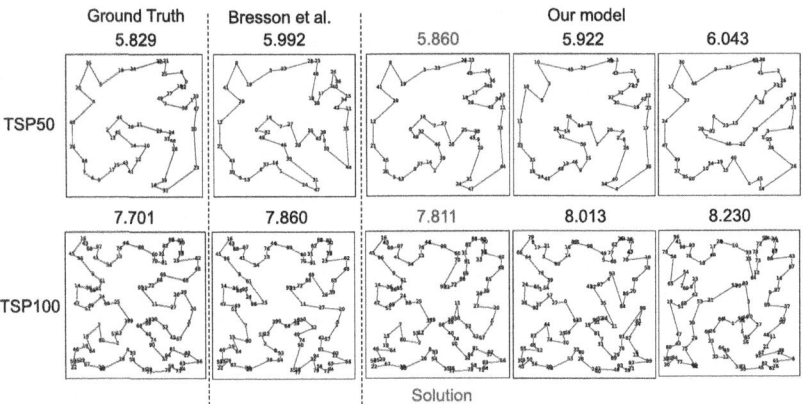

Fig. 3. Solution examples in TSP50 and TSP100. The number indicates the tour length. Our model assumes $M = 3$.

5   Experimental Results

5.1   Comparative Experiments with the Conventional TSP Solvers

Table 1 shows the average tour length ("Obj.") and optimality gap ("Gap.") on TSP50 and TSP100. "G" and "BS" stand for the greedy search and the beam search in the decoding process of the comparative models. The results show that our model outperforms two comparative models in both datasets and in both metrics. In particular, in TSP50, our method archives almost the same tour length as the exact solver.

As an ablation study, we evaluated our model without using the distance matrix. More specifically, we used the city coordinates $\{x_i\}$ instead of the column vectors $\{d_i\}$. We used positional encoding for x_i, and therefore, we still performed TTA in this ablation setup. (Note that this setup can be seen as an extended version of Bresson et al. [4] by introducing the same positional encoding and then TTA.) Our model shows much better performance than this ablation setup ("Our model w/o Distance Matrix"). Consequently, using the distance matrix as an ordered sequence of the column vectors $d_1, \ldots, d_N$ is proved to be useful in the TTA framework.

As another ablation study, we evaluated our model without TTA. The result in Table 1 shows that the performance of our model degrades drastically by the omission of TTA. In other words, this ablation study emphasizes how TTA is effective in our model. (Note that the performance drop by the ablation of TTA is far more serious than the drop by the above ablation of the distance matrix.) As described before, random city index permutation affects the distance matrix in both the column and row directions. This large effect of permutation helps TTA find a better solution. We also evaluated our model without TTA but with beam search; its performance is still much lower than our model with TTA. This difference also proves the effectiveness of TTA with the distance matrix.

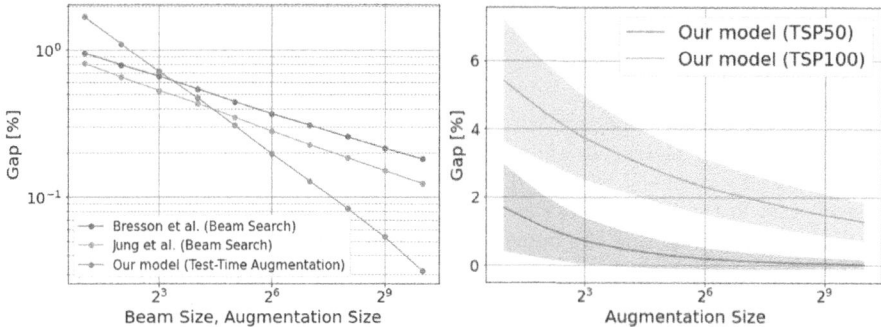

Fig. 4. Left: Effect of the augmentation size M on the performance of our model on TSP50. The performance of the latest models [4,17] under different beam width B is also plotted. Right: Performance variations in all instances in TSP50 or TSP100. Shaded regions indicate the standard deviation intervals. Note that the vertical axis of the left plot is logarithmic, whereas that of the right plot is not.

Figure 3 illustrates the predicted tour for a specific instance of TSP50 and TSP100. Our model predicted different tours by TTA. Compared with Bresson et al. [4], the tour shown in the center of Fig. 3 by our model resembles the ground-truth by Concorde and thus gives a shorter (i.e., better) tour.

5.2   Effect of Augmentation Size

Figure 4(Left) shows how the augmentation size M affects the performance of our model on TSP50. By increasing M, the performance (Gap) of our model is monotonically improved. This result also proves the effectiveness of TTA in improving the accuracy. The computational cost increases with M. Therefore, this result shows that we have a simple trade-off between the computational cost and accuracy. This property is important for the practical use of our model. (In other words, if this curve is non-monotonic, it is difficult to determine the best compromise between computations and accuracy.)

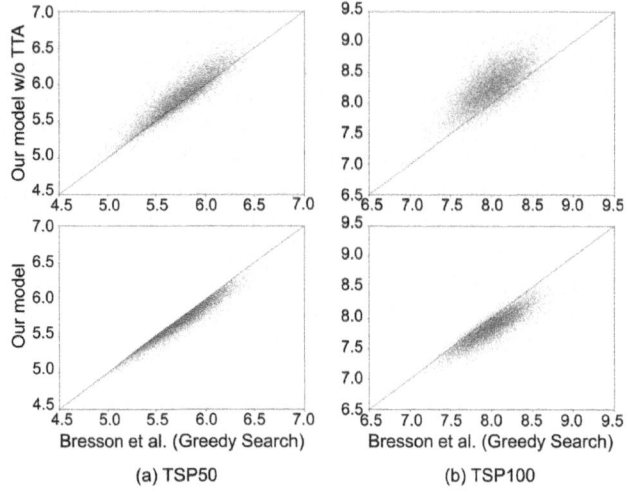

(a) TSP50   (b) TSP100

Fig. 5. Instance-level comparison with Bresson et al. [4]. Here, the tour length is used as the performance metric. In the upper plots, the vertical axes correspond to the performance of our model without TTA, whereas the horizontal axes correspond to Bresson et al. In the lower plots, the vertical axes show the performance of our model (with TTA). By comparing the upper and lower plots, we can also observe the effect of TTA on our model.

Figure 4(Left) also shows the performance of the latest models [4,17] under different beam width B. Like our model, these models also show monotonic performance improvements according to B. However, the degree of their improvements is not as large as ours. Note that, in this graph, all three models show a near-linear relationship between the gap (i.e., accuracy) and the augmented size M (or the beam size B) in this logarithmic plot. This means that the gap $\propto e^{-\alpha M}$ (or $e^{-\beta B}$), where α (or β) is a positive constant representing the slope of the graph. In addition, since the graph shows $\alpha > \beta$, the increase of M is more effective than that of B, although it is not meaningful to compare M and B directly.

Figure 4(right) shows the performance variations in all instances of TSP50 and TSP100 by our model. As the augmentation size M increases, the variation decreases; this means that more accurate paths are obtained in most instances. Note that the curve of TSP100 does not show performance saturation. Larger M will give better tours even for more difficult tasks with more cities, N.

5.3   Instance-Level Performance Evaluation

Figures 5(a) and (b) illustrate the instance-level performance on TSP50 and TSP100, respectively. In these plots, each dot corresponds to a TSP instance. The horizontal and vertical axes are the tour length by Bresson et al. and our model, respectively. About our model, the upper plots show the case without TTA and the lower with TTA. The upper plot of (a) shows that Bresson et al. outperformed our model without TTA in most instances; however, the lower plot shows that our model with TTA outperformed Bresson et al. in almost all instances. The same observation can be made in (b). Consequently, from these plots, we can confirm that TTA is beneficial for our model, and our model with TTA is better than Bresson et al. regardless of the distributions of cities.

6   Conclusion, Limitation, and Future Work

In this study, we introduce test-time data augmentation (TTA) to a machine learning-based traveling salesperson problem (TSP) solver. For the introduction, we reformulated TSP from a set-to-sequence task to a sequence-to-sequence task by (re)employing positional encoding. Moreover, we represent a TSP instance by a distance matrix; more specifically, we represent a TSP instance as a sequence of the column vectors of the matrix. Then, we apply random permutation of city order, which causes variants of the original distance matrix in both the column and row directions. By choosing the best solution among the solutions for the individual variants, we finally have an accurate solution for the instance. Through various experiments, we confirmed the effectiveness of TTA. Moreover, we also confirmed that our model outperforms the latest solvers [4,17].

The limitation of the current model is that it assumes a fixed city number by following the tradition of machine learning-based TSP solvers. Research on approximation methods using machine learning for the TSP often assumes variable numbers of cities [8,15]. However, the current model relies on the fixed size of the distance matrix because we assume the dimension of the column vector d_i is fixed. (Note that the basic architecture of the Transformer encoder and decoder can accept variable N.) Therefore, to deal with the arbitrary number of cities, N, we need to employ some module (such as a Transformer) that can convert $\boldsymbol{d}_i$ (with an arbitrary dimension N) into a fixed dimensional vector.

There are several other future works. First, one may introduce a learning scheme to find the optimal TTA strategies. In this paper, we use TTA based on uniform random permutations. In contrast, we can find trainable TTA strategies, especially in recent image processing research fields [27,32]. Second, one may develop a more computationally efficient TTA. Through the experimental results in this paper, we found that the accuracy improves along with the augmentation size M. Therefore, we want to increase M as many as possible, but the computational cost increases linearly with M. To relax this trade-off, we can introduce some algorithmic methods, as well as the trainable TTA strategies mentioned above.

Acknowledgments. This work was supported by JST-JPMJAX23CR and JSPS-JP23KJ1723, JP21K18312, JP22H05172, JP22H05173 and JP24K22308.

References

1. Applegate, D., Bixby, R., Chvatal, V., Cook, W.: Concorde tsp solver (2006). https://www.math.uwaterloo.ca/tsp/concorde.html
2. Bello, I., Pham, H., Le, Q.V., Norouzi, M., Bengio, S.: Neural combinatorial optimization with reinforcement learning. In: preprint arXiv:1611.09940 (2016)
3. de Berg, M., Bodlaender, H.L., Kisfaludi-Bak, S., Kolay, S.: An ETH-Tight exact algorithm for euclidean TSP. SIAM J. Comput. **52**(3), 740–760 (2023)
4. Bresson, X., Laurent, T.: The transformer network for the traveling salesman problem. In: preprint arXiv:2103.03012 (2021)
5. Christofides, N.: Worst-case analysis of a new heuristic for the travelling salesman problem. Operat. Res. Forum **3** (1976)
6. Deudon, M., Cournut, P., Lacoste, A., Adulyasak, Y., Rousseau, L.M.: Learning heuristics for the tsp by policy gradient. In: Integration of AI and OR Techniques in Constraint Programming (2018)
7. Dufter, P., Schmitt, M., Schütze, H.: Position information in transformers: an overview. Comput. Linguist. **48**(3), 733–763 (2022)
8. Fu, Z.H., Qiu, K.B., Zha, H.: Generalize a small pre-trained model to arbitrarily large tsp instances. In: Proceedings of the AAAI Conference on Artificial Intelligence, vol. 35, pp. 7474–7482 (2021)
9. Held, M., Karp, R.M.: A dynamic programming approach to sequencing problems. In: ACM National Meeting (1962)
10. Helsgaun, K.: An effective implementation of the lin-kernighan traveling salesman heuristic. Eur. J. Oper. Res. **126**, 106–130 (2000)
11. Helsgaun, K.: An extension of the lin-kernighan-helsgaun tsp solver for constrained traveling salesman and vehicle routing problems. Roskilde: Roskilde University **12**, 966–980 (2017)
12. Hudson, B., Li, Q., Malencia, M., Prorok, A.: Graph neural network guided local search for the traveling salesperson problem. In: International Conference on Learning Representations (2022)
13. Johnson, D.S.: Local optimization and the traveling salesman problem. In: International Colloquium on Automata, Languages and Programming (1990)

14. Johnson, D.S., McGeoch, L.A.: The traveling salesman problem: a case study. Local search in combinatorial optimization, pp. 215–310 (1997)
15. Joshi, C.K., Cappart, Q., Rousseau, L.M., Laurent, T.: Learning the travelling salesperson problem requires rethinking generalization. Constraints **27**(1), 70–98 (2022)
16. Joshi, C.K., Laurent, T., Bresson, X.: An efficient graph convolutional network technique for the travelling salesman problem. In: preprint arXiv:1906.01227 (2019)
17. Jung, M., Lee, J., Kim, J.: A lightweight CNN-transformer model for learning traveling salesman problems. Appl. Intell. 1–12 (2024)
18. Kaempfer, Y., Wolf, L.: Learning the multiple traveling salesmen problem with permutation invariant pooling networks. In: preprint arXiv:1803.09621 (2018)
19. Khalil, E.B., Dai, H., Zhang, Y., Dilkina, B.N., Song, L.: Learning combinatorial optimization algorithms over graphs. In: preprint arXiv:1704.01665 (2017)
20. Kim, I., Kim, Y., Kim, S.: Learning loss for test-time augmentation. Adv. Neural Inform. Process. Syst. (2020)
21. Kimura, M.: Understanding test-time augmentation. In: Mantoro, T., Lee, M., Ayu, M.A., Wong, K.W., Hidayanto, A.N. (eds.) ICONIP 2021. LNCS, vol. 13108, pp. 558–569. Springer, Cham (2021). https://doi.org/10.1007/978-3-030-92185-9_46
22. Kool, W., van Hoof, H., Welling, M.: Attention, learn to solve routing problems! In: International Conference on Learning Representations (2019)
23. Kwon, Y.D., Choo, J., Kim, B., Yoon, I., Gwon, Y., Min, S.: Pomo: policy optimization with multiple optima for reinforcement learning. Adv. Neural. Inf. Process. Syst. **33**, 21188–21198 (2020)
24. Kwon, Y.D., Choo, J., Yoon, I., Park, M., Park, D., Gwon, Y.: Matrix encoding networks for neural combinatorial optimization. Adv. Neural. Inf. Process. Syst. **34**, 5138–5149 (2021)
25. Lin, S.: Computer solutions of the traveling salesman problem. Bell Syst. Tech. J. **44**, 2245–2269 (1965)
26. Lin, S., Kernighan, B.W.: An effective heuristic algorithm for the traveling-salesman problem. Oper. Res. **21**, 498–516 (1973)
27. Lyzhov, A., Molchanova, Y., Ashukha, A., Molchanov, D., Vetrov, D.: Greedy policy search: a simple baseline for learnable test-time augmentation. In: Conference on Uncertainty in Artificial Intelligence, pp. 1308–1317. PMLR (2020)
28. Moshkov, N., Mathe, B., Kertesz-Farkas, A., Hollandi, R., Horvath, P.: Test-time augmentation for deep learning-based cell segmentation on microscopy images. Sci. Rep. **10**(1), 5068 (2020)
29. Nazari, M., Oroojlooy, A., Snyder, L.V., Takác, M.: Reinforcement learning for solving the vehicle routing problem. Neural Inform. Process. Syst. (2018)
30. Nowak, A.W., Villar, S., Bandeira, A.S., Bruna, J.: A note on learning algorithms for quadratic assignment with graph neural networks. In: preprint arXiv:1706.07450 (2017)
31. Papadimitriou, C.H.: The euclidean travelling salesman problem is np-complete. Theoret. Comput. Sci. **4**(3), 237–244 (1977)
32. Shanmugam, D., Blalock, D., Balakrishnan, G., Guttag, J.: Better aggregation in test-time augmentation. In: Proceedings of the IEEE/CVF International Conference on Computer Vision, pp. 1214–1223 (2021)
33. Vaswani, A., et al.: Attention is all you need. Adv. Neural Inform. Process. Syst. **30** (2017)
34. Vinyals, O., Fortunato, M., Jaitly, N.: Pointer networks. Adv. Neural Inform. Process. Syst. **28** (2015)

35. Wang, G., Li, W., Aertsen, M., Deprest, J., Ourselin, S., Vercauteren, T.: Aleatoric uncertainty estimation with test-time augmentation for medical image segmentation with convolutional neural networks. Neurocomputing **338**, 34–45 (2019)
36. Williams, R.J.: Simple statistical gradient-following algorithms for connectionist reinforcement learning. Mach. Learn. **8**, 229–256 (1992)
37. Xiao, Y., et al.: Reinforcement learning-based non-autoregressive solver for traveling salesman problems. In: preprint arXiv:2308.00560 (2023)

Novel Neural Architectures

Architectural Architecture

Resonator-Gated RNNs

Robert Deibel[1], Shahram Eivazi[1], Matrin V. Butz[2], and Sebastian Otte[3(✉)]

[1] Autonomous Systems Lab, Department of Computer Science, University of
Tübingen, Sand 14, 72076 Tübingen, Germany
[2] Neuro-Cognitive Modeling, Department of Computer Science, University of
Tübingen, Sand 14, 72076 Tübingen, Germany
[3] Adaptive AI Lab, Institute for Robotics and Cognitive Systems, University of
Lübeck, Ratzeburger Allee 160, 23562 Lübeck, Germany
sebastian.otte@uni-luebeck.de

Abstract. Detecting repetitive and periodic temporal patterns is essential for accurate predictions and informed decision-making in various domains of sequence learning. In RNN-based approaches to sequence learning, gated RNNs, such as long short-term memory networks (LSTMs) and gated recurrent units (GRUs), are the *de facto* standard for these predictions. While adept at capturing longer-term dependencies, gated RNNs still sometimes struggle with periodic data components because their gating mechanism is designed to prioritize retaining transient relevant information. As a result, these networks are often challenged by periodicity in the data. We present a novel memory unit that incorporates a simple resonator circuit. The circuit facilitates the recognition of periodic data patterns, focusing on data-specific time scales and respective frequencies. Moreover, it enables the forward propagation of information through resonating dynamics, while stably channeling the gradient backwards. We show that our resonator-gated RNN (RG-RNN) accelerates the training convergence on multiple sequence classifications tasks. Moreover, it significantly outperforms vanilla LSTMs on four out of five benchmark tasks in terms of accuracy. We conclude that resonator-based gating offers a new inductive bias to gated-RNNs, focusing learning on the detection and processing of periodic data patterns.

Keywords: Sequence Learning · RNN · LSTM · Resonators · Time Series · ECG · MNIST · Speech Commands · Transformer

1 Introduction

Sequence learning involves various domains such as, speech recognition [30], biomedical signal analysis [24], or time series forecasting in climate science [10]. Models in these domains are mainly used for prediction, classification, generation, and comprehensive learning of and from time sequence information [12,30]. Among the most popular techniques are recurrent neural networks (RNNs) which include long short-term memories (LSTMs) [17] or gated recurrent units (GRUs) [6].

© The Author(s), under exclusive license to Springer Nature Switzerland AG 2024
M. Wand et al. (Eds.): ICANN 2024, LNCS 15016, pp. 211–225, 2024.
https://doi.org/10.1007/978-3-031-72332-2_15

Data in this area of research often contains repetitive and periodic temporal patterns, e.g., seasonality in climate data. Regardless of the underlying RNN architectures, the ability to generate or differentiate these signals requires models to learn seasonal patterns, fundamental frequencies, and shifts therein. While gated RNNs are good at capturing longer-term dependencies over time, they often fail to recognize reoccurring or periodic patterns as they operate on fixed time scales and lack the necessary inductive biases [22, 23]. Therefore, targeting periodic signals can be a viable approach to boost performance in sequence learning tasks [18, 23].

Resonators may offer just the right kind of inductive bias. In contrast to integrators, resonators have the capability of responding to spike patterns that arise at specific frequencies rather than within a short period of time. The same input experienced at different times can unfold either an inhibitory or an excitatory effect. This enables these neurons to naturally extract frequency patterns within the time domain and enables bridging the temporal information gap when relevant information is sparsely but regularly scattered across time [19, 20].

While the value of resonator neurons in physiological neural networks has been discussed [1, 15, 20, 31], the applicability in ANNs has not yet been investigated deeply. Examples of RNN extensions that facilitate the detection of periodic signals indicate the demand for such capabilities [18, 23]. Similarly, recent advances in sequence models pushed the performance boundaries of RNNs [13, 26, 28].

We believe that the integration of resonators in ANNs and RNNs, in particular, can be of high interest for sequence learning, since their inductive bias has the potential to overcome the above mentioned shortcomings of gated-RNNs. As a possible solution to fill this gap, we here propose the resonator-gated RNN (RG-RNN), a novel memory unit that incorporates a simple resonator circuit that facilitates the recognition of periodic data patterns. We implement the RG-RNN by using an LSTM as the carrier network. This enables us to directly compare the models with almost identical hyper parameters and uses an established model with clear intuition as the signal carrier.

2   Background

Resonator neurons respond favorably to incoming spike patterns matching their resonating frequency. This behavior can be modeled by a complex-valued state $z \in \mathbb{C}$ or equivalently by a two dimensional state vector $z \in \mathbb{R}^2$, where $z = [v, u]^T$, as opposed to the scalar state of an integrator neuron. The integration inside the resonator is coupled to the position of z around its origin [19, 20].

In [20] the resonate-and-fire neuron is defined by two differential state equations.

$$\frac{dv}{dt} = bv - \omega u \tag{1}$$

and

$$\frac{du}{dt} = \omega v + bu, \tag{2}$$

where $b \in \mathbb{R}_{\leq 0}$ and $\omega \in \mathbb{R}_{\geq 0}$ are parameters that represent the dampening and the resonating frequency of the resonator neuron, respectively. Input signals to the resonator can be handled in different ways. [20] uses a simple summation in the real axis of the complex state. The resonators response can be measured by its excitation, i.e., the magnitude of z.

Via Euler integration the differential state Eqs. 1 and 2 can be transferred into discrete time with time step $t \in \mathbb{N}_0$ and step size $\delta \in \mathbb{R}_{\geq 0}$:

$$v^t = v^{t-1} + \delta(bv^{t-1} - \omega u^{t-1} + x^t) \tag{3}$$

and

$$u^t = u^{t-1} + \delta(\omega v^{t-1} + bu^{t-1}), \tag{4}$$

where x^t is an input signal at time step t, that is added to v^t. The dependence on the previous state v^{t-1} and u^{t-1} acts as a local recurrence similar to the internal cell state of an LSTM or the membrane potential in regular spiking neurons.

Parameterizations of the resonate-and-fire neuron will express different behavior towards (periodic) signals including the capability of "skipping" the resonating mechanism entirely, thus being able to emulate integrators. Figure 1 shows the response of a resonator to an arbitrary input signal and the resonating frequency of the resonator through a Bode diagram. The resonating frequency generates the strongest response—strongest growth—in the resonator state. Because the integration of the differential Eqs. 1 and 2 are directly dependent on δ, it is a key parameter in this method.

RNNs utilize their recurrent connections to capture long-term dependencies in data. In practice, LSTMs [17] and GRUs [6] are the *de facto* standard for gated RNNs. The incorporated gating allows information to be selectively passed through the control flow of the unit or forgotten if the need arises. The internal mechanisms are trained with back-propagation through time (BPTT) [35], which is largely trivialized in modern automatic differentiation frameworks [27].

The internal function of an LSTM is usually denoted with four equations for the gate activation and two equations for the state update. The gate activations are as follows

$$g^t = \tanh\left(x^t W_{g,i} + b_{g,i} + h^{t-1} W_{g,h}\right) \tag{5}$$

$$i^t = \sigma\left(x^t W_{i,i} + b_{i,i} + h^{t-1} W_{i,h}\right) \tag{6}$$

$$f^t = \sigma\left(x^t W_{f,i} + b_{f,i} + h^{t-1} W_{f,h}\right) \tag{7}$$

$$o^t = \sigma\left(x^t W_{o,i} + b_{o,i} + h^{t-1} W_{o,h}\right) \tag{8}$$

where g^t, i^t, f^t, and o^t are the cell input, input gate, forget gate, and output gate, respectively. The current time step is denoted by t. The input is given by x^t and the last hidden state by h^{t-1}. The parameters are contained in matrices $W_{\text{gate,connection}}$. To calculate the activation and hidden state of the LSTM the following equations are used

$$c^t = f^t \odot c^{t-1} + i^t \odot g^t \tag{9}$$

$$h^t = o^t \odot \tanh(c^t) \tag{10}$$

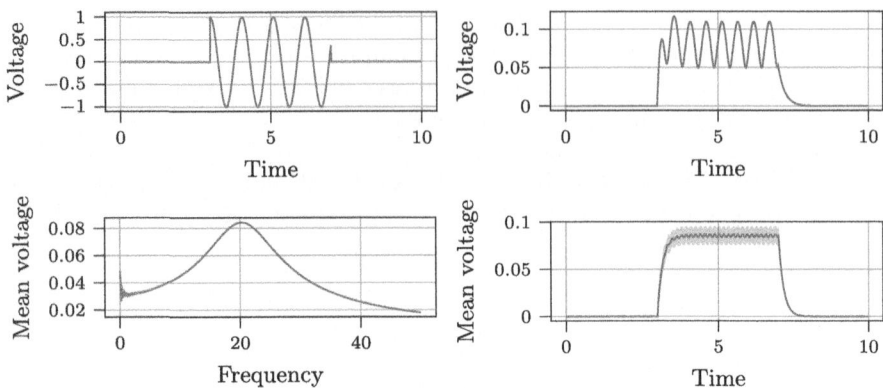

Fig. 1. Responses of a resonator ($b = -7.808$, $\omega = 18.966$, $\delta = 0.01$) to input signals. **(top-left)** Input signal to the resonator. The periodic part is arbitrarily chosen. **(top-right)** Norm of the state, i.e. activation of the resonator. The resonator responds to the periodic input with a pattern similar to a square wave with oscillations. **(bottom-left)** Bode diagram of the resonator over frequencies in $f \in [0, 50]$Hz. It shows the averaged response to an input of similar structure as **(top-left)** with frequency f. The peak indicates the resonating, i.e. most favourable, frequency. **(bottom-right)** Averaged response of the resonator at its resonating frequency when the phase of the input is shifted randomly between 0 and 2π for 200 samples. The small standard deviation (shaded area) demonstrates the invariance to phase shifts at the resonating frequency.

Here c^t denotes the internal activation of the LSTM. In Eqs. 6-8 σ refers to the sigmoid activation and tanh to the hyperbolic tangent activation functions, the output of the LSTM is calculated by iteratively processing each input step in the input sequence and evaluating the hidden state. The last hidden state or the full hidden state sequence can then be used as output. In multi-layer LSTMs the input of the next recurrent layer is the hidden state of the previous layer at the current time step.

3   Resonator-Gated RNN

Our definition and implementation of the resonator-gated RNN (RG-RNN) combines the properties and inductive biases of RNNs and resonator neurons by extending an LSTM with a resonator-gating unit. The resonator is derived and implemented based on Eqs. 3 and 4. Its state is calculated by

$$v^t = v^{t-1} + \delta(bv^{t-1} - \omega u^{t-1} + x_r^t) \tag{11}$$

$$u^t = u^{t-1} + \delta(\omega v^{t-1} + bu^{t-1}) \tag{12}$$

$$z^t = \begin{bmatrix} v^t & u^t \end{bmatrix} \tag{13}$$

and its activation by

$$r^t = |z^t| - \|\delta\|_2 \tag{14}$$

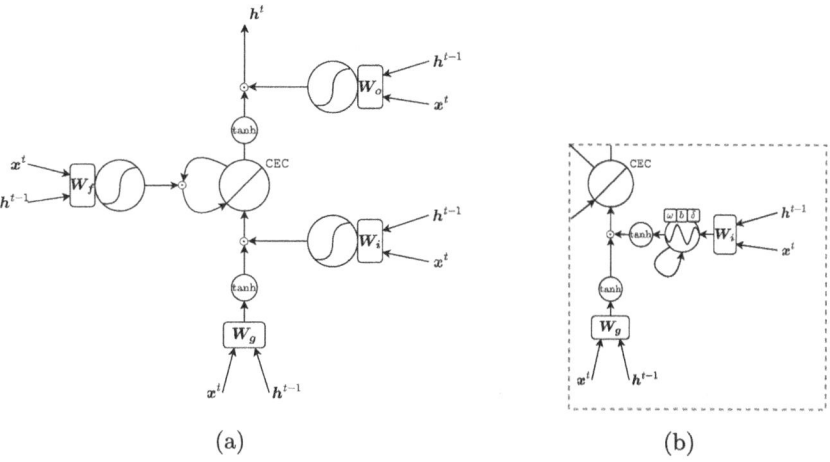

(a)   (b)

Fig. 2. (a) Schematic of the LSTM. It has four input paths for passing x^t and h^{t-1}. tanh transformations are indicated with tanh and sigmoid activations with a stylized "S" in the nodes. Small nodes labeled $\odot$ depict element wise multiplication (Hadamard product). During the computation of step t the CEC out-edge leading towards the forget gate holds the activation from the previous step $t-1$. (b) Extension of the LSTM with a resonator. Notation and schematic are similar to Fig. 2a. The resonator neuron is placed after the affine transformation of the input gate and before the activation. Because the resonator output is $>= 0$, the activation is replaced by a tanh activation to cover the range $[0, 1]$.

with $|z^t|$ being the magnitude of the two dimensional state and not the full vector.

The parameters b, ω, and δ are trainable and initialized randomly in the ranges $(0, 1)$ for b and ω, and $(0.01, 0.1)$ for δ. To guarantee that the parameters adhere to the numerical ranges described in Sect. 2 they need to be transformed accordingly. This was achieved by taking the absolute value of the real underlying parameters $\hat{b}$, $\hat{\omega}$, and $\hat{\delta}$ and further negating it in the case of b.

$$b = -|\hat{b}|, \ \omega = |\hat{\omega}|, \ \delta = |\hat{\delta}| \tag{15}$$

Because the resonator unit is derived from the Euler integrated differential equations, its stability greatly depends on the parametrization of δ. To aid the stability of the resonator, the magnitude of the resonator state is regularized by the magnitude of δ itself in Eq. 14. This reduces the potential for δ to grow during training, destabilizing the resonator. In initial tests, which omitted the regularization term, the state of the resonator was able to grow into regimes where the numerical range of floating point numbers was exhausted and even simple random input signals would drive the resonator into self-oscillation. This can already be the case when $\delta_i \geq 0.8$ is chosen.

The resonator unit is implemented as a differentiable module using PyTorch [27]. It can be inserted into the gating of an LSTM, where we can keep track of the resonator state and execute it as part of the forward pass of the LSTM. The control flow of the implemented unit is depicted in Fig. 2b.

The resulting RG-RNN modifies LSTM Eq. 6 as

$$i^t = \tanh\left(r^t\right) \tag{16}$$

with resonator input

$$x_r^t = x^t W_{i,i} + b_{i,i} + h^{t-1} W_{i,h} \tag{17}$$

The resonator sits after the affine transformation and before the non-linearity of the LSTM gate (Fig. 2b). The activation of the gate was converted to a tanh activation, because the output r is always positive. With a sigmoid activation the range would be restricted to $[0.5, 1]$ while providing a gradient even when $z = 0$.

Note that the integration of the resonator as a building block is of course not limited to LSTMs. It could, for instance, also be integrated into the GRU model replacing the update gate. Moreover, it can be reasonable to implement the resonator as a standalone layer, which could be flexibly arranged within RNN architectures.

4   Experiments

The RG-RNN was evaluated against an LSTM baseline as well as a Transformer [32]. For each experiment and model type five models were trained from scratch and evaluated using the mean loss and mean accuracy of the five runs. The training was based on a per-time step evaluation i.e. the models calculated a prediction after each presented time step. The labels were either uniform across all time steps or a segmentation, in the case of the QT Database (QTDB[1]). We found that using a single-label approach, where the training is only based on the last time step, did not deviate greatly from the results of the per-time step train- ing. We used the cross-entropy loss as the training criterion and RMSprop [16] as the optimizer during training. Between one and two recurrent layers followed by a single linear layer were used for the RNN-based models. The input to the linear layer was the hidden state sequence of the recurrent layers which was input per time step. The Transformer architecture we chose was a simple linear embedding layer followed by the transformer encoding blocks, which fed into a linear output layer. We did not utilize transformer decoding blocks because the experiments did not demand the transformation of the input into a domain with different shape. Such a transformation is common in text translation where input and output dimensions vary [30]. Further, we chose the number of encoding blocks and hidden layer sizes such that the total number of parameters was close to

---

[1] QT referres to the QT waveform interval in the ECG.

the number of parameters used in the RNN-based models. This way we ensure that the model capacity was on par with the other models. A consequence of this was that the number of encoding blocks was restricted to one or two layers. We used eight attention heads and a positional encoding on the inputs before feeding into the model.

4.1   Datasets

To evaluate the implementation of the RG-RNN we trained LSTMs and RG-RNNs and the transformer on three sequence datasets chosen from the benchmarks presented in [37] and a dataset of time-series sensor data. The datasets were sequential MNIST (SMNIST), including permuted sequential MNIST (PMNIST) [21], Google Speech Commands V2 (GSC) [34], QTDB [9], and a multivariate gait analysis (MGA) dataset from [14].

SMNIST and PMNIST. MNIST is a dataset consisting of hand written digits collected by the US postal service. Each sample is a 28 by 28 pixel gray scale image. The associated label of each sample represents the digit.

The SMNIST variation of MNIST is a popular benchmark for sequence models. In SMNIST the 28×28 pixel images are converted into sequences of length $T = 28 \cdot 28 = 784$ where the rows (or columns) are concatenated one after another.

PMNIST raises the complexity of the problem by permuting the 784 steps of the sequence in a random fashion. While the permutations are chosen randomly it is necessary to fix the permutation for each run since learning the problem would otherwise be impossible.

The periodicity in SMNIST is expressed through the regular arrangement of signals in the samples. The data is condensed in the center of each image which leads to patterns when concatenating the rows into sequences. In PMNIST this periodicity is deliberately disrupted through the random permutation. This removes the original causal relationship between signal and placement in the sequence.

While the loss is calculated based on each time step the accuracy is calculated using only the last prediction of the model.

GSC. The GSC dataset contains one second audio recordings. The recordings contain word utterances recorded across a wide range of devices. Each sample is labeled with one of 21 classes representing the uttered word. The dataset includes utterances that do not correspond to a class label but are instead placed into the *unknown* class. The classes are: *unknown, yes, no, up, down, left, right, on, off, stop, go, zero, one, two, three, four, five, six, seven, eight, nine.*

The task of the models is to predict the class of the utterance. As a preprocessing step the raw 16000 Hz audio signals are converted into MFCC frames. The length of the sequences would otherwise prevent the models from learning anything meaningful in a feasible amount of time [29]. The conversion was performed using `torchaudio`. The training setup was roughly based on [4]. We modified the parameters of the MFCC transformation to produce sequences of length 801 because we estimated that this would be most beneficial for the learning of meaningful temporal dynamics.

The conversion into MFCC frames transforms the problem such that the underlying frequencies are not immediately apparent. This poses an additional challenge for RG-RNNs. In its un-processed form, GSC presents itself as an interesting study of inherently periodic data that is unfortunately computationally infeasible. To leverage the resonator, repeating patterns in frequency bands have to be present and extracted by the RG-RNN from the MFCC.

While the loss is calculated based on each time step, the accuracy is calculated using only the last prediction of the model.

QTDB. The last dataset from [37] is generated from QTDB [9]. QTDB holds ECG records collected from different resources for a total of 105 patients. The ECG signals are two channel 15 min recordings, recorded at or converted to a 250 Hz sampling rate and annotated by two experts for a selection of the beats in each recording. In addition to the expert annotations machine annotations were calculated for each sample. Similar to [37] we use sequences of length 1300. We omit the termination signal that is included in [37]. To ensure the best possible quality of the training data we only considered the expert annotated samples of expert one in our experiment. Records from *Sudden Death* were excluded in the training similar to [24] since they show vast differences to all other signals. Additionally, record *sel232* was excluded due to the use of the undocumented label *A*.

The remaining signals were cut into 1300 long sequences starting from the first annotation, until the end of the sequence. The last segment was omitted if it was shorter than 1300 steps, additionally all segments that only contained 0-label steps, i.e. unannotated baseline signals, were also omitted to reduce class imbalance. Lastly, there were cases were the annotations had large gaps, which can be observed in sample *sel104*. This lead to challenges to automate the pre-processing and selection of samples, because the signal which contained the proper wave forms was not fully annotated. This would lead to problems in training. For these cases the samples were manually sighted and selected to exclude such cases. The total number of samples in the resulting dataset was 379 which was further split into a train and test set with a 80/20 split.

The task was to predict the current label that represents the phase of the ECG signal, i.e. p, N, t, or u representing the corresponding complex [3].

ECGs hold the possibility to learn both the periodic components of the signal itself and the patterns of the recurring complexes across the sequence. Both the loss and the accuracy are calculated per-time step as opposed to the other datasets.

MGA. For an additional diversification of the experiments the MGA dataset was analysed. It consists of the measured joint angles collected from ten subjects during walking on a treadmill. Each subject performed ten repetitions, i.e., step cycles and was measured under three conditions: (1) unbraced, (2) with a knee brace, and (3) with an ankle brace. The samples were constructed by concatenating the angles for a subject under a repetition into a matrix $x \in \mathbb{R}^{6 \times 101}$ where 6 is the number of joint angles (three per leg) and 101 is the number of time steps in a repetition. The task is to classify the condition of the subject.

4.2 Results

The training results of the RG-RNN follow a similar pattern on all experiments. We observe a fast decline in the loss and rise in the accuracy, immediately superseding the LSTM in both measures. This can be observed in all experiments in Fig. 3 during the first epochs. Additionally, the training stability is vastly improved which is indicated by the small error bands around the training trajectory. The difference compared to the LSTM is most clearly visible in the SMNIST and PMNIST experiment in Fig. 3 The overall performance also seems to benefit from the addition of the resonator. The model performance is recorded in Table 1 for all measures. While it is not possible for the RG-RNN to outperform the LSTM in the GSC or the Transformer in the MGA experiments the other experiments show improvements in both accuracy and loss. The experiment on MGA shows a similar convergence profile as the previously discussed experiments. The results suggest that the problem itself is easily handled by both LSTM and RG-RNN.

The Transformers performance varied across the tasks. In the SMNIST and PMNIST tasks the training curves show a close resemblance to the curves of the RG-RNN. The performance in the QTDB and GSC experiments was clearly worse than the RNN-based models. But the Transformer outperformed the RNN-based apporaches on the last MGA experiment.

5   Discussion

We were pleasantly surprised to see the RG-RNNs convergence and performance on PMNIST to follow a similar trend as in SMNIST, where information is more

Table 1. Average test results of all experiments after the final training epoch.

Benchmark	LSTM	Transformer	RG-RNN (ours)
SMNIST	98.44%	97.42%	**98.51%**
P-SMNIST	89.49%	**96.19%**	93.55%
QTDB	72.86%	55.77%	**84.46%**
GSC	**92.30%**	88.12%	91.79%
MGA	96.83%	**100%**	96.62%

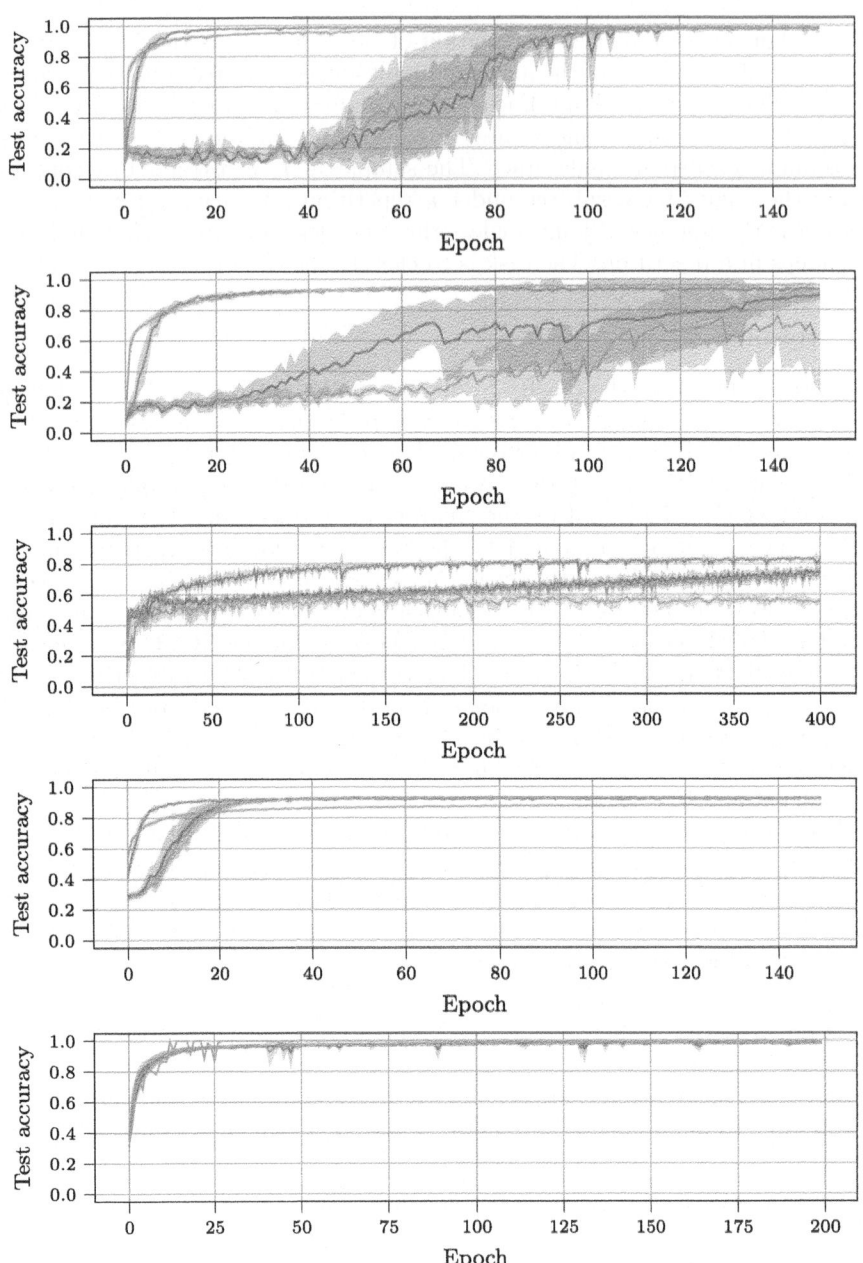

Fig. 3. Averaged test accuracy of the main experiments as described in Sect. 4:(from the top) **(1st)** SMNIST, **(2nd)** PMNIST, **(3rd)** QTDB, **(4th)** GSC, and **(5th)** MGA. The orange curve depicts the RG-RNN, the blue curve the LSTM, the green curve is a variation on the LSTM with an additional hidden unit, and the pink curve is the Transformer. The shaded area is the standard deviation between recorded runs. The curves show similar patterns across all four experiments. There is a sharp rise in performance during the first few epochs with an early convergence for the RG-RNN and the Transformer, compared to the LSTM. Additionally, the variance between each run of the RG-RNN is almost non-existing, while the LSTM shows large deviations in the curves.

clearly structured. While the regular structure of the samples is disrupted, the RG-RNN is still able to quickly converge to a level that is almost on-par with current SotA-RNNs on PMNIST [7,36]. Additionally, we suspect that the use of the sequence-labeled training regime (Sect. 4) leads to faster propagation of information compared to a conventional single-label training. Nevertheless, because the recorded performance of the LSTM is comparable to the values recorded in literature [2] we consider our approach as a valid alternative.

The improved convergence and performance on QTDB demonstrates the practical applicability of resonators and the RG-RNN on real-world data. The added inductive bias appears to facilitate the detection of key frequencies in the ECG samples. At the same time, the convergence behavior of the RG-RNN on GSC reveals a limited advantage here. We suspect that this is due to the fact that the preprocessing already extracts frequency components in form of short-term spectra. Repetitive patterns have to be extracted from within these spectrograms and the time signal directly. This suggests that, although computationally demanding, future experiments should also be performed on raw audio data to correctly assess the impact of resonators on speech detection.

These results were achieved with only a small addition of trainable parameters. Note, however, that an LSTM model with a slightly larger parameter count is not able to achieve similar performance. This further suggests that the inductive bias encoded by the resonator-gating is responsible for the improved performance rather that just an increase in parameters as is often suspected [8].

In addition to the capability to respond to specific frequency ranges within the data, another advantage of RG-RNNs is that the resonator effectively pursues an anticipatory forward propagation of information through time via its local recurrence. This shields the information from outer disturbances and noise. At the same time, the resonator provides a "gradient highway" in backward direction similar to CEC-forget gate dynamics in LSTMs but with a constant period, which seems to stabilize the overall training behavior further.

Over the last decade advances in sequence learning gave rise to new models that differ from the classic gated RNNs, namely Transformers [33], temporal convolutional networks (TCNs) [25], and, more recently, state space models (SSMs) such as S4 [13], S5 [28], and similar. While transformers are powerful due to the use of the attention mechanism and their ability to process sequences in parallel, the computation can be costly as they scale in the order $\mathcal{O}(n^2)$ with the input size [13,26,33]. TCNs utilize dilated convolutions and deep networks to build powerful embeddings [25]. In S4 and S5 the model is derived from SSMs and leverages HiPPO theory and the diagonalization of complex matrices to achieve high performance on long range dependency tasks without the use of attention [13]. Interestingly, slight parallels can be drawn from the recurrent matrix in S4 to the resonator when rewriting the state in Eq. 13 with complex parameters. If and which relation there is to SSMs can be an interesting approach to explore in future works that focuses on the theoretical embedding of resonators. Lastly, the recently published linear recurrent unit (LRU) [26] aims to solve known issues in RNNs by applying insights from S4-based models to linear RNNs. LRUs bring a number of enhancements into the vanilla RNN framework that increase the parallelization and especially performance on long-range dependencies. This promising advancement in sequence learning and in particular RNN literature can pose an opportunity for research to focus more on applications that employ RNNs and may offer an option to integrate resonators into LRUs. Future work could extend on the possibility of integrating the RG-RNNs inductive bias into LRUs as well as conduct comprehensive performance studies of the RG-RNN against current baselines.

As an initial performance comparison we implemented and trained a Transformer model on the same experiments as the RNN-based models (Fig. 3). A complete superiority of the Transformer model could not be found during our experiments. On the one hand, it showed fast convergence and even better performance in the PMNIST and MGA experiments. On the other hand, it clearly performed worse than the LSTM baseline in QTDB and GSC. These outcomes may be due to the fact that we limited the model to a similar parameter count and therefore could not use deep encoding blocks, which are often integral to the performance success of transformer models. Nevertheless, the experiments show that a RG-RNN with similar capacity to a Transformer can achieve comparable results to this modern ANN type.

Especially the experiments on SMNIST and QTDB have shown a great response of the resonator to highly regular and inherently periodic inputs. The regular structure of the samples in MNIST, which places the information into predictable intervals corresponding to the center of the image, and the combination of periodic signals that form ECGs are captured, mirrored and anticipated by the rotating state of the resonator circuit. This can be exploited by the resonator leading in principle to a fast training result and fast convergence.

5.1   Limitations

The time required to train the RG-RNNs during the performed experiments was greatly higher than of the more established methods. Because RNN output is evaluated in sequence along the time dimension speedups from parallelization via GPUs is limited. Additionally, we did not perform optimizations that are in place in `PyTorch` implementations of RNNs where major parts of the implementation are externalized into custom C++ and CUDA modules. With similar optimizations we expect an only marginally slower runtime from our additional computations.

Even though the training stability showed improvements in RG-RNNs it is yet unclear how long a time context can be chosen without the need to truncate gradients [11]. Because time sequence length is a major limitation in vanilla RNNs it is very well possible that similar effects are encountered at some point in RG-RNNs.

The lack of (great) improvement on GSC can be seen as an archetype for experiments on data where frequency information is already encoded in the input. In such cases the use of a resonator-based model may be seen as redundant.

Because the RG-RNNs performance did not worsen in our experiment a RG-RNN may often be a substitute that is in practice worth trying.

6   Conclusion

We have introduced the resonator as a means to add to the inductive bias of RNNs by enabling the detection and response to periodic signals in sequential data. The implemented RG-RNN shows improved convergence on all tested benchmarks and better performance in four out of five experiments against vanilla LSTMs with comparable performance to a transformer baseline.

We conclude that resonator-based gating offers a strong, new inductive bias for gated RNNs that is applicable to a variety of problems, but excels at detecting patterns and periodicity in the input sequence. This can be attributed to the addition of an additional internal loop, which enables the interplay of rotation and decay and opens the possibility to tune into periodicity in the data.

The resonators introduce an anticipatory forward propagation, effectively bridging information gaps across time, which we propose to be beneficial for the overall stability of the training procedure. In addition, the internal dynamics of the resonator appear to allow the RG-RNN to learn more effectively without the utilization of a memory. In the realm of follow-up research, it could be promising to blend the resonator-gating principle with the concept of state update skipping [5]. The latter introduces sparsity to state updates but also fortifies the shielding of the internal state, thereby enhancing the preservation of gradients. This fusion holds the potential for even more efficient and robust sequence learning models.

As further future work we intend to explore possible synergies of RG-RNNs embedded within larger-scale recurrent architectures, balancing resonating and non-resonating components, as well as the use of smaller resonator units without the need to be embedded in LSTMs. Additionally, a rigorous comparison of performance of RG-RNNs with current S4 and LRUs, as well as the integration of resonators into these models as well as transformers will reveal the relevance of resonator-gated and resonator-enhanced models in practice.

References

1. AlKhamissi, B., ElNokrashy, M.N., Bernal-Casas, D.: Deep spiking neural networks with resonate-and-fire neurons. CoRR abs/ arXiv:2109.08234 (2021)
2. Arjovsky, M., Shah, A., Bengio, Y.: Unitary evolution recurrent neural networks. CoRR abs arxiv: 1511.06464 (2015)
3. Ashley, E.A., Niebauer, J.: Cardiology Explained. Chapter 3, Conquering the ECG (2004). https://www.ncbi.nlm.nih.gov/books/NBK2214/
4. Bernardo, P.P., Gerum, C., Frischknecht, A., Lübeck, K., Bringmann, O.: Ultra-trail: a configurable ultralow-power tc-resnet ai accelerator for efficient keyword spotting. IEEE Trans. Comput. Aided Des. Integr. Circuits Syst. **39**(11), 4240–4251 (2020). https://doi.org/10.1109/TCAD.2020.3012320

5. Campos, V., Jou, B., Giró-i Nieto, X., Torres, J., Chang, S.F.: Skip rnn: learning to skip state updates in recurrent neural networks. In: International Conference on Learning Representations (2018)

6. Cho, K., van Merrienboer, B., Gülçehre, Ç., Bougares, F., Schwenk, H., Bengio, Y.: Learning phrase representations using RNN encoder-decoder for statistical machine translation. CoRR abs/ arXiv:1406.1078 (2014)

7. Cooijmans, T., Ballas, N., Laurent, C., Courville, A.C.: Recurrent batch normalization. CoRR arXiv: 1603.09025 (2016)

8. Frankle, J., Carbin, M.: The lottery ticket hypothesis: Training pruned neural networks. CoRR arXiv: 1803.03635 (2018)

9. Goldberger, A., et al.: Physiobank, physiotoolkit, and physionet: components of a new research resource for complex physiologic signals. Circulation **101**(23), e215–e220 (2000)

10. Granata, F., Di Nunno, F.: Forecasting evapotranspiration in different climates using ensembles of recurrent neural networks. Agric. Water Manag. **255**, 107040 (2021)

11. Graves, A.: Supervised Sequence Labelling with Recurrent Neural Networks. SCI, vol. 385. Springer (2012). https://doi.org/10.1007/978-3-642-24797-2

12. Graves, A.: Generating sequences with recurrent neural networks. CoRR abs/ arXiv: 1308.0850 (2013)

13. Gu, A., Goel, K., Ré, C.: Efficiently modeling long sequences with structured state spaces. CoRR abs/arXiv: 2111.00396 (2021)

14. Helwig, N., Hsiao-Wecksler, E.: Multivariate gait data. UCI Mach. Learn. Repository (2022). https://doi.org/10.24432/C5861T

15. Higuchi, S., Kairat, S., Bohte, S., Otte, S.: Balanced resonate-and-fire neurons. In: Forty-first International Conference on Machine Learning (2024)

16. Hinton, G.: Neural networks for machine learning, lecture 6e (2018). https://www.cs.toronto.edu/~tijmen/csc321/slides/lecture_slides_lec6.pdf, slide 26

17. Hochreiter, S., Schmidhuber, J.: Long short-term memory. Neural Comput. **9**(8), 1735–1780 (1997). https://doi.org/10.1162/neco.1997.9.8.1735

18. Huang, B., Zheng, H., Guo, X., Yang, Y., Liu, X.: A novel model based on da-rnn network and skip gated recurrent neural network for periodic time series forecasting. Sustainability **14**(1) (2022). https://doi.org/10.3390/su14010326

19. Izhikevich, E.M.: Neural excitability, spiking and bursting. Inter. J. Bifurcation Chaos **10**(06), 1171–1266 (2000)

20. Izhikevich, E.M.: Resonate-and-fire neurons. Neural Netw. **14**(6–7), 883–894 (2001). https://doi.org/10.1016/S0893-6080(01)00078-8

21. Le, Q.V., Jaitly, N., Hinton, G.E.: A simple way to initialize recurrent networks of rectified linear units. CoRR arXiv: abs/1504.00941 (2015)

22. Liu, Z., Hartwig, T., Ueda, M.: Neural networks fail to learn periodic functions and how to fix it. In: Larochelle, H., Ranzato, M., Hadsell, R., Balcan, M., Lin, H. (eds.) Advances in Neural Information Processing Systems 33: Annual Conference on Neural Information Processing Systems 2020, NeurIPS 2020, 6-12 December 2020, virtual (2020)

23. Neil, D., Pfeiffer, M., Liu, S.C.: Phased lstm: accelerating recurrent network training for long or event-based sequences. NIPS (2016). https://doi.org/10.5167/uzh-149394

24. Nurmaini, S., et al.: Electrocardiogram signal classification for automated delineation using bidirectional long short-term memory. Inform. Med. Unlocked **22**, 100507 (2021). https://doi.org/10.1016/j.imu.2020.100507

25. van den Oord, A., et al.: Wavenet: A generative model for raw audio. CoRR abs/ arXiv: 1609.03499 (2016)
26. Orvieto, A., et al.: Resurrecting recurrent neural networks for long sequences (2023)
27. Paszke, A., et al.: Pytorch: An imperative style, high-performance deep learning library. CoRR abs/ arXiv: 1912.01703 (2019)
28. Smith, J.T.H., Warrington, A., Linderman, S.W.: Simplified state space layers for sequence modeling (2023)
29. Sutskever, I.: Training Recurrent Neural Networks. Ph.D. thesis, University of Toronto, Canada (2013)
30. Sutskever, I., Vinyals, O., Le, Q.V.: Sequence to sequence learning with neural networks. CoRR abs/ arXiv: 1409.3215 (2014)
31. Tolmachev, P., Dhingra, R.R., Pauley, M., Dutschmann, M., Manton, J.H.: Modeling the respiratory central pattern generator with resonate-and-fire izhikevich-neurons. In: Neural Information Processing, vol. 11301, pp. 603–615. Cham (2018)
32. Vaswani, A., et al.: Attention is all you need. In: Guyon, I., Luxburg, U.V., Bengio, S., Wallach, H., Fergus, R., Vishwanathan, S., Garnett, R. (eds.) Advances in Neural Information Processing Systems, vol. 30. Curran Associates, Inc. (2017)
33. Vaswani, A., et al.: Attention is all you need. CoRR abs/ arXiv: 1706.03762 (2017)
34. Warden, P.: Speech Commands: A Dataset for Limited-Vocabulary Speech Recognition. ArXiv e-prints (Apr 2018)
35. Werbos, P.J.: Generalization of backpropagation with application to a recurrent gas market model. Neural Netw. **1**(4), 339–356 (1988). https://doi.org/10.1016/0893-6080(88)90007-X
36. Wisdom, S., Powers, T., Hershey, J., Le Roux, J., Atlas, L.: Full-capacity unitary recurrent neural networks. In: Lee, D., Sugiyama, M., Luxburg, U., Guyon, I., Garnett, R. (eds.) Advances in Neural Information Processing Systems, vol. 29. Curran Associates, Inc. (2016)
37. Yin, B., Corradi, F., Bohté, S.M.: Accurate and efficient time-domain classification with adaptive spiking recurrent neural networks. Nat. Mach. Intell. **3**(10), 905–913 (2021). https://doi.org/10.1038/s42256-021-00397-w

Towards a Model of Associative Memory with Learned Distributed Representations

Matej Fandl$^{(\boxtimes)}$ ⓘ and Martin Takáč ⓘ

Centre for Cognitive Science, Faculty of Mathematics, Physics and Informatics, Comenius University, Bratislava, Slovakia
matej.fandl@fmph.uniba.sk, martin.takac@fmph.uniba.sk
http://cogsci.fmph.uniba.sk/

Abstract. Associative memory is a device capable of storage of data and its retrieval from incomplete or noisy probes. This article describes a neural model of associative memory inspired by continuous Modern Hopfield networks. The proposed learning procedure produces distributed representations of the fragments of input data which collectively represent the stored memory patterns, governed by the activation dynamics of the network. This allows for effective storage of data without the need to grow the network. In comparison to training by error back propagation, the training procedure is relatively fast (few-shot learning).

Keywords: Associative memory · Hopfield networks · Competitive learning · Distributed representations

1 Introduction

Associative memory is a system that allows access to stored content with partial or noisy probes. This is different from traditional retrieval techniques in computers, such as reading data from a particular address or using queries entered as text. The associative memory allows searching by the content alone, hence it is also called content-addressable.

Computational models of associative memory have been an active research field for decades [4,5,16]. Probably the best known is the Hopfield network [4]—a recurrent fully-connected attractor neural network capable of storage and retrieval of binary patterns. It allows for one-shot Hebbian learning [3] of new memories and does not require synchronized firing of the units, but suffers from low storage capacity [4,6,12].

Recently the Hopfield networks came to the spotlight again thanks to a discrete model called Dense Associative Memory (DAM) [6]. DAM is a generalization of the original Hopfield model that solves its capacity problems. Its updated energy function uses an interaction function that can help emphasize the differences between patterns that are close together and avoid spurious attractors. This work triggered a new line of research about a new generation of models of associative memory that are now called Modern Hopfield networks. A different

© The Author(s), under exclusive license to Springer Nature Switzerland AG 2024
M. Wand et al. (Eds.): ICANN 2024, LNCS 15016, pp. 226–241, 2024.
https://doi.org/10.1007/978-3-031-72332-2_16

interaction function that further increased the capacity of the discrete model was introduced in [2]. A continuous version that is fully differentiable and therefore usable in end-to-end trained deep learning architectures was derived in [12] and was proven equivalent to self-attention mechanism in transformer neural networks [15]. A general framework that can be used to construct wide variety of these models was proposed in [11] and [7] described a way to construct its arbitrarily deep variants.

This paper explores a novel method of training models from the Modern Hopfield networks family. Related works either directly store the whole patterns in the model weights, or use backpropagation [6]. The former training method is one-shot, but the space complexity of the model and computational complexity of memory retrieval grow with the number of stored patterns. The latter method is slow and not considered biologically plausible [10]. Our approach aims to induce division of labor between partial representations in the network's weights so that they collectively represent the patterns to be stored. This can help keep the model's space and time complexity reasonably small, which leads to lower energy demands and possibility to use the model on wider range of consumer-grade computational devices.

2   Continuous Modern Hopfield Networks

The model proposed in [12] is capable of storage and reconstruction of continuous patterns of length d. After setting the state of the network $\boldsymbol{\xi} \in \mathbf{R}^d$ to values of a probe (an incomplete or noisy pattern), the networks state falls into an attractor basin of one of its point attractors. These point attractors are points in the state space where the memories are stored and towards which the state of the network evolves. The evolution of the networks state is governed by its activation dynamics defined by the following state update rule:

$$\boldsymbol{\xi}^{new} = \boldsymbol{W} \operatorname{softmax} \left(\beta \boldsymbol{W}^\top \boldsymbol{\xi} \right), \tag{1}$$

where $\boldsymbol{W} \in \mathbf{R}^{d \times N}$ is a matrix, each column $\boldsymbol{W}_{:,i} = \boldsymbol{w}_i$ of which corresponds to one stored memory and β is an inverse temperature parameter. From now on, we will refer to these stored memories as *memory vectors*.

This model can be interpreted as a network with two layers—the visible layer of input/feature neurons and the hidden layer of memory neurons [7]. Weight vectors of the individual neurons are the memory vectors mentioned above.

After the network is probed with an input $\boldsymbol{\xi}$, the neural activations on the visible layer are projected to the hidden layer via the weight matrix $\boldsymbol{W}$. The term $\boldsymbol{W}^\top \boldsymbol{\xi}$ expresses the similarity between the input and weight vectors of the individual memory neurons. The softmax function introduces competition between hidden neurons (as it makes their activations sum to 1). The parameter β regulates how much will the most similar neurons be emphasised and the less similar ones suppressed. Multiplying the weight matrix $\boldsymbol{W}$ by the softmax term can be interpreted as a top-down projection of the memory vectors weighted by their activations back to the visible layer.

For higher values of β, this update rule leads to a state very close to a particular memory vector—a stable state; for lower values it leads to a mixture of multiple memory vectors—a meta-stable state [12]. As we will show below, these meta-stable states are important for our work.

The simplest (one-shot) method for storing patterns in the Modern Hopfield network consists in adding a new memory neuron for each pattern we want to store and making the weights of this neuron equal to the pattern [7]. With this procedure, however, the computational complexity of the model grows with every pattern we want to store and the space demands of the model grow as well, which is not desirable.

3   Our Model

Our model retains the architecture described in [7] in that it consists of a visible layer of feature neurons and a hidden layer of memory neurons, but it differs in the state update rule and the method for storing patterns, which we will now describe in turn.

3.1   Activation Dynamics

Same as in Eq. 1, we want the activation dynamics to produce reconstructions that are linear combinations of the weight vectors of memory neurons. However, in order to use the memory capacity effectively, the weight vectors should not necessarily represent the whole patterns, but distributed fragments (see Fig. 1). The function computing the similarity between the weight vector of the memory neuron and the input should ensure that the memory neuron works as a detector of a fragment it represents in the input pattern, effectively ignoring the rest of the input. We achieve this by interpreting the input features (and the corresponding weight vectors) in a special way: each component represents the presence (1) or absence (−1) of a feature. Continuous values between −1 and 1 represent the degree of certainty about the presence/absence of the feature, with 0 representing a complete ambiguity. This, in combination with using cosine similarity for vector comparisons, will ensure that zeros or very small values in the weight vector will serve as a wildcard, i.e. giving a low importance to values of corresponding components in the input. Thus, different memory neurons can specialize in detecting and representing different input features.

Zero components in the input pattern can be interpreted as lack of information about the corresponding feature, and serve for representing masked probes or partial memory cues. Retrieval of a complete pattern from a partial cue is the most typical application of associative memory. The similarity s_i of the i-th weight vector $\boldsymbol{w}_i$ with the input $\boldsymbol{v}$ is computed as

$$s_i\left(v\right) = \operatorname{cossim}\left(\boldsymbol{w}_i \circ \boldsymbol{v}, \boldsymbol{v} \circ \boldsymbol{v}\right), \tag{2}$$

where cossim is cosine similarity between its two arguments and $\circ$ is Hadamard (element-wise) product of two vectors. The element-wise multiplication of both

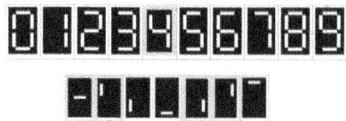

Fig. 1. Schematic of distributed representations of digits. In the second row, there is 7 manually created parts that can be used to assemble any of the digits in the first row. Highlighted are parts that form number 4. This is an example to illustrate what we want to achieve. The aim of our work is to design an algorithm capable of automatic formation of these parts.

w_i and v with v before applying cosine similarity effectively erases the masked (close-to-zero) input components from the weight vector argument, causing the weight vectors only compete in their non-masked parts, which is crucial for successful reconstruction of the masked content. Cosine similarity has a desirable property of its value being bounded between –1 and 1, which helps to keep the weights from growing significantly during training (see Sect. 3.2).

After computing similarity between the input pattern and the memory neurons, the state is updated using the rule

$$\boldsymbol{\xi}^{new} = \boldsymbol{W} \boldsymbol{s}^{\circ\beta}, \tag{3}$$

where $\boldsymbol{W}$ is a column matrix of weight vectors $\boldsymbol{w}_i$, $\boldsymbol{s} = \begin{bmatrix} s_0 \ s_1 \ \cdots \ s_m \end{bmatrix}^\top$ is a vector of similarities and $s_i = \mathrm{cossim}\left(\boldsymbol{w}_i \circ \boldsymbol{\xi}, \boldsymbol{\xi} \circ \boldsymbol{\xi}\right)$. The $\circ$ operator is Hadamard power—element-wise exponentiation. β is used to emphasize or reduce the differences between individual memories (β is required to be an odd number in case the sign of s_i is expected to be preserved).

This update rule allows us to also utilize memories that are far from the current state. While a positive similarity value determines the degree with which we want to include the weights of this neuron in reconstruction, a negative value determines how much we do *not* want them in the reconstruction. The units with negative similarity values participate in the reconstruction by pushing it away from the undesired states. This can be thought of as "wanting the opposite of this memory vector".

3.2   Training

Unlike in the naïve training procedure that consists in adding a new neuron with the whole pattern in its weights each time we want to store a new pattern, we keep the number of hidden neurons fixed. The goal of the proposed training procedure is to achieve division of labor in that the neurons learn to represent frequently occurring fragments rather than full patterns. We achieve this by asynchronous training of a selected subset of hidden neurons for a presented

---

Algorithm 1 One epoch of training procedure.

---
for *pattern* **in** *allPatterns* **do**
   residue ← *pattern* − *reconstruction*
   for *iteration* ← 0 **to** *maxIterations* **do**
      cw ← *closestWeightsTo* (*residue*)
      updateWeights (*cw*, *residue*)
      residue ← *residue* − *cw* · $s^{\circ\beta}$
      if *norm* (*residue*) < *threshold* **then**
         break
      end if
   end for
end for

---

input pattern. Hence, we distinguish between training *epochs* and *iterations*. One epoch consists of presenting all input patterns in a training set. For every presented pattern, a number of update iterations is performed, selecting a different neuron to be updated in each iteration.

The pseudocode of the training procedure is described in Algorithm 1. When a new input pattern is presented, we first compute its reconstruction using Eq. 3 ($\beta_{training}$ is used as the β parameter in the reconstruction during training). The difference between the pattern and the reconstruction is the *residue*. The goal of the training is to eliminate the residue. The division of labor between the neurons is achieved by training them to represent the residue (i.e., the remaining underrepresented fragments), rather than the whole pattern. We select the neuron with weights most similar to the residue (see details below) and move them even closer. Then we present the residue as the new input and the next iteration starts. In each iteration, we find the memory vector most similar to the current residue (Eq. 2) and update it so it represents it better (Eq. 5).

The whole procedure is repeated until the residue becomes small enough (the *residue norm threshold* parameter) or until a predefined maximum number of iterations (the *max iterations* parameter) has been reached.[1] Then we present the next input pattern from the training set.

The top-down pass in Eq. 3 can be split into individual steps for the weights of memory neurons: $s_0^\beta \boldsymbol{w_0} + s_1^\beta \boldsymbol{w_1} + \cdots + s_M^\beta \boldsymbol{w_M}$. The proposed algorithm mimics the reverse of this process by finding and updating the closest neurons one by one and removing their contribution $s_i^\beta \boldsymbol{w_i}$ from the actual residue.

Neuron Selection. Selection of the neuron to be updated happens in one or two passes. In the first pass, we simply find the neuron whose weight vector is the most similar to the residue. If the similarity is sufficient (bigger than the *similarity threshold* parameter), we update the weights of this neuron. Otherwise,

---

[1] During iterative updates for one input pattern, the neurons are selected *without repetition*. This means that the upper bound on the number of iterations is the total number of memory neurons, however, the predefined limit for maximum number of iterations is set to a smaller value to ensure sparsity of the representations.

we prefer to imprint (with a higher learning rate) the residue to a different, ideally not trained yet, neuron. In that case, a second search is attempted where neurons that have been less updated so far are favored. We do this by tracking the amount of updating of each memory vector with a value $\phi_i \mid 0 \leq \phi_i \leq 1$. This value is set to 0 in the beginning of training for each memory vector and is increased toward 1 proportionally to the neuron's updates with

$$\phi_i = (1 - \alpha_i') \cdot \phi_i + \alpha_i', \tag{4}$$

where α_i' is a neuron specific learning rate (Eq. 6). The second pass is searching for $\mathrm{argmax}_i(s_i(1 - \phi_i))$. A similar mechanism to favor unused neurons was used in [14].

The Weight Update Rule. Weight changes are computed with the update rule

$$\boldsymbol{w_i}^{t+1} = \boldsymbol{w_i}^t + \alpha_i' \left(\boldsymbol{v} - \boldsymbol{w_i}^t \right), \tag{5}$$

where $\boldsymbol{v}$ is the target vector we are moving the weight $\boldsymbol{w_i}$ closer to, and α_i is a neuron specific learning rate, which is computed as follows

$$\alpha_i' = \begin{cases} \alpha^I \cdot s_i & \text{selected in the second pass,} \\ \alpha \cdot s_i(1 - \phi_i) & \text{otherwise,} \end{cases} \tag{6}$$

where α is a fixed learning rate parameter and α^I is a value higher than α that is used in case we want to imprint the current pattern.

The value $(1 - \phi_i)$ in this context is effectively the "willingness to learn" of a neuron—a form of self-inhibition. This mechanism is similar to conscience and habituation mechanisms used in Self-Organizing Maps [13] or the adaptive threshold in the BCM learning rule [1].

4   Experiments

Here we report two pilot experiments with our model. First, we tested the performance of our method on a subset of the MNIST dataset of handwritten digits [9], then on a more complex Omniglot dataset [8].

4.1   The MNIST Dataset

The MNIST dataset [9] consists of 70 000 examples, 60 000 in the training set and 10 000 in the test set. Our training set comprises of 1000 randomly sampled exemplars from the training set—100 for each digit. Images – 28×28 matrices – were reshaped to vectors of length 784 and their component values remapped to the interval $[-1, 1]$. The network used in this experiment had 784 input neurons and 100 hidden neurons. The parameters of the model are in Table 1. Initial memory vectors were sampled from Gaussian distribution with $\mu = 0$ and $\sigma = 0.1$.

Table 1. Model parameters used during both the MNIST and Omniglot experiments.

Training Parameter	Value
α	0.2
α_I	1.0
$\beta_{training}$	1.0
max iterations	10
residue norm threshold	0.1
similarity threshold	0.1

Training Progress Although our algorithm does not learn one-shot, we hypothesised that it would reach a good performance after a very few epochs of training. That is why we did not use a stopping condition but ran it for a fixed number (100) of epochs[2] and we examined the reconstruction quality after each epoch of training for various values of β. As a measure of reconstruction quality we took a standard cosine similarity between the input pattern and its 1-step reconstruction using Eq. 3 averaged for all input patterns from the training set in an epoch. Figure 2 shows that, for values of $\beta = 1$ and $\beta = 3$, the learned representations are already useful after the first epoch and reach the peak performance in epoch 2 for $\beta = 3$, then after a slight drop $\beta = 3$ reconstruction quality plateaus around the same value as $\beta = 1$ (approx. 0.93). Higher values of β show steep drop in performance in the first epoch and then plateau as well.

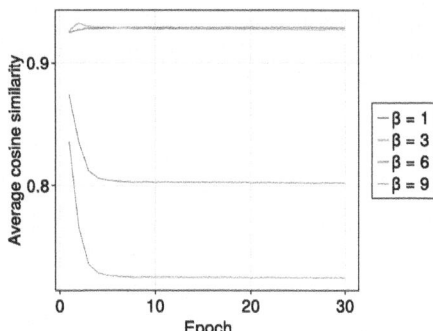

Fig. 2. The evolution of reconstruction quality during training in 30 epochs and the effect of β.

To get a better idea of how good are the values of achieved reconstruction similarity, we calculated average values of cosine similarity between exemplars

---

[2] Figure 2 only shows the first 30 epochs, because after that there was very little change for all values of β.

from different classes (Table 2). For values of $\beta = 1$ and $\beta = 3$, our algorithm reaches the reconstruction values above 0.92, while similarities between different exemplars from the same class are rarely above 0.8.

Table 2. Average cosine similarities between exemplars from different classes in the MNIST subset used in our experiments (intra-class similarities between different exemplars of the same digit are on the main diagonal). These values are properties of our dataset, unrelated to our algorithm.

	0	1	2	3	4	5	6	7	8	9
0	0.72	0.65	0.64	0.65	0.65	0.67	0.67	0.66	0.66	0.65
1	0.65	0.87	0.73	0.74	0.74	0.75	0.74	0.76	0.77	0.77
2	0.64	0.73	0.73	0.69	0.69	0.68	0.70	0.68	0.70	0.69
3	0.65	0.74	0.69	0.76	0.69	0.73	0.67	0.70	0.72	0.71
4	0.65	0.74	0.69	0.69	0.77	0.71	0.73	0.75	0.71	0.77
5	0.67	0.75	0.68	0.73	0.71	0.74	0.72	0.72	0.73	0.74
6	0.67	0.74	0.70	0.67	0.73	0.72	0.79	0.70	0.71	0.74
7	0.66	0.76	0.68	0.70	0.75	0.72	0.70	0.81	0.72	0.77
8	0.66	0.77	0.70	0.72	0.71	0.73	0.71	0.72	0.77	0.74
9	0.65	0.77	0.69	0.71	0.77	0.74	0.74	0.77	0.74	0.81

As another naïve baseline, we computed the average cosine similarity of training patterns with the vector obtained by averaging all the patterns in the training set. An algorithm that would always return this average as reconstruction, would achieve the similarity 0.845 for the MNIST subset.

Because the main role of the associative memory is to reconstruct distorted or incomplete patterns, we further examined the effect of three kinds of distortion of patterns on retrieval: additive Gaussian noise, missing/masked parts in patterns and reconstruction of patterns not used in training. We used the best performing model (reconstructing with $\beta = 3$ after 2 epochs of training) for each of these tasks.

Noise. In this test, the probes were generated from the training set exemplars by adding noise sampled from Gaussian distribution with $\mu = 0$ and $\sigma \in \{0, 0.1, 0.2, \cdots, 1\}$ to each component. We measured the cosine similarity of the reconstruction with the original training set exemplar (i.e., without the noise). Figure 3 shows the reconstruction similarity for different values of β.[3] The reconstruction is very robust—increasing σ up to the value of 1 only had slightly detrimental effect on the performance of the model. The performance of $\beta = 1$ shows the slowest decrease and exceeds the performance of $\beta = 3$ for $\sigma \geq 0.7$.

---

[3] Note the difference between $\beta_{training}$ and β. While the former was fixed during training, we varied the latter during the reconstruction tests.

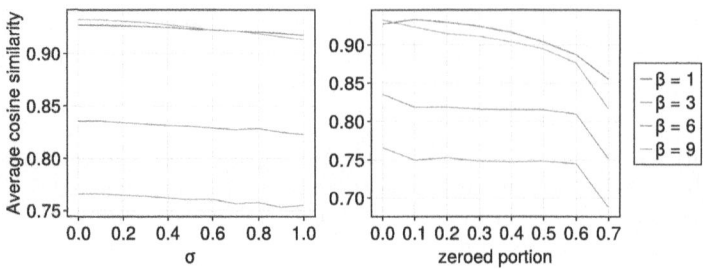

Fig. 3. The effect of noise (left) and masking (right) on reconstruction quality.

Incomplete Patterns. In this test, the probes were generated by setting a proportion of components of each training pattern to 0, thus masking/hiding a part of the pattern. The mask was a continuous block of zeroes from right end to the left. The effect of this kind of distortion is visible in Fig. 3. Interestingly, masking 10 % of the pattern slightly increases performance when reconstructing with $\beta = 1$ and this value achieves the best overall performance. The reconstruction similarity is still over 0.9 even if 50% of each input pattern has been masked. Masking more than 60% of the pattern yields a steep decrease in performance.

Unseen Digits. In the third test we probed the model with test set exemplars not present in the training set, but sampled from the same distribution as the training set. The expectation behind this test was that the digits of some category, while not being in the training set, would contain roughly the same features as the digits used to train the network and the reconstruction of such digits should not be significantly worse. Probes were generated by picking 100 random *test* set exemplars from each category.

Table 3. Reconstruction quality of seen and unseen digits. The reported values are cosine similarities between the probe and the reconstruction without an added noise or any other kind of distortion.

β	training set	test set
1	0.927	0.920
3	0.933	0.923
6	0.836	0.831
9	0.766	0.760

Table 3 shows that the drop in cosine similarity when reconstructing unseen probes does not exceed 0.01 for all reported values of β.

Distributed Representations. We further analysed the stored weight vectors and their contributions to reconstruction to verify whether the hidden neurons really divide the labor and represent the input patterns collectively. Figure 4 shows the contribution of different memory vectors to reconstruction for values of $\beta = 1$ and $\beta = 3$. There is significant difference in the number of contributing units. While for $\beta = 1$ almost all vectors contribute somehow, for $\beta = 3$ there is just a couple of neurons participating. Majority of the neurons participate positively. Negative contributions are negligible. Because low values of β achieved the best performance, the reported results hint at formation of distributed representations. The higher the value of β, the lower amount of neurons participate in reconstruction – this effect is caused by $^{\circ \beta}$ in Eq. 3.

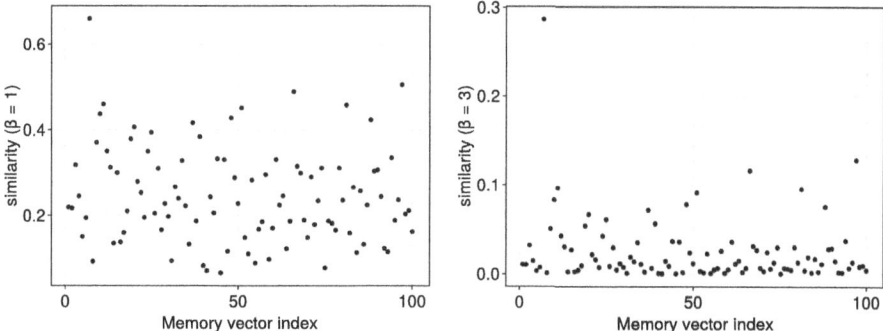

Fig. 4. A significant difference in the amount of neurons participating in reconstruction $(s^{\circ \beta})$ for $\beta = 1$ (left) and $\beta = 3$ (right) for a randomly selected training pattern. While in the former case almost all the neurons participate in reconstruction, in the latter case a lot fewer units effectively contribute.

Figure 5 shows the memory vectors from epoch 2 reshaped to dimensions 28×28 and plotted as images (with red pixels encoding negative values, blue pixels encoding positive values and gray pixels encoding values close to zero). All memory vectors are fragments; none represents a complete training pattern. Most of the memory vectors resemble individual digits, where negative values surround the positive values, but large parts of the background are close to 0. The negative components on the boundary of the positive ones ensure that the memory vector encodes a specific shape. A couple of neurons learned what looks like features of digits (a line shared by numbers 6 and 8, a cross in number 7 with a line in the middle). This corroborates our hypothesis that the stored representations are to a large extent distributed.

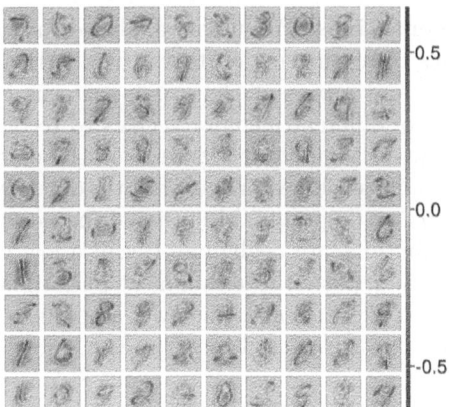

Fig. 5. Memory vectors formed after training on a subset of the MNIST dataset ($\beta =$ 3, epoch 2). Some memory vectors resemble the whole exemplars, others represent fragments.

Figure 6 shows the reconstructions of undistorted (left) and distorted (right) training patterns for different values of β. In all reported cases ($\beta \in \{1, 3, 6, 9\}$) the reconstructions resemble the probe and seem to fall into the correct category. Higher values of β lead to lower background component values in reconstructions, but the digits are perceptible. The reason for this is the aforementioned lower amount of neurons participating in reconstruction for higher values of β and the fact that the individual memory vectors do not encode full background.

4.2 Omniglot Experiments

While we consider the results on MNIST dataset promising and encouraging for further work on our model, MNIST only contains 10 classes with relatively low variation between the individual members. To see how our model performs on a more challenging dataset, we performed a second pilot experiment on a subset of Omniglot [8]. Our subset consisted of 964 exemplars, where each exemplar was from a different character class. Same as for MNIST, we computed a naïve baseline as an average cosine similarity between the training patterns and their average. This yielded the value 0.853, comparable to the value for MNIST (0.845). In Omniglot, however, the mean background to foreground ratio is 0.92 compared to 0.809 for MNIST. Thus, the background in Omniglot exemplars contributed more to the measured average similarity with an average Omniglot pattern. The Omniglot patterns were 105×105, hence the model had 11025 input neurons and 100 hidden ones (same as in the MNIST experiment). The training algorithm in both MNIST and Omniglot experiments used the same values of hyperparameters (Table 1).

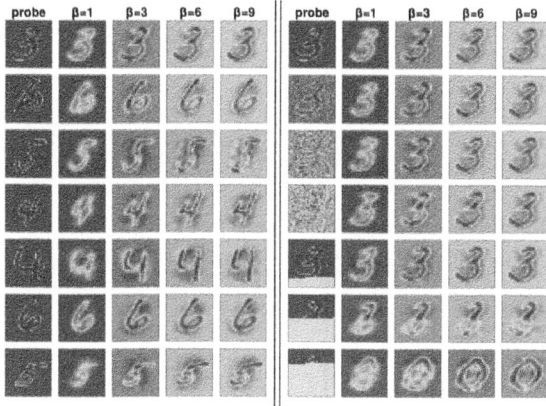

Fig. 6. Seven selected training set exemplars (left) with their reconstructions and examples of noise and masking effect on reconstruction quality (right). Columns are organized such that β grows from left to right. The amount of negative components is degrading with the growing value of β. The noise σ values used (top to bottom) are 0, 0.2, 0.5 and 1. The masked portion ratios were 0.2, 0.5 and 0.7.

Results. Figure 7 summarizes the results of the Omniglot experiment. The learned representations have structure similar to the ones from experiments on MNIST (Fig. 5): memory vectors are diverse, visually interpretable and represent fragments of the training patterns. The highest average reconstruction quality on clean patterns is 0.923. Effects of noise and masking on model performance are also similar. Noise up to the studied $\sigma = 1$ does not have a strong effect on the reconstruction quality, but the similarity of masked patterns drops more rapidly than in the case of MNIST. In summary, the performance of the proposed algorithm on the two datasets is similar.

5   Discussion

While the experiments we performed so far are limited and deeper understanding of the model is needed, these preliminary results are encouraging.

In the reported experiments we attempted to store 1000 MNIST patterns and 964 Omniglot patterns in a model with 100 hidden neurons. This value is lower by an order of magnitude than the cardinality of our training sets.

The experimental results support our hypothesis that the proposed learning rule will lead to division of labor between units on the hidden layer. The model learns visually interpretable distributed representations of exemplars from both training datasets. The reconstructed patterns are identifiable as members of the correct category for relatively high values of noise and for slightly masked patterns. The learned representations already appear useful after the first training epoch for certain reconstruction configurations.

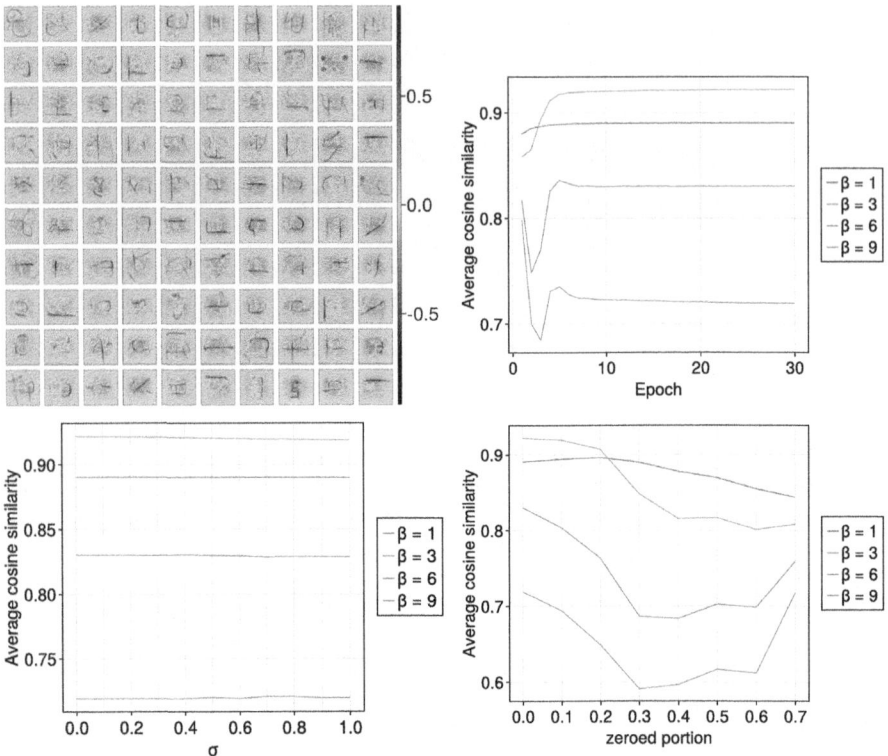

Fig. 7. Results of Omniglot experiments. Top left: learned memory vectors. Top right: the reconstruction quality during training. Bottom left: reconstruction of noisy patterns. Bottom right: reconstruction of masked patterns.

5.1 Limitations

The presented results are preliminary. While we consider them to be sufficiently promising to continue our work, the experiments themselves are rather limited to draw solid conclusions about the behavior of our model. More thorough experiments including ablation studies and exploration of sensitivity to different parameter values are needed. It will be important to investigate training on more complex datasets and also on datasets that do not fall into visual domain. In order to understand the learning dynamics and properties of the proposed model, it would also be beneficial to theoretically analyze whether the proposed algorithm minimizes the modern Hopfield energy function.

The achieved reconstruction similarity is higher than the naïve baseline (mean exemplar similarity with the average of the whole training set) and is sufficient for visually verifying that the reconstruction is similar to the probe in most cases, but the reconstruction is far from perfect. The resulting patterns might miss details important for real world tasks. For exemplars from the visual domain, we ideally aim for reconstruction visually indistinguishable from the probe.

The reason for the performance plateauing during longer training (Fig. 2) is not understood at the moment. Perhaps it can be improved by finding appropriate parameter values, but understanding causal relationships between the training dynamics, model performance and effects of individual parameters need to be further analysed, both theoretically and empirically.

In the presented experiments we aimed to get a reconstruction with *direction* as close as possible to the corresponding training pattern, ignoring its norm. Currently, the norm of the reconstruction depends on the number of participating neurons (Fig. 6). Further investigation into updating the activation dynamics to account for this, i.e. by employing hyperbolic tangent, is needed.

6 Conclusion

We introduced a new model of associative memory inspired by Continuous Modern Hopfield networks [12]. The model is designed with the goal to learn distributed representations of training data with a fixed number of neurons in the hidden layer. Reconstructions are assembled from parts stored in memory vectors. We proposed a competitive training procedure that emphasizes gradual learning of parts of the training patterns by multiple units. We evaluated the performance of the model trained on a subset of the MNIST dataset by measuring the reconstruction quality of original as well as distorted patterns, seeing a good robustness in the presence of noise and incomplete patterns. An important novel contribution of this work is the speed of training (several epochs in comparison to usually much slower backpropagation training) and lower memory capacity demands in comparison with models explicitly storing the whole memory patterns.

The analysis of memory vectors participating in the assembly of reconstructions shows that the representations are really distributed. Important next steps are better understanding of the impact of various parameters on formation of the hidden representations, theoretical analysis of the model's energy function and increasing overall reconstruction quality.

Acknowledgments. This research was supported by Slovak Grant Agency for Science (VEGA) project 1/0373/23 and Comenius University grant UK/1154/2024.

Competing Interests. The authors have no competing interests to declare that are relevant to the content of this article.

References

1. Bienenstock, E.L., Cooper, L.N., Munro, P.W.: Theory for the development of neuron selectivity: orientation specificity and binocular interaction in visual cortex. J. Neurosci. Off. J. Soc. Neurosci. **2**(1), 32–48 (1982). https://doi.org/10.1523/JNEUROSCI.02-01-00032.1982
2. Demircigil, M., Heusel, J., Löwe, M., Upgang, S., Vermet, F.: On a model of associative memory with huge storage capacity. J. Stat. Phys. **168**(2), 288–299 (2017). https://doi.org/10.1007/s10955-017-1806-y
3. Hebb, D.O.: The Organization of Behavior; A Neuropsychological Theory. Wiley, Hoboken (1949)
4. Hopfield, J.J.: Neural networks and physical systems with emergent collective computational abilities. Proc. Natl. Acad. Sci. **79**(8), 2554–2558 (1982). https://doi.org/10.1073/pnas.79.8.2554
5. Kanerva, P.: Sparse Distributed Memory. MIT Press, Cambridge (1988)
6. Krotov, D., Hopfield, J.J.: Dense associative memory for pattern recognition. In: Lee, D.D., Sugiyama, M., von Luxburg, U., Guyon, I., Garnett, R. (eds.) Advances in Neural Information Processing Systems 29: Annual Conference on Neural Information Processing Systems 2016, Barcelona, Spain, 5–10 December 2016, pp. 1172–1180 (2016)
7. Krotov, D., Hopfield, J.J.: Large associative memory problem in neurobiology and machine learning. In: 9th International Conference on Learning Representations, ICLR 2021, Virtual Event, Austria, 3–7 May 2021. OpenReview.net (2021)
8. Lake, B.M., Salakhutdinov, R., Tenenbaum, J.B.: The Omniglot challenge: a 3-year progress report. Curr. Opin. Behav. Sci. **29**, 97–104 (2019). https://doi.org/10.1016/j.cobeha.2019.04.007
9. LeCun, Y., Cortes, C., Burges, C.: MNIST handwritten digit database. ATT Labs, vol. 2 (2010). http://yann.lecun.com/exdb/mnist
10. Lillicrap, T.P., Santoro, A., Marris, L., Akerman, C.J., Hinton, G.: Backpropagation and the brain. Nat. Rev. Neurosci. **21**(6), 335–346 (2020). https://doi.org/10.1038/s41583-020-0277-3
11. Millidge, B., Salvatori, T., Song, Y., Lukasiewicz, T., Bogacz, R.: Universal Hopfield networks: A general framework for single-shot associative memory models. In: Chaudhuri, K., Jegelka, S., Song, L., Szepesvári, C., Niu, G., Sabato, S. (eds.) International Conference on Machine Learning, ICML 2022, Baltimore, Maryland, USA, 17–23 July 2022. Proceedings of Machine Learning Research, vol. 162, pp. 15561–15583. PMLR (2022)
12. Ramsauer, H., et al.: Hopfield networks is all you need. In: 9th International Conference on Learning Representations, ICLR 2021, Virtual Event, Austria, 3–7 May 2021. OpenReview.net (2021)
13. Rizzo, R., Chella, A.: A comparison between habituation and conscience mechanism in self-organizing maps. IEEE Trans. Neural Netw. **17**(3), 807–810 (2006). https://doi.org/10.1109/TNN.2006.872354
14. Sacouto, L., Wichert, A.: Competitive learning to generate sparse representations for associative memory. Neural Netw. **168**, 32–43 (2023). https://doi.org/10.1016/j.neunet.2023.09.005

15. Vaswani, A., et al.: Attention is all you need. In: Guyon, I., et al. (eds.) Advances in Neural Information Processing Systems 30: Annual Conference on Neural Information Processing Systems 2017, Long Beach, CA, USA, 4–9 December 2017, pp. 5998–6008 (2017)
16. Willshaw, D.J., Buneman, O.P., Longuet-Higgins, H.C.: Non-holographic associative memory. Nature **222**(5197), 960–962 (1969). https://doi.org/10.1038/222960a0

Neural Architecture Search

Accelerated NAS via Pretrained Ensembles and Multi-fidelity Bayesian Optimization

Houssem Ouertatani[1,3]([✉]), Cristian Maxim[1], Smail Niar[2],
and El-Ghazali Talbi[3]

[1] IRT SystemX, Palaiseau, France
`houssem.ouertatani@irt-systemx.fr`
[2] UPHF & LAMIH UMR CNRS, Valenciennes, France
[3] Univ. Lille, CNRS, Inria, Centrale Lille, UMR 9189 CRIStAL, 59000 Lille, France

Abstract. Bayesian optimization (BO) is a black-box search method particularly valued for its sample efficiency. It is especially effective when evaluations are very costly, such as in hyperparameter optimization or Neural Architecture Search (NAS). In this work, we design a fast NAS method based on BO. While Gaussian Processes underpin most BO approaches, we instead use deep ensembles. This allows us to construct a unified and improved representation, leveraging pretraining metrics and multiple evaluation fidelities, to accelerate the search. More specifically, we use a simultaneous pretraining scheme where multiple metrics are estimated concurrently. Consequently, a more general representation is obtained. A novel multi-fidelity approach is proposed, where the unified representation is improved both by high and low quality evaluations. These additions significantly accelerate the search time, finding the optimum on NAS-Bench-201 in an equivalent time and cost to performing as few as 50 to 80 evaluations. The accelerated search time translates to reduced costs, in terms of computing resources and energy consumption. As a result, applying this NAS method to real-world use cases becomes more practical and not prohibitively expensive. We demonstrate the effectiveness and generality of our approach on a custom search space. Based on the MOAT architecture, we design a search space of CNN-ViT hybrid networks. The search method yields a better-performing architecture than the baseline in only 70 evaluations.

Keywords: Neural Architecture Search · deep ensembles ·
simultaneous pretraining · multi-fidelity BO

1 Introduction

The design of high-quality neural network architectures has been a major driver of the impressive advances seen in deep learning and its applications. Many of the introduced improvements in widely-used architectures come from years of iterations and trial-and-error, guided by a set of general best practices and intuitions.

M. Wand et al. (Eds.): ICANN 2024, LNCS 15016, pp. 245–260, 2024.
https://doi.org/10.1007/978-3-031-72332-2_17

Neural architecture search (NAS) [1] seeks to automate this design process. The goal is to find the best performing architecture(s), according to one or many criteria, in a large search space. NAS can be useful in adapting a widely used architecture to a specific use case, or in building new architectures.

NAS consists of 3 main components: The search space, the search method, and the architecture performance evaluation method. Many search methods have been successfully used to perform NAS, including reinforcement learning [9,30], evolutionary algorithms [4,13,14], local search [26], and Bayesian optimization [25]. The evaluation of potential architectures is usually expensive, as it consists of a training phase and a test phase. With considerations of cost, energy consumption, and practicality, there is a special focus to explore these search spaces efficiently. This can be achieved by speeding up each evaluation, by making as few evaluations as possible, or a combination of the two.

1.1   Bayesian Optimization

Surrogate model-based optimization methods offer a viable solution for searching in large search spaces when evaluations are expensive. Instead of performing many costly evaluations, a model approximates the expensive objective function. In particular, Bayesian optimization (BO) relies on a probabilistic model to guide the search by suggesting which points to evaluate next at each iteration. This makes BO an effective blackbox search approach, particularly noted for its good sample efficiency. As a result, it is a useful method in Neural Architecture Search.

In Bayesian optimization literature, the model is most often a Gaussian Process (GP) [7]. However, GPs have certain shortcomings. A commonly mentioned limitation is scalability: as they scale in $O(n^3)$ w.r.t the number of datapoints, they can become computationally expensive. In addition, GP-based Bayesian optimization can be difficult to apply for use cases like Neural Architecture Search. More specifically, the required kernel function and distance function can be non-trivial for complicated spaces like neural network architecture spaces. A number of kernels and distance measures have been suggested, including optimal transport metrics for architectures of neural networks (OTMANN) [11], Weisfeiler-Lehman kernels [17], edit-distance neural network kernels [10], and phenotypic distance [8].

However, it is possible to use alternative models for Bayesian Optimization, such as random forests and Bayesian neural networks [7]. For NAS in particular, models like graph convolutional networks or Bayesian graph NNs have yielded good results [16,21].

1.2   Deep Ensembles and NAS

Ensembling methods [6] involve aggregating the predictions of a set of diverse models. Using an aggregate prediction is more robust and reliable than using an individual prediction model. The models can compensate mutual weaknesses and offer complementary strengths. A deep ensemble (DE) is the ensemble of a number of independently-trained neural networks.

DEs can reliably quantify uncertainty in many scenarios [12], and are effective in out-of-distribution cases. With a large number of parameters in neural networks, there are many low-loss regions which can approximate the training data. Deep ensembles can produce a set of diverse networks representing different low-loss regions. In practice, compared to other models such as Bayesian NNs, DEs are also easier to implement and to parallelize.

The Trieste BO-based search framework [20] presents an example of using a deep ensemble as a probabilistic model for Bayesian optimization. In the NAS context, the most prominent example is BANANAS [25], where the authors thoroughly analyzed many components of BO applied to architecture search: predictor choice, acquisition functions, etc. This led to the design of a successful NAS method based on an ensemble of networks, coupled with a novel architecture encoding scheme: path encoding. This work clearly demonstrated the potential of deep ensemble-based NAS.

Compared to GPs, the previously mentioned methods eliminate the need for kernel and distance functions. Flexibility in the choice of architecture for the ensemble networks is another advantage. For example, a GNN-based predictor can be used for graph-based search spaces. The ensembling allows an ordinarily non-probabilistic model to be an effective predictor for BO.

1.3   Contributions

Our NAS method is built on the idea of deep-ensembles as a predictor for BO-based NAS, focusing specifically on the capacity and advantages of **weights reuse**. In fact, DEs inherently enable the use of their weights to approximate many prediction targets. As a result, instead of only relying on the expensive evaluations of the objective function, we can leverage less expensive data to improve the internal representations of the ensemble. The idea is to incorporate useful information from cheaper-to-evaluate sources in the weights of the ensemble, with the aim of reducing search time and cost. We illustrate this with two complementary ideas: **simultaneous pretraining** on zero-cost metrics, and **multi-fidelity search**.

While pretraining is widely-used in machine learning, our main insight in the **simultaneous pretraining** section is how to efficiently combine multiple pretraining targets. Instead of pretraining the ensemble on one metric at a time, we use a shared embedding to predict all the metrics at once. This ensures the internal representation is more generic, and results in a significant speedup of the BO search using the pretrained ensemble.

The notion of shared embeddings also allows us to share information between different fidelity levels, allowing us to trade a few high quality evaluations for many lower quality evaluations. This **multi-fidelity BO** approach also considerably accelerates the search.

Our experiments on a NAS benchmark have shown that these two additions to deep ensemble-based NAS significantly reduce the search cost and boost the performance. In order to test our method's generality and applicability outside of NAS benchmarks, we construct a custom search space of ViT-CNN networks.

The search space is based on an existing family of architectures, the MOAT family [29]. Our search method quickly yields an improved architecture.

We summarize our contributions as follows:

– Simultaneous pretraining on multiple zero-cost metrics, which outperforms classic pretraining
– Multi-fidelity BO using shared weights
– An application example on a custom search space

2   NAS Method Description

2.1   Bayesian Optimization with DEs

At a glance, the Bayesian optimization (BO) procedure with deep ensembles is similar to BO with Gaussian processes (Fig. 1). At each iteration, the probabilistic model is used to select the best points to be evaluated, and the evaluations are in turn used to update the model and improve it in preparation for the next iterations.

The acquisition function is the criteria used to select the most interesting points to evaluate. It relies on the predictions and uncertainty measures provided by the probabilistic model. Different acquisition functions strike various balances between exploration and exploitation.

The surrogate model we use is a deep ensemble, made up of a number of neural networks which take an architecture (described by an encoding), and predict its performance. Owing to the different initializations, the networks in the ensemble occupy different low-loss regions in the loss landscape. As a result, their predictions for a particular candidate architecture are different, and combining them yields a more robust prediction.

We used the probability of improvement (PI) as acquisition function. It selects the points with the highest probability of improving over the current best observed value. Although in GP-based BO other acquisition functions (e.g. Expected Improvement) usually outperform PI, our tests showed PI to result in slightly faster search times. We use the predictions computed by the ensemble networks as MC samples to approximate the acquisition function PI.

At each iteration, the newly acquired observations are added to the training set of the deep ensemble, and the networks of the DE are trained to incorporate the new information.

2.2   Simultaneous Pretraining

With the performance gains it unlocks, pretraining is a widely-used technique in machine learning. Pretraining gives the model a head-start, by allowing it to learn general patterns in the data beforehand, which boosts its training and performance later.

In Bayesian Optimization, there are methods to pretrain GPs, such as HyperBO [23]. For Neural Architecture Search, FBNetV3 [2] is a notable

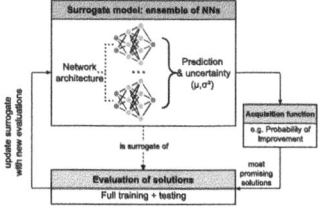

Fig. 1. Overview of Bayesian optimization with deep ensembles

example of successfully using a pretrained predictor in the search process. In a first step, architecture statistics are used to pretrain a predictor, which is then used as a proxy for the objective function in a predictor-based evolutionary search. The pretraining procedure led to an important boost in the sample efficiency of the predictor.

In the context of Bayesian optimization, the use of deep ensembles significantly simplifies this pretraining step. As opposed to GPs and other probabilistic models, pretraining a deep ensemble is simple: it trivially consists of pretraining its component networks. The impact of pretraining in general, and the way to pretrain the ensemble, is well established. Instead, our main insight is related to the pretraining data we use, and especially to the manner in which the pretraining on many metrics is performed.

For neural networks in the search space, we can quickly calculate a number of metrics to use as pretraining data. We focus on zero-cost metrics, i.e. metrics whose computation or estimation doesn't require prior training of the neural network in question. These metrics depend only on the architecture, or hardware, but not on whether it is trained or not. Among such zero-cost metrics, we use the number of trainable parameters, the average inference latency, the number of floating point operations (FLOPs) and memory footprint.

With access to multiple pretraining metrics, the main question we focused on is **which pretraining method is the best way to fully leverage these multiple metrics**.

Our idea is to force the common representation generated by the networks to remain as generic as possible, while retaining relevant information. Since these metrics are not our real prediction target, simultaneous pretraining is used to keep the embedding from specializing on any one metric to the detriment of our real prediction target. Practically, this is achieved by using a common embedding to predict all the metrics simultaneously.

In practice, each ensemble network is composed of a section of n layers $(L_1, .., L_n)$, followed by prediction head(s). This first section is mutualized to all the pretraining metrics. In the forward step, this shared section produces a common embedding $\mathbf{y}$ of the input data $\mathbf{x}$. The embedding $\mathbf{y}$ is subsequently used to predict each of the m pretraining metrics, using a separate prediction head f_i for each metric.

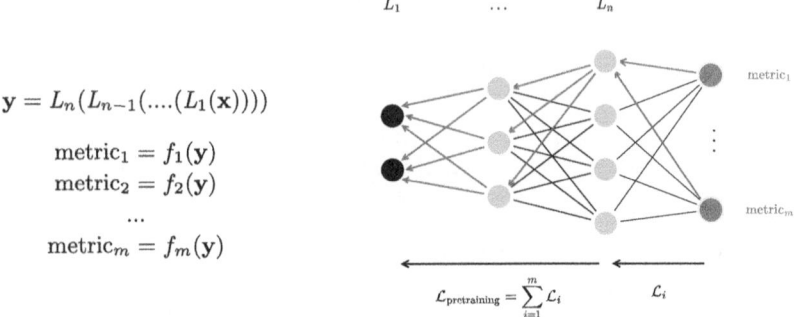

$$\mathbf{y} = L_n(L_{n-1}(....(L_1(\mathbf{x}))))$$

$$\mathrm{metric}_1 = f_1(\mathbf{y})$$
$$\mathrm{metric}_2 = f_2(\mathbf{y})$$
$$...$$
$$\mathrm{metric}_m = f_m(\mathbf{y})$$

Fig. 2. Backpropagation step of the simultaneous pretraining scheme, with a separate prediction head for each zero-cost metric

In other words, all the metrics are predicted using a common vector $\mathbf{y}$ representing the input architecture. The loss is the sum of losses of each pretraining metric, as illustrated in Fig. 2. As a result of this setup, the backpropagation step updates the weights of the shared section using combined loss information from all the zero-cost metrics.

In our experiments, we measure the impact of correlations between the metrics, and the impact of the number of metrics used. We also compare this pretraining scheme to performing no pretraining and to sequential pretraining.

2.3   Multi-fidelity Search

The motivation for multi-fidelity search is to replace a low number of high quality evaluations, with a high number of lower quality evaluations. For example, we can use the validation score after fully training the model (N epochs) as the high-fidelity evaluation. The low-fidelity evaluation is then the validation score after a smaller number n of training epochs ($n < N$).

The practical implementation is similar to simultaneous pretraining: a common embedding is used to estimate both fidelity levels of the evaluation. As a result, in a similar way to the multi-head pretraining method, our ensemble networks have a shared section, followed by a specialized section (a dense layer and prediction head) for each level of fidelity. Figure 3 illustrates this architecture.

The training procedure is as follows:

- At each iteration, perform p low-fidelity evaluations.
- Every f iterations, perform q high-fidelity evaluations.

Both of these steps update the shared sections of the ensemble networks, which results in this section being updated more frequently than in the mono-fidelity case. The high-fidelity evaluations are given more importance than the low-fidelity evaluations, by training the ensemble on their data for more epochs.

In subsequent experiments, we compare this training procedure to the mono-fidelity approach, and illustrate whether its impact is additive or contradictory with the impact of simultaneous pretraining.

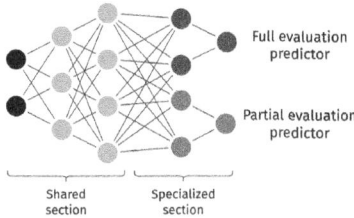

Fig. 3. Ensemble network architecture for multi-fidelity search

3   Experiments and Results

We use an MLP-based architecture for the deep ensemble networks, both in the benchmark and the custom search space (Sect. 4).

3.1   Simultaneous Pretraining Experiments

In Fig. 4, we compare the evolution of the best value and the Spearman rank correlation with and without the simultaneous pretraining, as we advance along the search procedure. The Spearman rank correlation designates how well the model ranks the architectures in the benchmark: the correlation between the true ranking, and the ranking based on the model's prediction. The x-axis tracks the number of evaluations performed.

In Fig. 5, we report the average search time, i.e. the number of evaluations needed to reach the optimum, for CIFAR10 and ImageNet16-120. We also report the Spearman rank correlation values ρ after 512 evaluations.

The results show that pretraining has a significant impact on the predictive performance of the ensemble: the increase in rank correlations means the model suggests better points to be evaluated, and as a result the best value plot shows a significant advantage for the pretrained DE.

Impact of the Pretraining Method
To specifically evaluate the impact of **simultaneous pretraining**, we compare it to **sequential pretraining**, where pretraining is performed on $metric_1$, then $metric_2$, and so on.

We varied the number of pretraining epochs to account for the different hyperparameters each pretraining mode might require. We reported the average search time in Fig. 6. The results show a clear trend where the search is significantly accelerated by simultaneous pretraining.

Pairwise Correlations and Number of Pretraining Metrics
We used the following zero-cost metrics: number of trainable parameters, FLOPs, average inference time. The NAS-Bench-201 [3] benchmark allows us to calculate some pairwise correlations between the different metrics and the evaluations at high and low fidelities (Table 1). We also measured the impact of the number of used metrics on the search time (Table 2). The search time corresponds here to

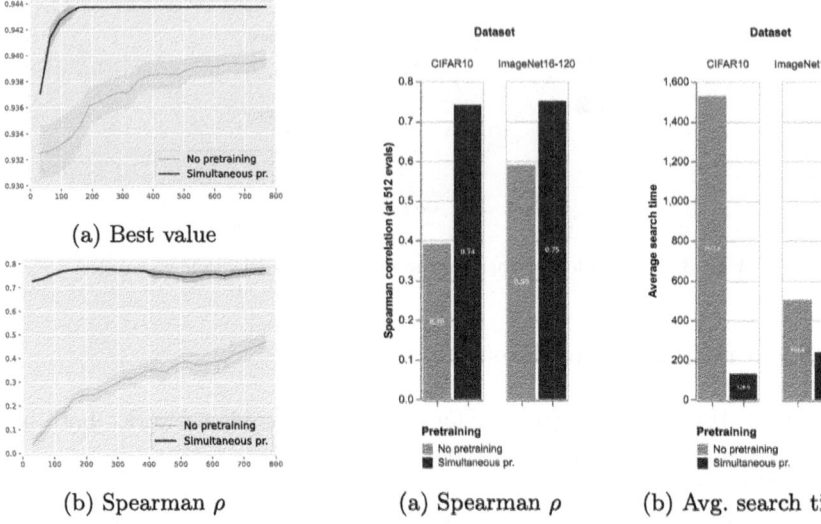

(a) Best value

(b) Spearman ρ

Fig. 4. Best value and correlation plots (CIFAR10 - 20 run averages)

(a) Spearman ρ

(b) Avg. search time

Fig. 5. Impact of pretraining on correlation and search time (C10 and ImageNet16-120)

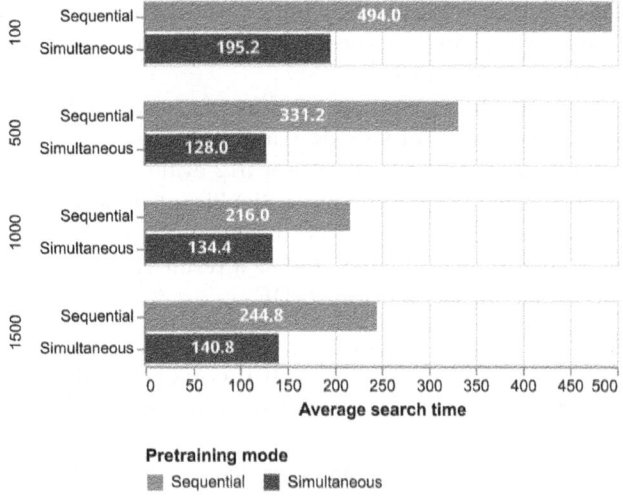

Fig. 6. Average search times using sequential and simultaneous pretraining. The rows indicate the number of pretraining epochs.

Table 1. Pairwise correlations of pretraining metrics

	N.params	Latency	Hi-Fi acc.	Lo-Fi acc.
N.params	1	0.68	0.32	0.38
Latency	0.68	1	0.38	0.48

Table 2. Effect of the number of metrics on search time

N.metrics	1	2	3
Search time (in evaluations)	113.4	96	78.9

the average number of evaluations needed to find the optimal architecture in the benchmark.

While testing with 2 metrics, we found that the differences in pairwise correlations have a small impact on the search time. The number of metrics has a much more significant impact. Therefore, using more metrics, even if that includes highly correlated metrics (e.g. number of parameters and FLOPs), yields the fastest search times.

3.2 Multi-fidelity Search Experiments

NAS-Bench-201 includes validation scores after 12 epochs and after 200 epochs of training, which we use respectively as the low-fidelity and the high-fidelity evaluations. We use values of $p = 5$, $q = 3$, $f = 10$, i.e. 3 full evaluations are performed for every 50 partial evaluations.

In Figs. 7 and 8, we plot the average best values found and Spearman rank correlation with and without multi-fidelity training. For CIFAR10, we also include the values without any pretraining or multi-fidelity training, to illustrate the combined impact of these two procedures. On the x-axis, we track the current cost in terms of "equivalent full evaluations", to be able to compare results with the mono-fidelity approach. We calculate this as the total number of epochs at that point, divided by the number of epochs for a full evaluation.

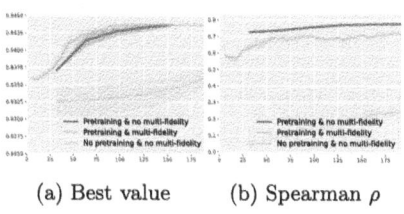

(a) Best value   (b) Spearman ρ

Fig. 7. CIFAR10

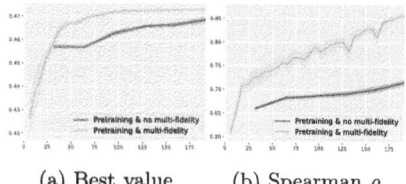

(a) Best value   (b) Spearman ρ

Fig. 8. ImageNet120-16

Our experiments show that multi-fidelity search with shared weights improves the search speed with or without pretraining, and that combining both produces much faster search speeds. The next section provides an overview of the method, and its main results on NAS-Bench-201.

3.3   NAS Method Overview and Results

Figure 9 sums up the NAS approach in its two steps: the simultaneous pretraining, and the multi-fidelity BO loop. Figure 10 provides an overview of the impact simultaneous pretraining and multi-fidelity search have on search time.

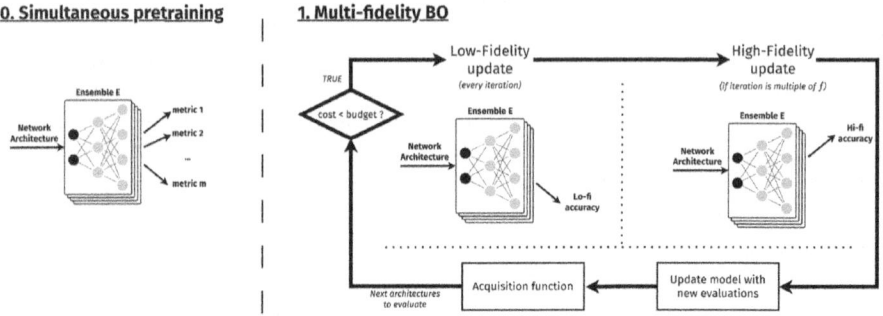

Fig. 9. General overview of the NAS method. Ensemble E's weights are pretrained in **step 0**, then updated both by high fidelity and low fidelity evaluation data. It is used at each iteration to suggest the next points to evaluate.

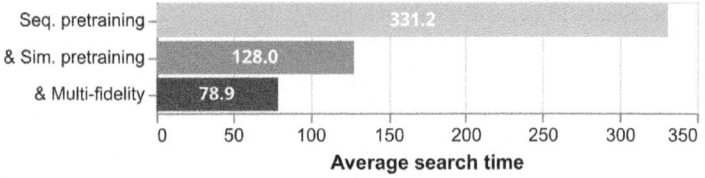

Fig. 10. Combined effect of simultaneous pretraining and multi-fidelity search on the search time (CIFAR10, 20 run average)

Table 3. Statistics about the number of evaluations to reach optimum (20-run averages)

		Min	Max	Mean	Std
CIFAR10	Random search	280	13690	7717.0	4498.8
	Evolution	20	1000	250.0	247.41
	BANANAS	20	470	128.5	118.59
	Local search	30	190	99.0	56.82
	Ours	36	132	**78.9**	25.45
CIFAR100	Random search	520	14690	7141.5	4137.08
	Evolution	40	650	247.5	154.2
	BANANAS	40	310	101.0	54.49
	Local search	20	490	118.5	96.4
	Ours	18	96	**53.7**	24.03

In Table 3, we report statistics about the number of evaluations needed to reach the optimum on NAS-Bench-201 datasets. We compare to other NAS methods in the literature where it was possible to access this data, most notably using the NASZilla library [24], which includes BANANAS [25] and the local search approach [27].

In Table 4, we report the best accuracy value found by the search method, after respectively 100 and 200 evaluations (or queries) on NAS-Bench-201. We compare this with other NAS methods in the literature, either using our own tests or reported results from their respective papers.

Table 4. Best value after 100 and 200 epochs (20-run averages)

NAS Method	CIFAR10	CIFAR100	ImageNet16-120	Queries
Optimum*	**94.37**	**73.51**	**47.31**	
Arch2vec & BO [28]	94.18	73.37	46.27	100
AG-Net [15]	94.24	73.12	46.20	100
Random search	93.70	70.94	45.44	100
Local search	94.26	73.15	46.09	100
Evolution	94.20	72.70	46.03	100
BANANAS	94.15	73.28	46.46	100
Ours	**94.36**	**73.51***	**47.19**	100
Random search	93.86	71.78	45.70	200
Local search	**94.37***	73.48	46.21	200
Evolution	94.24	73.07	46.30	200
BANANAS	94.20	73.43	46.51	200
AG-Net	**94.37***	**73.51***	46.43	192
Ours	**94.37***	**73.51***	**47.31***	200

NB: There are of course better-performing networks on these datasets in the literature in general, but here we are interested in how fast the different methods perform the search, and are restricted to the architectures contained in the benchmark. As a result, the accuracies have an upper bound (first line).

On the 100 evaluation budget, our method finds architectures with the top value in all 3 benchmark datasets, finding the optimum for C100. By the 200 evaluations mark, our method find all 3 optima for the NAS-Bench-201 dataset.

4   Application to a Custom Search Space

The architecture of the ensemble networks we used are generic, not tailored to the graph-based search space of the benchmarks. To further test the generality of this method, we decided to apply it to a very different search space, and kept the same ensemble architecture and hyperparameters.

4.1   Search Space Design

MOAT [29] is a family of hybrid CNN-ViT networks for vision tasks. They are built on a hybrid block which effectively merges mobile convolution with transformer blocks: the MOAT block. The design of the MOAT block involves replacing the MLP in the transformer block with a mobile convolution, and rearranging the components.

Inspired by the architecture of the MOAT block, we built a search space of hybrid CNN-ViT networks. It contains purely mobile convolution-based networks including MobileNetV2 [19], purely transformer-based networks including ViT [22], and various hybrid configurations including MOAT.

It is a cell-based design, made up of 4 stages, where the image resolution is divided by 2 at each new stage. An architecture is described by which cell design is selected for each section of the network. Our optimization objective is to find the optimal sequence of cell types associated with the stages. This way, we allow sufficient flexibility, but avoid having an unreasonably large space potentially full of ineffective or unfeasible architectures, keeping it at a similar size to benchmark search spaces like NAS-Bench-201.

The possible cell designs were constructed by leveraging an architectural pattern common to the MOAT block, the ViT block and the MBConv block. They can all be described using a succession of two mini-blocks, each with a skip connection. Figure 11 illustrates this pattern.

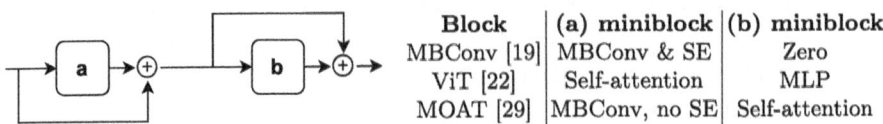

Block	(a) miniblock	(b) miniblock
MBConv [19]	MBConv & SE	Zero
ViT [22]	Self-attention	MLP
MOAT [29]	MBConv, no SE	Self-attention

Fig. 11. Meta-architecture of a cell in the search space

The cell types are described by specifying the miniblocks (a) and (b) from a list of possible miniblocks: {*MBConv with Squeeze-Excitation (SE), MBConv with no SE, Self-attention, MLP, Zero*}. We eliminate unfeasible designs, resulting in a total of 7 possible cell designs. We add two experimental cell types, with miniblocks (a) and (b) in parallel.

4.2   Experiment and Results

The objective function is the validation score on Imagewoof [5], a challenging 10-class subset of ImageNet [18]. The high-fidelity evaluation is computed after 200 epochs of training epochs, and the low-fidelity evaluation is after 15 training epochs. We compute 4 pretraining metrics using PyTorch Profiler: FLOPs, inference time, GPU memory footprint and number of trainable parameters.

We use the `tiny-moat-0` configuration and number of channels as a baseline. It is a small network of around 2.7M parameters, and the ImageNet results should be interpreted in comparison to this baseline network and not performances of much bigger networks.

The baseline architecture **tiny-moat-0** is as follows:

```
(MBConv) | (MBConv) | (MOAT block)| (MOAT block)
```

The search procedure yielded the following architecture after approximately 70 evaluations:

```
(MBConv no SE) | (MOAT block) | (MBConv & SE + MLP) | (MOAT block & SE)
```

We report the validation results on Imagewoof and ImageNet-1k (with no pretraining) in Table 5. We also test a bigger version of the baseline and the searched architecture, using the `MOAT-0` channel sizes. The search procedure has produced a better-performing architecture than the baseline we started with.

Table 5. Performance comparison of the searched architecture and the baseline MOAT architecture

Experiment	Baseline	Ours
Imagewoof - 200 epochs (Objective function)	81.13	83.76
Imagewoof - 300 epochs	84.34	86.54
Imagewoof - 300 epochs (MOAT-0 channels)	89.82	90.86
ImageNet-1k - 90 epochs	61.84	67.74

5   Conclusion

Deep ensembles are an effective probabilistic model for Bayesian optimization applied to NAS. They present a number of upsides, especially the flexibility to leverage additional sources of information to improve the predictive performance. We showcase this in practice with simultaneous pretraining on zero-cost metrics and multi-fidelity search. These additions lead to significant speedups of the search procedure, respectively accelerating the search 2.5 and 1.6 times on NAS-Bench-201 (CIFAR10). We also highlight how our method can be applied more generally outside the scope of usual NAS benchmarks, by successfully applying it on a fully custom and complex hybrid CNN-ViT search space, outperforming the baseline on ImageNet-1k. As a conclusion, our NAS method is a practical and effective way to search for architectures in customized and novel search spaces.

Acknowledgements. This work has been supported by the French government under the "France 2030" program, as part of the SystemX Technological Research Institute within the Confiance.ai project.

Experiments presented in this paper were carried out using the Grid'5000 testbed, supported by a scientific interest group hosted by Inria and including CNRS, RENATER and several Universities as well as other organizations (see https://www. grid5000.fr).

This work was partially supported by the EXAMA (Methods and Algorithms at Exascale) project under grant ANR-22-EXNU-0002.

References

1. Benmeziane, H., El Maghraoui, K., Ouarnoughi, H., Niar, S., Wistuba, M., Wang, N.: Hardware-aware neural architecture search: survey and taxonomy. In: Proceedings of the Thirtieth International Joint Conference on Artificial Intelligence, IJCAI-2021, pp. 4322–4329 (8 2021)
2. Dai, X., et al.: Fbnetv3: joint architecture-recipe search using predictor pretraining. In: Proceedings of the IEEE/CVF Conference on Computer Vision and Pattern Recognition, pp. 16276–16285 (2021)
3. Dong, X., Yang, Y.: Nas-bench-201: extending the scope of reproducible neural architecture search. In: International Conference on Learning Representations (ICLR) (2020). https://openreview.net/forum?id=HJxyZkBKDr
4. Elsken, T., Metzen, J.H., Hutter, F.: Efficient multi-objective neural architecture search via lamarckian evolution. arXiv preprint arXiv:1804.09081 (2018)
5. Fast.ai: Imagewoof (2019). https://github.com/fastai/imagenette. Accessed 20 July 2023
6. Ganaie, M., Hu, M., Malik, A., Tanveer, M., Suganthan, P.: Ensemble deep learning: a review. Eng. Appl. Artif. Intell. **115**(C) (2022). https://doi.org/10.1016/j. engappai.2022.105151
7. Garnett, R.: Bayesian Optimization. Cambridge University Press, Cambridge (2023)
8. Hagg, A., Zaefferer, M., Stork, J., Gaier, A.: Prediction of neural network performance by phenotypic modeling. In: Proceedings of the Genetic and Evolutionary Computation Conference Companion, GECCO 2019, pp. 1576–1582 (2019). Association for Computing Machinery, New York (2019). https://doi.org/10.1145/ 3319619.3326815
9. Hsu, C.H., et al.: Monas: multi-objective neural architecture search using reinforcement learning. arXiv preprint arXiv:1806.10332 (2018)
10. Jin, H., Song, Q., Hu, X.: Auto-keras: an efficient neural architecture search system. In: Proceedings of the 25th ACM SIGKDD International Conference on Knowledge Discovery & Data Mining, pp. 1946–1956 (2019)
11. Kandasamy, K., Neiswanger, W., Schneider, J., Poczos, B., Xing, E.P.: Neural architecture search with bayesian optimisation and optimal transport. Adv. Neural Inf. Process. Syst. **31** (2018)
12. Lakshminarayanan, B., Pritzel, A., Blundell, C.: Simple and scalable predictive uncertainty estimation using deep ensembles. In: Proceedings of the 31st International Conference on Neural Information Processing Systems, NIPS 2017, pp. 6405-6416. Curran Associates Inc., Red Hook (2017)

13. Lu, Z., Deb, K., Goodman, E., Banzhaf, W., Boddeti, V.N.: NSGANetV2: evolutionary multi-objective surrogate-assisted neural architecture search. In: Vedaldi, A., Bischof, H., Brox, T., Frahm, J.-M. (eds.) ECCV 2020. LNCS, vol. 12346, pp. 35–51. Springer, Cham (2020). https://doi.org/10.1007/978-3-030-58452-8_3

14. Lu, Z., et al.: Nsga-net: neural architecture search using multi-objective genetic algorithm. In: Proceedings of the Genetic and Evolutionary Computation Conference, pp. 419–427 (2019)

15. Lukasik, J., Jung, S., Keuper, M.: Learning where to look-generative nas is surprisingly efficient. In: Avidan, S., Brostow, G., Cisse, M., Farinella, G.M., Hassner, T. (eds.) ECCV 2022, vol. 13683, pp. 257–273. Springer, Heidelberg (2022). https://doi.org/10.1007/978-3-031-20050-2_16

16. Ma, L., Cui, J., Yang, B.: Deep neural architecture search with deep graph bayesian optimization. In: IEEE/WIC/ACM International Conference on Web Intelligence, pp. 500–507 (2019)

17. Ru, B., Wan, X., Dong, X., Osborne, M.: Interpretable neural architecture search via bayesian optimisation with weisfeiler-lehman kernels. In: International Conference on Learning Representations (2021). https://openreview.net/forum?id=j9Rv7qdXjd

18. Russakovsky, O., Deng, J., Su, H., Krause, J., Satheesh, S., Ma, S., Huang, Z., Karpathy, A., Khosla, A., Bernstein, M., Berg, A.C., Fei-Fei, L.: ImageNet large scale visual recognition challenge. Int. J. Comput. Vision (IJCV) **115**(3), 211–252 (2015). https://doi.org/10.1007/s11263-015-0816-y

19. Sandler, M., Howard, A., Zhu, M., Zhmoginov, A., Chen, L.C.: Mobilenetv2: inverted residuals and linear bottlenecks. In: Proceedings of the IEEE Conference on Computer Vision and Pattern Recognition, pp. 4510–4520 (2018)

20. Secondmind-Labs: Bayesian optimization with deep ensembles (2020). https://secondmind-labs.github.io/trieste/1.0.0/notebooks/deep_ensembles.html. Accessed Aug 2023

21. Shi, H., Pi, R., Xu, H., Li, Z., Kwok, J., Zhang, T.: Bridging the gap between sample-based and one-shot neural architecture search with bonas. Adv. Neural. Inf. Process. Syst. **33**, 1808–1819 (2020)

22. Vaswani, A., et al.: Attention is all you need. Adv. Neural Inf. Process. Syst. **30** (2017)

23. Wang, Z., et al.: Pre-trained gaussian processes for bayesian optimization. arXiv preprint arXiv:2109.08215 (2021)

24. White, C., Neiswanger, W., Nolen, S., Savani, Y.: A study on encodings for neural architecture search. Adv. Neural Inf. Process. Syst. (2020)

25. White, C., Neiswanger, W., Savani, Y.: Bananas: bayesian optimization with neural architectures for neural architecture search. In: Proceedings of the AAAI Conference on Artificial Intelligence, vol. 35, pp. 10293–10301 (2021)

26. White, C., Nolen, S., Savani, Y.: Exploring the loss landscape in neural architecture search. In: Uncertainty in Artificial Intelligence, pp. 654–664. PMLR (2021)

27. White, C., Nolen, S., Savani, Y.: Exploring the loss landscape in neural architecture search. In: Uncertainty in Artificial Intelligence. PMLR (2021)

28. Yan, S., Zheng, Y., Ao, W., Zeng, X., Zhang, M.: Does unsupervised architecture representation learning help neural architecture search? Adv. Neural. Inf. Process. Syst. **33**, 12486–12498 (2020)

29. Yang, C., et al.: Moat: alternating mobile convolution and attention brings strong vision models. arXiv preprint arXiv:2210.01820 (2022)

30. Zoph, B., Le, Q.: Neural architecture search with reinforcement learning. In: International Conference on Learning Representations (2017). https://openreview.net/forum?id=r1Ue8Hcxg

Feature Activation-Driven Zero-Shot NAS: A Contrastive Learning Framework

Di Wang[1] , Xunzhi Xiang[1] , Kun Jing[2(✉)] , and Jungang Xu[1(✉)]

[1] University of Chinese Academy of Sciences, Beijing, China
xujg@ucas.ac.cn
[2] Anhui University, Hefei, Anhui, China
jingkun@ahu.edu.cn

Abstract. The main bottleneck in current neural architecture search (NAS) algorithms is the inability to efficiently evaluate various neural architectures. While existing performance predictor-based approaches have significantly reduced the time required to assess alternative architectures, they still necessitate the training of a substantial number of architectures. Unfortunately, this training process consumes considerable computational resources. To address this challenge, we introduce a novel training-free evaluation metric rooted in the principles of contrastive learning. This innovative metric evaluates architecture performance by analyzing the differential responses elicited from positive and negative samples within the deep architecture. By leveraging the insights of contrastive learning, it offers a more resource-efficient solution for assessing neural architectures. We prove the superiority of our method with evaluation benchmarks such as NAS-bench-101 and NAS-bench-201. This method combined with the search strategy finally achieved an accuracy of 97.5 % on CIFAR-10. The code for our work is available at https://github.com/wdi-nancy/FADZS-NAS.

Keywords: Neural architecture search · Training-free proxies · Deep feature activation · Contrastive learning

1 Introduction

Neural Architecture Search (NAS) has recently attracted considerable attention for automatically designing architectures. Early researchers evaluated the alternative architecture through standard training. Due to the low cost of computing resources, predictor-based NAS methods [1–5] and one-shot NAS methods [6–9] have become popular. Performance predictor evaluation strategy maps the architecture space to absolute or relative performance values. One-shot NAS merges all possible architectures from the search space into a supernet and thus only needs to train the supernet once. However, such methods can be very computationally expensive as they require numerous training resources. To address

---

D. Wang and X. Xiang—Equal contribution.

M. Wand et al. (Eds.): ICANN 2024, LNCS 15016, pp. 261–276, 2024.
https://doi.org/10.1007/978-3-031-72332-2_18

this limitation, zero-shot (training-free) NAS has emerged recently, which ranks alternative architectures based on their performance on a specific property in a neural architecture.

The first proposed zero-shot proxy is NN-Mass [10], which theoretically links how the architecture topology influences gradient propagation and model performance. Some methods use the number of architecture parameters and FLOPs as naive performance proxies to evaluate the architecture and have achieved relatively good results. However, these methods often fall short in probing the underlying causes behind architecture performance, a gap that might compromise the efficacy of zero-shot NAS approaches. For instance, the role of features and gradients present before the activation layer could be pivotal in determining the final performance of the architecture. Simultaneously, various pre-training tasks demonstrate notable success across different models, underscoring their significance [11]. These methods typically involve calculating the loss function by utilizing the deep features of the architecture for training purposes. Such practices verify that the feature vectors from the architecture accurately mirror its overall performance. Contrastive learning measures the final performance of the architecture by performing similarity calculations on features in the architecture layer. For example, CLIP [12] uses a contrastive learning framework to achieve multi-modal representation learning of images and text by training models in a shared embedding space. Current works [11,13] demonstrate that the ability of architectures to differentiate between positive and negative sample features indicates their ultimate performance on downstream tasks.

Inspired by these works, we propose a novel zero-shot proxy: Feature Activation Driven Zero-Shot NAS: A Contrastive Learning Framework (FADZS-NAS). In evaluating architecture performance, we initially select features just before the activation layer, as these are typically not yet saturated in the early stages of training. We then project these features onto a hypersphere and calculate a feature similarity matrix by determining the cosine similarity between them. Finally, we apply the InfoNCE loss [14] to this matrix, incorporating the value of its determinant, to derive a comprehensive score for the architecture performance. This approach guarantees a detailed assessment of the architecture learning capabilities and efficiency in feature representation. We verify our proxy's performance on both NAS-bench-101 [15] and NAS-bench-201 [16] using Spearman ρ as the performance testing metric. The experimental results demonstrate that it is feasible and effective to evaluate architecture performance by combining contrastive learning, resulting in a stronger proxy method.

We summarize our main contributions as follows:

1. We propose a novel zero-shot proxy FADZS-NAS for NAS. The proposed FADZS-NAS is computationally efficient and has experimentally been proved to be insensitive to how the neural architecture is initialized.
2. We explore how to calculate the performance score through the similarity matrix and perform different calculation for different search spaces.

3. We integrate FADZS-NAS with different search strategies, and remarkably, the architecture derived from this combination achieves exceptional results on CIFAR-10 in only a handful of GPU hours.

2   Related Work

2.1   Zero-Shot NAS

Traditional neural architecture search methods typically involve numerous training iterations of alternative architectures to identify the optimal architecture. Zero-shot NAS aims to design a suitable proxy to rank the accuracy of various candidate neural architectures without training. Existing accuracy proxies are divided into two categories based on whether gradients are involved in the proxy calculation: gradient-based accuracy proxy and gradient-free accuracy proxy.

The gradient norm reflects the convergence and performance of the architecture, SNIP [17] multiplies the value of each parameter and its gradient to measure parameter importance in forward inference and gradient propagation. Fisher [18] approximates the Fisher information of a neural architecture by the square of the activation value and its gradient to evaluate the importance of each neuron/channel of a given architecture. Beyond gradients, various approaches use alternative metrics to assess architecture effectiveness. The number of parameters and FLOPs reflect the performance upper limit and computing power of the architecture, so some work uses them as naive proxies to evaluate the architecture. Zen-score measures model expressivity by averaging the Gaussian complexity under randomly sampled input and a small input perturbation.

Guided by these studies, we propose from a contrastive learning viewpoint that an architecture's skill in discerning positive from negative samples may also indicate its ultimate performance capabilities.

2.2   Contrastive Learning

Contrastive learning is an unsupervised learning approach that learns the representation of data by comparing the similarity and difference between two different samples. SimCLR [13] uses large batches and data augmentation to maximize the similarity between positive samples and minimize the similarity between negative samples. MoCo [11] introduces a momentum update strategy to compare online updated target samples with historical samples. CLIP [12] applies contrastive learning to multi-modal tasks.

Inspired by these studies, we generate pairs of positive and negative samples using various data augmentation methods and then conduct a preliminary evaluation of architecture performance by comparing the correlation of activation features between different sample pairs. The data augmentation methods we use include: RandomCrop, RandomHorizontalFlip, ColorJitter, and Lighting.

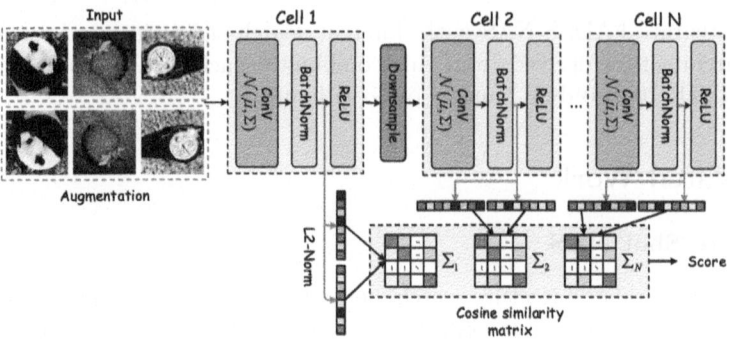

Fig. 1. Overall Framework of FADZS-NAS.

3 Method

In this section, we first describe the theoretical basis for why features should be mapped to the hypersphere. Then, we present the method for calculating the cosine similarity of activation features based on contrastive learning. We employ two distinct approaches for proxy score calculation: the softmax form and the determinant calculation form. Finally, we combine this proxy metric with a search strategy to search architectures in different search spaces.

3.1 Overall Framework

Figure 1 shows the overall framework of our proposed method called FADZS-NAS. We define a sample's activation as the feature representation generated in the hidden layer during forward inference with various sample inputs in the architecture. In untrained neural architectures, the nonlinear activation function typically resides in a saturated state, potentially leading to substantial information loss. For instance, feature representations preceding the ReLU nonlinear activation function often contain numerous negative values. These negative elements are converted to zero upon passing through the ReLU function, culminating in the loss of information contained within them. Consequently, we choose to utilize the features before the activation layer for calculating the similarity matrix and the proxy score, to mitigate this loss of information.

Specifically, given a neural architecture with N_a nonlinear activation functions and a batch of randomly sampled inputs with batch size of N, we collect the activation vector of the n-th sample before the i-th nonlinear activation function through architecture forward inference and calculate their cosine similarity matrix. This process can be described as:

$$a_n^i = l2\_normalize(f(x_n)) \tag{1}$$

$$b_n^i = l2\_normalize(f(y_n)) \tag{2}$$

$$\Sigma^i = A^i \cdot (B^i)^T \tag{3}$$

where x_n denotes the n-th sample of the input, and y_n denotes the n-th sample after weak data augmentation, a_n^i, b_n^i denotes the features of the sample pair before passing through the i-th activation layer of the architecture f ($n = 1, 2, ..., N$ and $i = 1, 2, ..., N_a$), $A^i = \{a_1^i, a_2^i \cdots a_n^i\}$ and $B^i = \{b_1^i, b_2^i \cdots b_n^i\}$ denotes the feature set of a batch.

3.2   Calculation of Zero-Shot Proxy Metrics

The similarity matrix is designed to quantify the feature similarities across various samples. Our objective is to ensure that this matrix is as close as possible to a diagonal matrix, thereby enhancing the model's ability to distinguish positive and negative samples. To meet the requirements of diverse search spaces, we apply two distinct computational approaches to the acquired cosine similarity matrix, thereby deriving the proxy score.

Contrastive Learning-Based Feature Similarity Calculation. In the NAS-bench-101 search space, the absence of residual connections and zeroing operations means that the model relies more heavily on convolution operations, thereby enhancing its capability to extract features. This implies a potentially sparser distribution of the computed cosine similarity. In light of this observation, we adopt the calculation approach of InfoNCE to determine the model's proxy score. This is due to the exponential calculation method's high sensitivity to the similarity values among different samples. This process can be described as:

$$\text{FADZS-NAS} = \sum_{i=1}^{N_a} \sum_{n=1}^{N} log \left(\frac{exp(\Sigma_{nn}^i)}{\sum_{m=1}^{N} exp(\Sigma_{nm}^i)} \right) \tag{4}$$

where Σ_{nm}^i denotes the similarity value in the n-th row and m-th column of the similarity matrix before the i-th activation layer, N_a de notes the number of activation function layers in the architecture under evaluation.

Exponential-Based Proxy Score Calculation. In contrast, the NAS-bench-201 search space features an increased prevalence of residual connections and zeroing operations. When the model is untrained, the feature of architecture extraction capability remains relatively weak. Consequently, the data distribution in the resulting cosine similarity matrix tends to be more compact. When the value distribution of the similarity matrix is compact, using the exponential form to calculate will result in a small distribution variance. This lower variance is less favorable for evaluation purposes. Hence, we employ the determinant calculation method to map the similarity matrix to the architecture score. This process can be described as:

$$\Sigma = \sum_{i=1}^{N_a} abs(\Sigma^i) \tag{5}$$

$$\text{FADZS-NAS} = log(det(\Sigma)) \tag{6}$$

where Σ^i denotes the similarity matrix before the i-th activation layer, *abs* means absolute value computation, and *det* denotes the computation of the determinant.

Algorithm 1: Training-free search algorithm based on evolutionary algorithm

---

Input: Search space A, training-free proxy P, the number of neural architectures to explore N, the number of architectures to initialize the population I

1 Initialize the population to an empty queue
2 Initialize the history to an empty set
3 **while** *population* $< I$ **do**
4 | Randomly sample a neural architecture a from the search space A
5 | Calculate the score of the architecture by P
6 | Add a to the end of queue and history
7 **end**
8 **while** *history* $< N - I$ **do**
9 | Randomly sample S neural architectures from the population to form candidate contenders C
10 | Select the model with the highest proxy score in C as the parent
11 | Mutate the parent architecture to obtain a child and calculate its proxy score
12 | Add child architecture to the end of queue and history
13 **end**
Output: The highest scoring neural architecture in history

3.3 Search Strategy

To demonstrate the applicability of our agent score calculation method in real-world search process for identifying superior architectures, we conducted experiments in different search spaces using evolutionary algorithm. The evolutionary algorithm obtains the optimal architecture through Tournament in the population and performs mutation operations on a randomly selected candidate to generate the next generation. The process is described in Algorithm 1.

4 Experiments

4.1 Search Spaces of Experiments

To demonstrate that our proxy method is versatile and effective in both constrained and expansive search spaces, we rigorously tested its performance across three benchmarks: NAS-Bench-101, NAS-Bench-201, and DARTS.

1. The NAS-Bench-101 search space consists of 423K unique convolutional architectures, and the space is "operations on nodes" (OON). Each architecture is restricted to the cell space. Each cell can have a maximum of 7 nodes

and 9 edges. For each node, the candidate operations include 1×1 convolution, 3×3 convolution, and 3×3 max pooling. NAS-Bench-101 provides the validation accuracy and the test accuracy on CIFAR-10.

2. The NAS-Bench-201 search space consists of 15625 unique convolutional architectures, and the space is "operations on edges" (OOE). Each cell is composed of 4 nodes and 6 edges. The candidate operations on each edge include zeroize, skip connection, 1×1 convolution, 3×3 convolution, and 3×3 average pooling. NAS-Bench-201 provides three different results for each architecture (CIFAR-10, CIFAR-100, and ImageNet).

3. The architectures in the DARTS search space are composed of normal cells and reduced cells. Each cell consists of 7 nodes and 14 edges. The DARTS search space is a large space containing approximately 10^{18} structures and is also OOE. The candidate operations on each edge include 3×3 and 5×5 separable convolutions, 3×3 and 5×5 dilated separable convolutions, 3×3 max pooling, 3×3 average pooling, and skip connection.

4.2   Experiment Setup

We consistently utilize the Spearman correlation coefficient ρ to evaluate the predictive performance of our method. Except for the ablation studies that explore different calculation methods, we apply the determinant calculation method for scoring candidate architectures. Furthermore, to ensure consistency with past experiments and to guarantee time efficiency and fair comparisons across all experiments, we randomly select 1,000 neural architectures for testing, maintaining consistency in the random number seeds used. To demonstrate the effectiveness of our method in the DARTS search space, we train the derived architecture on CIFAR-10 from scratch. The training process involves these key hyperparameters: learning rate of 0.025, momentum of 0.9, weight decay of 0.0003, 600 epochs, 36 channels, 20-layer model, and a batch size of 96. All experiments utilize a V100 GPU for enhancing computational efficiency and accuracy.

4.3   Proxy Evaluation Experiment

In order to verify the robustness of our method and reduce the experimental verification of the entire search space, we first conduct proxy evaluation experiment on NAS-bench-201. We randomly sample 1000 architectures in the NAS-bench-201 search space and calculate the Spearman coefficient ρ of the corresponding proxy method. As shown in Table 1, The true (computed using all neural architectures) Spearman rank correlation coefficients for all proxies and the estimated (computed using 1000 random neural architectures) Spearman rank correlation coefficients deviate within 10%, which is acceptable for fast proxy evaluation and ablation studies. Specially, Δ denotes the deviation between the true and estimated Spearman rank correlation coefficients, % denotes the deviation rate. ValAcc denotes the validation accuracy achieved by the model following conventional training. This metric is not a score generated from a training-free proxy, but an authentic measure of the architecture's performance. Therefore, ValAcc

Table 1. Comparison between the real/estimated Spearman ρ of different proxies

proxy	PNorm [19]	PGNorm [19]	SNIP [17]	GraSP [20]	SynFlow [21]	Fisher [18]
estimation	0.684	0.591	0.593	0.543	0.743	0.506
reality	0.684	0.594	0.596	0.515	0.739	0.503
Δ	0.000	−0.003	−0.003	0.028	0.004	0.003
%	0.00	−0.51	−0.50	5.44	0.54	0.60
proxy	JacCor [22]	EPE [23]	Phi [24]	Zen [24]	BT [25]	BndlEnt [26]
estimation	0.726	0.693	0.703	0.257	0.510	0.199
reality	0.726	0.694	0.723	0.237	0.472	0.188
Δ	0.000	−0.001	−0.020	0.020	0.038	0.011
%	0.00	−0.14	−2.77	8.44	8.05	5.85
proxy	LyDynIso [27]	DynIso [27]	IptSens	BCLR [22]	FADZS-NAS†	ValAcc
estimation	0.633	0.662	0.686	0.779	**0.787**	0.842
reality	0.632	0.674	0.685	0.780	**0.790**	0.811
Δ	0.001	−0.012	0.001	−0.001	−0.002	0.000
%	0.16	−1.78	0.15	−0.13	−0.25	0.00

is the theoretical upper limit for all training-free proxies. Since there is not much difference in the final experimental results between evaluating the entire search space and evaluating 1000 randomly sampled architectures, we choose to sample 1000 architectures in subsequent experiments.

Table 2. Comparison between the estimated Spearman ρ of different proxies in different search spaces

proxy	PNorm [19]	PGNorm [19]	SNIP [17]	GraSP [20]	SynFlow [21]	Fisher [18]
NAS-bench-101	0.561	0.248	0.157	0.357	0.429	0.281
NAS-bench-201	0.684	0.591	0.593	0.543	0.743	0.506
Δ	0.123	0.343	0.436	0.186	0.314	0.225
proxy	JacCor [22]	EPE [23]	Phi [24]	Zen [24]	BT [25]	BndlEnt [26]
NAS-bench-101	0.362	0.008	0.016	0.200	0.054	0.012
NAS-bench-201	0.726	0.693	0.703	0.257	0.510	0.199
Δ	0.364	0.685	0.687	0.057	0.456	0.187
proxy	LyDynIso [27]	DynIso [27]	IptSens	BCLR [22]	FADZS-NAS†	ValAcc
NAS-bench-101	0.486	0.281	0.302	0.410	0.537	0.647
NAS-bench-201	0.633	0.662	0.686	0.779	**0.787**	0.842
Δ	0.147	0.381	0.384	0.369	0.251	0.195

4.4   Performance in Different Search Spaces

Table 2 presents the Spearman ranking correlation coefficients for various proxies across different search spaces, including NAS-bench-101 and NAS-bench-201.

It can be found that the Spearman coefficient of most proxy methods on NAS-bench-101 dropped significantly. A notable decline in the Spearman coefficient for most proxy methods is observed in the NAS-bench-101, potentially due to its considerably higher complexity compared to NAS-bench-201. This variation results from the distinct sets of alternative operations within the architectures of the two operation spaces. PNorm is a proxy calculation method related to the number of architecture parameters, so there is not much difference in performance between the two search spaces. Similarly, the final performance of our proxy method on NAS-bench-101 also drops, yet it surpasses most other proxies and outperforms PNorm, especially when employing the InfoNCE calculation method. These experimental details are further elaborated in Sect. 4.7.

4.5   Performance of the Architecture Under Different Initializations

In order to prove that our method is not sensitive to the initialization of the neural architecture, we use different initialization methods for verification. These include Kaiming Uniform Initialization, Kaiming Normal Initialization and Standard Normal Initialization. As shown in Table 3, The average value of the Spearman coefficient of different initialization methods is close to 0, and at the same time it has a low standard deviation, which is the same as the existing state-of-the-art method BCLR. Therefore, our proposed proxy does not need to carefully choose the architecture initialization method during calculation to obtain better performance.

Table 3. Comparison between the estimated Spearman ρ of different proxies with different initializations

proxy	PNorm [19]	PGNorm [19]	SNIP [17]	GraSP [20]	SynFlow [21]	Fisher [18]
uniform	0.684	0.591	0.593	0.543	0.743	0.506
normal	0.692	0.559	0.570	0.451	0.738	0.480
n01	0.725	0.369	0.518	0.106	0.743	0.453
avg.	0.700	0.506	0.560	0.367	0.741	0.480
std.	0.018	0.098	0.031	0.188	0.002	0.022
proxy	JacCor [22]	EPE [23]	Phi [24]	Zen [24]	BT [25]	BndlEnt [26]
uniform	0.726	0.693	0.703	0.257	0.510	0.199
normal	0.739	0.698	0.670	0.036	0.558	0.028
n01	0.733	0.719	0.657	0.352	0.568	0.040
avg.	0.733	0.703	0.677	0.215	0.545	0.089
std.	0.005	0.011	0.019	0.215	0.594	0.089
proxy	LyDynIso [27]	DynIso [27]	IptSens	BCLR [22]	FADZS-NAS[†]	ValAcc
uniform	0.633	0.662	0.686	0.779	**0.787**	0.842
normal	0.612	0.587	0.641	0.777	0.775	0.842
n01	0.717	0.358	0.625	0.777	0.766	0.842
avg.	0.654	0.536	0.651	0.778	0.776	0.842
std.	0.045	0.739	0.503	0.001	0.009	0.000

Table 4. Comparison between the estimated Spearman ρ of different proxies with different data augmentation strategies

Augmentation	Same	RandomCrop	RandomHorizontalFlip	ColorJitter	Lighting
NAS-bench-101	0.537	0.029	0.457	0.594	0.571
NAS-bench-201	0.787	0.237	0.651	0.735	0.791

4.6 Performance of Different Data Augmentation Strategies

To explore the impact of different positive and negative sample constructions on experimental results, we use different data augmentation strategies on NAS-bench-101 and NAS-bench-201. As shown in Table 4, when strong data augmentation strategies, such as random cropping, are employed, the performance of the proxy method declines sharply. This aligns with the conclusion that architectures exhibit weaker differentiation capabilities for various samples prior to training. Conversely, with mild data augmentation, the model recognizes positive samples and distinguish negative ones, yielding better results. It can also be observed that using weak data augmentation in NAS-bench-201 leads to a decline of performance, whereas the opposite effect is noted in NAS-bench-101. This demonstrates that the architectures within the NAS-bench-101 search space have a stronger discriminatory ability across different samples. This is attributed to the absence of residual connections and zeroizing operations in the NAS-bench-101 search space, which aligns with previous conjectures.

4.7 Performance of Different Similarity Calculation Methods

In this section, our experiments focus on the computation of proxy scores, encompassing methods like the InfoNCE form and determinant calculation, as delineated in Table 5. Notably, the InfoNCE form demonstrates superior performance in the NAS-bench-101 search space, it can be attributed to the absence of non-convolution operations in the models within the NAS-bench-101, which enhances their capability to differentiate between samples. The findings from these experiments align with the conclusions drawn in the Sect. 3.1.

Table 5. Comparison between the Spearman ρ of different proxies with different calculation methods

Calculation Methods	InfoNCE	Determinant	ValAcc
NAS-bench-101	0.588	0.537	0.674
NAS-bench-201	0.695	0.787	0.809

4.8 Performance on the Top 20% of High-Performance Neural Architectures

In neural architecture search tasks, NAS algorithms predominantly aim to identify high-performance neural architectures, as the primary objective is to unearth optimal configurations. This subsection outlines the evaluation results of various proxies on the top 20% of high-performance neural architectures within the NAS-Bench-201 search space. We conducted an experiment of the correlation coefficient ranking between 1000 randomly sampled architectures and the top 20% of the search space. The data in Table 6 reveal a significant reduction in Spearman rank correlation coefficients for all proxies when assessing against the upper tier of high-performance architectures. While our FADZS-NAS proxy exhibits a marginally lower performance compared to SynFlow and BT in this high-performing segment, it still retains a competitive result.

4.9 Architecture Search and Evaluation Experiments

Search Experiment of NAS-Bench-101. To demonstrate the efficacy of our proxy method in conjunction with a search strategy for identifying optimal neural architectures, we performed search experiments on the NAS-bench-101 benchmark, utilizing a determinant-based approach for architecture scoring. As shown in Table 7, our method consumes only a minimal amount of time to obtain competitive architectures, improving search efficiency by approximately 50 times. The resulting architecture has an average test accuracy of 93.80% and a standard deviation of 0.08, which demonstrates the efficiency and stability of our method.

Table 6. Comparison between the Spearman ρ of different proxies on different architectures

proxy	PNorm [19]	PGNorm [19]	SNIP [17]	GraSP [20]	SynFlow [21]	Fisher [18]
1000	0.684	0.591	0.593	0.543	0.743	0.506
top-20%	0.146	0.110	0.105	0.078	0.466	0.133
Δ	0.538	0.481	0.488	0.465	0.277	0.373
proxy	JacCor [22]	EPE [23]	Phi [24]	Zen [24]	BT [25]	BndlEnt [26]
1000	0.726	0.693	0.703	0.257	0.510	0.199
top-20%	0.126	0.073	0.044	0.134	0.242	0.190
Δ	0.600	0.620	0.659	0.123	0.268	0.009
proxy	LyDynIso [27]	DynIso [27]	IptSens	BCLR [22]	FADZS-NAS[†]	ValAcc
1000	0.633	0.662	0.686	0.779	0.787	0.842
top-20%	0.094	0.026	0.073	0.124	0.113	0.440
Δ	0.539	0.636	0.613	0.655	0.674	0.402

Table 7. Comparison of different training-free NAS algorithms on NAS-Bench-101

			CIFAR-10 Accuracy	
Architecture	Costs(s)	Validation(%)	Test(%)	
RS-ValAcc	142980	93.92±0.11	93.42±0.02	
RL-ValAcc	157375	94.23±0.33	93.72±0.27	
RE-ValAcc	152146	94.28±0.24	93.78±0.16	
Ours	**3088**	93.80±0.08	93.22±0.08	

Search Experiment of NAS-Bench-201. To further validate the adaptability of our method across various search spaces, we extend our experiment to NAS-bench-201, also utilizing a determinant-based approach for architecture scoring. Table 8 illustrates that our method, when integrated with evolutionary algorithms, delivers outstanding performance on the CIFAR-10, CIFAR-100 and ImageNet16-120 datasets. Our approach surpasses all existing methods on these datasets, with the exception of GDAS and TE-NAS. Moreover, our method notably exceeds others like GDAS in computational efficiency, almost by a factor of 10, underscoring its superior efficacy.

Table 8. Comparison of different training-free NAS algorithms on NAS-Bench-201

Architecture	Costs(s)	CIFAR-10(%)	CIFAR-100(%)	ImageNet16(%)
ENAS [6]	13315	54.30±0.00	15.61±0.00	16.32±0.00
RSPS [28]	7587	87.66±1.69	58.33±4.34	31.14±3.88
DARTS-V1 [29]	10890	54.30±0.00	15.61±0.00	16.32±0.00
DARTS-V2 [29]	29902	54.30±0.00	15.61±0.00	16.32±0.00
GDAS [30]	28926	93.51±0.13	70.61±0.26	41.84±0.90
SETN [31]	31010	86.19±4.63	56.87±7.77	31.90±4.07
NASWOT [22]	–	91.78±1.45	67.05±2.89	37.07±6.39
TE-NAS [32]	–	93.90±0.47	71.24±0.56	42.38±0.46
Ours	**2412**	93.61±0.91	68.74±0.87	43.97±0.86

Search Experiment of DARTS. To show the efficacy of our approach in an expansive and open search domain, we meticulously selected 1500 neural architectures within the DARTS search space, employing a sophisticated evolutionary algorithm for this purpose. As shown in Table 9, our method achieves 97.50% accuracy on CIFAR-10, surpassing manually designed neural architectures and neural architectures based on predictor search. As shown in Fig. 2, the architecture discovered in this chapter has more parameters and is deeper than the architectures discovered by other NAS algorithms. In addition, our method only requires 0.3 GPU days to obtain the corresponding architecture, which proves the efficiency of our method.

Table 9. Comparison of different NAS algorithms on DARTS

Architecture	Params. (M)	Accuracy (%)	Costs (G · D)
VGG-19 [33]	–	95.13	–
DenseNet [34]	25.6	96.54	–
Swin-S [35]	–	94.17	–
Nest-S [36]	–	96.97	–
AmoebaNet [37]	3.2	96.66	3150
ENAS [6]	4.6	97.11	0.5
PNAS [1]	3.2	96.59	–
GHN [38]	5.7	97.16	0.8
D-VAE [39]	–	94.80	–
NGE [40]	–	97.40	0.1
BONAS-A [41]	3.45	97.31	2.5
CATE [4]	–	97.45	3.3
CTNAS [42]	3.6	97.41	0.3
TNASP [2]	3.6	97.43	0.3
Ours	4.7	**97.50**	0.3

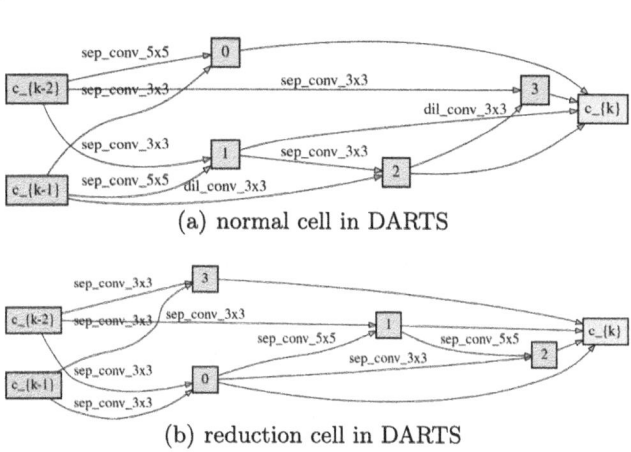

(a) normal cell in DARTS

(b) reduction cell in DARTS

Fig. 2. Our best searched normal cell and reduction cell.

5   Conclusion

In this paper, we introduce FADZS-NAS, a novel approach for Zero-Shot NAS using Feature Activation-Driven Contrastive Learning. We evaluate the model's performance by analyzing feature distributions before the activation layer on the hypersphere. Our proposed methods utilize feature cosine similarity matrices for proxy score calculation, validated experimentally on NAS-bench-101 and NAS-bench-201. We believe our proxy can improve search efficiency and architecture performance when integrated with various search strategies. Our future work will enhance the proxy's performance in broader search spaces and develop more efficient proxies.

References

1. Liu, C., Zoph, B., Neumann, M., Shlens, J., Hua, W., et al.: Progressive neural architecture search. In: European Conference on Computer Vision, ECCV (2018)
2. Lu, S., Li, J., Tan, J., Yang, S., Liu, J.: TNASP: a transformer-based NAS predictor with a self-evolution framework. In: Conference on Neural Information Processing Systems, NeurIPS (2021)
3. Lu, S., Hu, Y., Wang, P., Han, Y., Tan, J., et al.: PINAT: a permutation invariance augmented transformer for NAS predictor. In: AAAI Conference on Artificial Intelligence, AAAI (2023)
4. Yan, S., Song, K., Liu, F., Zhang, M.: CATE: computation-aware neural architecture encoding with transformers. In: International Conference on Machine Learning, ICML, Machine Learning Research (2021)
5. Jing, K., Xu, J., Li, P.: Graph masked autoencoder enhanced predictor for neural architecture search. In: International Joint Conference on Artificial Intelligence, IJCAI (2022)
6. Pham, H., Guan, M., Zoph, B., Le, Q., Dean, J.: Efficient neural architecture search via parameters sharing. In International Conference on Machine Learning (ICML) (2018)
7. Chu, X., Zhang, B., Xu, R.: Fairnas: rethinking evaluation fairness of weight sharing neural architecture search. In: IEEE International Conference on Computer Vision, ICCV (2021)
8. Xie, S., Zheng, H., Liu, C., Lin, L.: SNAS: stochastic neural architecture search. In: International Conference on Learning Representations, ICLR (2019)
9. Guo, Z., et al.: Single path one-shot neural architecture search with uniform sampling. In: Vedaldi, A., Bischof, H., Brox, T., Frahm, J.-M. (eds.) ECCV 2020. LNCS, vol. 12361, pp. 544–560. Springer, Cham (2020). https://doi.org/10.1007/978-3-030-58517-4_32
10. Bhardwaj, K., Li, G., Marculescu, R.: How does topology influence gradient propagation and model performance of deep networks with densenet-type skip connections? In: IEEE Conference on Computer Vision and Pattern Recognition, CVPR (2021)
11. He, K., Fan, H., Wu, Y., Xie, S., Girshick, R.: Momentum contrast for unsupervised visual representation learning. In: IEEE Conference on Computer Vision and Pattern Recognition, CVPR (2020)
12. Radford, A., Kim, J.W., Hallacy, C., Ramesh, A., Goh, G., et al.: Learning transferable visual models from natural language supervision. In: Meila, M., Zhang, T. (eds.) International Conference on Machine Learning, ICML (2021)
13. Chen, T., Kornblith, S., Norouzi, M., Hinton, G.E.: A simple framework for contrastive learning of visual representations. In: International Conference on Machine Learning, ICML (2020)
14. Wan, C., Zhang, T., Xiong, Z., Ye, H.: Representation learning for fault diagnosis with contrastive predictive coding. In: Symposium on Fault Detection, Supervision, and Safety for Technical Processes, SAFEPROCESS (2021)

15. Ying, C., Klein, A., Christiansen, E., Real, E., Murphy, K., et al.: Nas-bench-101: towards reproducible neural architecture search. In: International Conference on Machine Learning, ICML (2019)
16. Dong, X., Yang, Y.: Nas-bench-201: extending the scope of reproducible neural architecture search. In: International Conference on Learning Representations, ICLR (2020)
17. Lee, N., Ajanthan, T., Torr, P.H.S.: Snip: single-shot network pruning based on connection sensitivity. In: International Conference on Learning Representations, ICLR (2019)
18. Liu, L., Zhang, S., Kuang, Z., Zhou, A., Xue, J.H., et al.: Group fisher pruning for practical network compression. In: International Conference on Machine Learning, ICML (2021)
19. Abdelfattah, M.S., Mehrotra, A., Dudziak, L., Lane, N.D.: Zero-cost proxies for lightweight NAS. In: International Conference on Learning Representations, ICLR (2021)
20. Wang, C., Zhang, G., Grosse, R.B.: Picking winning tickets before training by preserving gradient flow. In: International Conference on Learning Representations, ICLR (2020)
21. Tanaka, H., Kunin, D., Yamins, D.L.K., Ganguli, S.: Pruning neural networks without any data by iteratively conserving synaptic flow. In: Conference on Neural Information Processing Systems, NeurIPS (2020)
22. Mellor, J., Turner, J., Storkey, A.J., Crowley, E.J.: Neural architecture search without training. In: International Conference on Machine Learning, ICML (2021)
23. Lopes, V., Alirezazadeh, S., Alexandre, L.A.: EPE-NAS: efficient performance estimation without training for neural architecture search. In: Farkaš, I., Masulli, P., Otte, S., Wermter, S. (eds.) ICANN 2021. LNCS, vol. 12895, pp. 552–563. Springer, Cham (2021). https://doi.org/10.1007/978-3-030-86383-8_44
24. Lin, M., Wang, P., Sun, Z., Chen, H., Sun, X., et al.: Zen-nas: a zero-shot NAS for high-performance deep image recognition. CoRR (2021)
25. Zbontar, J., Jing, L., Misra, I., LeCun, Y., Deny, S.: Barlow twins: self-supervised learning via redundancy reduction. In: International Conference on Machine Learning, ICML (2021)
26. Peer, D., Stabinger, S., Rodríguez-Sánchez, A.J.: Auto-tuning of deep neural networks by conflicting layer removal. CoRR (2021)
27. Lee, N., Ajanthan,T., Gould, S., Torr, P.H.S.: A signal propagation perspective for pruning neural networks at initialization. In: International Conference on Learning Representations, ICLR (2020)
28. Li, L., Talwalkar, A.: Random search and reproducibility for neural architecture search. In: Conference on Uncertainty in Artificial Intelligence, UAI (2019)
29. Liu, H., Simonyan, K., Yang, Y.: DARTS: differentiable architecture search. In: International Conference on Learning Representations, ICLR (2019)
30. Dong, X., Yang, Y.: Searching for a robust neural architecture in four GPU hours. In: IEEE Conference on Computer Vision and Pattern Recognition, CVPR (2019)
31. Dong, X., Yang, Y.: One-shot neural architecture search via self-evaluated template network. In: IEEE International Conference on Computer Vision, ICCV (2019)
32. Chen, W., Gong, X., Wang, Z.: Neural architecture search on imagenet in four GPU hours: a theoretically inspired perspective. In: International Conference on Learning Representations, ICLR (2021)
33. Simonyan, K., Zisserman, A.: Very deep convolutional networks for large-scale image recognition. In: International Conference on Learning Representations, ICLR (2015)

34. Huang, G., Liu, Z., van der Maaten, L., Weinberger, K.Q.: Densely connected convolutional networks. In: IEEE Conference on Computer Vision and Pattern Recognition, CVPR (2017)
35. Liu, Z., Lin, Y., Cao, Y., Hu, H., Wei, Y., et al.: Swin transformer: hierarchical vision transformer using shifted windows. In: IEEE International Conference on Computer Vision, ICCV (2021)
36. Zhang, Z., Zhang, H., Zhao, L., Chen, T., Arik, S.Ö., et al.: Nested hierarchical transformer: towards accurate, data-efficient and interpretable visual understanding. In: AAAI Conference on Artificial Intelligence, AAAI (2022)
37. Real, E., Aggarwal, A., Huang, Y., Le, Q.V.: Regularized evolution for image classifier architecture search. In: AAAI Conference on Artificial Intelligence, AAAI (2019)
38. Zhang, C., Ren, M., Urtasun, R.: Graph hypernetworks for neural architecture search. In: International Conference on Learning Representations, ICLR (2019)
39. Zhang, M., Jiang, S., Cui, Z., Garnett, R., Chen, Y.: D-VAE: a variational autoencoder for directed acyclic graphs. In: Conference on Neural Information Processing Systems, NeurIPS (2019)
40. Li, W., Gong, S., Zhu, X.: Neural graph embedding for neural architecture search. In: AAAI Conference on Artificial Intelligence, AAAI (2020)
41. Shi, H., Pi, R., Xu, H., Li, Z., Kwok, J.T., et al.: Bridging the gap between sample-based and one-shot neural architecture search with BONAS. In: Conference on Neural Information Processing Systems, NeurIPS (2020)
42. Chen, Y., Guo, Y., Chen, Q., Li, M., Zeng, W., et al.: Contrastive neural architecture search with neural architecture comparators. In: IEEE Conference on Computer Vision and Pattern Recognition, CVPR (2021)

NAS-Bench-Compre: A Comprehensive Neural Architecture Search Benchmark with Customizable Components

Di Wang[1], Kun Jing[2]([⊠]), and Jungang Xu[1]([⊠])

[1] University of Chinese Academy of Sciences, Beijing, China
xujg@ucas.ac.cn
[2] Anhui University, Hefei, Anhui, China
jingkun@ahu.edu.cn

Abstract. Neural Architecture Search (NAS) aims to discover the optimal architectures for various tasks, which developed rapidly but disorderly in the past few years. Therefore, the proposal of NAS benchmark is necessary and significant for the development of NAS. NAS has three components (search space, search strategy, and estimation strategy) to be evaluated. However, the existing benchmarks mainly evaluate search strategy only, lacking consideration of the other two components. Thus, it is difficult for current benchmarks to comprehensively evaluate the three components of NAS. In this work, we propose NAS-Bench-Compre, a comprehensive benchmark for NAS that allows users to customize search spaces, search strategies, and evaluation strategies for evaluation. Additionally, we provide essential user interfaces and implementations for each decoupled and customized component to ensure fairness and usability of NAS evaluation. We support the evaluation of the three components of the prominent NAS algorithms. Currently, we implement evaluations for DARTS, GMAENAS, OFA, AutoFormer, and Zero-shot NAS. All codes are publicly available at https://github.com/kunjing96/NAS-Bench-Compre.

Keywords: Neural architecture search · NAS benchmark · Customizable component · User-friendly interface

1 Introduction

In recent years, deep learning has demonstrated remarkable learning capabilities and achieved breakthrough success in various tasks. The performance of deep neural networks is heavily reliant on architectural design, which often demands substantial human efforts. In this context, Neural Architecture Search (NAS) [1] has been introduced to automate the design of high-performance neural architectures based on target tasks, utilizing limited computational resources and minimizing human intervention. With the increasing focus of researchers, many efficient NAS methods for neural architecture have been proposed. Differentiable

© The Author(s), under exclusive license to Springer Nature Switzerland AG 2024
M. Wand et al. (Eds.): ICANN 2024, LNCS 15016, pp. 277–291, 2024.
https://doi.org/10.1007/978-3-031-72332-2_19

architecture search (DARTS) [2] introduces continuous relaxation to construct a differentiable search space, simplifying the NAS problem into a differentiable learning process of neural architecture parameters. Once-for-all (OFA) [3] decouples parameter training and architectural optimization into two distinct stages, attracting more researchers to study the two-step NAS. As a representative of the predictor-based NAS, BANANAS [4] maps the architectural space to absolute or relative performance values, reducing reliance on training architectures and improving efficiency. ZeroCost [5] employs zero-shot proxy metrics to evaluate architectures, improving the efficiency of evaluation strategies.

The emergence of diverse NAS algorithms emphasizes the growing necessity of a fair evaluation. NAS consists of three components: search space, search strategy, and evaluation strategy, and has achieved success in various tasks. A few works have proposed benchmarks of NAS, enabling scientifically sound evaluations of NAS to some extent. However, some existing benchmarks only consider the evaluation of different search strategies, for example, NAS-Bench-101 [6], NAS-Bench-201 [7], and NAS-Bench-301 [8] serve as NAS benchmarks to evaluate image classification tasks. NAS-Bench-ASR [9] is utilized to evaluate speech recognition tasks, and TransNAS-Bench-101 [10] aims at the evaluation of object classification and other six visual tasks. NAS-Bench-NLP [11] focuses on evaluating the natural language processing tasks. Apart from this, NAS-Bench-Suite [12], NAS-Bench-Suite-Zero [13], and BenchENAS [14] gather different benchmarks as a collection to mitigate the issue caused by the varied performance of NAS algorithms across different benchmarks, focusing on evaluating the three components of NAS. However, they need a decoupled and user-friendly interface.

To address these issues, we propose a comprehensive, fair, and simple benchmark for NAS called NAS-Bench-Compre, which provides interfaces for individually customizing search space, search strategy, and evaluation strategy separately. The implementations are independent and decoupled, which brings convenience to users. Moreover, NAS-Bench-Compre improves the comprehensiveness of evaluation by shifting the focus from solely evaluating search strategy to considering the evaluation of both search space and evaluation strategy. Based on NAS-Bench-Compre, we have implemented evaluations for DARTS [2], GMAE-NAS [15], OFA [3], AutoFormer [16], and Zero-shot NAS currently. Among them, Zero-shot NAS is a NAS algorithm based on a zero-shot evaluation metric.

In summary, the contributions of this work are as follows.

- We propose NAS-Bench-Compre, a comprehensive, fair, and user-friendly benchmark for NAS, which enables evaluating the three NAS components separately or interactively. Additionally, we implement a simple and user-friendly interface that allows users to individually customize search space, search strategy, and estimation strategy for evaluation.
- We evaluate existing NAS algorithms, including DARTS [2], GMAENAS [15], OFA [3], AutoFormer [16], and Zero-shot NAS, providing valuable insights into the comparison of different algorithms.

Table 1. Summary of different NAS benchmarks.

Benchmark	Tasks/Datasets	Size	Search spaces	Search strategies	Estimation strategies
NAS-Bench-101 [6]	CIFAR-10 [17]	423k		√	
NAS-Bench-201 [7]	CIFAR-10/100 [17], ImageNet [18]	15.6k		√	
NAS-Bench-301 [8]	CIFAR-10 [17]	60k		√	
TransNAS-Bench-101 [10]	7 visual tasks	7k		√	
NAS-Bench-NLP [11]	PTB [19], WikiText-2 [19]	14k		√	
NAS-Bench-ASR [9]	ASR on TIMIT [20]	8k		√	
NAS-Bench-Graph [21]	9 Graph Datasets	26k		√	
NAS-Bench-Macro [22]	CIFAR-10 [17]	6.5k		√	
Blox [23]	CIFAR-100 [17]	91k		√	
NAS-Bench-360 [24]	10 different domain datasets	15.6k		√	
HW-NAS-Bench [25]	hardware cost on 6 devices	46k		√	
EC-NAS [26]	energy consumption	423k		√	
NAS-Bench-Suite [12]	A collection of several benchmarks		√	√	√
NAS-Bench-Suite-Zero [13]	A collection of 13 Zero-cost proxies on 28 tasks		√	√	√
BenchENAS [14]	A collection of nine representative evolutionary search algorithms			√	
NAS-Bench-1Shot1 [27]	Introduce 3 different search spaces			√	
NAS-Bench-MR [28]	Designs a multi-branch multi-resolution search space			√	
NAS-Bench-x11 [29]	Introduces 3 different search spaces			√	
NAS-Bench-Compre (Ours)	User-customized search spaces,search strategies and estimation strategies		√	√	√

2   Related Work

There are two key reasons why NAS benchmarks are essential. First, it is difficult to compare NAS algorithms due to the variation in training pipelines and settings among them. NAS benchmark establishes a standardized and fair evaluation framework for various NAS algorithms. Second, NAS benchmarks enable more researchers to participate in advancing these algorithms since they improve efficiency and allow researchers to focus on enhancing the algorithms. In Table 1, we briefly summarize different NAS benchmarks. The last three columns of Table 1 respectively demonstrate whether the corresponding NAS benchmarks include these evaluations.

From the perspective of the search space, benchmarks can be categorized into two types. One category constrains a fixed search space with one or more tasks/datasets, termed as naive benchmark, while another aggregates different benchmarks of different search spaces, tasks, or datasets, referred to as comprehensive benchmark.

2.1   Naive Benchmark

Most NAS benchmarks belong to this type, and they can be divided into two groups based on the task type: visual task-related and non-visual task-related. Among these, NAS-Bench-101 [6], NAS-Bench-201 [7], NAS-Bench-301 [8], TransNAS-Bench-101 [10], NAS-Bench-Macro [22], and Blox [23] are visual task-related benchmarks, whereas NAS-Bench-NLP [11], NAS-Bench-ASR [9], and NAS-Bench-Graph [21] are non-visual related benchmarks. Besides, NAS-bench-360 [22]evaluates NAS algorithms in 10 different domain datasets, from visual tasks to DNA sequence predictions. HW-NAS-Bench [25] evaluates the hardware cost of NAS on six devices. EC-NAS [26] evaluates the energy consumption of NAS. While they have advanced benchmark research and contributed to the development of NAS, their approach of constraining the search space and primarily evaluating search strategy hinders the achievement of a comprehensive and fair evaluation. NAS-Bench-Compre provides a user-driven evaluation of the three components of NAS by allowing users to individually customize the search space, search strategy, and evaluation strategy according to their specific requirements. It enables users to evaluate both the search space and the evaluation strategy instead of solely focusing on the evaluation of search strategy. NAS-Bench-Compre provides a more flexible evaluation approach and a more comprehensive benchmark.

2.2   Comprehensive Benchmark

The strong performance of NAS algorithms on one or a few benchmarks does not guarantee similar performance on other benchmarks [12]. Meanwhile, it is inconvenient for researchers to gather information on different search spaces and datasets to evaluate the NAS algorithms [30]. NAS-Bench-Suite [12] can mitigate the above issues since it is a collection of several benchmarks. NAS-Bench-Suite-Zero [13], a benchmark collection of 13 Zero-cost proxies on 28 tasks, is used to analyze the generalizability information. BenchENAS [14] consists of nine representative evolutionary search algorithms. NAS-Bench-MR [28] designs a multi-branch, multi-resolution search space. NAS-BENCH-1SHOT1 [27] aims to track the trajectory and performance of the discovered architectures computationally cheaply and introduces three different search spaces. NAS-Bench-x11 [29] introduces three different search space, NAS-Bench-111, NAS-Bench-311, and NAS-Bench-NLP11, respectively. Although they have gathered different benchmarks and enhanced the comprehensiveness and fairness of the benchmark, the interface they provide lacks the simplicity and convenience comparable to NAS-Bench-Compre. Moreover, most of them can only evaluate different search strategy instead of all three components of NAS. NAS-Bench-Suite [12] and NAS-Bench-Suite-Zero [13] allow users to evaluate the three components of NAS, but the NAS algorithms they can evaluate are relatively single. NAS-Bench-Compre allows users to customize the three components of NAS: search space, search strategy, and estimation strategy to evaluate a broader range of NAS algorithms. By implementing simple interfaces for these three components, NAS-Bench-Compre enables the evaluation of each component, thereby facilitating more convenient and fair NAS evaluations.

NAS consists of search space, search strategy, and estimation strategy. Thus, a comprehensive evaluation of these three aspects is essential. Given the diversity in training pipelines and search spaces employed by NAS algorithms, prudent consideration for an overall evaluation approach is necessary. We implement a comprehensive, fair, and simple evaluation of NAS algorithms by decoupling these components, enabling users to customize their search space, search strategy, and evaluation strategy.

3   NAS-Bench-Compre Methodology

We provide a comprehensive, fair, and user-friendly NAS benchmark called NAS-Bench-Compre. In contrast to other benchmarks that confine themselves to one or more search space to evaluate the performance of search strategy predominantly, NAS-Bench-Compre offers the flexibility to customize search space, search strategy, and evaluation strategy to facilitate a more comprehensive NAS evaluation.

3.1   NAS Problem Definition

The overall framework of NAS is shown in Fig. 1. Neural architecture search typically starts with a predefined set of operations. Second, it uses a search strategy to obtain many candidate neural architectures from the search space based on the set of operations. Third, train and validate candidate neural architectures to obtain performance evaluations of these candidate architectures. Fourth, adjust the search strategy based on the ranking information of the candidate neural architectures to obtain a new set of candidate neural architectures. At the end of the search, use the most promising neural architecture as the final optimal neural architecture and select that neural architecture for the final performance evaluation.

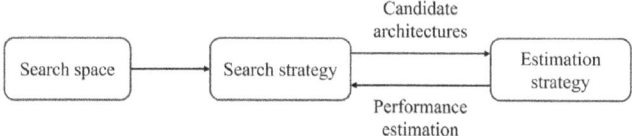

Fig. 1. The overall framework of NAS. The NAS algorithm uses a search strategy to obtain numerous candidate network architectures from the search space and then evaluates the candidates using an estimation strategy. We utilize multiple performance estimations to update the search strategy and generate next-generation candidates iteratively, eventually acquiring the final architecture.

3.2   Overall Design

As mentioned in the problem definition, NAS has three components, each of which contributes to the performance of NAS algorithms. In NAS-Bench-Compre, any of these three components can be defined by users as long as the interface code is written properly (see Sect. 3.6). We can use each component of the existing NAS algorithm implemented in the benchmark to achieve a fair evaluation of NAS. We implement DARTS [2], GMAENAS [15], OFA [3], Auto-Former [16], and Zero-shot NAS. The details of each available NAS component are shown in Table 2.

Table 2. Summary of supported NAS algorithms components.

Components	Available Candidates
Search space	AutoFormer [16], DARTS [2], OFA [3]
Search strategy	AgingEvolution [31], DARTS [2], Random, PredictorBasedRandom
Estimation strategy	AutoFormer [16], DARTS [2], OFA [3], StandardTraining, Zero-shot NAS

The evaluation process can proceed once the search space, search strategy, and evaluation strategy are completely defined (whether user-defined or from existing algorithms). Figure 2 demonstrates the design details of NAS-Bench-Compre. Specifically, the search space needs to be passed into the estimation strategy as a parameter. The search space and the estimation strategy should be considered as parameters of the search strategy. The design details are explained in Sect. 3.6.

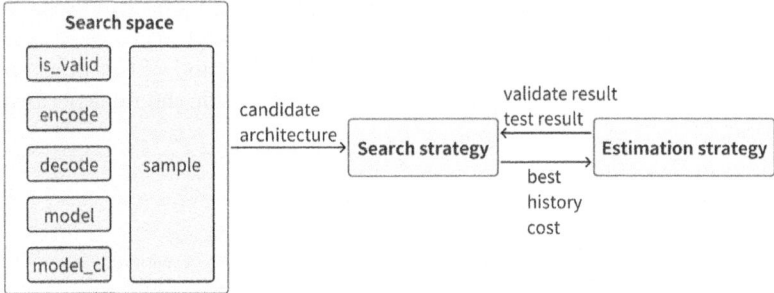

Fig. 2. The design details of each component of NAS. In the search space, in order to obtain candidate architectures, attributes such as *model* and *model_cls* should be defined during initialization. Additionally, functions such as *is_valid*, *encode*, *decode*, and *sample* are required implementations. The search strategy passes the parameters of candidate architectures, such as *best*, *history*, and *cost*, to the estimation strategy. Subsequently, the estimation strategy returns parameters including *validation result* and *test result* to the search strategy for further adjustments.

3.3   Search Spaces

To acquire candidate architectures from the search space, the model and the class of model to be used should be declared. It is necessary to individually implement the architecture validity assessment, architecture encoding, architecture decoding, and final sampling of candidate architectures within their respective functions. We implement the search space of AutoFormer [16], DARTS [2], and OFA [3]. The authors of AutoFormer design a large transformer search space that includes five variable factors in transformer building blocks: embedding dimension, Q-K-V dimension, number of heads, MLP ratio, and network depth [16].

We implement three versions of the search space of AutoFormer, which are base, small, and tiny. The search space of DARTS is cell-level and continuous; a cell is a directed acyclic graph consisting of an ordered sequence of N nodes where the node is a latent representation, and the edge represents some operation that transforms node [2]. The authors of NAG expand the search space of DARTS; NAG can generate various architectures after adversarial training [32]. OFA provides one model but supports many sub-networks of different sizes, covering four important dimensions of the convolutional neural network (CNN) architectures [3]. We implement four versions of the search space of OFA, which are MBV3, MBV3L, Proxyless, and ReNet50.

3.4   Search Strategies

During initialization, predefined search space and estimation strategy are parameters. In this part, values of search history, best architecture, and time cost should be returned. We implement search strategies of AgingEvolution [31], DARTS [2], Random, and PredictorBasedRandom. AgingEvolution simulates the operations of species propagation in the natural world [31]. AgingEvolution randomly generates some network architectures as a population and selects the ones with the highest accuracy as parents. Mutate operations are performed on these parents to obtain new network architectures, which are then trained and added to the population. Subsequently, the earliest added architectures are removed from the population. This process is iterated for a specified number of times. The search strategy of DARTS entails optimizing a differentiable hypernetwork to discover the optimal neural network architecture [2]. Random search attempts to find the optimal architecture by randomly selecting candidates within the search space and evaluating their objective function values. PredictorBasedRandom selects a subset of candidates randomly. It employs a predictor to estimate the performance of the remaining candidates. These candidates are then incorporated into the search space in descending order based on their predicted results. The optimal structure is subsequently identified by computing their objective function values.

3.5   Estimation Strategies

During initialization, predefined search space is a parameter. Both the validation and test results of candidate architectures need to be returned for further adjustments. The specific implementation details depend on the specific algorithm. We implement the estimation strategy of AutoFormer [16], DARTS [2], OFA [3], StandardTraining, and TrainingFreeProxy. Candidate architectures are evaluated according to the manager of the evolution algorithm [16]. In DARTS, selected architectures are trained from scratch with initialized weights to obtain their evaluation results [2]. OFA constructs accuracy and latency table. Structure selection can be executed by referencing the above tables [3]. Zero-shot NAS is a NAS algorithm based on a zero-shot evaluation metric.

3.6   System Design

We offer decoupled implementations for search spaces, search strategies, and estimation strategies, granting users significant flexibility in evaluating NAS performance.

The Implementation of API. Once the search space, evaluation strategy, and search strategy are defined, all the predefined components are obtained at this point of API invocation for evaluating the NAS algorithm. It is important to note that the search space needs to be passed to the evaluation strategy, and both the search space and evaluation strategy need to be passed to the search strategy.

When implementing the API for NAS-Bench-Compre, the hyperparameter dictionary for search space, search strategy, and estimation strategy are initialized.

The Implementation of Search Space. While implementing the search space, attributes *model* and *model_cls* need to be defined during initialization. *model* is the model you use, such as Vision_TransformerSuper from AutoFormer, ofa_net from OFA, and others. *model_cls* is the model class, for example, Network, NASNetworkImageNet, NASNetworkCIFAR, and others. Additionally, the function *is_valid* is employed to assess the validity of the structures (generally used for controlling model parameters, FLOPs, and latency). The architectures should be encoded and decoded in function *encode* and *decode*, respectively. Lastly, obtain candidate architectures in function *sample*.

The Implementation of Search Strategy. Search space and estimation strategy are needed when initializing the search strategy. The historical result, best result, and time cost should be returned. Users must define a subclass such as RANDOM that inherits from superclass, which is *Base* in NAS-Bench-Compre, and implement specific functions; other functions should be written properly if necessary to acquire the history result, best result, and time cost.

The Implementation of Estimation Strategy. Search space is needed when initializing the estimation strategy. The *validate result* and *test result* should be returned. Users should define a subclass that inherits from *base* superclass and implement specific functions. Other functions to acquire validate results and test results should be written properly if necessary.

3.7   Case Study

This section will describe the process of evaluating new models using NAS-Bench-Compre. Our description will be divided into three parts based on the three components of NAS: search space, search strategy, and evaluation strategy.

For search space, subclass *YourSearchSpace* inherited from superclass *Base* should be defined, and codes to implement specific functions need to be added. Please note that the decorator *@_register* should be written at the beginning.

```
@_register
class YourSearchSpace(Base):

def __init__(self, config):
    super(YourSearchSpace, self).__init__(config)
    # add your initialization

def is_valid(self, arch):
    # add code to determine whether the architecture
    # is valid, generally used for controlling model parameters, FLOPs,
        and latency.
    return True

def encode(self, decoded_arch):
    arch = None
    # add code to encode architectures
    return arch

def decode(self, arch):
    decoded_arch = None
    # add code to deconde architectures
    return decoded_arch

def sample(self):
    arch = list()
    # sample an architecture
    return tuple(arch)
```

For estimation strategy, subclass *YourEstimationStrategy* inherited from superclass *Base* should be defined, and codes to implement specific functions to obtain *validate result* and *test result* need to be added. Please note that the decorator *@_register* should be written at the beginning.

```
@_register
class YourEstimationStrategy(Base):

def __init__(self, config):
    super(YourEstimationStrategy, self).__init__(config)
    # add your initialization

def __call__(self):
    val_res = None # validate result
    test_res = None # test result
    # define your estimation strategy
    return val_res, test_res
```

For search strategy, subclass *YourSearchStrategy* inherited from superclass *Base* should be defined, and codes to implement specific functions to obtain *best, history* and *cost* need to be added. Please note that the decorator *@_register* should be written at the beginning.

```
@_register
class YourSearchStrategy(Base):

def __init__(self, config):
    super(YourSearchStrategy, self).__init__(config)
    # add your initialization

def __call__(self):
    history = None # search history
    best = None # best architecture
```

```
cost = None # time cost
# define your search strategy
return best, history, cost
```

4   Experiment

After presenting the methodology of NAS-Bench-Compre, we proceed to intro-duce the experimental setup and discuss the results in this section. Since NAS-Bench-Compre enables users to individually customize search spaces, search strategies, and estimation strategies, our discussion will prove the flexibility and effectiveness of evaluating different components of NAS.

This section will introduce and analyze four experiments. The first three experiments cover the experimental setup and analysis of evaluating three com-ponents of NAS with NAS-Bench-Compre. The fourth experiment is conducted to validate the reliability of NAS-Bench-Compre.

4.1   Experiments on Different Search Spaces

To prove that NAS-Bench-Compre could effectively evaluate different search spaces, we use Random as the search strategy and StandardTraining as the estimation strategy. Subsequently, we evaluate AutoFormerSmall and OFA (two different search spaces) on ImageNet. We run all experiments on Nvidia Gefore GTX TITAN V. All search algorithms search in ten architectures during the experimental process. Since our experimental goal is to observe the performance differences of different NAS algorithms under the same configuration, we did not use the configurations mentioned in the original paper.

Table 3. The best accuracy and time cost of NAS on ImageNet with different search spaces using NAS-Bench-Compre.

Search space	Search strategy	Estimation strategy	Acc. (%)	Cost (h)
AutoFormerSmall	Random	StandardTraining	57.06	4.85
OFA	Random	StandardTraining	70.29	4.11

Table 3 presents an overview of NAS performance on the ImageNet with different search spaces, it shows the best accuracy and time cost of NAS. NAS-Bench-Compre distinguishes the performance of various search spaces of NAS algorithms, demonstrating the effectiveness of NAS-Bench-Compre. The combi-nation of OFA, Random, and StandardTraining achieves the best accuracy of 70.29%. Furthermore, when using OFA as the search space, searching architec-tures takes less time compared to using AutoFormerSmall. That is because OFA is based on CNN, while AutoFormer is based on transformers. Since they use the same hyperparameters, CNN converges faster and achieves higher accuracy.

4.2   Experiments on Different Search Strategies

Like the experimental setup in Sect. 4.1, we use Nvidia Gefore GTX TITAN V to evaluate AgingEvolution, PredictorBasedRandom, and Random (three different search strategies) by NAS-Bench-Compre, we use AutoFormerSmall as search space and AutoFormer as Estimation strategy. Subsequently, we search in ten architectures and record the best accuracy. For AgingEvolution, we set mutate probability as 1. The reason we chose AutoFormer instead of OFA as the search space is that the transformer is currently the mainstream model and is more widely used. Therefore, we conducted evaluation experiments on different search strategies for AutoFormer search space.

Table 4 demonstrates the best accuracy and time cost of NAS using different search strategies on ImageNet. The statistical variance of these results is 0.27, which shows that NAS-Bench-Compre can effectively distinguish the performance of various search strategies. As for accuracy, Random attains the highest accuracy of 81.58% among these three search strategies. In terms of time cost, PredictorBasedRandom needs the longest time of 1.52 h since it requires a predictor to search for architectures.

Table 4. The best accuracy and time cost of NAS on dataset ImageNet with different search strategies using NAS-Bench-Compre.

Search space	Search strategy	Estimation strategy	Acc. (%)	Cost (h)
AutoFormerSmall	AgingEvolution	AutoFormer	80.43	0.89
AutoFormerSmall	PredictorBasedRandom	AutoFormer	81.46	1.52
AutoFormerSmall	Random	AutoFormer	81.58	0.85

4.3   Experiments on Different Estimation Strategies

To validate the effectiveness of NAS-Bench-Compre in evaluating different evaluation strategies, given that the DARTS search space is a commonly used NAS search space, and many works have experimented on it, we use DARTS space as the search space and AgingEvolution as the search strategy. We utilize various zero-shot evaluation strategies [5] to search in ten architectures on the CIFAR-10. The best-performing architecture's accuracy is recorded for analysis. Our search spans three epochs with a specified learning rate of 0.025, weight decay of 3e-4, and a dropout probability of 0.2. We search in ten architectures.

Table 5 shows the best accuracy and time cost of NAS using different estimation strategies on CIFAR-10. The statistical variance of these results is 85.69, proving that NAS-Bench-Compre could evaluate various estimation strategies effectively. Among all these estimation strategies, the top three search strategies in terms of accuracy are as follows: plain achieves 63.3%, phi_score achieves 59.86%, and jacob_cor_logdet achieves 58.36%. The average search time across these evaluation strategies is 7.33 h. The plain proxy computes the sum of the absolute gradients of the neural network weights, phi_score calculates the L1-norm of the difference between the outputs of two networks, one with noisy inputs and the other with original inputs, and jacob_cor_logdet computes the correlation of the Jacobian matrices for different inputs.

Table 5. The best accuracy and time cost of NAS on CIFAR-10 with different search strategies using NAS-Bench-Compre.

Search space	Search strategy	Estimation strategy	Acc. (%)	Cost (h)
DARTS	AgingEvolution	act_grad_cor_weighted	36.63	7.26
DARTS	AgingEvolution	bundle_entropy	56.20	7.21
DARTS	AgingEvolution	epe_score	57.46	7.36
DARTS	AgingEvolution	feature_distance	48.84	7.30
DARTS	AgingEvolution	fisher	37.73	7.28
DARTS	AgingEvolution	grad_cor	33.72	7.13
DARTS	AgingEvolution	grad_hamming	39.71	7.42
DARTS	AgingEvolution	grad_norm	40.88	7.25
DARTS	AgingEvolution	jacob_cor_logdet	58.36	7.29
DARTS	AgingEvolution	jacob_cor	57.45	7.46
DARTS	AgingEvolution	l2_norm	36.87	7.55
DARTS	AgingEvolution	layerwise_dyn_isometry	37.63	7.03
DARTS	AgingEvolution	phi_score_with_data	56.19	7.33
DARTS	AgingEvolution	phi_score	59.86	7.14
DARTS	AgingEvolution	plain	63.03	7.26
DARTS	AgingEvolution	resp_sens	39.15	7.31
DARTS	AgingEvolution	snip	42.95	7.37
DARTS	AgingEvolution	zen_score_cossim	49.80	7.42
DARTS	AgingEvolution	zen_score_with_data_cossim	50.74	7.59
DARTS	AgingEvolution	zen_score	51.36	7.58
DARTS	AgingEvolution	zen_score_with_data	51.84	7.54

4.4   Experments on Reliability Validation

To affirm the reliability of NAS-Bench-Compre, we compare the results from the original papers with those generated by NAS-Bench-Compre. In order to establish a direct comparison, we use the same experimental parameters as the original work. For AutoFormerSmall, we search in 1000 architectures, and set the batch size to 64.

Table 6 presents the results of the AutoFormersmall as reported in the original paper and the evaluation results from NAS-Bench-Compre. These two results are close, thus proving the reliability and effectiveness of NAS-Bench-Compre.

Table 6. Accuracy of autoformer in original paper and generated from NAS-Bench-Compre.

Search space	Search strategy	Estimation strategy	Original paper	NAS-Bench-Compre
AutoFormerSmall	AgingEvolution	AutoFormer	81.70	81.59

5   Conclusion

We discuss the two challenges of the NAS benchmark. First, the existing benchmarks primarily focus on evaluating search strategies, neglecting the evaluation of search spaces and evaluation strategies. Second, the existing benchmarks lack a simple and convenient interface for users to evaluate the three components of NAS separately. To mitigate the above problems, we propose NAS-Bench-Compre, a more comprehensive, fair, and user-friendly benchmark, which provides the flexibility for users to customize three components of NAS for evaluation. We elaborate on how NAS-Bench-Compre achieves this goal by decoupling search strategies, search spaces, and evaluation strategies. Some experimental results are shown and discussed to validate the effectiveness of NAS-Bench-Compre. We also provide details of the system design and interface function design of NAS-Bench-Compre. In the future, our work will include two main aspects. First, we will broaden the range of supported NAS algorithms for evaluation. Second, we will conduct more comprehensive tests and analyses on NAS-Bench-Compre.

References

1. Zoph, B., Le, Q.V.: Neurals architecture search with reinforcement learning. In: International Conference on Learning Representations, ICLR (2017)
2. Liu, H., Simonyan, K., Yang, Y.: DARTS: differentiable architecture search. In: International Conference on Learning Representations, ICLR (2019)
3. Cai, H., Gan, C., Wang, T., Zhang, Z., Han, S.: Once-for-all: train one network and specialize it for efficient deployment. In: International Conference on Learning Representations, ICLR (2020)
4. White, C., Neiswanger, W., Savani, Y.: BANANAS: bayesian optimization with neural architectures for neural architecture search. In: AAAI Conference on Artificial Intelligence, AAAI (2021)
5. Abdelfattah, M.S., Mehrotra, A., Dudziak, L., Lane, N.D.: Zero-cost proxies for lightweight NAS. In: International Conference on Learning Representations, ICLR (2021)

6. Ying, C., Klein, A., Christiansen, E., Real, E., Murphy, K., et al.: NAS-Bench-101: towards reproducible neural architecture search. In: International Conference on Machine Learning, ICML (2019)
7. Dong, X., Yang, Y.: Nas-bench-201: extending the scope of reproducible neural architecture search (2020)
8. Siems, J., Zimmer, L., Zela, A., Lukasik, J., Keuper, M., et al.: NAS-Bench-301 and the case for surrogate benchmarks for neural architecture search. arXiv preprint arXiv:2008.09777 (2022)
9. Mehrotra, A., Ramos, A.G.C.P., Bhattacharya, S., Dudziak, L., Vipperla, R., et al.: NAS-Bench-ASR: reproducible neural architecture search for speech recognition. In: International Conference on Learning Representations, ICLR (2021)
10. Duan, Y., Chen, X., Xu, H., Chen, Z., Liang, X., et al.: TransNAS-Bench-101: improving transferability and generalizability of cross-task neural architecture search. In: IEEE Conference on Computer Vision and Pattern Recognition, CVPR (2021)
11. Klyuchnikov, N., Trofimov, I., Artemova, E., Salnikov, M., Fedorov, M.V., et al.: NAS-Bench-NLP: neural architecture search benchmark for natural language processing. IEEE Access **10**, 45736–45747 (2022)
12. Mehta, Y., White, C., Zela, A., Krishnakumar, A., Zabergja, G., et al.: NAS-bench-suite: NAS evaluation is (now) surprisingly easy. In: International Conference on Learning Representations, ICLR (2022)
13. Krishnakumar, A., White, C., Zela, A., Tu, R., Safari, M., et al.: NAS-bench-suite-zero: accelerating research on zero cost proxies. In: Koyejo, S., Mohamed, S., Agarwal, A., Belgrave, D., Cho, K., Oh, A. (eds.) Conference on Neural Information Processing Systems, NeurIPS (2022)
14. Xie, X., Liu, Y., Sun, Y., Yen, G.G., Xue, B., Zhang, M.: BenchENAS: a benchmarking platform for evolutionary neural architecture search. IEEE Trans. Evol. Comput. **26**(6), 1473–1485 (2022)
15. Jing, K., Xu, J., Li, P.: Graph masked autoencoder enhanced predictor for neural architecture search. In: International Joint Conference on Artificial Intelligence, IJCAI (2022)
16. Chen, M., Peng, H., Fu, J., Ling, H.: AutoFormer: searching transformers for visual recognition. In: International Conference on Computer Vision, ICCV (2021)
17. Krizhevsky, A.: Learning multiple layers of features from tiny images (2009)
18. Chrabaszcz, P., Loshchilov, I., Hutter, F.: A downsampled variant of imagenet as an alternative to the cifar datasets. arXiv preprint arXiv:1707.08819 (2017)
19. Mikolov, T., Karafiát, M., Burget, L., Cernocký, J., Khudanpur, S.: Recurrent neural network based language model. In: Annual Conference of the International Speech Communication Association (2010)
20. Garofolo, J.S.: Timit acoustic phonetic continuous speech corpus. In: Linguistic Data Consortium (1993)
21. Qin, Y., Zhang, Z., Wang, X., Zhang, Z., Zhu, W.: NAS-bench-graph: benchmarking graph neural architecture search. In: Conference on Neural Information Processing Systems, NeurIPS (2022)
22. Su, X., Huang, T., Li, Y., You, S., Wang, F., et al.: Prioritized architecture sampling with monto-carlo tree search. In: IEEE Conference on Computer Vision and Pattern Recognition, CVPR (2021)
23. Chau, T., Dudziak, L., Wen, H., Lane, N.D., Abdelfattah, M.S.: BLOX: macro neural architecture search benchmark and algorithms. In: Conference on Neural Information Processing Systems (2022)

24. Tu, R., Roberts, N., Khodak, M., Shen, J., Sala, F., et al.: NAS-Bench-360: benchmarking neural architecture search on diverse tasks. In: Conference on Neural Information Processing Systems (2022)
25. Li, C., Yu, Z., Fu, Y., Zhang, Y., Zhao, Y., et al.: HW-NAS-bench: hardware-aware neural architecture search benchmark. In: International Conference on Learning Representations, ICLR (2021)
26. Bakhtiarifard, P., Igel, C., Selvan, R.: Energy consumption-aware tabular benchmarks for neural architecture search. arXiv preprint arXiv:2210.06015 (2022)
27. Zela, A., Siems, J., Hutter, F.: NAS-bench-1Shot1: benchmarking and dissecting one-shot neural architecture search. In: International Conference on Learning Representations, ICLR (2020)
28. Ding, M., et al.: Learning versatile neural architectures by propagating network codes. In: The Tenth International Conference on Learning Representations, ICLR (2022)
29. Yan, S., White, C., Savani, Y., Hutter, F.: NAS-Bench-x11 and the power of learning curves. In: Conference on Neural Information Processing Systems (2021)
30. Chitty-Venkata, K.T., Emani, M., Vishwanath, V., Somani, A.K.: Neural architecture search benchmarks: insights and survey. IEEE Access **11**, 25217–25236 (2023)
31. Real, E., Aggarwal, A., Huang, Y., Le, Q.V.: Regularized evolution for image classifier architecture search. In: AAAI Conference on Artificial Intelligence (2019)
32. Jing, K., Jungang, X., Zhang, Z.: A neural architecture generator for efficient search space. Neurocomputing **486**, 189–199 (2022)

NAVIGATOR-D3: Neural Architecture Search Using VarIational Graph Auto-encoder Toward Optimal aRchitecture Design for Diverse Datasets

Kazuki Hemmi[1,2(✉)], Yuki Tanigaki[3], and Masaki Onishi[2]

[1] University of Tsukuba, Tsukuba, Japan
[2] National Institute of Advanced Industrial Science and Technology, Tokyo, Japan
`{henmi-kazuki,onishi-masaki}@aist.go.jp`
[3] Osaka Institute of Technology, Osaka, Japan
`yuki.tanigaki@oit.ac.jp`

Abstract. Neural architecture search (NAS) is an automated machine learning method that optimizes neural network architectures depending on the dataset or its purpose. With the advances in NAS, high-accuracy neural network architectures can be built for a specific dataset without any expert skills. However, NAS is an expensive, time-consuming, and resource-intensive technique. Therefore, searching for the optimal architecture from scratch for each new dataset is inefficient. To accommodate the expected future increase in datasets, a technique is required that directly predicts the optimized architecture for unknown datasets. Therefore, we propose a framework that generates architectures for unknown datasets by mapping adequate architectures for existing datasets into the latent feature space. A variational graph autoencoder (VGAE) is utilized for latent feature mapping. Our experimental results indicate that the architecture generated by the proposed method from the information of previously obtained high-accuracy architectures performs effectively for new datasets.

Keywords: Neural Architecture Search · Automated Machine Learning · Deep Learning · Graph Neural Network

1 Introduction

Neural architecture search (NAS) has come to occupy a pivotal position in automated machine learning (AutoML), enabling the automatic optimization of neural network architectures based on specific datasets or tasks. Before NAS, the construction and refinement of neural architectures were performed manually and repeatedly to improve accuracy. Specialized knowledge is required for constructing neural architectures, which is a protracted process. However, NAS has democratized the creation of high-accuracy neural network architectures,

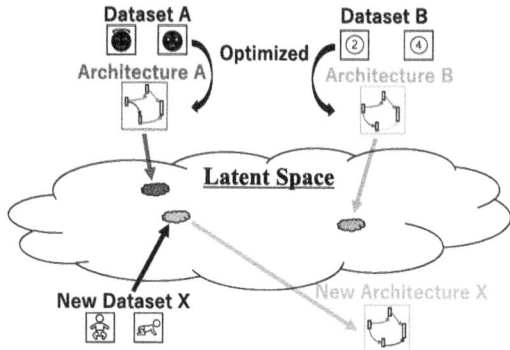

Fig. 1. Concept of NAVIGATOR-D3. The architectures A and B are optimized based on datasets A and B, respectively. Optimized architectures A and B are mapped to the latent space. When a new dataset X is born, the optimal architecture X is instantly generated from the latent space, considering the characteristics of the dataset.

even without specialist knowledge. Nevertheless, NAS has a major disadvantage which remains unaddressed the time-consuming searches for network architectures. This search for the optimal architecture from scratch for each new dataset is inefficient. To accommodate the imminent expansion of dataset types, a method is required that directly predicts architectures optimized for unknown datasets. Generating architectures accelerates and reduces the search time by predicting the optimal architecture. Generation methods, such as repeating random generation multiple times or changing a part of the existing architecture are available. However, the multiple architectures cannot be considered to generate, and the difficulty of searching for optimal points is due to the discrete method. To solve this issue, we use latent feature space by mapping the architecture into a continuous space. We have assumed that mapping the architectures of high-quality solutions in the latent feature space would help predict and generate optimal architectures from neighboring features. Thus, we proposed a Neural Architecture search using VarIational Graph AuTO-encodeR (NAVIGATOR) as a framework that maps the architecture of the adequate existing datasets into the latent feature space.

Mapping the latent features utilizing the encoder-decoder transformation model is based on variational graph autoencoders (VGAE) [11]. VGAE extracts the graph features using the encoder and decoder and maps the NAS architecture represented by a graph structure. By reducing the loss between the input and output architectures, the latent feature in the latent space represents the feature of the architecture. Figure 1 shows the concept of NAVIGATOR-D3, which builds the latent space from VGAE training onto suitable architectures, thus designing optimal architectures for diverse datasets (D3). The proposed method can generate architectures that are as accurate as existing NAS methods with less search time. The performance of the architectures neighboring the optimal architecture is also investigated. Considering the generation of architecture through

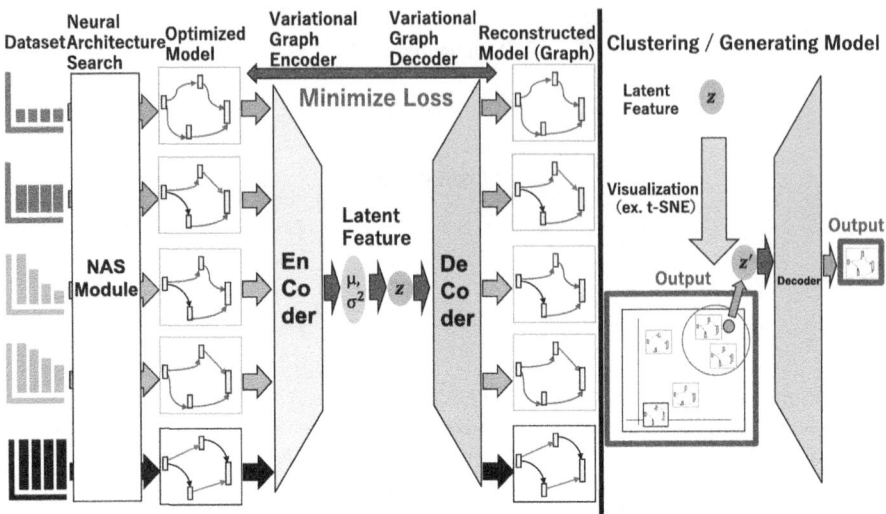

Fig. 2. Schematic of the proposed approach

the latent space as an information transfer onto the architecture, the distribution of structures in the latent space significantly affects the proposed method's performance. Figure 2 shows a schematic overview of the proposed method and details of the encoder and decoder are shown in Fig. 3.

The paper is organized as follows. Section 2 explains NAS, graph neural network (GNN), and related approaches. Section 3 discusses the details of the proposed approach. Section 4 presents the experimental settings and results obtained using the proposed approach and Sect. 5 presents our conclusions.

2   Related Works

This section explains the NAS and GNN methods, which were used in this study.

2.1   Neural Architecture Search (NAS)

NAS is an AutoML optimization method that automatically searches for neural network architectures used in deep learning. Since NAS is a costly approach, an efficient representation of architectures has been sought. Cell-based NAS emerges as a prevalent solution. Among cell-based NAS, DARTS [18] and PC-DARTS [40] are the most common methods that allow efficient architecture searches. The network architecture is an assembly of compact modules, each depicted as a directed acyclic graph. This facilitates creating large-scale architectures within a simple search space. Recently, many highly accurate architectures using cell-based NAS have been discovered. The reuse of those high-accuracy architectures is proposed [39] for efficient architecture generation.

2.2   Graph Neural Network (GNN)

GNN [30] incorporate deep learning mechanisms into graph-structured data and perform tasks such as graph classification and edge prediction (link prediction) and can operate on any structure represented by nodes and edges. These characteristics are applied, in, amongst many others, to molecular chemistry [8] and particle simulations in physics [29]. In this study, we employed the GNN method because the output of NAS can be represented as a graph structure and as an input.

Several GNN methods use graph convolution networks (GCN) [10] for learning the structure of local graphs using adjacency and order matrices (1).

$$\mathbf{F_N}^{(l+1)} = h(\mathbf{D}^{-1/2}\mathbf{A}\mathbf{D}^{-1/2}\mathbf{F_N}^{(l)}\mathbf{W}^{(l)}) \tag{1}$$

Here, $\mathbf{F_N}$ is the input or output of the layer and represents the node features, $\mathbf{A}$ is the adjacency matrix of the undirected graph that depicts edge connections and the unit matrix, $\mathbf{W}$ is the trainable weight matrix, $\mathbf{D}$ is the degree matrix indicating the node relationships, and h is an activation function (separately using softmax and ReLU). Another popular GNN is the graph attention network (GAT) [35]. This alternative to the GCN improves expressivity by adding multiple attention layers to the GCN, thereby focusing on the importance of nodes. VGAE [11] is a GNN method that can extract graph features using encoders and decoders. VGAE is an extension of the VAE [9] which uses images to represent graphs. Details of the encoder and decoder are presented as follows.

2.3   Related Approaches

Our proposed method has the following five characteristics:

(a) Alleviates computational cost and search time
(b) Builds the latent feature space using VGAE
(c) Capable of handling continuous architectures
(d) Directly generated architectures
(e) Enables information transfer in existing datasets

The expensive search time and computational cost of NAS have been widely discussed, and researchers have attempted to find possible solutions [12,24,42]. Related methods [3,6,15,16,20,22,26,34,41] for extracting features from the architecture and constructing the latent feature space have been studied. Among these, methods [3,6,16,20,34] of extracting features using VGAE, like the proposed method can be differentiated from the others from the point of view (b). VGAE enables the capture of fine-grained architectural features and properties via graphs. Although there are related methods [1,4,5,16,21,26,32,36,37,41] that construct the predictor, the proposed method directly generates the architecture from the latent feature space in (d). Additionally, it can generate both discrete and continuous architectures by targeting architectures in (c) that have already been discovered and simultaneously learning their embedded representations. In transfer learning [19,33,38,39] promising information is used to improve

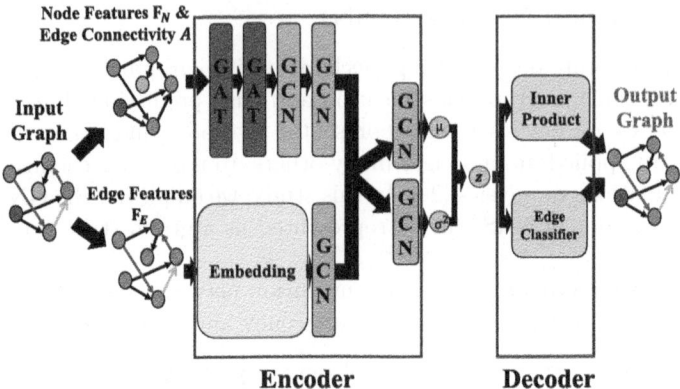

Fig. 3. Details of encoder & decoder. GAT: graph attention network; GCN: graph convolution network; edge classifier: two fully connected layers with a ReLU.

performance (e). Transfer learning methods focus mainly on dataset information transitions, whereas the proposed method specializes in information transitions in the architecture itself.

Although methods with one or several of these features exist, the proposed method has all five characteristics, (a)–(e), which differentiate it from related approaches. All five characteristics are crucial in architecture generation. Additionally, our study uniquely employed an improved VGAE to capture the detailed characteristics of selected architectures. Most related methods are primarily focused on improving accuracy, while an objective of this study was to investigate the architectures on various datasets. Specifically, this study investigated the kind of architecture obtained and the performance achieved by generating models from the neighborhoods of the optimal architectures in the latent feature space.

3   Proposed Approach NAVIGATOR

We propose a Neural Architecture search using VarIational Graph AuTO-encodeR (NAVIGATOR), which combines NAS and VGAE.

3.1   Overview of NAVIGATOR

Figure 2 shows a schematic overview of the proposed method and details of the encoder and decoder are in Fig. 3. The algorithm for the proposed method is given in Alg. 1. NAVIGATOR consists of three components, NAS, VGAE, and the generating model, and is executed in the Alg. 1 pipeline (NAS: Lines 1–5, VGAE: Lines 6–15, Generating model: Line 16–19). Initially, NAS is run on the given dataset to obtain optimized architectures. The number of architectures depends on the types of datasets and the number of searches with different

---

Algorithm 1. NAVIGATOR: NAS using VGAE (1 trial)

Input: Dataset D
Output: Latent Feature Space z , New Architecture $\mathbf{A_r'}$
1: Prepare dataset D
2: **for** Epoch $= 1$ to Specified Epochs in NAS **do**
3:    Update architecture and evaluate on D
4:    Retain architecture $\mathbf{A_r}$ with best performance
5: **end for**
6: Convert A_r into graph format $\mathbf{F_N}, \mathbf{A}, \mathbf{F_E}$
7: **for** Epoch $= 1$ to Specified Epochs in VGAE **do**
8:    *Encoder* $\mathbf{F_N}, \mathbf{A}, \mathbf{F_E}$
9:    μ_i, σ_i^2 *Encoder*
10:    $\mathbf{z}$ Sample from μ_i, σ_i^2
11:    *Decoder* $\mathbf{z}$
12:    $\mathbf{A'}, \mathbf{F_E'}$ *Decoder*
13:    $\boldsymbol{Loss}$ = Difference $(\mathbf{A}, \mathbf{F_E}$, $\mathbf{A'}, \mathbf{F_E'})$
14:    Update weights w of *Encoder* and *Decoder* by descending ∇_w $\boldsymbol{Loss}$
15: **end for**
16: μ_i, σ_i^2 Trained *Encoder* with $\mathbf{F_N}, \mathbf{A}, \mathbf{F_E}$
17: Latent Feature Space $\mathbf{z}$ Sample from μ_i, σ_i^2
18: Generate $\mathbf{z'}$ using $\mathbf{z}$ and the generating model algorithm (**Section 3.5**)
19: New Architecture $\mathbf{A_r'}$ Trained *Decoder* with $\mathbf{z'}$

---

seeds. Latent features were obtained using VGAE's Encoder and Decoder for the optimized architectures. As shown in Fig. 2, NAVIGATOR allows clustering by visualization methods or can generate new models after acquiring latent features. For visualization, the dimensionality reduction algorithms PCA [27] are leveraged.

3.2   Encoder

Assume that μ is the mean of the latent variable, and σ^2 is the variance of the latent variable. The encoder calculates μ and $\log \sigma$ in (2). The input to the proposed method is a network architecture searched using the NAS method.

$$\mu = \text{GNN}_\mu(\mathbf{F_N}, \mathbf{A}, \mathbf{F_E})$$
$$\log \sigma = \text{GNN}_\sigma(\mathbf{F_N}, \mathbf{A}, \mathbf{F_E}) \tag{2}$$

where $\mathbf{F_N}$ contains the node features, $\mathbf{A}$ is the edge adjacency matrix, and $\mathbf{F_E}$ is the edge feature. The node features $\mathbf{F_N}$ consist of the input node, middle node, and output node in NAS and the three node types are input as features. And the input simultaneously trains the graph edge features $\mathbf{F_E}$. The edge feature $\mathbf{F_E}$ depends on the NAS search space. Each operator is mapped onto an edge feature $\mathbf{F_E}$, and an embedding layer passes through the edge feature $\mathbf{F_E}$. The embedding layers are trained simultaneously. The embedding layer, based on the edge features of $\mathbf{F_E}$, allows the relationship between operators to be considered such that the information of each operator is more probably reflected in the latent space. The proposed method adds to change original VGAE to represent both the graph structure and the edge operations in the latent space. The details of the GNN with $\mathbf{F_N}$ and $\mathbf{A}$ as inputs are present in two types of layers: one that uses only the GCN and the other that combines the GAT and GCN, as shown in Fig. 3. In the early phase, the GAT layer trains specific relationships among nearby nodes, and in the later phase, the GCN layer trains the local graph information between nearby nodes. Therefore, combining the two types of layers improves the expressiveness of the mechanism.

The latent feature $\mathbf{z}$ is assumed to follow a normal distribution represented by the calculated μ and $\log \sigma$, and $\mathbf{z}$ is obtained using (3).

$$q(\mathbf{z}|\mathbf{F_N}, \mathbf{A}, \mathbf{F_E}) = \prod_{i=1}^{N} q(\mathbf{z}_i|\mathbf{F_N}, \mathbf{A}, \mathbf{F_E}) \tag{3}$$

$$\text{with } q(\mathbf{z}_i|\mathbf{F_N}, \mathbf{A}, \mathbf{F_E}) = Norm(\mathbf{z}_i|\mu_i, \sigma^2{}_i)$$

where N is number of nodes. The encoder is used to convert to parameters of normal distribution, and the product of the normal distribution for each node is the overall distribution.

3.3 Decoder

The decoder reconstructs the graph structure from the latent feature $\mathbf{z}$ using the inner product of (7), where $\hat{\mathbf{A}}$ denotes the adjacency matrix of the generated graphs. $\hat{\mathbf{F_E}}$ is a feature of the generated and represented type of edge, which is predicted from latent feature $\mathbf{z}$ by traversing it through the edge classifier.

The latent features of the target graph can be extracted by optimizing the graph reconstructed by the decoder, which includes the inner product and edge classifier. The original graph is input to the encoder to obtain the same structure by reducing the loss.

3.4 Loss Function

The loss function of the proposed method comprises the reconstruction loss (Recon Loss) of VGAE, edge-class loss, and the Kullback-Leibler divergence loss (KL Loss) of VGAE. Specifically, this function is expressed in (4), where W_{Edge} is the importance of the edge class loss and W_{KL} is a hyperparameter that adjusts the importance of the KL Loss (typically set to 1.0).

$$Loss = L_{Recon} + W_{Edge} * L_{Edge} + W_{KL} * L_{KL} \tag{4}$$

where L_{Recon} is the reconstruction loss of the VGAE obtained from (5) using positive and negative losses. The role of L_{Recon} is to index the graph and approximate architectures. W_P is the importance of the positive loss, and W_N is a hyperparameter that adjusts the importance of the negative loss.

$$L_{Recon} = \sum_{(i,j) \in \mathbf{A}} W_P * A_{i,j} \log(\hat{A}_{i,j}) + W_N * (1 - A_{i,j}) \log(1 - \hat{A}_{i,j}) \tag{5}$$

where $A_{i,j}$ is the true graph and $\hat{A}_{i,j}$ is the ith row jth column element of the adjacency matrix of the generated graph. L_{Edge} represents the edge class loss, which predicts the edge type and is computed using the cross-entropy loss. L_{Edge} indexes the edge type and approximate architectures. Another loss is also used in the original VGAE, but L_{Edge} is added to apply fine features to NAS. L_{KL} represents the KL divergence of the normal and prior distributions of the latent variable given by (6). The KL divergence represents the difference between two probability distributions. L_{KL} makes the latent space continuous and stabilizes learning. In (6), N is the number of dimensions of the latent variable.

$$L_{KL} = -\frac{1}{2} \sum_{i=1}^{N} (1 + \log(\sigma_i^2) - \mu_i^2 - \sigma_i^2) \tag{6}$$

The sum of these losses constitutes the loss function, which is the objective function of the proposed method.

$$p(\hat{\mathbf{A}}, \hat{\mathbf{F}}_\mathbf{E}|\mathbf{z}) = \prod_{i=1}^{N} \prod_{j=1}^{N} p(\hat{A}_{i,j}, \hat{F}_{Ei,j}|\mathbf{z}_i, \mathbf{z}_j) \tag{7}$$

$$p(\hat{A}_{i,j} = 1|\mathbf{z}_i, \mathbf{z}_j) = Sigmoid(\mathbf{z}_i^T, \mathbf{z}_j)$$

3.5   Generating Model

The architecture generation method is based on the following steps:

1. The user identifies the promising latent feature $\mathbf{z}$ for the target task based on, e.g., the similarity of the dataset.
2. The identified latent feature $\mathbf{z}$ is normalized to the range of [0,1] for each dimension considering the other latent features.
3. New latent features in the neighborhood of the optimal point are generated by scaling the generated direction vectors and adding them to the normalized latent feature corresponding to an optimal point.
4. The new latent features generated in Step 3 are inverse normalized using the inverse transform of Step 2 and the new latent feature is input to the decoder of the trained VGAE.
5. The new network architecture $\mathbf{z}'$ is finally generated.

Methods for identifying promising latent features include drawing analogies from the features of the dataset. In this study, assuming that the best latent feature can be estimated, it is manually specified as the architecture that performed best- on the selected dataset. In scaling steps 2–4, new latent feature $\mathbf{z}'$ is generated by reducing them by a factor, as expressed in (8). Sgn is a sign function that returns the sign of the input value to generate a direction vector with a value of -1 or 1 for each dimension from a uniform random number $\mathcal{U}(-0.5, 0.5)$.

$$\mathbf{z}_{\text{normalize}} = \frac{\mathbf{z} - \mathbf{z}_{\min}}{\mathbf{z}_{\max} - \mathbf{z}_{\min}}, \quad \mathbf{d} = \text{Sgn}\left(\mathcal{U}(-0.5, 0.5)\right)$$
$$\mathbf{z}' = (\mathbf{z}_{\text{normalize}} + \gamma \cdot \mathbf{d})(\mathbf{z}_{\max} - \mathbf{z}_{\min}) + \mathbf{z}_{\min} \tag{8}$$

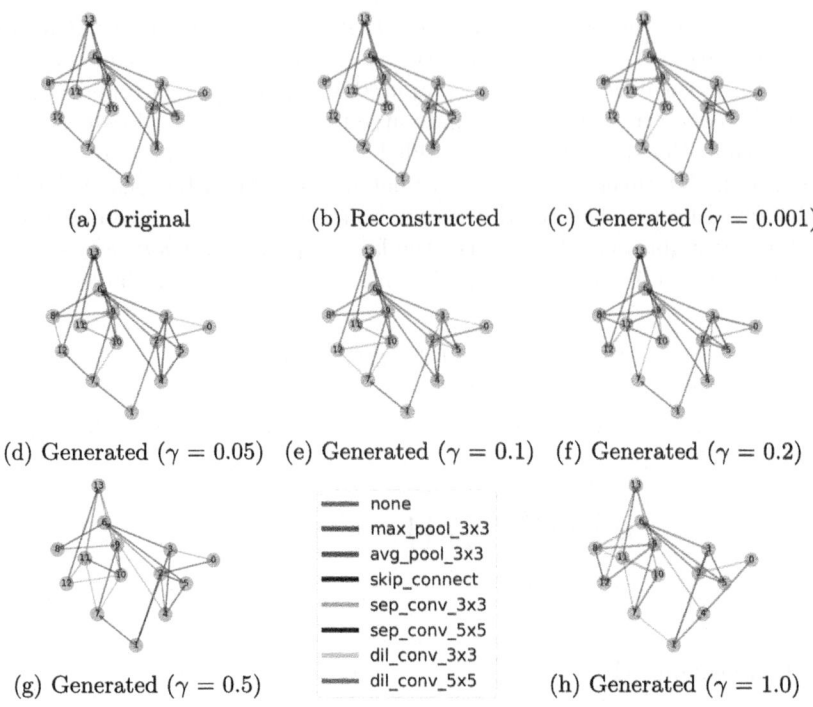

(a) Original   (b) Reconstructed   (c) Generated ($\gamma = 0.001$)

(d) Generated ($\gamma = 0.05$) (e) Generated ($\gamma = 0.1$) (f) Generated ($\gamma = 0.2$)

(g) Generated ($\gamma = 0.5$)

- none
- max_pool_3x3
- avg_pool_3x3
- skip_connect
- sep_conv_3x3
- sep_conv_5x5
- dil_conv_3x3
- dil_conv_5x5

(h) Generated ($\gamma = 1.0$)

Fig. 4. Graphs generated from input graphs and latent features

In Step 3, uniform random numbers and γ are used to provide minor perturbations to obtain new latent features in the neighborhood of the original latent
features. Introducing minor perturbations to repeat multiple trial experiments
allowed us to obtain good values. While the direction of optimization must be
considered, the proposed method's generating model can be adjusted. γ is a
hyperparameter. Keeping the distance of the direction vectors constant in the
normalized latent feature space, the new latent features take their absolute distances according to the value of γ. This process posits that the newly generated
network architectures share similar characteristics with the network architectures corresponding to the original optimal points when γ is sufficiently small.
The architectures generated through experiments were evaluated and validated
in the following section.

4   Experiments

We evaluated the effectiveness of the proposed method through experiments on
multiple datasets. PC-DARTS was utilized to search for the appropriate architecture for each dataset. Otherwise expressed, this study used PC-DARTS for
the NAS module of NAVIGATOR. The PC-DARTS search space contains 10^{25}

architectures, which is sufficient for experiments owing to the search efficiency and relatively large size. This experiment was run using one NVIDIA Tesla V100 SXM2 16 GB per trial (ImageNet requires four GPUs in parallel).

4.1 Experiment I: Visualizing the Output from Latent Features

In Experiment I, we examined the possibility of generating network architectures with similar features from latent features space acquired by NAVIGATOR. The results of adding noise to verify the shape of the model architecture distributed around the specified features are described below. Experiment I used the CIFAR-10 and MNIST datasets. CIFAR-10 [13] is a commonly used dataset in deep learning and NAS experiments, featuring 60,000 images. MNIST dataset [14] contains handwritten numbers. Because latent features encapsulate the characteristics of the dataset to be optimized, we adjusted the latent features (refer to the generating model of the proposed approach) extracted using NAVIGATOR from the optimized architecture for CIFAR-10. This adaptation demonstrated the feasibility of generating various architectures similar to an output reconstruction graph. Additionally, we used an architecture connecting two types of cells for more detailed feature extraction and readily visualized graphs by learning an architecture connecting two types of cells instead of the entire graph.

Figure 4 presents a graph of the input architecture and the architecture generated from latent features with different γ values. The input nodes are blue, the middle nodes orange, and the output nodes green. The number of edges was set to be identical to that in the input graph.

First, the original graph in (a) and the reconstructed graph in (b) are identical, which confirms that NAVIGATOR's encoder and decoder can correctly reconstruct the architecture.

Next, regarding the small perturbations in γ, two edge connections change when $\gamma = 0.05$ in (d), but the architecture was almost identical until the other half (original graph in (a)—$\gamma = 0.05$ in (d)).

As shown in the generating model algorithm Sect. 3.5, when generating a new latent feature, normalization is performed once per dimension in the range of [0,1] for each dimension in consideration of the other latent features in the latent space. The vector over which the gammas were multiplied was fixed to a length of 1. Therefore, when given small perturbations up to $\gamma = 0.05$, maximally 5 % from the original latent features in the latent feature space was moved, which is why almost the same architecture was generated as shown in Fig. 4.

Finally, after $\gamma = 0.1$ in (e), the architecture gradually changed while retaining the features of the original graph, and the different computation labels of the edges are represented by indicating the colors that the edge features also changed. Therefore, increasing the γ value freely generated architectures with features distant from the original graph.

The proposed method was constrained to perform in NAS when generating the architecture so that no more than two edges can be connected to each middle node. In addition, because the two types of cells are combined into a single graph, fixed edges (such as 1 to 7 or edges that are definitely connected to the output edge) were included, and constraints were added to ensure that no other edges were connected across each cell other than the fixed edges. This rule was the constraint added to reproduce the same search space as the DARTS and PC-DARTS architectures with constraints, matching the input architecture to ensure a fair comparison. The proposed method can be generated by the removal of the constraint in the generation of architectures that have a wide range of sizes.

4.2 Experiment II: Experimental Settings for the Latent Feature Visualization of Architecture Searched in CIFAR-10 and MNIST

In Experiment II, we leveraged NAS to search for the optimal architecture for different types of datasets with varying content characteristics and visualized the latent features of the network architectures optimized for each dataset. Experiment II verifies this hypothesis that NAVIGATOR's input models in VGAE are distributed in different areas of the latent feature space depending on the dataset.

Each architecture included edge connectivity relationships and edge types, and the number of nodes was fixed. Five different PC-DARTS trials were executed on the MNIST and CIFAR-10 datasets, with 10 optimized architectures. For the search phase, we used 50 training epochs, the learning rate followed a cosine schedule from 0.1 to 0, the learning rate for architecture parameters was 0.0003, and random cropping and horizontal flipping was used for data augmentation. In configuring the VGAE, we used 500,000 training epochs, the learning rate followed a cosine schedule from 0.001 to 0.00025, the optimizer was Adam, for the dimension of the 4 latent features, the seed value was 2024, and the loss function weights, W_P, W_N, W_{Edge}, and W_{KL}, were set respectively to 2.0, 1.0, 0.5, and 1.0. The hyper-parameters were selected based on the NAS method and the original VGAE study.

4.3 Experiment II: Results and Discussion

Figure 5 visualizes the latent features acquired using the proposed method. It is a dimensionality-reduced graph obtained via PCA, where the axes correspond to the coordinate axes of the result of compressing the multidimensional data of the latent features. We assumed that NAS could search for architectures that better extract the characteristics of the dataset because CIFAR-10 and MNIST possess different image sizes and dataset characteristics (RGB images for CIFAR-10 and grayscale images for MNIST). CIFAR-10 has an image size of 32×32 pixels, whereas MNIST has an image size of 28×28 pixels. Therefore, the input image sizes are distinct. Our results validated that CIFAR-10 and MNIST exhibit different trends, with each obtaining approximate latent.

NAVIGATOR's VGAE training time was 14.52 h (500,000 epochs with 10 architectures). The approximate training time for NAS was 2.4 h $\times$ the number of architectures, identical to using the original PC-DARTS. The advantage and essence of the proposed method are that the architecture can be generated for a

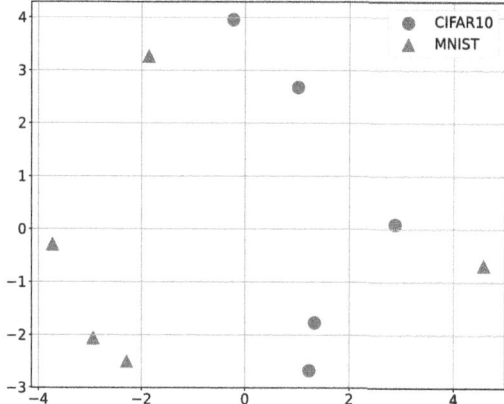

Fig. 5. Visualization of latent features acquired using the proposed method. The axes represent the coordinate axes on a two-dimensional PCA plot.

new dataset with minimal time to prepare a large latent space that has already been optimized for multiple datasets.

4.4 Experiment III: Experimental Settings for Performance Test of Generated Architecture in NAVIGATOR

In Experiment III, we compared the architectures generated via NAVIGATOR with randomly generated architectures on the USPS and the ImageNet dataset to evaluate the proposed method's ability to generate advantageous architectures. Moreover, we investigated possible improvements by comparing it with random architectures. The newly used United States Postal Service (USPS) dataset [7] features 16 - 16 grayscale images of handwritten digits ranging from 0 to 9. Similar to the MNIST dataset, the USPS dataset is generally subjected to domain adaptation [31]. Considering this, we verified whether an architecture optimized for a similar dataset could perform similarly to Experiment III. We evaluated the accuracy by retraining the four architecture types. The first architecture type was constructed from latent features adjacent ($\gamma = 0.1$) to the MNIST obtained in Experiment II. The second type is an architecture built from the latent features adjacent ($\gamma = 0.1$) to CIFAR-10. The third and fourth are architectures constructed from latent features in the acquired MNIST ($\gamma = 0.0$), and architectures constructed from random latent features. The $\gamma = 0.1$ was set as a parameter to construct the surrounding latent features since its architecture is partially different while capturing the features of the architecture with $\gamma = 0.0$, such as in Experiment I. MNIST and CIFAR-10 each have five latent feature architectures, and the model was constructed by four times generating random numbers obtained during direction vector generation. In total 50 architectures (five Random models, five MNIST models ($\gamma = 0.0$), 20 CIFAR-10 models ($\gamma = 0.1$), and 20 MNIST models ($\gamma = 0.1$)) were evaluated for the five types. For

Table 1. Results of Experiment III using the USPS dataset

Architecture	Model Size (Mean ± SD) (M)	Best Test Acc. (Max ± SD) (%)	Search Time (h)
CENAS [33]	0.57	89.94	7.4
ASED [25]	–	97.75	153.6
NASDA [17]	2.70	98.00	7.2
NAVIGATOR (Random model)	2.15 ± 0.28	97.96 ± 0.10	0
NAVIGATOR (CIFAR-10 model, $\gamma = 0.1$)	3.32 ± 0.63	98.01 ± 0.11	0
NAVIGATOR (MNIST model, $\gamma = 0.0$)	3.37 ± 0.60	98.01 ± 0.09	0
NAVIGATOR (MNIST model, $\gamma = 0.1$)	**3.16 ± 0.66**	**98.11 ± 0.13**	0

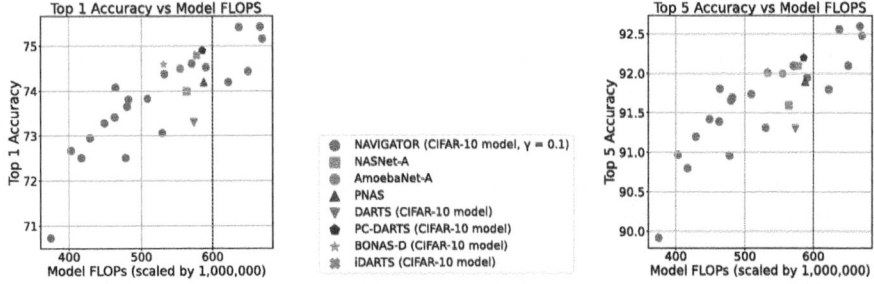

Fig. 6. Results of Experiment III using the ImageNet dataset (left: Top 1 Accuracy, right: Top 5 Accuracy), dashed line: 600M (mobile setting)

retraining on the USPS dataset, we used 600 training epochs, a batch size of 128, an initial learning rate of 0.025 (reduced by cosine schedule to 0), a momentum of 0.9, a weight decay of 3×10^{-4}, a drop path probability of 0.3, and cutout augmentation.

The USPS is a small dataset at 16×16. Therefore, it is advisable to verify whether NAVIGATOR works well on datasets with large image sizes. Additional experiments using the ImageNet dataset were performed in Experiment III. ImageNet [28] is a dataset commonly used for image identification, and the model searched in CIFAR-10 is also used in the base PC-DARTS as a metastasis. Experiment III used architectures (20 architectures) constructed from latent features around CIFAR-10 and was evaluated using ImageNet. RandomHorizontalFlip which randomly performs left-right flipping, and ColorJitter which randomly changes brightness and contrast, were selected for data augmentation. For retraining in ImageNet, we used 250 training epochs, a batch size of 768, an initial learning rate of 0.5 (reduced by linear schedule to 0), a momentum of 0.9, and a weights decay of 3×10^{-5}. Other settings include adopting label smoothing and using warm-up in the first five Epochs. Auxiliary Loss was not used to simplify the experiment. Hyper-parameters were selected for retraining in ImageNet and USPS regarding the original study of PC-DARTS.

4.5   Experiment III: Results and Discussion

Table 1 presents the results of each NAS. The MNIST model ($\gamma = 0.1$) secured the highest accuracy, with a Best Test Accuracy value of 98.11%. This outcome emphasizes that architectures optimized through latent features from kindred datasets yield superior performance. The MNIST model performed better than the Random Model and the CIFAR-10 model, and among the MNIST models, $\gamma = 0.1$ performed better than $\gamma = 0.0$. This is because $\gamma = 0.0$ are the optimized architectures for the MNIST dataset, which is similar to, and different from the USPS dataset. The advantages of the proposed method of generating architectures by adding perturbations are confirmed. Moreover, this approach can be adapted to new datasets, thereby eliminating the need for thorough NAS and significantly reducing the search time.

Figure 6 presents the results of each NAS in the experiment using ImageNet. NAVIGATOR (CIFAR-10 model, $\gamma = 0.1$) secured the highest accuracy, with a Best Test Accuracy value of 75.43% in Top1 and 92.60% in Top5. When sliced by Mobile Setting with 600M FLOPs, PC-DARTS is the most accurate, but it can be confirmed that the architecture generated by NAVIGATOR exhibits identical performance. Therefore, the proposed approach can generate a set of architectures that are non-inferior to the compared architectures even in cases where FLOPs are relatively small. Furthermore, considering that the results were shown without any search time, the performance was superior. The experimental results indicate that the proposed method is efficient for datasets with large image sizes and large architectures with transposed Cells.

5   Conclusion

We focused on reducing the search time by implementing NAVIGATOR, a framework that combines NAS and VGAE. The next challenge is to generate similar architectures in search spaces other than the trained ones. Since there are methods [2,23] for measuring the distance between datasets, we plan to combine the proposed method to quantify the distance between datasets and improve it to develop an optimization method in the latent feature space. This approach promises to navigate toward advances in neural architecture search and fosters deeper insights into the realm of network design and optimization.

References

1. Agiollo, A., Omicini, A.: Gnn2gnn: graph neural networks to generate neural networks. In: Uncertainty in Artificial Intelligence, pp. 32–42. PMLR (2022)
2. Alvarez-Melis, D., Fusi, N.: Geometric dataset distances via optimal transport. In: Advances in Neural Information Processing Systems, vol. 33, pp. 21428–21439. Curran Associates, Inc. (2020)
3. Chatzianastasis, M., et al.: Graph-based neural architecture search with operation embeddings. In: Proceedings of the IEEE/CVF International Conference on Computer Vision Workshops, pp. 393–402 (2021). https://doi.org/10.1109/ICCVW54120.2021.00048

4. Chen, Y., et al.: Contrastive neural architecture search with neural architecture comparators. In: Proceedings of the IEEE/CVF Conference on Computer Vision and Pattern Recognition, pp. 9502–9511 (2021). https://doi.org/10.1109/CVPR46437.2021.00938

5. Dudziak, L., et al.: Brp-nas: prediction-based nas using gcns. Adv. Neural. Inf. Process. Syst. **33**, 10480–10490 (2020)

6. Friede, D., et al.: A variational-sequential graph autoencoder for neural architecture performance prediction. arXiv preprint (2019). https://doi.org/10.48550/arXiv.1912.05317

7. Hull, J.J.: A database for handwritten text recognition research. IEEE Trans. Pattern Anal. Mach. Intell. **16**(5), 550–554 (1994). https://doi.org/10.1109/34.291440

8. Jiang, D., et al.: Could graph neural networks learn better molecular representation for drug discovery? a comparison study of descriptor-based and graph-based models. J. Cheminf. **13**(1), 1–23 (2021). https://doi.org/10.1186/s13321-020-00479-8

9. Kingma, D.P., Welling, M.: Auto-encoding variational bayes. arXiv preprint (2013). https://doi.org/10.48550/arXiv.1312.6114

10. Kipf, T.N., Welling, M.: Semi-supervised classification with graph convolutional networks. arXiv preprint (2016). https://doi.org/10.48550/arXiv.1609.02907

11. Kipf, T.N., Welling, M.: Variational graph auto-encoders. arXiv preprint (2016). https://doi.org/10.48550/arXiv.1611.07308

12. Krishnakumar, A., et al.: Nas-bench-suite-zero: accelerating research on zero cost proxies. Adv. Neural. Inf. Process. Syst. **35**, 28037–28051 (2022)

13. Krizhevsky, A., Hinton, G.: Learning multiple layers of features from tiny images. Master's thesis, Department of Computer Science, University of Toronto (2009)

14. LeCun, Y., et al.: Gradient-based learning applied to document recognition. Proc. IEEE **86**(11), 2278–2324 (1998). https://doi.org/10.1109/5.726791

15. Lee, H., Hyung, E., Hwang, S.J.: Rapid neural architecture search by learning to generate graphs from datasets. arXiv preprint (2021). https://doi.org/10.48550/arXiv.2107.00860

16. Li, J., et al.: Neural architecture optimization with graph vae. arXiv preprint (2020). 10.48550/arXiv.2006.10310

17. Li, Y., Peng, X.: Network architecture search for domain adaptation. arXiv preprint (2020). https://doi.org/10.48550/arXiv.2008.05706

18. Liu, H., et al.: Darts: differentiable architecture search. arXiv preprint (2018). https://doi.org/10.48550/arXiv.1806.09055

19. Lu, Z., et al.: Neural architecture transfer. arXiv preprint (2020). https://doi.org/10.48550/arXiv.2005.05859

20. Lukasik, J., et al.: Smooth variational graph embeddings for efficient neural architecture search. In: International Joint Conference on Neural Networks, pp. 1–8 (2020)

21. Lukasik, J., et al.: Learning where to look - generative NAS is surprisingly efficient. CoRR arxiv:2203.08734 (2022)

22. Luo, R., et al.: Neural architecture optimization. Adv. Neural Inf. Process. Syst. **31** (2018)

23. Mathisen, B.M., et al.: Learning similarity measures from data. Prog. Artif. Intelli. **9** (2019). https://doi.org/10.1007/s13748-019-00201-2

24. Mellor, J., et al.: Neural architecture search without training. In: International Conference on Machine Learning, vol. 139, pp. 7588–7598. PMLR (2021). https://doi.org/10.1109/ACCESS.2021.3052996

25. Muravev, A., et al.: Neural architecture search by estimation of network structure distributions. IEEE Access **9** (2021). https://doi.org/10.1109/ACCESS.2021.3052996

26. Ning, X., Zheng, Y., Zhao, T., Wang, Yu., Yang, H.: A generic graph-based neural architecture encoding scheme for predictor-based NAS. In: Vedaldi, A., Bischof, H., Brox, T., Frahm, J.-M. (eds.) ECCV 2020. LNCS, vol. 12358, pp. 189–204. Springer, Cham (2020). https://doi.org/10.1007/978-3-030-58601-0_12

27. Pearson, K.: Liii. on lines and planes of closest fit to systems of points in space. Lond. Edinburgh Dublin Phil. Maga. J. Sci. **2**(11) (1901). https://doi.org/10.1080/14786440109462720

28. Russakovsky, O., et al.: Imagenet large scale visual recognition challenge. Int. J. Comput. Vision **115**, 211–252 (2015). https://doi.org/10.1007/s11263-015-0816-y

29. Sanchez-Gonzalez, A., et al.: Learning to simulate complex physics with graph networks. In: Proceedings of the International Conference on Machine Learning, pp. 8459–8468 (2020)

30. Scarselli, F., et al.: The graph neural network model. IEEE Trans. Neural Netw. **20**(1), 61–80 (2008). https://doi.org/10.1109/TNN.2008.2005605

31. Schrod, S., et al.: FACT: federated adversarial cross training. arXiv preprint (2023). https://doi.org/10.48550/arXiv.2306.00607

32. Shi, H., et al.: Bridging the gap between sample-based and one-shot neural architecture search with bonas. Adv. Neural. Inf. Process. Syst. **33**, 1808–1819 (2020)

33. Singamsetti, M., et al.: Conceptual expansion neural architecture search (cenas). arXiv preprint (2021). DOI: https://doi.org/10.48550/arXiv.2110.03144

34. SuchopÃrovÃ, G., Neruda, R.: Graph embedding for neural architecture search with input-output information. In: AutoML Conference Workshop Track (2022)

35. Velickovic, P., et al.: Graph attention networks. In: International Conference on Learning Representations (2018). https://doi.org/10.17863/CAM.48429

36. Wen, W., et al.: Neural predictor for neural architecture search. In: Proceedings of European Conference on Computer Vision (2019)

37. White, C., et al.: Bananas: bayesian optimization with neural architectures for neural architecture search. In: Proceedings of the AAAI Conference on Artificial Intelligence (2021)

38. Wistuba, M.: XferNAS: transfer neural architecture search. In: Hutter, F., Kersting, K., Lijffijt, J., Valera, I. (eds.) ECML PKDD 2020. LNCS (LNAI), vol. 12459, pp. 247–262. Springer, Cham (2021). https://doi.org/10.1007/978-3-030-67664-3_15

39. Wong, C., et al.: Transfer learning with neural automl. Adv. Neural Inf. Process. Syst. (2018)

40. Xu, Y., et al.: PC-DARTS: partial channel connections for memory-efficient architecture search. In: International Conference on Learning Representations (2020)

41. Yan, S., et al.: Does unsupervised architecture representation learning help neural architecture search? Adv. Neural Inf. Process. Syst. (2020)

42. Zoph, B., et al.: Learning transferable architectures for scalable image recognition. Proceedings of the IEEE/CVF Conference on Computer Vision and Pattern Recognition, pp. 8697–8710 (2018). DOI: https://doi.org/10.1109/CVPR.2018.00907

ResBuilder: Automated Learning of Depth with Residual Structures

Julian Burghoff[1]([✉]), Matthias Rottmann[2], Jill von Conta[3],
Sebastian Schoenen[1], Andreas Witte[4], and Hanno Gottschalk[5]

[1] Research and Development, Control Expert, Langenfeld, Germany
{j.burghoff,s.schoenen}@controlexpert.com
[2] Department of Mathematics, University of Wuppertal, Wuppertal, Germany
rottmann@uni-wuppertal.de
[3] Research and Development, Control Expert (at time of research), Langenfeld,
Germany
jillmignon.vonconta@uk-essen.de
[4] Control Expert, Langenfeld, Germany
a.witte@controlexpert.com
[5] Institute of Mathematics, TU-Berlin, Berlin, Germany
gottschalk@math.tu-berlin.de

Abstract. In this work, we develop a neural architecture search algorithm, termed Resbuilder, that develops ResNet architectures from scratch that achieve high accuracy at moderate computational cost. It can also be used to modify existing architectures and has the capability to remove and insert ResNet blocks, in this way searching for suitable architectures in the space of ResNet architectures. In our experiments on different image classification datasets, Resbuilder achieves close to state-of-the-art performance while saving computational cost compared to off-the-shelf ResNets. Noteworthy, we once tune the parameters on CIFAR10 which yields a suitable default choice for all other datasets. We demonstrate that this property generalizes even to industrial applications by applying our method with default parameters on a proprietary fraud detection dataset.

Keywords: automated machine learning · neural architecture search · residual neural networks

1 Introduction

Machine learning is a key technology for data-driven automation that achieved intriguing success in many applications, such as image recognition [1] and natural language processing [2]. However, the application of machine learning to ever new tasks requires data scientists with the skill to design problem specific machine learning algorithms. In particular this applies when working with deep neural networks [3] and the question arises how to choose network architectures and select the hyper parameters to control the training process. Especially for small

© The Author(s), under exclusive license to Springer Nature Switzerland AG 2024
M. Wand et al. (Eds.): ICANN 2024, LNCS 15016, pp. 308–323, 2024.
https://doi.org/10.1007/978-3-031-72332-2_21

and medium size enterprises, this constitutes a major obstacle in the application of machine learning in their processes or products. Automated Machine Learning (AutoML) [4,5] is a research area that studies hyperparameter optimization [6] as well as neural architecture search (NAS) [7].

Many of the existing methods in the field of network architecture search (NAS) modify existing network architectures [4]. In many cases, runtime optimization of a baseline neural network while preserving accuracy of predictions is a motivation for many of the works in this field [8].

Another field closely related to NAS is the task of network pruning [9,10] where an existing network is made sparser in order to reduce the computational burden. More drastically, complete layers are removed in [11,12].

In this work, we introduce ResBuilder that tackles the problem of constructing ResNet [13] architectures from scratch. The goal is to find a ResNet architecture in an automated way that achieves high accuracy for a given problem while not exceeding a predefined computational budget. More precisely, we utilize MorphNet [14] as a baseline which performs channel pruning as well as layer removal during training. We introduce a method to add and remove layers during training, dynamically controlling the network's capacity while balancing test accuracy and computational expense.

To achieve this, we focus on ResNet blocks that support layer insertion and removal during training in a natural way. Due to the skip connection, layers with weights close to zero almost act as identity layers. Such layers can thus be inserted during training without undoing the previous training progress. Similarly, layers with weights small in magnitude can be seamlessly removed during training with undoing previous progress. During architecture search, we utilize a layer LASSO approach in order to identify unnecessary layers of the parameters.

We demonstrate the efficiency of ResBuilder on six datasets, namely Animals10 [15], CIFAR10 [16], CIFAR100 [16], MNIST [17], FashionMNIST [18], EMNIST [19], and study its hyperparameters in-depth. It turns out that ResBuilder easily builds ResNet architectures achieving close to state-of-the-art performance (without pre-training) on a variety of image classification benchmarks while saving computational cost compared to off-the-shelf ResNets.

Our contribution can be summarized as follows:

- We develop ResBuilder which constructs ResNet architectures from scratch and modifies existing ones to achieve high accuracy at a given computational budget.
- We achieve close to state-of-the-art accuracy on various academic benchmark data sets with a default hyperparameter setting that was tuned once for CIFAR10.
- We showcase the effectiveness of ResBuilder for the industrial application of detecting manipulated images. Specifically, we focus on its application within insurance companies, where the authenticity of images is crucial for automated processing in claims management. We provide evidence of ResBuilder's capability to successfully solve the NAS task using our default hyperparameters.

The remainder of this work is structured as follows. In Sect. 2 we discuss the related work and position ourselves in the field of neural architecture search. In Sect. 3 we describe the methodology behind ResBuilder. This is followed by numerical experiments in Sect. 4 and finally a conclusion in Sect. 5.

2 Related Work

AutoML is an active field of research. While the survey [7] approaches NAS focusing on its different structural aspects, such as comparing different works regarding their search spaces, search strategy and performance estimation on cell structured network architectures where each cell can be arbitrary connected to any of the previous cells and each different cell type stands for a different kind of layer.

The survey [4] provides an overview of NAS approaches, providing a general and comprehensive view on the field of AutoML. The presented methods are classified into different categories, such as reinforcement learning, evolution-based algorithms, as well as gradient descent, random search and surrogate model-based optimization. Regarding NAS and its performance on image classification tasks, they compare the different methods especially with respect to their accuracy on the CIFAR10 and ImageNet datasets.

NAS and Reinforcement Learning. The authors of [20] present an approach where they use reinforcement learning in order to predict hyperparameters like the number of filters, filter height/width and stride for every layer up to a chosen maximum of layers using a recurrent neural network as meta model. Another reinforcement learning approach for NAS is introduced in [21] where the authors develop the NASNet search space, in which for a given problem (here CIFAR10 and ImageNet) a certain number of pooling layers are given, between which any number of feature map size-preserving blocks can be inserted. This search space is then iterated using a controller recurrent neural network to design a problem-specific architecture, while our search strategy is simpler and based on a penalization approach.

Penalization-Based NAS. Our work is also very related to MorphNet [14] which optimizes the number of channels of convolutional layers. A group Lasso regularization term penalizes the weights, such that channels with weights below a chosen threshold are removed. The threshold and the penalization are chosen such that the computational cost is below a given budget. In an iterative fashion, this step alternates with a expansion step wherein remaining computational budget is re-distributed proportionally to the different layers of the network.

Similar to the regularisation terms of BatchNormalisation in MorphNet, gatekeeper variables are introduced in [22] that are multiplied by the output of each channel of each layer. The peculiarity here is that the actual weights are not trained during the process of finding the gatekeeper variables, but are still in the state of their random initialisation, whereby the importance of the layers is calculated before the actual training of the weights starts. To keep the gatekeeper variables low, an additional regularisation term is introduced depending on the cost of the layer and the status of the gatekeeper variables.

Another penalization-based approach regarding NAS can be seen in [23] where the authors present their Transformable Architecture Search (TAS). Instead of the approach of alternating training and pruning of a network architecture, the authors consider the possibility of training a large network with an excessive number of weights and then transferring the knowledge learned to another network for which the depth and width have been determined independently of the first network. The loss function also contains a penalty term that penalises the size of the resulting network.

NAS and ResNets. ResBuilder works on the search space of ResNet architectures [13]. In [24], the authors also focus on ResNets, in particular the links between the different residual blocks, and define masks such that (in case of ResNet) the input of each residual block can take the outputs of every previous residual block.

The authors of [25] have a similar idea of a modular architecture search space with residual structures as we do, but develop an active-learning approach "incremental neural architecture search (iNAS)", which can be combined with any query strategy. They then interpret their search space as a directed acyclic graph in which they start with the smallest possible architecture in order to find a suitable, problem-specific residual architecture.

In contrast to most of these examples, our ResBuilder can be adapted to new data sets or tasks quickly, easily and without adapting further hyperparameters, which simplifies its use for small and medium-sized companies.

3   Methods

ResBuilder operates on the search space of ResNet architectures [13]. It modulates a given ResNet by optimally dropping layers while randomly inserting layers. At the same time, it is able to shrink and expand the number of filters per layer. For the latter, we utilize MorphNet which relies on a group Lasso regularization term, where the groups refer to weights corresponding to a given channel, as outlined in the previous section. We now introduce the MorphNet algorithm and afterwards wrap our architecture search method around it. Our notation is inspired by [14].

For a given image of size $H \times W$ and N the number of classes, let $F : \mathbb{R}^{H \cdot W} \to \mathbb{R}^N, H, W, N \in \mathbb{N}$ be a network that assigns class labels to Images:

$$F = \mathrm{SM} \circ \mathrm{FC} \circ f \circ \mathrm{CL}$$

for CL a convolutional, FC a fully connected and SM a softmax layer as well as a backbone $f : \mathbb{R}^{d_1} \to \mathbb{R}^{d_{2L}}$, $d_1, d_{2L} \in \mathbb{N}$, $L \in \mathbb{N}$, which consists of $\ell = 1, \ldots, L$

ResNet blocks B_ℓ, i.e.,

$$f = B_L \circ B_{L-1} \circ \ldots \circ B_1$$

It should be noted that also pooling layers can be part of the architecture, which are positioned in between of two residual blocks of convolutional layers. However, these are omitted from the notation for the sake of simplicity.

Although our method, in principle, is able to handle residual blocks of other sizes, in this paper we only consider the possibility of inserting a block that contains two convolutional layers as they are used in smaller types of residual networks (e.g. ResNet18, ResNet34 [13]) and are shown as the green block of layers in Fig. 1b. Because of this, we consider that each $B_\ell : \mathbb{R}^{d_{2\ell-2}} \to \mathbb{R}^{d_\ell}$ is composed of two convolutional layers with weight tensors $\theta_j, j \in \{2\ell-1, 2\ell\}$. For $x_{\ell-1} \in \mathbb{R}^{d_{\ell-1}}$, the operations of these layers are described as

$$x_\ell = x_{\ell-1} + \sigma(BN(\theta_{2\ell} \cdot \sigma(BN(\theta_{2\ell-1} \cdot x_{\ell-1}))))$$

where σ is the ReLU activation $\sigma(t) = \max\{0, t\}$ and BN is the batch normalization process:

$$BN(z_{i,j,\cdot}) = \left(\frac{z_{i,j,\cdot} - m_\ell(z)}{s_\ell(z)} \right) \gamma_\ell + \beta_\ell, z \in \mathbb{R}^{u_\ell \times v_\ell \times c_\ell}$$
$$\forall i \in \{1,\ldots,u_\ell\}, \forall j \in \{1,\ldots,v_\ell\} \forall \ell = 1,\ldots,2L$$

with $\gamma_\ell, \beta_\ell \in \mathbb{R}^{c_\ell}$ and the mean (standard deviation) $m_\ell(z)$ $(s_\ell(z))$ of the tensor z alongside its channels c_ℓ. When BN is applied to the tensor z, it is done by applying it to each $z_{i,j,\cdot} \forall i, j$ seperately.

Any of the $j = 1,\ldots,2L$ convolutions maps from $\mathbb{R}^{d_j}$ to $\mathbb{R}^{d_j+1}$, where the dimension is the product $d_j = u_j v_j c_j$ of the spatial dimensions u_j, v_j and the number of channels c_j. The application of the jth convolution requires

$$C_j = 2s_j u_j v_j c_j c_{j+1}$$

floating point operations, where $s_j = \text{size}(\theta_j)/(c_j c_{j+1})$ denotes the filter size of the convolution's kernel.

The MorphNet regularization term $\mathcal{G}_M(j)$ for a given layer j is defined as:

$$\mathcal{G}_M(\theta, j) = C_j \sum_{p=1}^{c_{j-1}} |\gamma_{j-1,p}| \sum_{k=1}^{c_j} \mathbb{1}^O_{\{\gamma_{j,\cdot} > \tau_M\}} + C_j \sum_{p=1}^{c_j} \mathbb{1}^I_{\{\gamma_{j-1} > \tau_M\}} \sum_{k=1}^{c_{j-1}} |\gamma_{j,p}| \quad (1)$$

with $\mathbb{1}^I$ ($\mathbb{1}^O$) the indicator function for a given statement on the input (output) channels. For further information on how the MorphNet regularization term is calculated have a look in [14]. The overall MorphNet regularization term $\mathcal{G}_M$ is then defined by:

$$\mathcal{G}_M = \sum_{\ell=1}^{2L} \mathcal{G}_M(\theta, \ell) \quad (2)$$

The Morphnet algorithm (presented in [14]) is given in Algorithm 1, where c'_ℓ denotes the channel width of Layer ℓ, $\mathcal{F}$ is in our case the number of floating

Algorithm 1. The MorphNet algorithm

1: Train the network to find

$$\theta^* = \operatorname{argmin} \theta \{\mathcal{L}(\theta) + \lambda_M \mathcal{G}_M(\theta)\}$$

for suitable λ_M.
2: Find the new widths $c'_{1:2L}$ induced by θ^*.
3: Find the largest ω, such that $\mathcal{F}(\omega \cdot c'_{1:2L}) \leq \zeta$.
4: Repeat from Step 1 for as many times as desired, setting $c^0_{1:2L} = \omega \cdot c'_{1:2L}$.

point operations (FLOPs) the net currently uses and ζ the maximum capacity of FLOPs the net should have. Furthermore, we denote $c_{1:2L} = (c_1, \ldots, c_{2L})$.

Our LayerLasso method also uses a $L1$-regularization term $\mathcal{G}_\Lambda$ which is based on regularization of the weights θ:

$$\mathcal{G}_\Lambda = \sum_{\ell=1}^{2L} \|\theta_\ell\|_1. \tag{3}$$

It should be noted that this regularisation term is only applied to layers in the residual blocks and does neither penalise the weights of the initial layer nor the fully connected layers.

Besides the regularization terms, our loss function also consists of a cross entropy loss $\mathcal{L}_{CE}$:

$$\mathcal{L}_{CE} = -\sum_i^q y_i \log F_i(x). \tag{4}$$

In summary, our loss function consists of three different parts:

- The weight-optimizing part $\mathcal{L}_{CE}$ including a default L2-regularization term
- The MorphNet-L1-regularization $\mathcal{G}_M$ multiplied by the MorphNet regularization strength λ_M
- The LayerLasso-L1-regularization $\mathcal{G}_\Lambda$ multiplied by the MorphNet regularization strength λ_Λ

The complete loss term $\mathcal{L}$ can then be defined as:

$$\mathcal{L} = \mathcal{L}_{CE} + \lambda_\Lambda \mathcal{G}_\Lambda + \lambda_M \mathcal{G}_M \tag{5}$$

3.1   Inserting ResNet Blocks

Due to the construction of ResNet blocks, in particular their identity given by the addition of $x^{(B_i)}$, it is natural to initialize $\theta^{(B_i,j)}$, $j = 1, 2$, close to zero (but randomly) in order to enable a seamless continuation of training from the current state of the network. Initialization close to zero implies $x^{(B_i+1)} \approx x^{(B_i)}$. Randomness is required to avoid symmetries in the weights that occur in case of zero initialization.

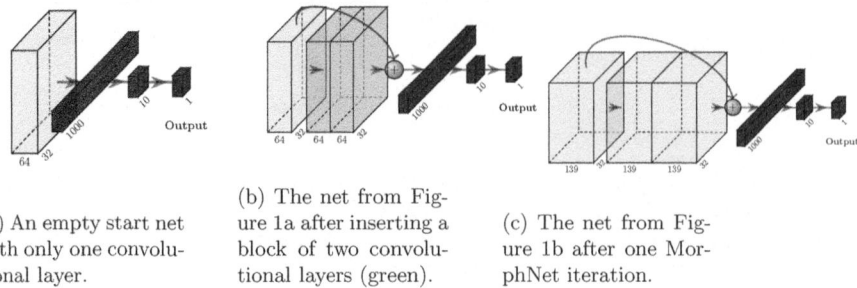

(a) An empty start net with only one convolutional layer.

(b) The net from Figure 1a after inserting a block of two convolutional layers (green).

(c) The net from Figure 1b after one MorphNet iteration.

Fig. 1. Example of structural changes the *ResBuilder* method uses

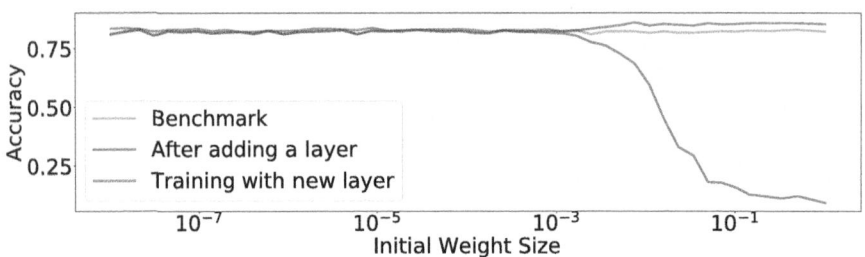

Fig. 2. Effect of weight initialization of an inserted layer block to a network on the CIFAR10 dataset. The green line indicates the benchmark of the startnet trained normally without a new block of layers inserted. Red indicates the accuracy after adding one of these blocks to our startnet but without further training while violet shows the accuracy after further training with the new layerblock inserted. (Color figure online)

In order to obtain the most effective trade-off between an initialisation close to 0 and the avoidance of undesired symmetries in the network's weights, we show in Fig. 2 how much the accuracy of the network suffers from the insertion of residual blocks with different initial weights, but also how strong the constraints of unbroken symmetries would be when continuing to train.

Therefore, we insert a layer block B_k behind a given block B_i into our network f according to our insertion strategy (see Sect. 3.3) like it can be seen in Fig. 1. We choose its initial weights such that the full potential of the new block can be exploited but the damage to the current knowledge of the network remains minimal. In our example, we therefore initialize the weights with an average initial weight of the order of $\theta_{init} = 10^{-2}$. From this, we get the new network architecture:

$$f' = B_n \circ \ldots \circ B_{i+1} \circ B_k \circ B_i \circ \ldots \circ B_1$$

3.2   LayerLasso

In order to also have the possibility to delete residual blocks of layers from positions where the net does not use its capacity efficiently, we introduce the **LayerLasso**: after a certain number of epochs of training every block B_i that includes at least one layer whose sum of weights $\sum_{\theta_m \in \theta^{B_i,j}} \|\theta_m\|_1, j = 1, 2$ lays under the set threshold for layer deleting τ_Λ will be erased from the net:

$$f' = B_n \circ \ldots \circ B_{i+1} \circ B_{i-1} \circ \ldots \circ B_1 \quad \forall B_i \exists j : \sum_{\theta_m \in \theta^{B_i,j}} \|\theta_m\|_1 < \tau_\Lambda. \quad (6)$$

Because of the residual architecture, a layer with small weights resembles the identity, so we continue training after the removal of blocks from the resulting architecture with the weights for the remaining blocks equal to the values before the removal step.

3.3 Strategy of Inserting and Removing

In order to optimize the resulting network architecture we used a "random in - greedy out" optimization strategy: we insert residual blocks of convolutional layers at random positions in the net with the only constraints that the new block can not be inserted before the first convolutional layer or after the flattening of the feature maps. In order to evenly spread the blocks across the entire architecture we choose the pooling stage randomly and then a random block B_i from this stage after which a new residual block is inserted (this might also be directly behind the pooling layer).

3.4 Structure of the Method

Our NAS training pipeline thus is defined as depicted in Fig. 3. We do n_Λ insertion steps before we start the MorphNet subroutine. Besides the training with all regularization terms active

$$\theta_{reg} := \text{argmin} \left\{ \mathcal{L}_{\text{CE}}(\theta) + \lambda_M \mathcal{G}_M + \lambda_\Lambda \mathcal{G}_\Lambda \right\}$$

our training pipeline determines which elements $\mathcal{A}' \subset \mathcal{A}$ of the architecture search space $\mathcal{A}$ are considered, while we also train the actual network architecture without regularization $\theta_{noReg} := \text{argmin} \{\mathcal{L}_{\text{CE}}\}$ in order to achieve the best accuracy for the given net.

4 Numerical Experiments

In this section we give an overview of our experimental setup and provide hyperparameters we use. In this section we also present our numerical results on six academic data sets and for one real world industrial application.

4.1 Experimental Setup

Initial Architectures. In our experiments, we apply our ResBuilder method to the different data sets. In one set of experiments, a ResNet18 is used as the initial network (RB-R18), and in the second set, a minimal network (RB-0Net), as shown in Fig. 4.

Training Types. The experiments we conduct contain three training types of each neural network architecture that is considered:

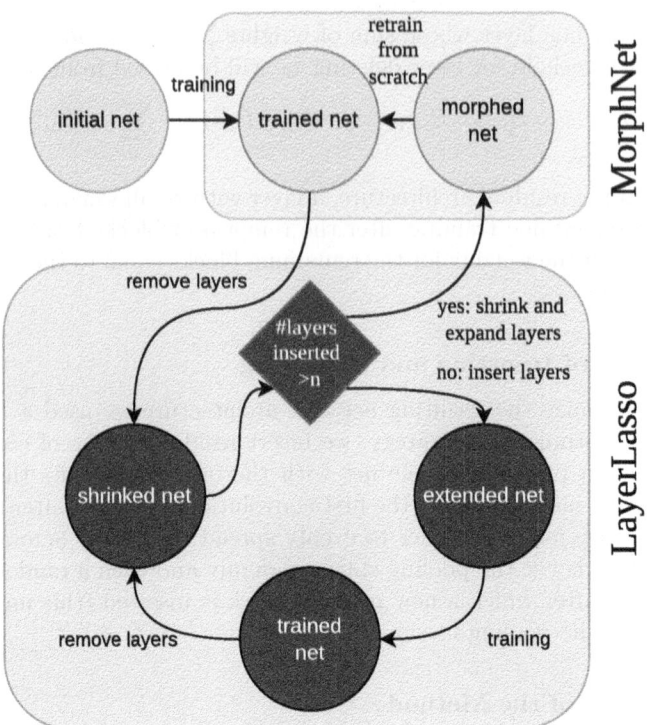

Fig. 3. Overview of the method

- Training Variant *With Reg*: With all possible regularisation terms.
- Training Variant *No Reg RI*: Without additional regularisation terms, starting from scratch, i.e. a random initialisation of the weights.
- Training Variant *No Reg WI*: Without additional regularisation terms, starting from the checkpoint induced by training *With Reg*.

The expression "without additional regularisation terms" used here means that the training takes place without additional regularisation by MorphNet or the LayerLasso, i.e. $\lambda_M = \lambda_\Lambda = 0$. An additional $L2$-regularisation term λ_0, which is intended to reduce overfitting, is nevertheless applied to the training.

Adjustable Hyperparameters. Unless otherwise specified, we use the default values given here for the hyperparameters, which we determined following extensive testing:

- $n_\Lambda = 4$ number of insertion steps before MorphNet step.
- $n_M = 7$ number of MorphNet steps.
- $\lambda_M = 10^{-7}$ the MorphNet regularization strength
- $\zeta = 100,000,000$ the aimed for FLOP costs, except for Animals10, for which ten times the amount of arithmetic operations was allowed ($\zeta =$

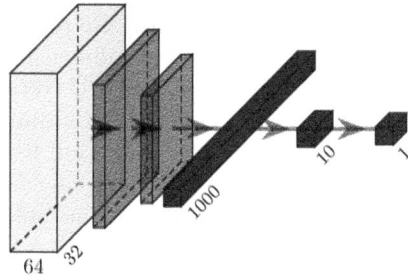

Fig. 4. This is the minimal initial startnet architecture with one convolutional layer (orange), two pooling layers (red) and one dense layer (violet) before the softmax and output layer. (Color figure online)

1,000,000,000), as the images in this dataset also have significantly higher resolution.

- $\lambda_\Lambda = 10^{-8}$ the LayerLasso regularization strength
- $\tau_\Lambda = 10^{-3}$ the threshold of our LayerLasso method which sets when a layer (block) is deleted (see Sect. 3.2)
- $\lambda_0 = 10^{-5}$ the additional $L2$-regularization strength
- *Adam* is used as optimizer.
- Data augmentation is active with 10% shifts in horizontal and vertical position as well as a possible horizontal flip.

Academic Datasets. In this work we evaluate our methods on the datasets listed in Table 1:

Table 1. Overview of datasets we use in this work

Name	Content	Size (px)	Classes
Animals10 [15]	Animals	300	10
CIFAR10 [16]	Misc.	32	10
CIFAR100 [16]	Misc.	32	100
MNIST [17]	Digits	28	10
FashionMNIST [18]	Clothing	28	10
EMNIST [19]	Letters	28	62

Data Preprocessing and Used Resources. For the Animals10 and CIFAR datasets, we apply a normalisation process to the data, which calculates for each pixel the difference between the current pixel value and the mean of all pixel values of all images in the current colour channel, and then divides it by the standard deviation. MNIST and FashionMNIST pixel values are simply scaled

from 0 to 1 and not pre-processed. We do also use data augmentation for our
trainings, where we allow a shift of 10% in both dimensions and also horizontal
flips. We used different packages like Tensorflow-GPU (version 1.14.0) [26], Keras
(version 2.2.4.) [27] or MorphNet (0.2.1) [14]. The visualization of neural nets
like Fig. 4 is based on the repository of [28]. For a full list of used packages see
the requirements.txt in our git repository. For the calculations, we used a Dell
Precision 7920 workstation with a Dual Intel Xeon Gold 6248R 3.0 GHz and
three Nvidia Quadro P 6000 graphic units with 24 GB VRAM each.

4.2   Numerical Results

Results on Academic Datasets. In one of our experimental setups, we ran
the ResBuilder method with the same hyper-parameters (described in Sect. 4)
for all our datasets under consideration and summarise these results in Table 2.
The column with the results of the run "RB-0Net" gives the accuracies we
achieve for the test series starting with a minimal network architecture as shown
in Fig. 4, without regularisation (*No Reg RI* and *No Reg WI*), as this would
be the later use case. The results in "RB-R18" were generated analogously to
the results of "RB-0Net", with the exception that a ResNet18 [13] was used
as the initial architecture. We benchmark our method against the accuracy we
achieve by training a ResNet18 on the specific dataset. The column "Res18"
therefore gives the accuracy when it is trained with the variant "No Reg RI"
which would be the typical way to go in order to train this architecture from
scratch. The "MorphNet"-column shows an ablation study where we omit our
additional LayerLasso routine from the architecture search process wherefore we
use a ResNet18 and perform the single MorphNet routine for the given number
of n_M. The training that is used for that, follows the variant "No Reg RI", as
our Res18 benchmark also does.

As can be seen in Table 2, our ResBuilder method performs better on all
datasets than both the standard ResNet18 architecture and the MorphNet approach without our depth-first search.

Table 2. Overview of achieved accuracies

Dataset	Res18	MorphNet	RB-R18	RB-0Net
Animals10	92.10%	92.02%*	**92.73%***	88.72%*
CIFAR10	85.50%	88.17%	88.32%	**89.92%**
CIFAR100	53.80%	59.78%	57.69%	**62.36%**
MNIST	99.14%	99.11%	99.17%	**99.34%**
FashionMNIST	92.81%	92.97%	93.55%	**93.71%**
EMNIST	86.48%	86.70%	86.86%	**86.95%**

In Fig. 5a we can see in the lower panel how the accuracy for the three
different training types over the iterations of architectures in the search space

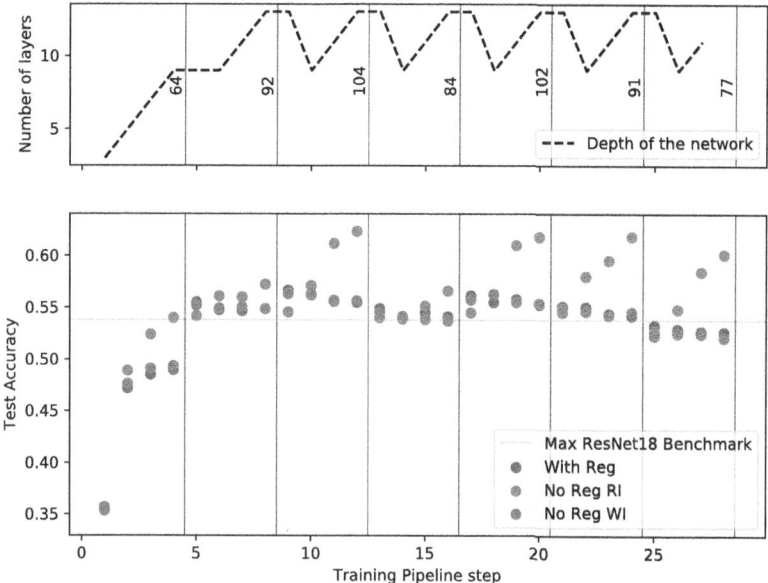

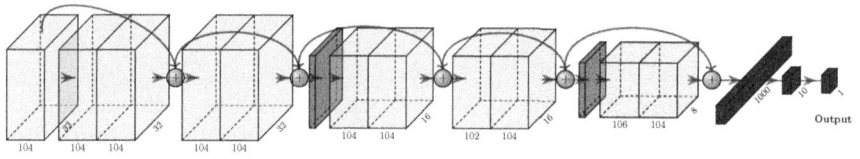

(a) History of accuracies for observed architectures.

(b) Best architecture (pipeline step 12)

Fig. 5. History of observed network architectures on CIFAR100 starting from minimal architecture (see Fig. 4)

(shown on the x-axis). Here we start searching from the minimal initial architecture (Fig. 4) on the CIFAR100 dataset. Meanwhile, the upper part of the figure shows how the depth of the network architecture while our ResBuilder method progresses, which is represented by the purple dashed line. Vertical red lines within the figure indicate MorphNet channel width optimization steps and the numbers next to the line in the upper part of the display the maximum channel size in the first pooling block in order to get a feeling for the width of the layer. Since this figure shows an experiment that started with a minimal initial architecture, the orange line shows the benchmark that was achieved by training a ResNet18. It should be noted that even at a fairly early stage of the ResBuilder, accuracies above this benchmark could be achieved. In this experiment, the best accuracy was achieved with the twelfth architecture, which is therefore shown in Fig. 5b. Furthermore, one can see that both the accuracy achieved and the depth and width of the network have converged very quickly in this example.

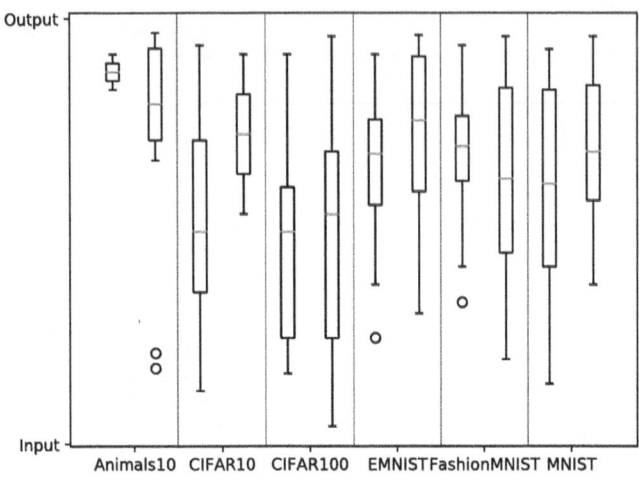

Fig. 6. Position of removed layerblocks relative to the network size. For each dataset the left plot shows the removed positions proceeding from the minimal initial architecture and the right one proceeding from the ResNet18 architecture

Removal Positions. In Fig. 6, we show at which positions of the network our method eliminates blocks of layers. The positions between input and output are always to be seen relative to the current network size. In particular, it can be observed that for CIFAR100, which has a relatively large number of classes (100), more layers tend to be thrown out of the front part of the net, which could result in a focus on the classification into the individual classes instead of optimizing encoding. The rather higher resolution of the Animals10 dataset, on the other hand, tends to throw out layer blocks at the back end of the architecture, which speaks for the importance of the front blocks for extracting information from the high resolution images.

Regularization Parameter Study. Figure 7 shows the accuracies of training different architectures with various regularization strengths. To ensure better visualisation, only the experiments with $\lambda_M = \lambda_\Lambda$ are considered in the figure. The shown accuracies all refer to the training variant *With Reg* without additional L2-regularization ($\lambda_0 = 0$). One can see that there is an accuracy trade-off between low FLOPs induced by a high regularization term and high FLOPs with a low regularization strength. For $\lambda_M = \lambda_\Lambda = 10^{-7}$ it can be also mentioned that the architecture broke down due to too high penalization applied.

Example of Industrial Application. In addition, the ResBuilder methodology has been challenged not only on standard benchmark datasets, but also

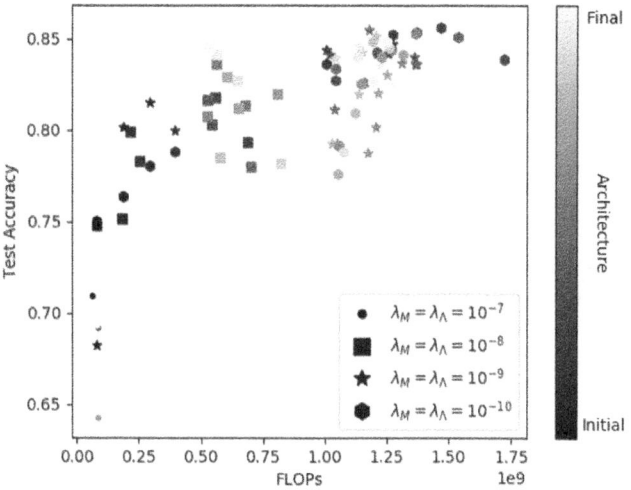

Fig. 7. Different regularization strengths for training on CIFAR10 data

on an independent set of data to assess its applicability in real-world scenarios. Specifically, the approach was applied to an insurance use case offered by ControlExpert (CE) [29], a company providing end-to-end motor claims management solutions. CE uses images, in particular of damaged vehicles, to process claims. For this, it must be ensured that these images are authentic, for instance by detecting manipulated images. This process can be described as a binary classification task. Ordinarily, a variety of architectures are scrutinized in these tasks to determine the optimal fit. These architectures are quite sophisticated, but are generally not developed specifically for a single classification task. Additionally, conducting the experiments to find an optimized architecture is time consuming. Hence, in an industry research process this might lead to competitive disadvantage. This motivated the application of the ResBuilder approach and its comparison with competitor networks like EfficientNet-b0 [30], EfficientNet-b4 [30], and ResNet18 [13]. The dataset utilized for this experiment was balanced, comprising 50% manipulated images and 50% non-manipulated images of damaged vehicles. The image manipulations were performed manually using various forgery methods such as splicing and copy-move. ResBuilder demonstrated a superior performance in terms of accuracy, achieving an enhancement of 1.2%-points[1] compared to the most effective competitor network, EfficientNet-b0. Moreover, it attained this while maintaining a reduction of 97.86% in parameter quantity. The findings suggest that ResBuilder is capable of generating efficient architectures for a real-world scenario, exhibiting not only an improved accuracy for the use case but also a significant memory efficiency. Another benefit of the ResBuilder approach is its capability to automate the architecture search process, thus reducing

---

[1] Absolute values cannot be disclosed due to ControlExpert's intellectual property rights.

the manual efforts required in creating an efficient model, a factor of substantial importance in an industrial context in order to lower the time-to-market for the development of such AI models.

5 Discussion

In this paper, we introduced the ResBuilder method, which provides a NAS algorithm that can generate neural networks from scratch or from existing architectures using suitable regularisation techniques. On many datasets (mainly academic, but also industrial), results close to state-of-the-art (without pretraining) are achieved. To this end, ablation studies related to the omission of our LayerLasso method were conducted and a parameter study on the regularisation strength was performed.

References

1. Russakovsky, O., et al.: ImageNet large scale visual recognition challenge. Int. J. Comput. Vision **115**, 211–252 (2015)
2. Nakano, R., et al.: WebGPT: browser-assisted question-answering with human feedback. arXiv preprint arXiv:2112.09332 (2021)
3. Goodfellow, I., Bengio, Y., Courville, A.: Deep Learning. MIT Press, Cambridge (2016)
4. He, X., Zhao, K., Chu, X.: AutoML: a survey of the state-of-the-art. Knowl. Based Syst. **212**, 106622 (2021)
5. Choudhary, T., Mishra, V., Goswami, A., Sarangapani, J.: A comprehensive survey on model compression and acceleration. Artif. Intell. Rev. **53**, 5113–5155 (2020)
6. Yao, Q., et al.: Taking human out of learning applications: a survey on automated machine learning. arXiv preprint arXiv:1810.13306 (2018)
7. Elsken, T., Metzen, J.H., Hutter, F.: Neural architecture search: a survey. J. Mach. Learn. Res. **20**(1), 1997–2017 (2019)
8. Hsu, C.-H., et al.: MONAS: multi-objective neural architecture search using reinforcement learning. arXiv preprint arXiv:1806.10332 (2018)
9. Reiners, M., Klamroth, K., Heldmann, F., Stiglmayr, M.: Efficient and sparse neural networks by pruning weights in a multiobjective learning approach. Comput. Oper. Res. **141**, 105676 (2022)
10. He, Y., Zhang, X., Sun, J.: Channel pruning for accelerating very deep neural networks. In: Proceedings of the IEEE International Conference on Computer Vision, pp. 1389–1397 (2017)
11. Srinivas, S., Babu, R.V.: Data-free parameter pruning for deep neural networks. arXiv preprint arXiv:1507.06149 (2015)
12. Liu, Z., Sun, M., Zhou, T., Huang, G., Darrell, T.: Rethinking the value of network pruning. arXiv preprint arXiv:1810.05270 (2018)
13. He, K., Zhang, X., Ren, S., Sun, J.: Deep residual learning for image recognition. In: Proceedings of the IEEE Conference on Computer Vision and Pattern Recognition, pp. 770–778 (2016)
14. Gordon, A., et al.: MorphNet: fast & simple resource-constrained structure learning of deep networks, 18 November 2017. arXiv:http://arxiv.org/abs/1711. 06798v3 [cs.LG]

15. Animals10 dataset. https://www.kaggle.com/datasets/alessiocorrado99/animals10. Accessed 30 May 2022
16. Krizhevsky, A., Hinton, G., et al.: Learning multiple layers of features from tiny images (2009)
17. LeCun, Y., Bottou, L., Bengio, Y., Haffner, P.: Gradient-based learning applied to document recognition. Proc. IEEE **86**(11), 2278–2324 (1998)
18. Xiao, H., Rasul, K., Vollgraf, R.: Fashion-MNIST: a novel image dataset for benchmarking machine learning algorithms. CoRR, vol. abs/1708.07747 (2017). arXiv: 1708.07747, http://arxiv.org/abs/1708.07747
19. Cohen, G., Afshar, S., Tapson, J., Van Schaik, A.: EMNIST: extending MNIST to handwritten letters. In: 2017 International Joint Conference on Neural Networks (IJCNN), pp. 2921–2926 (2017)
20. Zoph, B., Le, Q.V.: Neural architecture search with reinforcement learning. arXiv preprint arXiv:1611.01578 (2016)
21. Zoph, B., Vasudevan, V., Shlens, J., Le, Q.V.: Learning transferable architectures for scalable image recognition. In: Proceedings of the IEEE Conference on Computer Vision and Pattern Recognition, pp. 8697–8710 (2018)
22. Wang, Y., et al.: Pruning from scratch. In: Proceedings of the AAAI Conference on Artificial Intelligence, vol. 34, pp. 12 273–12 280 (2020)
23. Dong, X., Yang, Y.: Network pruning via transformable architecture search. In: Advances in Neural Information Processing Systems, vol. 32 (2019)
24. Ahmed, K., Torresani, L.: MaskConnect: connectivity learning by gradient descent. In: Ferrari, V., Hebert, M., Sminchisescu, C., Weiss, Y. (eds.) ECCV 2018. LNCS, vol. 11209, pp. 362–378. Springer, Cham (2018). https://doi.org/10.1007/978-3-030-01228-1_22
25. Geifman, Y., El-Yaniv, R.: Deep active learning with a neural architecture search. In: Advances in Neural Information Processing Systems, vol. 32 (2019)
26. Abadi, M., et al.: TensorFlow: large-scale machine learning on heterogeneous systems, Software available from tensorflow.org (2015). https://www.tensorflow.org/
27. Chollet, F., et al.: Keras (2015). https://keras.io
28. Iqbal, H.: Harisiqbal88/plotneuralnet v1.0.0, December 2018. https://doi.org/10.5281/zenodo.2526396
29. Controlexpert. https://www.controlexpert.com/de-de/. Accessed 5 July 2023
30. Tan, M., Le, Q.: EfficientNet: rethinking model scaling for convolutional neural networks. In: International Conference on Machine Learning, PMLR, pp. 6105–6114 (2019)

Self-Organization

A Neuron Coverage-Based Self-organizing Approach for RBFNNs in Multi-class Classification Tasks

Alberto Ortiz[1,2,3]([✉]) [iD]

[1] Department of Mathematics and Computer Science, University of the Balearic
Islands (UIB), Palma, Spain
alberto.ortiz@uib.es
[2] IDISBA (Health Research Institute of the Balearic Islands), Palma, Spain
[3] IAIB (Artificial Intelligence Research Institute of the UIB), Palma, Spain

Abstract. Radial Basis Function Neural Networks (RBFNN) constitute
a distinct type of artificial neural network known for its unique architecture and ability to approximate complex functions, and as well lead
to effective modeling in regression/classification tasks. Within the context of multi-class classification, this work introduces NC-SORBFNN as
a self-organizing approach that adapts the structure of RBFNNs to the
complexity of each task. NC-SORBFNN comprises an initialization step
that makes use of a *Self-Organizing Map* (SOM) to give rise to a first set
of hidden neurons that is next tuned by a self-adjustment process that
goes through the data to make the network *grow and shrink* as required
on the basis of the *sample coverage* of each neuron. The experimental
results show the effectiveness of NC-SORBFNN, which in turn compares
well with other similar solutions, leading in general to high performance
for a low amount of hidden neurons.

Keywords: Multi-class Classification · RBF Neural Networks
(RBFNN) · Self-Organizing Neural Networks

1 Introduction

Radial Basis Function Neural Networks (RBFNN) constitute a distinct type of
artificial neural network known for its simpler architecture, but at the same time
strong non-linear function approximation and effective modeling capabilities [21].
This has permitted to address both regression and classification problems, in
fields such as non-linear modeling [17] and forecasting [25], non-linear control [2]
and pattern recognition [1], to name but a few.

In a multilayer perceptron (MLP), the function approximation is defined by
a nested set of weighted summations, according to the layered structure of the

This work has been partially supported by the EU-H2020 grant BUG-
WRIGHT2 (GA 871260) and by grant PID2022-139248NB-I00 (funded by
MICIU/AEI/10.13039/501100011033 and by ERDF/EU). This publication reflects
only the authors views and the EU is not liable for any use that may be made of
the information contained therein.

M. Wand et al. (Eds.): ICANN 2024, LNCS 15016, pp. 327–342, 2024.
https://doi.org/10.1007/978-3-031-72332-2_22

model, while, in an RBFNN, the approximation is defined by a single weighted sum, i.e. a single hidden layer network. This can be regarded as the basic structural difference between the two kinds of networks. Another fundamental difference is the fact that, in RBFNNs, the hidden units implement a set of radial-basis functions that constitute an arbitrary basis for the input samples when they are expanded into the hidden unit space. Besides, all this is achieved with a considerably low computational cost, suitable for energy-limited scenarios, e.g. edge computing devices.

A *radial-basis function* is a real-valued function ρ whose value depends only on the distance between input samples x and a specific point μ that is associated with each hidden unit, and it is known as its *center*, so that $\rho(x;\mu) = f(d(x,\mu))$, where $d(x,\mu)$ is the distance between x and μ. Apart from other RBFs [21], the RBF most usually adopted is the Gaussian function $G_\sigma(r) = \exp\left(-\frac{r^2}{2\sigma^2}\right)$, with $r^2 = \|x - \mu\|_2^2$ and σ as the *width* of the unit.

A main problem with standard RBFNNs (and with feed-forward neural networks in general) is the fact that the network structure, i.e. the number of hidden neurons, as well as its key parameters, i.e. the centers μ and widths σ, must be fixed *a priori*. This is a crucial factor to the successful application of RBFNNs, since the set of centers and the related widths are expected to match the complexity of the problem to solve. In this regard, a learning methodology that can adjust the structure of the network to the data turns out to be more appropriate. To this end, this paper describes the *Neuron Coverage-based Self-Organizing RBFNN* (NC-SORBFNN) approach. It applies the self-adjustment idea to the network building process from end to end, in order to adapt the structure of the network to the specific task.

In more detail, NC-SORBFNN comprises two stages, initialization and self-adjustment. The *initialization stage* adopts a single-layer *Self-Organizing Map* (SOM) to produce in a parameterless way a first set of centers from the training set, which are next used to estimate the widths of the neurons of the hidden layer to complete the initialization. Next, the *self-adjustment stage* makes the network grow and shrink in an iterative process which goes through the training data and adapts the network to the complexity of the task. To make neuron pruning and creation decisions, this work revisits the concepts of *orthogonality* and *integrity of a set of neurons* introduced by Han et al. in [12]. Finally, NC-SORBFNN is evaluated in the context of multi-category classification tasks using publicly available datasets, comparing well with other self-organizing solutions.

The main contributions of this work are enumerated next:

- NC-SORBFNN is proposed as a self-organizing approach that involves an adaptive expansion and pruning mechanism (AEPM) to adjust the structure of an RBFNN to the complexity of the task at hand.
- NC-SORBFNN comprises a SOM-based initialization step followed by a self-adjustment iterative process. The SOM-based initialization avoids the specification of the number of clusters, contrary to other solutions based on e.g. K-means, while the second step is parameterized by only two thresholds that determine the size of the hidden layer. Although the latter fact prevents

NC-SORBFNN from being a parameterless algorithm, in the end, there are only two main parameters to set up what makes NC-SORBFNN an easy-to-configure algorithm.

- NC-SORBFNN does not require the involvement of derivatives for network updates in any of its components, thus extending the range of radial basis functions that can be adopted.
- The experimental setup for evaluating NC-SORBFNN takes into consideration a total of 18 datasets comprising samples of a quite diverse nature/source. All in all, the experimental results obtained show competitive performance for, in general terms, a low number of neurons in the hidden layer, hence reducing the number of calculations during inference.

The rest of the paper is organized as follows: Sect. 2 outlines previous work related to RBFNNs and self-organizing approaches; the RBFNN concept is reviewed in Sect. 3; Sect. 4 describes NC-SORBFNN in detail; Sect. 5 reports on the effectiveness of NC-SORBFNN as well as on the performance attained by NC-SORBFNN on a number of publicly available datasets; finally, Sect. 6 concludes the paper summarizing the main findings and outlining future work.

2   Previous Work

Training RBFNNs is generally accomplished through two-stage methodologies [27]: the first step typically finds the centers and widths of the hidden neurons, while the second step determines the connection weights between hidden and output neurons. Clustering [37], orthogonal least squares [3] and gradient descent [31] are among the methods adopted for the first stage. The second stage is normally fulfilled by means of least squares [20] or by gradient descent-based approaches [7,22]. In this latter regard and specifically for RBFNNs, a variety of strategies have been adopted to counteract slow learning (due to vanishing gradient in the cost function), such as, among others, *natural gradient learning* [36], *adaptive gradient learning* [29] or *accelerated gradient learning* [11].

On the other side, the number and location of the network centers are key parameters mostly affecting the performance of an RBFNN model. This has lead to a number of *self-organizing approaches* aiming at adapting the network structure to the task at hand. A number of these solutions adopt an AEPM strategy, by which the network structure is iteratively self-adjusted by adding and removing neurons as needed according to specific criteria: Huang et al. design in respectively [18,19,35] the algorithms GAP-RBF, GGAP-RBF and FGAP-RBF, where they introduce the concept of the *significance* of a neuron by which the number of hidden neurons increases and decreases inside an on-line approach that involves a decoupled Extended Kalman Filter for updating the network parameters; Han et al. describe FS-RBFNN in [8] for water quality prediction, which makes use of the neurons *average firing rate* to determine whether new neurons have to be incorporated, while a mutual information-inspired process decides whether a neuron has to be pruned; Han et al. develop IOA-RRBFNN in [9], which defines the *information processing strength* of a neuron supported by an independent component analysis process, all to configure a recurrent RBFNN model (RRBFNN); Xie et al. propose SAS-RBFNN in [33] for on-line prediction of ferrous ion concentration, comprising an initialization step that clusters samples in accordance to their associated output, and neuron merging and splitting operations, respectively based on the distances between neurons and the widths of the neurons, for on-line adjusting the RBFNN structure; Han et al. present ASOL-SORBFNN in [12], defining the concepts of *orthogonality* and *integrity of a set of neurons* to decide on neuron merging and splitting, followed by the update of the network parameters by means of the Levenberg-Marquardt update rule after each merge/split operation; Yang et al. describe RBFNN-GP in [34], a self-adjusting algorithm that is concerned with the generalization capability of the resulting model instead of focusing on the approximation error.

Alternatively to the previous self-adjusting approaches, there has been a great deal of research devoted to the adjustment of neural network structures using metaheuristics such as e.g. genetic and evolutionary algorithms [16], differential evolution [4] or artificial fish swarm [28]. Inside this category, *particle swarm optimization* (PSO)-based strategies have become particularly popular, e.g. Fathi and Montazer present in [5] a PSO algorithm to optimize the centers and widths of the hidden units (though not the number of neurons), and involves the Optimum Steepest Descent (OSD) method to train the weights; Han et al. adopt *adaptive PSO* to adjust both the network size and the neuron parameters in [10]; Li et al. propose MGPSO-SORBF in [24] for aero-engine thrust estimation using a specifically developed version of PSO; Han et al. describe AGMOPSO to predict the biochemical oxygen demand in a wastewater treatment process in [15], using in this case *adaptive gradient multi-objective PSO*; in [32], Wu et al. present GA-IPSO-CSRBF to deal with imbalanced data in medical diagnosis by means of a genetic algorithm combined with *improved PSO* to optimize jointly and adaptively the structure and the network parameters. Within the same category of solutions, although adopting a bee colony-inspired approach instead of a particle swarm, [6] proposes BeeRBF as a variation of the OptBees algorithm; this solution determines automatically the number of hidden neurons and their locations, leaving the estimation of the other parameters outside the approach.

3   Preliminaries

A *Radial Basis Function Neural Network* (RBFNN) is a special class of artificial neural network [26], whose main differences with *Feed-Forward Neural Networks* (FFNN) are the fact that an RBFNN typically comprises, apart from the input and output layers, a single hidden layer, and the fact that the hidden neurons implement an instance of a radial basis function $\rho(x; \mu_j)$ instead of the usual dot product $w \cdot x$ between connection weights w and the neuron input x. The value of an RBF depends on the distance from x to the *center vector* or *prototype* μ_j associated to the neuron, i.e. $\rho(x; \mu_j) = f(d(x, \mu_j))$, and hence they become radially symmetric about μ_j. Common elections are the Euclidean norm for the distance between samples x and centers μ_j, and a Gaussian as the radial basis function:

$$\rho_j(x;\mu_j) = G_{\sigma_j}(r_{x,j}) = \exp\left[-\frac{r_{x,j}^2}{2\sigma_j^2}\right] \tag{1}$$

where $r_{x,j}^2 = \|x - \mu_j\|_2^2$ and σ_j is the *width* of the hidden unit j. In the context of a classification problem involving c classes, output Φ_k (for class $k = 1, \ldots, c$) of an RBFNN with n_h hidden neurons is then given by:

$$\Phi_k(x) = \sum_{j=1}^{n_h} w_{k,j}\, \rho_j(x;\mu_j) + b_k \tag{2}$$

with $w_{k,j}$ as the weight of the connection between the hidden neuron j and the output neuron k, and b_k as the bias of the output neuron k.

4   Self-organizing Approach for RBFNNs

NC-SORBFNN makes use of the self-organization idea from end to end: in a first step, training data are clustered in a parameterless way by means of a *Self-Organizing Map* (SOM) to locate areas in feature space populated by samples; a second step addresses *network initialization* by building a first single-hidden layer network using the cluster representatives as centers of the hidden neurons of the network, while the neuron widths are set on the basis of the inter-sample distances in the training data; finally, in a third step, a *network adjustment* procedure adapts the initial set of hidden units to the complexity of the data by means of neuron pruning and growing processes. A high-level graphical description of NC-SORBFNN is shown in Fig. 1, while the details can be found in the following sections.

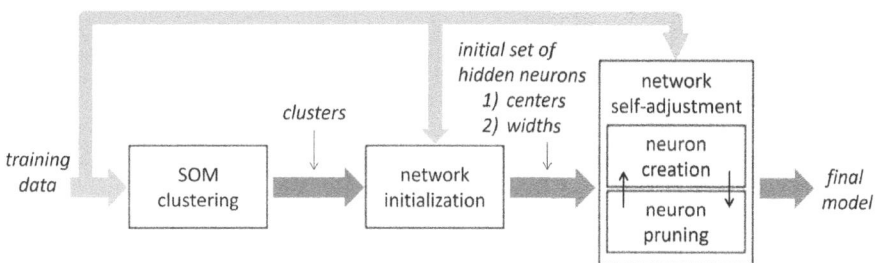

Fig. 1. High-level description of NC-SORBFNN.

4.1   Initialization

SOM Building and Training. A SOM is a special type of artificial neural network that comprises a single computational layer of neurons arranged in a lattice, i.e. each hidden neuron n_{ij} is assigned a cell (i,j) in a grid. Using a combination of competitive and cooperative training, each SOM neuron learns a *weight vector* ω_{ij} —with the same dimensionality as the input samples $x \in \mathbb{R}^n$—, so that the SOM model as a whole can deliver a set of *prototypes* that represent populated areas of the input space. After training, the network is expected to get organized in such a way that neighbouring samples in the input space map to the same or neighbouring cells in the lattice, hence the use of the SOM model for data clustering. (For more details, see [23] among many others).

In NC-SORBFNN, the SOM is defined as a 2D rectangular grid with 4-neighbour connectivity and a number of neurons n_{som} such that $n_{\text{som}} = 5\sqrt{N}$ for N training samples (in accordance to [23]). The size $n_1 \times n_2$ of the SOM grid is defined by the ratio of the two largest eigenvalues λ_1, λ_2 of the covariance matrix of the training set (i.e. $n_1 \times n_2 \approx n_{\text{som}}, \lambda_1/\lambda_2 = n_1/n_2$), while the initial weights $\omega_{ij}(0)$ for each neuron are determined using PCA. This defines a rectangular grid whose directions and extent match the two main directions and range of the data, so as to initialize the grid weights ω_{ij} with points of the feature space regularly spaced throughout the input data subspace, and hence get most advantage of the SOM [23]. Due to the way how the SOM is defined and trained, the initialization step of NC-SORBFNN does not require setting any critical parameters, in particular the number of clusters, unlike, for example, K-means and many others, or setting up any other parameter e.g. related to the density of samples (in DBSCAN and others).

From the SOM to the RBFNN. Once the SOM has been built, each training sample x is associated to the neuron whose weight vector $\omega_{ij}(t_{\max})$ is closest, what leads to *non-empty* and *empty cells* depending on whether any training sample has got associated to the cell or not. After this association process, each *non-empty* cell of the SOM gives rise to a cluster, and correspondingly to a neuron n_k in the hidden layer of the RBFNN. The neuron center μ_k is defined by the final weight vector $\omega_{ij}(t_{\max})$ of the SOM cell.

Regarding the neuron widths σ_j, inspired by [26], they are automatically defined as $\sigma_j = \alpha_w \, d_{\text{avg}}$, where $\alpha_w \in [1, 2]$ is a parameter and d_{avg} is the average of the distances between every center μ_j and its nearest center $\mu_{j'}$.

4.2   Network Self-adjustment

The initial configuration of the network that results from the SOM typically comprises more hidden neurons than necessary for the complexity of the task at hand. The next natural step is hence the simplification of the network trying to keep with the same or similar level of performance. On-line methods for RBFNNs typically incorporate pruning, merging and splitting processes to build the network in accordance to the samples as they are seen and processed. To this end, different criteria have been introduced: [18, 19, 35] handle the concept of *significance* to decide on neuron growing and pruning, being the significance a sort of average distance between the functions implemented by the network with and without that neuron; [12] refers to the *orthogonality* and the *integrity of a set of neurons* to make decisions on, respectively, pruning and growing relating both concepts to the performance of the basis functions underlying the Gaussian radial basis function; and [34] is concerned with the generalization ability of the network and performs a *sensitivity* analysis before deciding on neuron growing and pruning.

This work borrows the ideas of *orthogonality* and *integrity* from [12] to design a network self-adjustment process that adapts the network to the complexity of the task at hand. The key concept is the not necessarily symmetric matrix κ:

$$\kappa = \begin{bmatrix} 0.0 & \rho_1(\mu_2;\mu_1) & \cdots & \rho_1(\mu_{n_h};\mu_1) \\ \rho_2(\mu_1;\mu_2) & 0.0 & \cdots & \rho_2(\mu_{n_h};\mu_2) \\ \vdots & \vdots & \ddots & \vdots \\ \rho_m(\mu_1;\mu_{n_h}) & \rho_m(\mu_2;\mu_{n_h}) & \cdots & 0.0 \end{bmatrix} \qquad (3)$$

where ρ_i refers to the RBF implemented by the hidden neuron i.

Given the values taken by $\rho_i(\mu_j;\mu_i)$, $\kappa(i,j)$ is easily shown to capture the overlapping between the receptive fields of the set of hidden neurons, and hence the amount of redundancy: the closer κ is to a zero matrix $\mathbf{0}_{n_h \times n_h}$, the less redundant the set of neurons will be. Consequently, the *overlapping of the receptive fields of the set of neurons* can be measured as the maximum of matrix κ: the closer to 0, the less overlap exists between the receptive fields of the neuron set.

On the other side, the neuron set should be such that each training sample x is properly covered by the set of hidden neurons. This can be measured sample by sample by means of $\pi(x)$ defined as follows:

$$\pi(x) = \frac{1}{1 + \dfrac{\gamma_c}{\gamma_x}} = \frac{1}{1 + \dfrac{\max\limits_{i \neq j}\{\kappa(i,j)\}}{\rho_j(x;\mu_j)}} \ , \quad j = \arg\max_{j'}\{\rho_{j'}(x;\mu_{j'})\} \qquad (4)$$

where neuron j is the closest to sample x, and hence $\pi(x)$ relates the distance of x to neuron j versus the distance between neuron j and its closest neuron i, i.e. the *coverage of a sample x provided by the set of neurons*. Further, $0 \leq \pi(x) \leq 1$ and $\pi(x) = 0.5$ if $\gamma_x = \gamma_c$, $\pi(x) < 0.5$ if $\gamma_x < \gamma_c$ (i.e. the neuron closest to x is closer to its nearest neuron) and $\pi(x) > 0.5$ if $\gamma_x > \gamma_c$ (i.e. the neuron closest to x is farther to its nearest neuron). Correspondingly, the lower $\pi(x)$ is, the worse the coverage for sample x.

Given the aforementioned, the network adjustment approach of NC-SORBFNN consists in an iterative process of refinement that evaluates at each iteration the coverage that the set of neurons provides to the training set, to re-adjust the hidden layer accordingly to the complexity of the classification task. Specifically:

1. Neurons are pruned if redundancy is detected, i.e. excessive overlapping: $\exists i \neq j$ s.t. $\kappa(i,j) \geq \tau_O$, where $\tau_O \in [0,1]$ is a parameter. In such cases, one of neurons i or j is removed, that one with the smallest width σ, i.e. narrower coverage, and $n_h = n_h - 1$. The neuron pruning stage of NC-SORBFNN removes all neurons fulfilling the pruning condition at every iteration, unlike [12] that prunes one neuron per iteration as part of an on-line learning approach. Experimental evidence has shown this multi-neuron pruning operation to lead to the same network structure but reducing the number of iterations.

2. Neurons are added if there are training samples x with insufficient coverage from the set of neurons. Hence, the growing condition can be stated as: $\exists x$ s.t. $\pi(x) \leq \tau_I$, where $\tau_I \in [0,1]$ is a parameter. In such a case, the set of hidden neurons grows with a new neuron to cover better sample x. The new neuron is defined as $\mu_{\text{new}} = x$, $\sigma_{\text{new}} = \alpha_w d_{\text{avg}}$, where d_{avg} is the average of the distances between every center μ_j and its nearest center $\mu_{j'}$, $\alpha_w \in [1,2]$ is a parameter, and $n_h = n_h + 1$.

The adjustment process interleaves pruning and growing steps, and stops when the structure does not shrink or grow in the current iteration.

Summing up, step 1 above can be seen as a measurement of how well the centers are spread throughout the input space, so that centers are not close to each other and therefore redundant; on the other side, step 2 measures how well training samples x are covered by the set of hidden neurons.

5   Experimental Results

5.1   Evaluation Methodology and Experimental Setup

In the following, we evaluate the RBFNN model proposed within the context of multi-class supervised classification tasks using standard metrics and publicly available well-known datasets. Classification performance is reported in terms of accuracy $A = \frac{\text{TP}+\text{TN}}{\text{N}}$ and $F_1 = \frac{2\text{PR}}{\text{P}+\text{R}}$, which is the harmonic mean of the precision $P = \frac{\text{TP}}{\text{TP}+\text{FP}}$ and recall $R = \frac{\text{TP}}{\text{TP}+\text{FN}}$ metrics, where TP, TN, FP and FN are, respectively, the true positives, true negatives, false positives and false negatives, and N is the total number of samples. All the performance values reported in this section correspond to the average of the metric values after stratified 5-fold cross validation, using *weighted-averaging* for the multi-class datasets (that is to say, the weights match the support for each class, i.e. the number of true instances).

The evaluation of NC-SORBFNN considers a benchmark comprising several sorts of datasets, whose main features are summarized in Table 1, grouped in a first *basic set* (datasets 1–4), followed by a second set of *two-class (moderately) imbalanced* datasets [30] (datasets 5–9), datasets comprising *high-dimensional* samples (datasets 10–15) and a final subset with *highly imbalanced* classes

Table 1. Main features of the datasets considered. (IR stands for *imbalance ratio*, expressed by the quotient *no. samples majority class/no. samples minority class* [32].)

Id	Dataset	no. features	no. classes	no. samples	IR
1	Moons (synthetic dataset, see Fig. 2(left))	2	2	400	1.00
2	Iris	4	3	150	1.00
3	Penguins	6	3	333	2.15
4	Wine	13	3	178	1.50
5	Wisconsin breast cancer (original, WBC-org)	9	2	683	1.86
6	BUPA liver disorders	6	2	345	1.38
7	PIMA Indians Diabetes	8	2	768	1.87
8	Lower Back Pain Symptoms (LBPS)	11	2	310	2.10
9	Phoneme	5	2	5404	2.41
10	Segment	19	7	2310	1.00
11	Wisconsin breast cancer (WBC)	30	2	569	1.68
12	Satimage	36	6	6430	2.45
13	Digits	64	10	1797	1.05
14	DNA	180	2	3186	2.16
15	Sonar	60	2	207	1.16
16	Heart	13	5	297	12.31
17	Glass	9	6	214	8.44
18	Yeast	8	10	1484	92.6

(datasets 16–18). Except for *moons*, all datasets can be found in public repositories, e.g. UCI-ML, Kaggle or DataHub. All of them have been normalized (using *max-min* normalization) before training and building the RBFNN through NC-SORBFNN. The 2D *moons* dataset has been generated synthetically using the make_moons function from the *scikit-learn* Python library, being later rotated and translated (see Fig. 2[left]). In all cases, $\alpha_w = 1.2$ (Sect. 4.1), while the main parameters of NC-SORBFNN, the thresholds for neuron prunning τ_O and for growing the neuron set τ_I (Sect. 4.2), are set as indicated for every test.

5.2   Evaluation of NC-SORBFNN

For a start and by way of illustration of NC-SORBFNN, Fig. 2 shows classification results for fold 2 of the *moons* dataset. NC-SORBFNN has started with a SOM of $7 \times 15 = 105$ neurons that, after training, has given rise to an initial hidden layer of 69 neurons, which have finally become a hidden layer comprising 15 neurons (during the execution of NC-SORBFNN, 58 neurons have been pruned and 4 have been added). The final centers are indicated by black crosses in Fig. 2(right), while the final widths are $\sigma_j = 0.126, \forall j$. As can be observed, the classification performance achieved in this case is almost 100%.

Next, Fig. 3 shows graphically the performance achieved for datasets *iris* and *segment* for 16 combinations of values of $\tau_O = 0.6, 0.7, 0.8, 0.9$ and $\tau_I = 0.1, 0.2, 0.3, 0.4$. From the plots, one can observe that increasing τ_O gives rise to

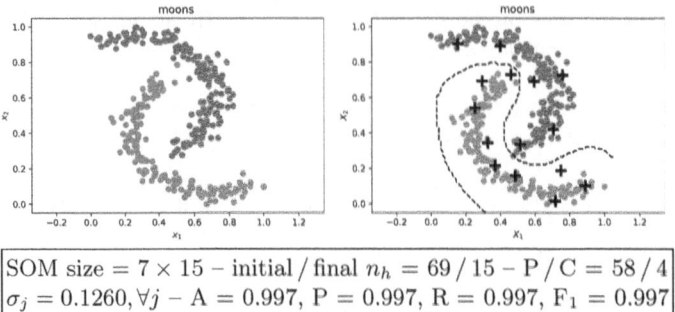

SOM size = 7×15 – initial / final $n_h = 69\,/\,15$ – P / C = $58\,/\,4$
$\sigma_j = 0.1260, \forall j$ – A = 0.997, P = 0.997, R = 0.997, F$_1$ = 0.997

Fig. 2. (top) *Moons* dataset. (bottom) Decision curve (dashed line) resulting from NC-SORBFNN for $\tau_O = 0.6$ and $\tau_I = 0.2$, final centers are plotted as black crosses. (bottom) Data from the RBFNN building process (P/C = number of neurons pruned/created).

more neurons, as there is less pruning activity, while keeping constant τ_O but increasing τ_I in general decreases the size of the hidden layer. In the case of the *iris* dataset, the classification performance in terms of accuracy (A) and F$_1$ score keeps almost unaltered across the different configurations, while for the *segment* dataset, of higher complexity, one can observe that models with less neurons leads to lower A and F$_1$ values.

Table 2 reports on the performance resulting from the evaluation of the previous 16 configurations for the 18 datasets (including the *iris* and *segment* datasets). In the table, for each dataset, the first row reports on the average F$_1$ values while the second row is for the average number of hidden neurons—F$_1$ values are used because of the class imbalance present in most datasets. Furthermore, values highlighted in red and blue correspond to, respectively, best and second best performance. The two last rows of the table summarize the number of times each configuration has given rise to the best or second best performance for F$_1$ or to the lowest/second lowest number of neurons. According to the results obtained, the configuration $(\tau_O, \tau_I) = (0.8, 0.4)$ gives rise to the highest amount of cases with high F$_1$ values [7 times as the best and 8 times as the second best]; another configuration with highest performance is $(0.9, 0.4)$ [8 times as the best and 6 times as the second best], though this is at the expense of the highest number of neurons for each dataset, so that $(0.8, 0.4)$ seems a good compromise between both classification performance and a smaller hidden layer. Regarding configurations leading to lower amounts of hidden neurons, as expected, they result from the lowest τ_O value; best cases are $(0.6, 0.1)$, and $(0.6, 0.4)$, being the latter the one that tends to give rise to higher F$_1$ values.

To finish, Table 3 compares NC-SORBFNN with other self-organizing approaches in terms of classification accuracy and number of hidden neurons: GAP-RBF [18], FGAP-RBF [35], AANN [14], CP-NN [13], BeeRBF [6] and GA-IPSO-CSRBF [32]. We include them in the comparison because they all have been published as self-adjusting schemes for single hidden layer neural

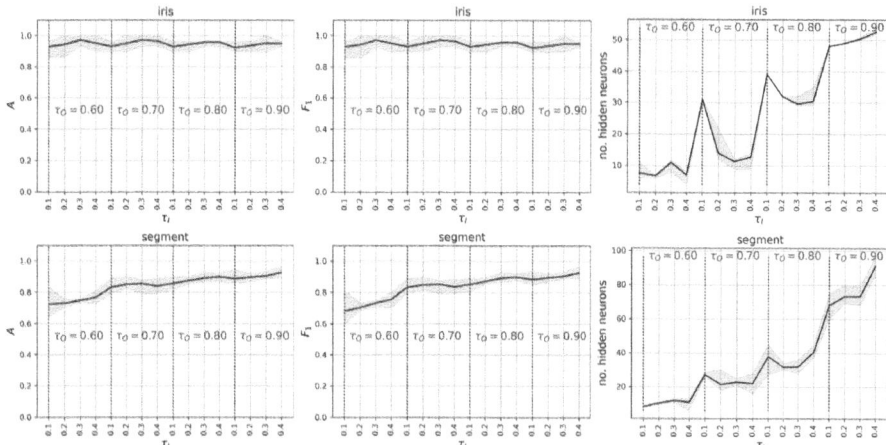

Fig. 3. Performance achieved by NC-SORBFNN for different combinations of values of τ_O and τ_I for (left) the *iris* dataset and (right) the *segment* dataset. (Shaded areas represent the interval between the minimum and maximum values for a specific pair (τ_O, τ_I) among the five folds, while the solid line corresponds to the average value).

networks focusing on classification (it must be noticed that this is not the typical orientation of self-organizing RBFNNs since, in general, these models focus on regression, e.g. [12]). On the other side, note also that the goal of AANN and CP-NN is the adjustment of multilayer perceptrons. Finally, the accuracy values that are reported for the aforementioned algorithms in Table 3 come directly from the original publications (F_1 values are not available in these papers). In the table, accuracy values are reported for four configurations of NC-SORBFNN: (C1) configuration achieving best F_1 score for that dataset; (C2) configuration leading to the smallest hidden layer for that dataset; (C3) configuration $(0.8, 0.4)$ as a compromise between classification performance and size of the hidden layer, prioritizing the classification performance; and (C4) configuration $(0.6, 0.4)$, as a compromise between classification performance and size of the hidden layer, prioritizing the size of the hidden layer. As can be observed, NC-SORBFNN compares well with the other solutions, being able to attain the highest performance on all except for the *glass* dataset. Moreover, the configurations deemed to be good compromises between performance and size of the hidden layer (C3 & C4), in particular the case $(0.8, 0.4)$, gives rise to a performance level which is the best or the second best in all cases, except for datasets *satimage*, *heart* and *glass*, though it gets very close to the best approach. Finally, configurations leading to smaller hidden layers (C2 & C4) achieve on a number of occasions, for a lower number of neurons, a performance level above the best competing solutions –datasets *wbc-org*, *sonar* and *heart*– or gets below but very close again –datasets *iris*, *pima* and *wbc*.

Table 2. Performance of NC-SORBFNN for different values of τ_O and τ_I.

$\tau_O \rightarrow$	0.6	0.6	0.6	0.6	0.7	0.7	0.7	0.7	0.8	0.8	0.8	0.8	0.9	0.9	0.9	0.9
$\tau_I \rightarrow$	0.1	0.2	0.3	0.4	0.1	0.2	0.3	0.4	0.1	0.2	0.3	0.4	0.1	0.2	0.3	0.4
(1) moons	0.99	1.00	0.99	0.99	0.99	1.00	1.00	1.00	0.99	1.00	1.00	1.00	0.99	1.00	1.00	0.99
	15.2	13.6	16.4	11.8	33.8	24.2	25.0	14.2	43.0	38.2	37.6	32.0	57.8	59.2	60.0	66.2
(2) iris	0.93	0.94	0.97	0.95	0.93	0.95	0.97	0.97	0.93	0.94	0.96	0.96	0.92	0.94	0.95	0.95
	7.6	6.8	11.0	7.0	31.0	14.0	11.4	12.8	39.0	32.0	29.6	30.4	48.0	49.0	50.4	52.6
(3) penguins	0.98	0.98	0.99	0.98	0.97	0.98	0.98	0.98	0.97	0.98	0.98	0.98	0.97	0.98	0.98	0.98
	9.6	10.6	11.4	10.8	21.2	22.4	23.4	29.4	43.2	35.2	37.8	48.2	80.4	79.2	87.0	100.0
(4) wine	0.96	0.98	0.96	0.82	0.97	0.97	0.97	0.98	0.95	0.97	0.96	0.98	0.95	0.97	0.97	0.97
	7.6	8.4	7.8	4.0	12.4	13.2	16.4	17.6	33.0	25.4	38.0	56.0	55.0	50.2	62.8	86.2
(5) wbc-org	0.97	0.98	0.97	0.97	0.97	0.97	0.97	0.97	0.97	0.97	0.98	0.98	0.96	0.97	0.97	0.97
	6.8	9.6	11.2	9.8	12.8	16.8	14.2	16.8	48.0	39.6	33.8	36.4	73.4	77.8	82.4	92.6
(6) bupa	0.59	0.56	0.55	0.55	0.62	0.66	0.66	0.66	0.66	0.66	0.70	0.69	0.67	0.67	0.67	0.67
	7.4	9.0	9.2	8.4	16.8	17.2	14.8	15.0	24.0	24.4	28.2	35.6	53.8	57.4	66.2	95.0
(7) pima	0.75	0.75	0.75	0.72	0.76	0.77	0.75	0.77	0.77	0.76	0.77	0.76	0.76	0.75	0.75	0.74
	15.6	17.0	13.6	6.0	28.2	28.4	23.6	15.0	40.6	51.4	65.2	67.4	54.4	73.2	94.6	135.0
(8) lbps	0.64	0.67	0.71	0.73	0.74	0.76	0.79	0.79	0.77	0.80	0.83	0.86	0.77	0.80	0.83	0.86
	15.4	15.8	19.4	14.6	21.4	30.0	34.2	42.0	40.8	52.8	65.0	80.8	50.6	60.6	74.2	90.6
(9) phoneme	0.81	0.80	0.80	0.75	0.81	0.81	0.83	0.82	0.84	0.85	0.86	0.88	0.86	0.87	0.88	0.91
	17.6	16.6	15.6	9.2	34.4	24.4	28.4	20.8	43.4	49.4	62.8	72.4	79.6	91.2	101.2	143.4
(10) segment	0.71	0.74	0.77	0.80	0.88	0.88	0.89	0.87	0.90	0.90	0.91	0.93	0.93	0.94	0.95	0.96
	8.2	10.4	12.0	11.0	27.2	21.4	22.8	22.0	37.8	31.6	31.8	40.4	68.0	73.2	73.2	91.2
(11) wbc	0.94	0.96	0.96	0.96	0.97	0.97	0.96	0.96	0.96	0.96	0.98	0.98	0.96	0.97	0.97	0.98
	18.0	16.4	21.0	17.4	24.0	27.4	32.6	31.8	35.2	45.0	59.8	72.4	70.0	70.8	93.0	131.6
(12) satimage	0.82	0.82	0.83	0.81	0.88	0.87	0.89	0.87	0.88	0.89	0.90	0.91	0.90	0.91	0.91	0.93
	20.4	22.4	25.2	24.6	43.8	41.6	51.4	45.0	56.6	67.2	85.2	91.8	96.6	103.0	121.0	169.6
(13) digits	0.91	0.92	0.93	0.83	0.95	0.97	0.98	0.97	0.98	0.98	0.99	0.99	0.98	0.99	0.99	0.99
	31.2	28.6	30.6	22.8	71.4	76.4	99.6	87.4	98.2	136.4	179.0	225.6	119.4	158.8	212.8	256.6
(14) dna	0.50	0.59	0.76	0.83	0.60	0.75	0.87	0.98	0.87	0.81	0.95	0.99	0.92	0.94	0.95	0.99
	17.0	23.2	32.8	36.8	46.0	64.8	95.0	173.2	71.2	102.4	158.6	304.0	71.4	103.4	158.6	304.0
(15) sonar	0.62	0.64	0.76	0.75	0.75	0.79	0.75	0.77	0.72	0.75	0.76	0.81	0.78	0.77	0.78	0.82
	9.4	11.6	11.0	12.6	19.8	19.8	27.6	31.8	30.4	35.4	45.2	63.6	36.2	43.2	55.8	78.8
(16) heart	0.50	0.48	0.49	0.51	0.53	0.51	0.54	0.50	0.51	0.54	0.51	0.50	0.56	0.53	0.51	0.50
	4.0	5.6	7.0	7.4	12.0	16.8	24.4	29.6	28.2	30.4	34.0	44.8	38.8	40.8	46.6	61.6
(17) glass	0.51	0.46	0.56	0.52	0.57	0.58	0.62	0.61	0.57	0.63	0.66	0.66	0.66	0.66	0.69	0.67
	10.4	8.8	7.4	9.2	12.6	13.2	12.4	12.2	12.8	16.0	20.2	21.0	27.4	31.2	36.8	40.2
(18) yeast	0.56	0.49	0.49	0.37	0.58	0.58	0.56	0.56	0.60	0.60	0.59	0.59	0.60	0.59	0.59	0.59
	15.2	7.8	9.6	4.6	23.6	31.4	21.4	18.0	58.0	78.2	78.4	78.0	81.8	100.2	127.6	196.4
F_1 #1/#2	1/3	3/1	2/2	1/3	0/5	2/4	2/5	1/3	2/3	2/6	0/4	7/8	2/3	2/6	3/9	8/6
n_h #1/#2	8/2	2/10	1/3	7/3	0/0	0/0	0/0	0/0	0/0	0/0	0/0	0/0	0/0	0/0	0/0	0/0

Table 3. Accuracy and number of hidden neurons n_h of NC-SORBFNN against other single-hidden layer self-organizing approaches.

dataset	GAP [18]	FGAP [35]	AANN [14]	CP-NN [13]	BeeRBF [6]	GAIPSO [32]	NC-SORBFNN C1	C2	C3	C4
(2) iris	–	–	–	0.95	0.96	–	0.97	0.94	0.96	0.95
	–	–	–	21.0	11.6	–	11.0	6.8	30.4	7.0
(4) wine	–	–	–	–	0.97	–	0.98	0.83	0.98	0.83
	–	–	–	–	18.7	–	8.4	4.0	56.0	4.0
(5) wbc-org	–	–	–	–	–	0.96	0.98	0.97	0.98	0.97
	–	–	–	–	–	–	9.6	6.8	36.4	9.8
(6) bupa	–	–	–	–	–	0.68	0.70	0.61	0.69	0.59
	–	–	–	–	–	–	28.2	7.4	35.6	8.4
(7) pima	–	–	–	–	0.77	0.74	0.77	0.73	0.76	0.73
	–	–	–	–	35.8	–	28.4	6.0	67.4	6.0
(8) lbps	–	–	–	–	–	0.81	0.86	0.75	0.86	0.75
	–	–	–	–	–	–	80.8	14.6	80.8	14.6
(9) phoneme	0.77	0.87	–	–	–	–	0.91	0.76	0.88	0.76
	150.4	996.5	–	–	–	–	143.4	9.2	72.4	9.2
(10) segment	0.89	0.96	–	–	–	–	0.96	0.75	0.93	0.80
	42.7	423.9	–	–	–	–	91.2	8.2	40.4	11.0
(11) wbc	–	–	0.67	0.66	0.97	0.96	0.98	0.96	0.98	0.96
	–	–	9.0	22.0	71.2	–	59.8	16.4	72.4	17.4
(12) satimage	–	0.92	–	–	–	–	0.93	0.84	0.91	0.83
	–	1455.0	–	–	–	–	169.6	20.4	91.8	24.6
(14) dna	–	0.93	–	–	–	–	0.99	0.57	0.99	0.83
	–	638.4	–	–	–	–	304.0	17.0	304.0	36.8
(15) sonar	–	–	–	–	0.59	–	0.82	0.66	0.81	0.75
	–	–	–	–	40.0	–	78.8	9.4	63.6	12.6
(16) heart	–	–	–	–	0.53	–	0.60	0.59	0.55	0.58
	–	–	–	–	11.4	–	38.8	4.0	44.8	7.4
(17) glass	–	–	–	–	0.73	–	0.72	0.59	0.70	0.55
	–	–	–	–	18.6	–	36.8	7.4	21.0	9.2
(18) yeast	–	–	0.55	0.52	–	–	0.61	0.44	0.60	0.44
	–	–	9.0	43.0	–	–	58.0	4.6	78.0	4.6

'–' means the value is not reported in the corresponding publication.

6   Conclusions and Future Work

This work has described NC-SORBFNN, a self-adjusting approach for designing RBFNNs with a very reduced number of parameters, i.e. τ_O and τ_I. NC-SORBFNN applies self-organization to the network building process at all steps of the training process, in order to adapt the structure of the network to the specific task at hand. It has been evaluated in the context of multi-class classification by means of a diverse benchmark, comprising up to 18 datasets featuring different imbalance ratios and number of dimensions. NC-SORBFNN has been shown to be able to attain a high level of performance and compares well with alternative self-organizing approaches, being capable of adjusting the structure of the network with, in general, a smaller hidden layer. A natural line of research that stems from this work is linking τ_O and τ_I with the achievable performance, as well as adapt and evaluate NC-SORBFNN for regression tasks.

References

1. Beltran-Perez, C., Wei, H.L., Rubio-Solis, A.: Generalized multiscale RBF networks and the DCT for breast cancer detection. Int. J. Autom. Comput. **17**(1), 55–70 (2020)
2. Chen, Z., Huang, F., Sun, W., Gu, J., Yao, B.: RBF-neural-network-based adaptive robust control for nonlinear bilateral teleoperation manipulators with uncertainty and time delay. IEEE Trans. Mechatron. **25**(2), 906–918 (2020)
3. Dachapak, C., Kanae, S., Yang, Z.J., Wada, K.: Orthogonal least squares for radial basis function network in reproducing Kernel Hilbert Space. IFAC Proceedings Volumes **37**(12), 847–852 (2004)
4. Dehuri, S., Jagadev, A.K., Cho, S.B.: Epileptic seizure identification from electroencephalography signal using DE-RBFNs ensemble. Procedia Comput. Sci. **23**, 84–95 (2013)
5. Fathi, V., Montazer, G.A.: An improvement in RBF learning algorithm based on PSO for real time applications. Neurocomputing **111**, 169–176 (2013)
6. Ferreira Cruz, D.P., Dourado Maia, R., da Silva, L.A., de Castro, L.N.: BeeRBF: a bee-inspired data clustering approach to design RBF neural network classifiers. Neurocomputing **172**, 427–437 (2016)
7. Ganapathy, K., Vaidehi, V., Chandrasekar, J.B.: Optimum steepest descent higher level learning radial basis function network. Expert Syst. Appl. **42**(21), 8064–8077 (2015)
8. Han, H.G., Li Chen, Q., Qiao, J.F.: An efficient self-organizing RBF neural network for water quality prediction. Neural Networks **24**(7), 717–725 (2011)
9. Han, H.G., Guo, Y.N., Qiao, J.F.: Self-organization of a recurrent RBF neural network using an information-oriented algorithm. Neurocomputing **225**, 80–91 (2017)
10. Han, H.G., Lu, W., Hou, Y., Qiao, J.F.: An adaptive-PSO-based self-organizing RBF neural network. IEEE Trans. Neural Netw. Learn. **29**(1), 104–117 (2018)
11. Han, H.G., Ma, M.L., Qiao, J.F.: Accelerated gradient algorithm for RBF neural network. Neurocomputing **441**, 237–247 (2021)
12. Han, H.G., Ma, M.L., Yang, H.Y., Qiao, J.F.: Self-organizing radial basis function neural network using accelerated second-order learning algorithm. Neurocomputing **469**, 1–12 (2022)
13. Han, H.G., Qiao, J.F.: A structure optimisation algorithm for feedforward neural network construction. Neurocomputing **99**, 347–357 (2013)
14. Han, H.G., Wang, L.D., Qiao, J.F.: Efficient self-organizing multilayer neural network for nonlinear system modeling. Neural Netw. **43**, 22–32 (2013)
15. Han, H.G., Wu, X., Zhang, L., Tian, Y., Qiao, J.F.: Self-organizing RBF neural network using an adaptive gradient multiobjective particle swarm optimization. IEEE Trans. Cybern. **49**(1), 69–82 (2019)

16. Hassanzadeh, Z., Kompany-Zareh, M., Ghavami, R., Gholami, S., Malek-Khatabi, A.: Combining radial basis function neural network with genetic algorithm to QSPR modeling of adsorption on multi-walled carbon nanotubes surface. J. Mol. Struct. **1098**, 191–198 (2015)
17. Henneron, T., Pierquin, A., Clénet, S.: Surrogate model based on the POD combined with the RBF interpolation of nonlinear magnetostatic FE model. IEEE Trans. Magn. **56**(1), 1–4 (2020)
18. Huang, G.B., Saratchandran, P., Sundararajan, N.: An efficient sequential learning algorithm for growing and pruning RBF (GAP-RBF) networks. IEEE Trans. Syst. Man Cybern. B **34**(6), 2284–2292 (2004)
19. Huang, G.B., Saratchandran, P., Sundararajan, N.: A generalized growing and pruning RBF (GGAP-RBF) neural network for function approximation. IEEE Trans. Neural Networks **16**(1), 57–67 (2005)
20. Kashiwao, T., Nakayama, K., Ando, S., Ikeda, K., Lee, M., Bahadori, A.: A neural network-based local rainfall prediction system using meteorological data on the Internet: a case study using data from the Japan meteorological agency. Appl. Soft Comput. **56**, 317–330 (2017)
21. Keller, J.M., Liu, D., Fogel, D.B.: Fundamentals of Computational Intelligence: Neural Networks, Fuzzy Systems, and Evolutionary Computation. Wiley - IEEE Press, New York (2016)
22. Kim, J.S., Jung, S.: Implementation of the RBF neural chip with the back-propagation algorithm for on-line learning. Appl. Soft Comput. **29**, 233–244 (2015)
23. Kohonen, T.: Self-Organizing Maps, vol. 30 in Series in Information Sciences, 3rd edn. Springer, Cham (2001). https://doi.org/10.1007/978-3-642-56927-2
24. Li, Z.Q., et al.: A proposed self-organizing radial basis function network for aero-engine thrust estimation. Aerosp. Sci. Technol. **87**, 167–177 (2019)
25. Liu, T., Chen, S., Liang, S., Gan, S., Harris, C.J.: Fast adaptive gradient RBF networks for online learning of nonstationary time series. IEEE Trans. Signal Process. **68**, 2015–2030 (2020)
26. Moody, J., Darken, C.J.: Fast learning in networks of locally-tuned processing units. Neural Comput. **1**(2), 281–294 (1989)
27. Perez-Godoy, M.D., Rivera, A.J., Carmona, C.J., del Jesus, M.J.: Training algorithms for Radial Basis Function Networks to tackle learning processes with imbalanced data-sets. Appl. Soft Comput. **25**, 26–39 (2014)
28. Shen, W., Guo, X., Wu, C., Wu, D.: Forecasting stock indices using radial basis function neural networks optimized by artificial fish swarm algorithm. Knowl.-Based Syst. **24**(3), 378–385 (2011)
29. Smith, J.S., Wu, B., Wilamowski, B.M.: Neural network training with Levenberg–Marquardt and adaptable weight compression. IEEE Trans. Neural Netw. Learn. **30**(2), 580–587 (2019)
30. Wang, S., Yao, X.: Multiclass imbalance problems: analysis and potential solutions. IEEE Trans. Syst. Man Cybern. B **42**(4), 1119–1130 (2012)
31. Wang, Y., Wang, T., Yang, X., Yang, J.: Gradient Descent-Barzilai Borwein-based neural network tracking control for nonlinear systems with unknown dynamics. IEEE Trans. Neural Netw. Learn. **34**(1), 305–315 (2023)
32. Wu, J.C., Shen, J., Xu, M., Liu, F.S.: An evolutionary self-organizing cost-sensitive radial basis function neural network to deal with imbalanced data in medical diagnosis. Int. J. Comput. Intell. Syst. **13**(1), 1608–1618 (2020)
33. Xie, Y., Yu, J., Xie, S., Huang, T., Gui, W.: On-line prediction of ferrous ion concentration in goethite process based on self-adjusting structure RBF neural network. Neural Netw. **116**, 1–10 (2019)

34. Yang, Y., Wang, P., Gao, X.: A novel radial basis function neural network with high generalization performance for nonlinear process modelling. Processes **10**(1), 140 (2022)
35. Zhang, R., Huang, G.B., Sundararajan, N., Saratchandran, P.: Improved GAP-RBF network for classification problems. Neurocomputing **70**(16), 3011–3018 (2007)
36. Zhao, J., Wei, H., Zhang, C., Li, W., Guo, W., Zhang, K.: Natural gradient learning algorithms for RBF networks. Neural Comput. **27**(2), 481–505 (2015)
37. Zheng, D., Jung, W., Kim, S.: RBFNN design based on modified nearest neighbor clustering algorithm for path tracking control. Sensors **21**(24), 8349 (2021)

Self-organising Neural Discrete Representation Learning à la Kohonen

Kazuki Irie[1]([✉]), Róbert Csordás[2]([✉]), and Jürgen Schmidhuber[3,4]

[1] Center for Brain Science, Harvard University, Cambridge, MA, USA
kirie@fas.harvard.edu
[2] Stanford University, Stanford, CA, USA
rcsordas@stanford.edu
[3] The Swiss AI Lab, IDSIA, USI & SUPSI, Lugano, Switzerland
juergen@idsia.ch
[4] AI Initiative, King Abdullah University of Science and Technology (KAUST),
Thuwal, Saudi Arabia

Abstract. Unsupervised learning of discrete representations in neural networks (NNs) from continuous ones is essential for many modern applications. Vector Quantisation (VQ) has become popular for this, in particular in the context of generative models, such as Variational Auto-Encoders (VAEs), where the exponential moving average-based VQ (EMA-VQ) algorithm is often used. Here, we study an alternative VQ algorithm based on Kohonen's learning rule for the Self-Organising Map (KSOM; 1982). EMA-VQ is a special case of KSOM. KSOM is known to offer two potential benefits: empirically, it converges faster than EMA-VQ, and KSOM-generated discrete representations form a topological structure on the grid whose nodes are the discrete symbols, resulting in an artificial version of the brain's *topographic map*. We revisit these properties by using KSOM in VQ-VAEs for image processing. In our experiments, the speed-up compared to *well-configured* EMA-VQ is only observable at the beginning of training, but KSOM is generally much more robust, e.g., w.r.t. the choice of initialisation schemes (Our code is public: https://github.com/IDSIA/kohonen-vae. The full version with an appendix can be found at: https://arxiv.org/abs/2302.07950).

Keywords: self-organizing maps · Kohonen maps · vector quantisation · VQ-VAE · discrete representation learning

1 Introduction

Internal representations in artificial neural networks (NNs) are continuous-valued vectors. As such, they are rarely identical in different contexts. In many scenarios, however, it is natural and desirable to treat some of these non-identical but similar vectors as representing a common discrete symbol from a fixed-size codebook/lexicon shared across various contexts [19,43]. Such *discrete representations* would allow us to express and manipulate hidden representations of NNs

---

K. Irie and R. Csordás—Equal contribution. Work done at IDSIA.

M. Wand et al. (Eds.): ICANN 2024, LNCS 15016, pp. 343–362, 2024.
https://doi.org/10.1007/978-3-031-72332-2_23

using a set of symbols. Unsupervised learning of such discrete representations is motivated in various contexts. For example, in certain algorithmic or reasoning tasks (e.g., [10,20,31]), learning of such representations may be a key for generalisation, since intermediate results in such tasks are inherently discrete.

Recently, discrete representation learning via *Vector Quantisation* (VQ) has become popular for practical reasons too. The Vector Quantised-Variational Auto-Encoders (VQ-VAEs) by van den Oord et al. [43] use VQ as a preprocessing step to *tokenise* high-dimensional data, such as images, i.e., to represent an image as a sequence of discrete symbols. The resulting sequence can then be processed by a powerful sequence processor, e.g., a Transformer variant [49,50,54]. Similar techniques have been applied to other kinds of data such as video [55,58] and audio [3,5,13,52]). Today, many large-scale text-to-image systems such as *DALL-E* [45], *Parti* [61], or *Latent Diffusion Models* [47], also use VQ in one way or another.

Here we study the learning rules of Kohonen Self-Organising Maps or Kohonen Maps [26,28] as the VQ algorithm for discrete representation learning in NNs. For shorthand, we refer to the corresponding algorithm as *KSOM* (reviewed in Sect. 2). In fact, KSOM is a classic VQ algorithm [41], and the exponential moving average-based VQ (EMA-VQ; [22,43,46,48]) commonly used today is a special case thereof (Sect. 3). KSOM is known to offer two potential benefits over EMA-VQ. First, KSOM is empirically reported to perform faster VQ (see, e.g., [11]). Second, discrete representations learned by KSOM form a *topological* structure in the pre-specified *grid* whose nodes represent the discrete symbols from the codebook. Such a grid is typically one- or two-dimensional, and symbols that are spatially close to each other on the grid represent features that are close to each other in the original input space. This *topological mapping* property has helped practitioners in certain applications to visualise/interpret their data (see, e.g. [51]). While such a property is arguably of limited importance in today's deep learning, Kohonen's algorithm is specifically designed to achieve this property which is known in the brain as *topographic organisation*. KSOM allows to naturally achieve an artificial version thereof, as a by-product of the VQ algorithm.

We explore these properties in modern NNs by using KSOM as the VQ algorithm in VQ-VAEs [43,46] for image processing. Importantly, we also revisit the configuration details of the baseline EMA-VQ (e.g., initialisation of EMAs). We show that proper configurations are crucial for EMA-VQ to optimally perform, while KSOM is robust, and performs well in all cases.

2   Background: Kohonen Maps

Here we provide a brief review of Kohonen's Self-Organising Maps (KSOMs).

2.1   (Online) Algorithm

Teuvo Kohonen (1934–2021)'s Self-Organising Map [26] is an unsupervised learning/clustering algorithm which achieves both *vector quantisation* (VQ) and *topological mapping* (Sect. 2.3). Let T, K, d_{in} and t denote positive integers. The algorithm requires to define a *distance function* $\delta : \mathbb{R}^{d_{\text{in}}} \times \mathbb{R}^{d_{\text{in}}} \rightarrow \mathbb{R}_{\geq 0}$ as well as a *neighbourhood matrix* $\boldsymbol{A} \in \mathbb{R}^{K \times K}$ with $0 \leq \boldsymbol{A}_{i,j} \leq 1$ for all $i, j \in \{1, ..., K\}$, whose roles are specified later. We consider T input vectors[1] $(\boldsymbol{x}_1, ..., \boldsymbol{x}_T)$ with $\boldsymbol{x}_t \in \mathbb{R}^{d_{\text{in}}}$ for $t \in \{1, ..., T\}$, and a weight matrix $\boldsymbol{W} \in \mathbb{R}^{K \times d_{\text{in}}}$ which we describe as a list of K weight vectors $(\boldsymbol{w}_1, ..., \boldsymbol{w}_K) = \boldsymbol{W}^{\mathsf{T}}$ with $\boldsymbol{w}_k \in \mathbb{R}^{d_{\text{in}}}$ for $k \in \{1, ..., K\}$ representing a *codebook* of size K. The algorithm clusters input vectors into K clusters where the prototype (or the centroid) of cluster $k \in \{1, ..., K\}$ is $\boldsymbol{w}_k$. At the beginning, these weight vectors are randomly initialised as $(\boldsymbol{w}_1^{(0)}, ..., \boldsymbol{w}_K^{(0)})$ where the super-script denotes the iteration step. The KSOM algorithm learns these weight vectors iteratively as follows.

For each step $t \in \{1, ..., T\}$, we process an input $\boldsymbol{x}_t$ by computing the index k^* (typically called *best matching unit*) of the weight vector that is the closest to the input $\boldsymbol{x}_t$ according to δ, i.e.,

$$k^* = \underset{1 \leq k \leq K}{\arg\min}\, \delta(\boldsymbol{x}_t, \boldsymbol{w}_k^{(t-1)}) \tag{1}$$

then the weights are updated; for all $k \in \{1, ..., K\}$,

$$\boldsymbol{w}_k^{(t)} = \boldsymbol{w}_k^{(t-1)} + \beta \boldsymbol{A}_{k^*,k}^{(t-1)}(\boldsymbol{x}_t - \boldsymbol{w}_k^{(t-1)}) \tag{2}$$

where $\beta \in \mathbb{R}_{>0}$ is the learning rate, and the super-script added to $\boldsymbol{A}_{k^*,k}^{(t)}$ indicates that it also changes over time.

Now, we need to specify the distance function δ in Eq. 1 and the neighbourhood matrix $\boldsymbol{A}^{(t)}$ in Eq. 2.

Distance function δ. A typical choice for δ is the Euclidean distance which we also use in all our experiments. Strictly speaking, δ does not have to be a metric; any kind of (dis)similarity function can be used, e.g., negative dot product is also a common choice (*dot-SOM* [28]).

Grid of Codebook Indices. In order to define the neighbourhood matrix $\boldsymbol{A}^{(t)}$, we first need to define a *grid* or lattice of the codebook indices on which *neighbourhoods* are defined. Let D denote a positive integer. We define a map $\mathcal{G} : \{1, ..., K\} \rightarrow \mathbb{N}^D$ which maps each codebook index $k \in \{1, ..., K\}$ to its Cartesian coordinates $\mathcal{G}(k) \in \mathbb{N}^D$ in the D-dimensional space representing the grid in question. Typically, the grid is 1D or 2D ($D = 1$ or 2). In the 1D case, the original index $k \in \{1, ..., K\}$ and its coordinates on the map $\mathcal{G}(k)$ are the same: $\mathcal{G}(k) = k$. In the 2D case, $\mathcal{G}(k)$ corresponds to the Cartesian (x, y)-coordinates on a 2D-rectangular grid formed by K nodes (assuming that K is chosen such that this is possible). Once $\mathcal{G}$ is defined, one can measure the "distance" between two codebook indices $i, j \in \{1, ..., K\}$ as their Euclidean distance on the grid, i.e., $\|\mathcal{G}(i) - \mathcal{G}(j)\|$. This finally allows us to define the neighbourhood: given a pre-defined threshold distance $d(\mathcal{G}) \in \mathbb{R}_{\geq 0}$, for $i, j \in \{1, ..., K\}$, i is within the neighbourhood of j on the map $\mathcal{G}$ if and only if $\|\mathcal{G}(i) - \mathcal{G}(j)\| \leq d(\mathcal{G})$.

---

[1] Here we use T as both the number of inputs and iterations.

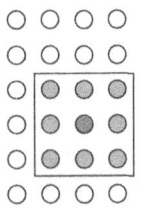
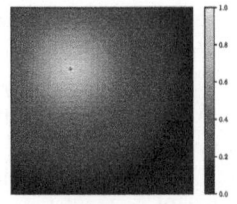
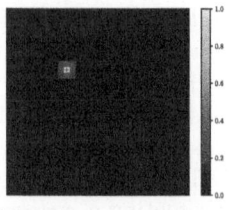

(a) Illustration of a hard neighborhood.

(b) Gaussian neighborhood: Early training.

(c) Gaussian neighborhood: Mid training.

Fig. 1. (a) Illustration of *hard* neighbourhoods (Eq. 3) in the 2D case. This is a 6×4 grid with $K = 24$ nodes. Considering the left-bottom corner node as the origin (0, 0), the eight neighbours of the node (2, 2) are highlighted. (b) and (c): Illustration of *Gaussian* neighbourhoods in the 2D case with shrinking (Eq. 5) at two different stages of training.

Neighbourhood Matrix $A^{(t)}$. In Eq. 2, the coefficient of the neighbourhood matrix $A^{(t)}_{k^*,k}$ has the role of adapting the weights/strengths of updates for each cluster k according to its distance from the best matching unit k^* on the map $\mathcal{G}$: as can be seen in Eq. 2, $\beta A^{(t)}_{k^*,k}$ is the effective learning rate for the update. Essentially, the best matching unit obtains the full update $A^{(t)}_{k^*,k^*} = 1$, while the updates for all others ($k \neq k^*$) are scaled down by $0 \leq A^{(t)}_{k^*,k} \leq 1$. In practice, there are various ways to define such an $A^{(t)}$. Here we focus on two variants: *hard* and *Gaussian* neighbourhoods.

Let us first define the *hard* variant *without* dependency on the time index t. In the *hard* variant, for all $i \in \{1, ..., K\}$, $A_{i,i} = 1$, and for all i, j such that $i \neq j$,

$$A_{i,j} = \begin{cases} 1 & \text{if } \|\mathcal{G}(i) - \mathcal{G}(j)\| \leq d(\mathcal{G}) \\ 0 & \text{otherwise} \end{cases} \tag{3}$$

In the 1D case, setting $d(\mathcal{G}) = 1$ yields two neighbour indices. Similarly, in the 2D case, $d(\mathcal{G}) = \sqrt{2}$ defines the eight surrounding nodes as the neighbours, which are illustrated in Fig. 1a.

In practice, we introduce an extra dependency on time t with the goal of *shrinking* the neighbourhood over time. For all $i \in \{1, ..., K\}$, $A^{(t)}_{i,i} = 1$, and for all i, j such that $i \neq j$,

$$A^{(t)}_{i,j} = \begin{cases} 1/(1 + t * \tau) & \text{if } \|\mathcal{G}(i) - \mathcal{G}(j)\| \leq d(\mathcal{G}) \\ 0 & \text{otherwise} \end{cases} \tag{4}$$

where $\tau \in \mathbb{R}_{>0}$ is a hyper-parameter representing the *shrinking step* which controls the speed of shrinking: larger τ implies faster shrinking. Shrinking in KSOM is important to obtain good performance at convergence [28].

In the alternative, *Gaussian* variant, $\boldsymbol{A}_{i,j}^{(t)}$ is expressed as an exponential function of the distance between indices on $\mathcal{G}$:

$$\boldsymbol{A}_{i,j}^{(t)} = \exp(-\|\mathcal{G}(i) - \mathcal{G}(j)\|^2/\sigma(t)^2) \tag{5}$$

where for $\sigma(t) \in \mathbb{R}$ we take $\sigma(t)^2 = 1/(1+t*\tau)$. Note that, also in this case, $\boldsymbol{A}^{(t)}$ reduces to an identity matrix when $t \to +\infty$. Figure 1 provides an illustration.

Relation to Hebbian Learning. We note that the online algorithm of Eqs. 1–2 can be also interpreted (see also [59] on this relation) as a variant of Hebbian learning [17]. Using the dot product-based similarity function for δ in Eq. 1, and by defining hardmax : $\mathbb{R}^K \to \mathbb{R}^K$ as the function that outputs 1 for the largest entry of the input vector, and 0 for all others, we can express Eqs. 1–2 in a fully matrix-form by defining a one-hot vector $\boldsymbol{y}_t \in \mathbb{R}^K$,

$$\boldsymbol{y}_t = \text{hardmax}(\boldsymbol{W}_{t-1}\boldsymbol{x}_t) \tag{6}$$

$$\boldsymbol{W}_t = \boldsymbol{W}_{t-1} + (\beta\boldsymbol{A}^{(t-1)}\boldsymbol{y}_t) \otimes (\boldsymbol{x}_t - \boldsymbol{W}_{t-1}^{\mathsf{T}}\boldsymbol{y}_t) \tag{7}$$

where $\otimes$ denotes outer product. With the neighbourhood reduced to zero, this is essentially the *winner-take-all* Hebbian learning. While this relation is not central to this work, we come back to this when we discuss the differentiable version of KSOM later in Sect. 5.

In passing, we also note that the last term in Eqs. 2 and 7 corresponds to Oja's [42] *forgetting term* which is not part of the original 1982 algorithm [26] but has been added later (see, e.g., [28]).

2.2   Batch Algorithm and Relation to K-Means

The algorithm described above is an *online* algorithm which updates weights after every input. The *batch* version thereof ([27,28]; see also [9]), which takes into account all data points $(\boldsymbol{x}_1, ..., \boldsymbol{x}_T)$ for a single update of $\{\boldsymbol{w}_1, ..., \boldsymbol{w}_K\}$, can be defined as follows.

At each iteration step t, we compute the best matching unit for each input x_i for all $i \in \{1, .., T\}$ (we now use the sub-script i to index data points not to confuse it with the iteration index t). Each input x_i is a *member* of one of the K clusters. The results can be summarised for each cluster $k \in \{1, .., K\}$ as the set $\mathcal{C}_k^{(t)}$ containing indices of its members at step t. We denote its cardinality by $|\mathcal{C}_k^{(t)}|$. The batch algorithm updates the weight vector $\boldsymbol{w}_k^{(t)}$ for cluster k as the average of its members and their neighbours weighted by the neighbourhood coefficients. That is, $\boldsymbol{w}_k^{(t)}$ is computed as the quotient of the sum $\boldsymbol{m}_k^{(t)}$ of all inputs belonging to the corresponding cluster k and their neighbours weighted by the neighbourhood coefficients, and the corresponding weighted count $N_k^{(t)}$, i.e.,

$$\boldsymbol{m}_k^{(t)} = \sum_{j=1}^{K} \boldsymbol{A}_{j,k}^{(t-1)} \sum_{i \in \mathcal{C}_j^{(t)}} \boldsymbol{x}_i \tag{8}$$

$$N_k^{(t)} = \sum_{j=1}^{K} A_{j,k}^{(t-1)} |\mathcal{C}_j^{(t)}| \tag{9}$$

$$w_k^{(t)} = m_k^{(t)} / N_k^{(t)} \tag{10}$$

Remark 1 (Relation to K-means). In the case where the *neighbourhood is reduced to zero*, i.e., for all t, $A_{i,i}^{(t)} = 1$ and for all i, j such that $i \neq j$, $A_{i,j}^{(t)} = 0$, Eqs. 8–10 reduce to the standard *K-means* algorithm [35]. Similarly, in such a case, Eqs. 1–2 reduce to an *online K-means* algorithm [36].

For deep learning applications, the algorithm needs to be both *online* and *mini-batch*; we discuss the corresponding extensions in Sect. 3.

2.3 Topographical Maps in the Brain as Motivation

Above we describe how KSOM performs clustering, i.e., *vector quantisation*. Here we discuss another property of this algorithm which is *topological mapping*.

It is known that there are multiple levels of *topographical maps* in the brain, e.g., different regions of the brain specialise to different types of sensory inputs (vision, audio, touch, etc.), and e.g., within the somatosensory part, regions that are responsible for different parts of the body are *ordered* according to the anatomical order in the body. A famous illustration of this is the "sensory homunculus."

The design of KSOM is inspired by such topographical maps. Many have proposed computational mechanisms to achieve such a property in the 1970s/80 s [2,37,38,56,57]. Kohonen [26] achieves this by introducing the concept of neighbourhoods between the (output) neurons. In the algorithm above (Sect. 2.1), all output neurons first compete against each other (Eq. 1) to yield a winner neuron. Then, the update is distributed to neurons that are spatially close to the winner through the coefficients of the neighbourhood matrix (Eqs. 2 and 8). As a result, clusters whose indices are spatially close on the grid are encouraged to store inputs that are close to each other in the feature space.

This is an unconventional feature for artificial NNs, since unlike the biological ones, artificial NNs do not have any physical constraints; there is no geometry nor distance between neurons. KSOM's neighbourhoods introduce such a structure. From the machine learning perspective, the resulting topological ordering has limited practical benefits. Even if it may potentially facilitate interpretation via direct visualisation, other embedding visualisation tools could fit the bill equally well. From the neuroscience perspective, however, it may be a property that contributes in filling the gap between artificial NNs and the biological ones (see also [8]).

3   Alternative VQ in VQ-VAEs

The general idea of this work is to replace the VQ algorithm used in standard VQ-VAEs by Kohonen's algorithm (Sect. 2). While the method can be applied to various data modalities, we focus on image processing as a representative example.

3.1   Background: VQ-VAEs

Let $d_{\mathrm{in}}, d_{\mathrm{emb}}, N, K$ be positive integers. A VQ-VAE [43] consists of an encoder NN, $\mathtt{Enc} : \mathbb{R}^{d_{\mathrm{in}}} \rightarrow \mathbb{R}^{N \times d_{\mathrm{emb}}}$, a decoder NN, $\mathtt{Dec} : \mathbb{R}^{N \times d_{\mathrm{emb}}} \rightarrow \mathbb{R}^{d_{\mathrm{in}}}$, and a codebook of size K whose weights are $(\boldsymbol{w}_1, ..., \boldsymbol{w}_K)$ with $\boldsymbol{w}_k \in \mathbb{R}^{d_{\mathrm{emb}}}$ for $k \in \{1, ..., K\}$. The encoder transforms an input $\boldsymbol{x} \in \mathbb{R}^{d_{\mathrm{in}}}$ to a sequence of N embedding vectors $(\boldsymbol{e}_1, ..., \boldsymbol{e}_N) = \boldsymbol{E}$ with $\boldsymbol{e}_i \in \mathbb{R}^{d_{\mathrm{emb}}}$ for $i \in \{1, ..., N\}$. Each of these embeddings is quantised to yield $(\mathtt{VQ}(\boldsymbol{e}_1), ..., \mathtt{VQ}(\boldsymbol{e}_N)) = \boldsymbol{E}'$ with $\mathtt{VQ}(\boldsymbol{e}_i) \in \mathbb{R}^{d_{\mathrm{emb}}}$ for $i \in \{1, ..., N\}$ where $\mathtt{VQ}$ denotes the VQ operation. The decoder transforms the quantised embeddings $\boldsymbol{E}'$ to a reconstruction of the original input $\hat{\boldsymbol{x}} \in \mathbb{R}^{d_{\mathrm{in}}}$. The corresponding operations can be expressed as follows.

$$(\boldsymbol{e}_1, ..., \boldsymbol{e}_N) = \mathtt{Enc}(\boldsymbol{x}) \tag{11}$$

$$\text{For all } i \in \{1, ..., N\}, \; k_i^* = \operatorname*{arg\,min}_{1 \le k \le K} \|\boldsymbol{e}_i - \boldsymbol{w}_k\|_2 \tag{12}$$

$$\boldsymbol{E}' = (\mathtt{VQ}(\boldsymbol{e}_1), ..., \mathtt{VQ}(\boldsymbol{e}_N)) = (\boldsymbol{w}_{k_1^*}, ..., \boldsymbol{w}_{k_N^*}) \tag{13}$$

$$\hat{\boldsymbol{x}} = \mathtt{Dec}(\boldsymbol{E}') \tag{14}$$

The parameters of the encoder and decoder are trained to minimise:

$$\|\boldsymbol{x} - \hat{\boldsymbol{x}}\|_2^2 + \lambda \frac{1}{N} \sum_{i=1}^{N} \|\mathtt{sg}(\boldsymbol{w}_{k_i^*}) - \boldsymbol{e}_i\|_2^2 \tag{15}$$

where the first term is the reconstruction loss, and the second term is the so-called *commitment loss*, weighted by a hyper-parameter $\lambda \in \mathbb{R}_{>0}$, which encourages the encoder to output embeddings that are close to their quantised counterparts ($\mathtt{sg}$ denotes "stop gradient" operation to prevent gradients to propagate through $\boldsymbol{w}_{k_i^*}$). As noted in [43], by assuming a uniform prior over the discrete latents, the standard KL term of the VAE loss [25] can be omitted as a constant. In the reconstruction term, as the quantisation operation is non-differentiable, the straight-through estimator [4,18] is used, i.e., gradients are directly copied from the decoder input to the encoder output.

The weights of the codebook prototypes $(\boldsymbol{w}_1, ..., \boldsymbol{w}_K)$ are trained by a variant of online mini-batch K-means algorithm which keeps track of exponential moving averages (EMAs) of two quantities for each cluster k: the sum of updates $m_k^{(t)}$, and the count of members in the cluster $N_k^{(t)}$. Their quotient yields the estimate of the weights at step t. Using the same notation $\mathcal{C}_k^{(t)}$ as in Sect. 2.2 for the set of encoder output indices (ranging from 1 to $N * B$ where B is a positive integer denoting the batch size) that are members of cluster k at step t, and by denoting encoder outputs in the batch as $(e_1, ..., e_{N*B})$, it yields:

$$m_k^{(t)} = (1 - \beta)m_k^{(t-1)} + \beta \sum_{i \in \mathcal{C}_k^{(t)}} e_i^{(t)} \tag{16}$$

$$N_k^{(t)} = (1 - \beta)N_k^{(t)} + \beta|\mathcal{C}_k^{(t)}| \tag{17}$$

$$w_k^{(t)} = m_k^{(t)}/N_k^{(t)} \tag{18}$$

where $(1 - \beta)$ is the EMA decay typically set to 0.99 (i.e., $\beta = 0.01$). For shorthand, we refer to this algorithm as EMA-VQ.

While it is also possible to train these codebook parameters through gradient descent by introducing an extra term in the loss of Eq. 15 (batching is omitted for consistency): $1/N \sum_{i=1}^{N} \|w_{k_i^*} - \mathrm{sg}(e_i)\|_2^2$, van den Oord et al. [43] recommend the EMA-based approach above, and many later works [44,46] follow this practice, while in VQ-GANs [14,60], the gradient-based approach is also commonly used.

3.2   Kohonen-VAEs

We use KSOM (Sect. 2) as the VQ algorithm to learn the codebook weights in VQ-VAEs (Sect. 3.1). Essentially, we replace the EMA-VQ algorithm of Eqs. 16–18 by an *online mini-batch* version of KSOM. Such an algorithm can be obtained by introducing exponential moving averaging into the batch version of KSOM (Eqs. 8–10) with a decay of $1 - \beta$. That is, we keep track of EMAs of both the weighted sum of the updates ($m_k^{(t)}$; Eq. 8) and the weighted count of members ($N_k^{(t)}$; Eq. 9) for each cluster $k \in \{1, ..., K\}$ (where weights are the neighbourhood coefficients). Their quotient yields the estimate of the weights of codebook prototypes at step t. Using the same notations as in Sect. 3.1, i.e., $\mathcal{C}_k^{(t)}$ denotes the set of indices of encoder outputs that are members of cluster k at step t, and $(e_1, ..., e_{N*B})$ denotes the encoder outputs in the batch, it yields:

$$m_k^{(t)} = (1 - \beta)m_k^{(t-1)} + \beta \sum_{j=1}^{K} A_{j,k}^{(t-1)} \sum_{i \in \mathcal{C}_j^{(t)}} e_i^{(t)} \tag{19}$$

$$N_k^{(t)} = (1 - \beta)N_k^{(t)} + \beta \sum_{j=1}^{K} A_{j,k}^{(t-1)}|\mathcal{C}_j^{(t)}| \tag{20}$$

$$w_k^{(t)} = m_k^{(t)}/N_k^{(t)} \tag{21}$$

All other aspects are kept the same as in the basic VQ-VAE (Sect. 3.1). We refer to this approach as *Kohonen-VAE*.

Remark 2 (Relation to EMA-VQ). If the neighbourhood is reduced to zero (see, Remark 1), this approach falls back to the standard EMA-VQ VAEs (Sect. 3.1).

3.3   Initialisation and Updates of EMAs

Initialisation. Both the basic EMA-VQ (Sect. 3.1) and KSOM (Sect. 3.2) require to initialise two EMAs: $\boldsymbol{m}_k^{(0)}$ (Eq. 16 and 19) and $N_k^{(0)}$ (Eq. 17 and 20). This is an important detail which is omitted in the common description of VQ-VAEs. Standard public implementations of VQ-VAEs, including the official one by van den Oord et al. [43], initialise $\boldsymbol{m}_k^{(0)}$ by a random Gaussian vector, and $N_k^{(0)}$ by 0. The latter is problematic for the following reason.

In fact, standard implementations apply the updates of Eqs. 16, 18 to all clusters including those that have no members in the batch—later, we show that this is another important detail for EMA-VQ. Since $N_k^{(0)}$ is initialised with 0, $N_k^{(1)}$ remains 0 at the first iteration for the clusters with no members in the first batch. To avoid division by zero, smoothing over counts $(N_1^{(t)}, ..., N_K^{(t)})$—typically additive smoothing; see, e.g., [6]—is applied to obtain smoothed counts $(\tilde{N}_1^{(t)}, ..., \tilde{N}_K^{(t)})$ such that $N_k^{(1)} = 0$ becomes $\tilde{N}_k^{(1)} = \epsilon$ where typically $\epsilon \sim 10^{-5}$. While this allows to avoid division by zero, the resulting multiplication of $\boldsymbol{m}_k^{(1)}$ by $1/\epsilon \sim 10^5$ in Eq. 18 also seems unreasonable. Our experiments (Sect. 4.1) show that this effectively results in certain sub-optimality in training. Instead, we propose $N_k^{(0)} = 1$ initialisation, which is consistent with non-zero initialisation of $\boldsymbol{m}_k^{(0)}$.

Updating EMAs of Clusters without Members. Given the update equations above (Eqs. 16–18 and 19–21), there remains the question whether the EMAs of clusters that have no members in the batch should be updated. As mentioned already in the previous paragraph, standard implementations *update all* EMAs. We conduct the corresponding ablation study in Section 4.1, and empirically show that, indeed, it is crucial to update all EMAs in the baseline EMA-VQ algorithm used in VQ-VAEs to achieve optimal performance.

Next, we'll also show that, unlike the standard EMA-VQ which is sensitive to these configurations, KSOM is robust, and performs well under any configurations.

4   Experiments

The goal of our experiments is to revisit the properties of KSOM when integrated into VQ-VAEs. In particular, we demonstrate its robustness, and analyse learned representations. Before that, we start with showing the sensitivity of the standard EMA-VQ w.r.t. various configuration details.

4.1   Sensitivity of the Baseline EMA-VQ

We first present two sets of experiments for the baseline EMA-VQ, which reveal its sensitivity to initialisation and update schemes of EMAs (discussed in Sect. 3.3).

Improving Initialisation. We start with evaluating $N_k^{(0)} = 1$ initialisation (instead of the standard $N_k^{(0)} = 0$) discussed in Sect. 3.3. Figure 2 shows the evolution of the validation reconstruction loss of VQ-VAEs trained with EMA-VQ on CIFAR-10 [29] for two runs of 10 seeds each with $N_k^{(0)} = 0$ (denoted by N=0/run1 and N=0/run2), and one run of 10 seeds with $N_k^{(0)} = 1$ (denoted by N=1). In both runs with N=0, we observe a plateau at the beginning of training. The variability of the results is also high: the performance of one of them (N=0/run1; the *blue* curve) remains above that of the N=1 case, even after the plateau, while the other one (N=0/run2, the *orange* curve) successfully reaches the performance of N=1. The final/best validation reconstruction losses $(1e^{-3})$ achieved by the respective configurations are: 148.7 ± 299.7 vs. 52.1 ± 0.6. In contrast, such a variability was not observed with $N_k^{(0)} = 1$.

Updating Clusters without Members. Now we also evaluate the effect of updating EMAs for the clusters that have no members in the batch. Table 1 shows the corresponding results. In all cases (with or without $N_k^{(0)} = 1$ initialisation), updating all clusters including those that have no member is crucial for good performance of the baseline EMA-VQ.

4.2   Reconstruction Performance and Convergence Speed

We evaluate the reconstruction performance and convergence speed of VQ-VAEs trained with KSOM using three datasets: CIFAR-10 [29], ImageNet [12], and a mixture of CelebA-HQ [23] and Animal Faces HQ (AFHQ; [7]). We use the basic VQ-VAE architecture [43] for CIFAR-10 and its extension VQ-VAE-2 [46] for ImageNet and CelebA-HQ/AFHQ without architectural modifications.[2]

Comparison to Optimised EMA-VQ. We first compare models trained with KSOM with those trained using carefully configured EMA-VQ (Sect. 4.1).

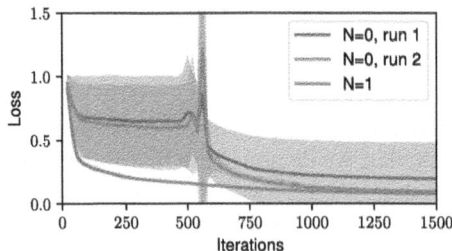

Fig. 2. Evolution of validation reconstruction loss on CIFAR-10 for baseline EMA-VQ with different initialisations $N_k^{(0)}$ in Eq. 9.

---

[2] Further experimental details are available through the link on page 1.

Table 1. Sensitivity of the baseline EMA-VQ. VQ-VAEs trained with EMA-VQ on CIFAR-10. "$N_k^{(0)}$" indicates its initialisation with 0 or 1. "Update-0" indicates whether to update clusters that have 0 members in the batch. "# Steps" is the number of training steps needed to achieve +10% of the final loss. Mean and std are computed using 10 seeds. For the $N_k^{(0)} = 0$ case, there is a high variability among results even with 10 seeds, as we report in Sect. 4.1: here, we report the result obtained using 10 *good* seeds.

$N_k^{(0)}$	Update-0	Loss ($1e^{-3}$)	# Steps ($1e^3$)
0	No	148.8 ± 11.0	13.5 ± 0.7
1	No	68.3 ± 0.8	16.3 ± 1.4
0	Yes	52.1 ± 0.6	6.8 ± 0.6
1	Yes	$\mathbf{51.9 \pm 0.2}$	$\mathbf{5.4 \pm 0.5}$

Table 2 summarises the results. We first observe that all methods achieve a similar validation reconstruction loss, with slight improvements obtained by KSOM over the baseline on CelebA-HQ/AFHQ. To compare the "speed of convergence," we measure the number of steps needed by each algorithm to achieve +10% and +20% of their final performance. Here "steps" correspond to the number of updates, and the batch size is the same for all methods. We observe that, indeed, KSOM tends to be faster than the basic EMA-VQ at the beginning of training, as can be seen in the column +20% (especially for the hard variant on CelebA-HQ/AFHQ). However, the baseline catches up later, and the difference becomes rather marginal at the +10% threshold: the corresponding speed up by KSOM is less than 5% relative compared to carefully configured EMA-VQ. In what follows, we show that KSOM is much more robust than EMA-VQ, and performs well under all configurations, including those that are sub-optimal for EMA-VQ.

Robustness of KSOM Against EMA-VQ Issues. Above we report that the performance gain (both in speed and reconstruction quality) by KSOM is rather marginal compared to our carefully configured EMA-VQ baseline obtained in Sect. 4.1. Here we compare the two approaches under various configurations. Table 3 shows the results. We observe that KSOM is remarkably robust: in all configurations, including those that are sub-optimal for EMA-VQ, KSOM achieves the same best validation loss as in the optimal configuration (Table 2). KSOM's neighbourhood updating scheme naturally fixes the problematic cases of the original EMA-VQ above. In these configurations, KSOM also generally converges faster than the baseline EMA-VQ. These results hold for both VQ-VAEs trained on CIFAR-10 and for VQ-VAE-2s trained on ImageNet and CelebA-HQ/AFHQ.[3]

---

[3] We also observed that KSOM generally improves the codebook utilisation. Further illustrations of this is available through the link provided on page 1.

Table 2. Validation reconstruction loss and number of steps needed to achieve +10% and +20% of the final loss for various Kohonen-VAEs. "None" in column "Neighbours" corresponds to our *well-configured* EMA-VQ (Sect. 4.1; "$N_k^{(0)} = 1$" and "Update-0, Yes" in Table 3). 2D grid is used for both "Hard" and "Gaussian" cases. Mean and standard deviations are computed with 10 seeds for CIFAR-10 and 5 seeds each for ImageNet and CelebA-HQ/AFHQ. Image resolution is 32×32 for CIFAR-10 and 256×256 for ImageNet and CelebA-HQ/AFHQ.

Dataset	Neighbours	Loss ($1e^{-3}$)	# Steps ($1e^3$) +10%	+20%
CIFAR-10	None (EMA-VQ)	51.9 ± 0.2	5.4 ± 0.5	3.7 ± 0.5
	Hard (KSOM)	52.1 ± 0.2	$\mathbf{5.2 \pm 0.6}$	$\mathbf{2.5 \pm 0.5}$
	Gaussian (KSOM)	$\mathbf{51.8 \pm 0.2}$	$\mathbf{5.2 \pm 0.6}$	3.0 ± 0.0
ImageNet	None (EMA-VQ)	$\mathbf{23.0 \pm 0.4}$	15.2 ± 1.6	8.4 ± 0.6
	Hard (KSOM)	23.5 ± 0.4	$\mathbf{13.6 \pm 2.2}$	$\mathbf{7.4 \pm 0.9}$
	Gaussian (KSOM)	23.2 ± 0.4	14.0 ± 1.4	7.6 ± 0.6
CelebA-HQ/AFHQ	None (EMA-VQ)	1.86 ± 0.10	$\mathbf{17.0 \pm 1.0}$	14.1 ± 1.4
	Hard (KSOM)	$\mathbf{1.73 \pm 0.01}$	$\mathbf{17.0 \pm 1.6}$	$\mathbf{10.8 \pm 1.3}$
	Gaussian (KSOM)	1.75 ± 0.03	17.8 ± 1.8	11.8 ± 1.3

4.3 Topologically Ordered Discrete Representations

Finally, we analyse the discrete representations learned by KSOM. We show that they are "topologically" ordered on the grid of indices, and consequently, reconstructed images remain close to the original ones even when we slightly shift their latent representations in the discrete index space.

Grid Visualisation. We first visualise how learned discrete representations are distributed over the grid. Here we use a VQ-VAE trained with KSOM (2D with hard neighbourhoods) on CIFAR-10, and proceed as follows. A discrete latent representation consists of N integers (Sect. 3.1). For each index in the codebook (corresponding to one of the nodes on the grid), we create a discrete latent representation whose N codes are all the same and equal to the corresponding index, and feed it to the decoder to obtain an output image. This results in a grid of images shown in Fig. 3. We observe that each code seems to correspond to some colour, we can effectively observe several local "islands" of colours which group colours that are visually close.

Impact of Perturbation in the Discrete Latent Space. To further illustrate the presence of neighbourhoods developed by KSOM in the discrete latent space, we show images obtained by perturbing the discrete latent code representing a proper image in the index space (by adding or subtracting an integer offset to each coordinates of the code indices). Here we use VQ-VAE-2 trained on

Table 3. Validation reconstruction loss and number of steps needed to achieve +10% of the final loss ("# Steps"), showing the **robustness of KSOM** w.r.t. configurations that are sub-optimal for EMA-VQ. "$N_k^{(0)}$" indicates its initialisation: 0 or 1. "Update-0" denotes whether to update clusters that have 0 members in the batch. Here we report the *hard* variant (results are similar for *Gaussian*).

Dataset	Method	$N_k^{(0)}$	Update-0	Loss ($1e^{-3}$)	# Steps ($1e^3$)
CIFAR-10	EMA-VQ	0	No	148.8 ± 11.0	13.5 ± 0.7
	KSOM			$\mathbf{51.8 \pm 0.2}$	$\mathbf{5.2 \pm 0.6}$
	EMA-VQ	1	No	68.3 ± 0.8	16.3 ± 1.4
	KSOM			$\mathbf{52.1 \pm 0.3}$	$\mathbf{4.8 \pm 0.9}$
	EMA-VQ	0	Yes	52.1 ± 0.6	6.8 ± 0.6
	KSOM			$\mathbf{51.9 \pm 0.2}$	$\mathbf{4.8 \pm 0.6}$
ImageNet	EMA-VQ	0	No	44.7 ± 2.6	13.5 ± 0.7
	KSOM			$\mathbf{24.4 \pm 1.3}$	$\mathbf{12.6 \pm 0.9}$
	EMA-VQ	1	No	27.9 ± 1.0	21.2 ± 1.5
	KSOM			$\mathbf{24.1 \pm 1.3}$	$\mathbf{12.2 \pm 1.3}$
	EMA-VQ	0	Yes	$\mathbf{23.6 \pm 0.9}$	15.4 ± 2.7
	KSOM			23.7 ± 0.7	$\mathbf{14.2 \pm 1.6}$
CelebA-HQ/AFHQ	EMA-VQ	0	No	3.73 ± 0.12	21.2 ± 3.9
	KSOM			$\mathbf{1.74 \pm 0.04}$	$\mathbf{17.6 \pm 3.6}$
	EMA-VQ	1	No	3.01 ± 0.28	22.6 ± 0.5
	KSOM			$\mathbf{1.74 \pm 0.02}$	$\mathbf{16.6 \pm 1.5}$
	EMA-VQ	0	Yes	1.80 ± 0.10	17.6 ± 1.5
	KSOM			$\mathbf{1.72 \pm 0.01}$	$\mathbf{16.8 \pm 1.9}$

ImageNet with 2D KSOM with hard neighbourhoods. The VQ-VAE-2 [46] has two levels of discretisation: we shift all of them by the same offset on both x and y axes of the 2D grid for KSOM or directly shift the code indices for EMA-VQ. Figure 4 shows the results. Obviously, with the baseline VQ-VAE-2 trained with EMA-VQ, the output images become complete noises under such perturbations, even with an offset of one. With the KSOM-trained representations, there is a certain degree of continuity in the space of indices (as illustrated in Fig. 3): the output images preserve the original contents though they become noisier as the offset increases. This illustrates the neighbourhoods learned by KSOM.

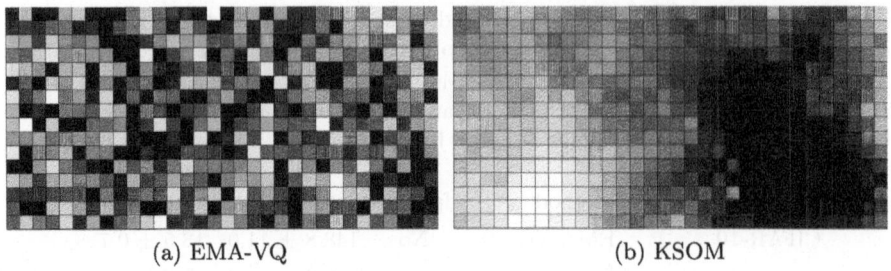

(a) EMA-VQ   (b) KSOM

Fig. 3. A visualization of codebook ($K = 512$) of VQ-VAEs trained on CIFAR10 with (a) EMA-VQ or (b) KSOM. The codebook of EMA-VQ obviously has no structure but serves as a reference. Similar visualisations for VQ-VAE-2 trained on ImageNet can be found on the link provided on page 1.

5   Discussion

Recommendations. Our first recommendation for any EMA-VQ implementation is to modify $N_k^{(0)} = 0$ to $N_k^{(0)} = 1$ (Sect. 4.1). For a more robust solution, we recommend using KSOM. Extending the standard implementation of EMA-VQ (Sect. 3.1) to KSOM (Sect. 3.2) is straightforward. While we introduce one extra hyper-parameter, shrink step τ (Eqs. 4-5), we found $\tau = 0.1$ to perform well across all tasks. Regarding the model variations, we recommend using the 2D variant with hard neighbourhoods. Virtually, any VQ implementation should benefit from these modifications.

Related Work. The most related work is Fortuin et al. [15]'s *SOM-VAE* which is also inspired by KSOM. The core difference between the SOM-VAE and our approach is that Fortuin et al. [15] train the codebook weights by gradient descent: all neighbour weight vectors are updated to be close to the encoder output (see the last paragraph of Sect. 3.1). While this is indeed inspired by Kohonen's neighbourhoods, none of Kohonen's learning rules (Sect. 2) is used. Also, the main focus of [15] is on modelling and interpreting time series. In fact, SOM-VAEs are extended by Manduchi et al. [39] for further applications to healthcare.[4]. Another type of VAEs exhibiting topographic properties can also be found in [24].

There are also several other works which attempt to improve the VQ algorithm. For example, Zeghidour et al. [62] propose to initialise all codebook

---

[4] Empirical results comparing KSOM and SOM-VAE is provided through the link on page 1. We find that the final reconstruction loss is similar for the EMA baseline, SOM-VAE, and KSOM, but KSOM converges the fastest (significantly faster than SOM-VAE). We also confirm that the EMA-based codebook learning outperforms the gradient-based one as noted by the original authors of VQ-VAE. We focus on KSOM because it is the natural extension of EMA which is recommended over the gradient-based variant (SOM-VAE is a natural extension of the gradient-based variant).

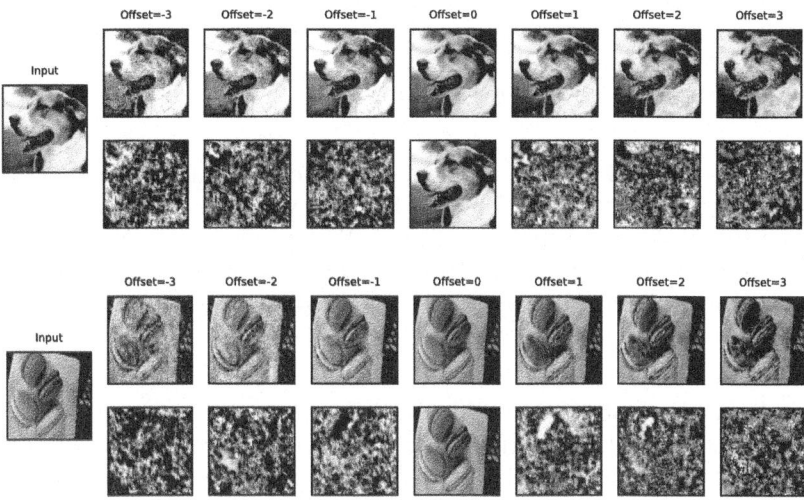

Fig. 4. Effects of perturbations to the discrete latent code on reconstructed images. For each image, the **top row** shows the results for KSOM, and the **bottom row** shows those for EMA-VQ. "Offset" indicates the offset added to the indices of the latent representations.

weights using examples in the first batch. Lee et al. [30] propose residual VQ that iteratively performs VQ for improved reconstruction quality (note that [30] use EMA-VQ which can be directly replaced by KSOM).

Applications of VQ. Following the VQ-VAE of [43], VQ has become popular across various modalities and applications. Besides the standard use case for high-dimensional data such as images, audio, and video, VQ has been also used for texts. For example, Kaiser et al. and Roy et al. [22, 48] downsample target sentences via VQ to speed up decoding in machine translation. Liu et al. [32] explore VQ to improve multi-lingual translation. Ozair et al. [44] applies VQ to model-based reinforcement learning/planning by quantising state-action sequences. Many text-to-image generation systems also include VQ components, e.g., Ramesh et al. [45] make use of discrete VAEs with Gumbel-softmax [21]), and Yu et al. [61] build upon VQ-GAN-2 [60]. Discrete representation learning is also motivated by out-of-distribution generalisation in certain tasks [33, 34, 53].

Differentiable Relaxation of KSOM. While working very well in practice, the use of the straight-through estimator (Sect. 3.1) to pass the gradients through the quantisation operation seems sub-optimal at first. For example, Agustsson et al. [1] show the possibility to perform discrete representation learning in fully differentiable NNs by using softmax with temperature annealing. In fact, we can also naturally derive a differentiable version of KSOM by replacing hardmax in Eq. 6 (presented in Section 2) by the softmax function. However, in our preliminary experiments, none of our differentiable variants with temperature annealing obtained a successful model that achieves a good reconstruction loss when the softmax is discretised for testing.

Semantic Codebook. While we demonstrate the emergence of neighbourhoods in the discrete space of codebook indices learned by KSOM, the features encoded by them remain low-level (mostly colours). While learning of more "semantic" discrete codes is out of scope here, we expect KSOM with such representations to be even more interesting, as it may potentially enable interpolation in the discrete latent space.

Other Variations of KSOM. Finally, there are several extensions of KSOM, e.g., *neural gas* [16,40]. Hopefully, our work will inspire further research on improving discrete representation learning in modern NNs through KSOM variations or other algorithms going beyond EMA-VQ.

6   Conclusion

We revisit the learning rule of Kohonen's Self-Organising Maps (KSOM) as the vector quantisation (VQ) algorithm for discrete representation learning in neural networks. KSOM is a generalisation of the exponential moving-average based VQ algorithm (EMA-VQ) commonly used in VQ-VAEs. We empirically demonstrate that, unlike the standard EMA-VQ, KSOM is robust w.r.t. initialisation and EMA update schemes. Our recipes can easily be integrated into existing code for VQ-VAEs. In addition, we show that discrete representations learned by KSOM effectively develop topological structures. We provide illustrations of the learned neighbourhoods in the image domain.

Acknowledgments. This research was partially funded by ERC Advanced grant no: 742870, project AlgoRNN, and by Swiss National Science Foundation grant no: 200021_192356, project NEUSYM. We are thankful for hardware donations from NVIDIA and IBM. The resources used for this work were partially provided by Swiss National Supercomputing Centre (CSCS) project s1145 and s1154.

References

1. Agustsson, E., et al.: Soft-to-hard vector quantization for end-to-end learning compressible representations. In: Proceedings of Advances in Neural Information Processing Systems (NIPS), pp. 1141–1151 (2017)
2. Amari, S.I.: Topographic organization of nerve fields. Bull. Math. Biol. **42**(3), 339–364 (1980)
3. Baevski, A., Schneider, S., Auli, M.: vq-wav2vec: self-supervised learning of discrete speech representations. In: International Conference on Learning Representations (ICLR). Virtual only (2020)

4. Bengio, Y., Léonard, N., Courville, A.: Estimating or propagating gradients through stochastic neurons for conditional computation. Preprint arXiv:1308.3432 (2013)

5. Borsos, Z., et al.: AudioLM: a language modeling approach to audio generation. Preprint arXiv:2209.03143 (2022)

6. Chen, S.F., Goodman, J.: An empirical study of smoothing techniques for language modeling. Comput. Speech Lang. **13**(4), 359–393 (1999)

7. Choi, Y., Uh, Y., Yoo, J., Ha, J.: StarGAN v2: diverse image synthesis for multiple domains. In: Proceedings of IEEE Conference on Computer Vision and Pattern Recognition (CVPR), pp. 8185–8194. Virtual only (2020)

8. Constantinescu, A.O., O'6Reilly, J.X., Behrens, T.E.: Organizing conceptual knowledge in humans with a gridlike code. Science **352**(6292), 1464–1468 (2016)

9. Cottrell, M., Olteanu, M., Rossi, F., Villa-Vialaneix, N.: Self-organizing maps, theory and applications. Revista de Investigacion Operacional **39**(1), 1–22 (2018)

10. Csordás, R., Irie, K., Schmidhuber, J.: CTL++: Evaluating generalization on never-seen compositional patterns of known functions, and compatibility of neural representations. In: Proceedings of Conference on Empirical Methods in Natural Language Processing (EMNLP), Abu Dhabi, UAE (2022)

11. De Bodt, E., Cottrell, M., Letremy, P., Verleysen, M.: On the use of self-organizing maps to accelerate vector quantization. Neurocomputing **56**, 187–203 (2004)

12. Deng, J., Dong, W., Socher, R., Li, L., Li, K., Fei-Fei, L.: ImageNet: a large-scale hierarchical image database. In: Proceedings of IEEE Conference on Computer Vision and Pattern Recognition (CVPR), Miami, Florida, USA, pp. 248–255 (2009)

13. Dhariwal, P., Jun, H., Payne, C., Kim, J.W., Radford, A., Sutskever, I.: Jukebox: a generative model for music. Preprint arXiv:2005.00341 (2020)

14. Esser, P., Rombach, R., Ommer, B.: Taming transformers for high-resolution image synthesis. In: Proceedings of IEEE Conference on Computer Vision and Pattern Recognition (CVPR), pp. 12873–12883. Virtual only (2021)

15. Fortuin, V., Hüser, M., Locatello, F., Strathmann, H., Rätsch, G.: SOM-VAE: interpretable discrete representation learning on time series. In: International Conference on Learning Representations (ICLR), New Orleans, LA, USA (May 2019)

16. Fritzke, B.: A growing neural gas network learns topologies. In: Proceedings of Advances in Neural Information Processing Systems (NIPS), Denver, CO, USA, pp. 625–632 (1994)

17. Hebb, D.O.: The organization of behavior; a neuropscholocigal theory. Wiley Book Clin. Psychol. **62**, 78 (1949)

18. Hinton, G.: Neural networks for machine learning. Coursera, video lectures (2012)

19. Hu, W., Miyato, T., Tokui, S., Matsumoto, E., Sugiyama, M.: Learning discrete representations via information maximizing self-augmented training. In: Proceedings of International Conference on Machine Learning (ICML), Sydney, Australia, pp. 1558–1567 (2017)

20. Hupkes, D., Singh, A., Korrel, K., Kruszewski, G., Bruni, E.: Learning compositionally through attentive guidance. In: Proceedings of International Conference on Computational Linguistics and Intelligent Text Processing, La Rochelle, France (2019)

21. Jang, E., Gu, S., Poole, B.: Categorical reparameterization with gumbel-softmax. In: International Conference on Learning Representations (ICLR), Toulon, France (2017)

22. Kaiser, L., Bengio, S., Roy, A., Vaswani, A., Parmar, N., Uszkoreit, J., Shazeer, N.: Fast decoding in sequence models using discrete latent variables. In: Proceedings

of International Conference on Machine Learning (ICML), Stockholm, Sweden, pp. 2395–2404 (2018)

23. Karras, T., Aila, T., Laine, S., Lehtinen, J.: Progressive growing of GANs for improved quality, stability, and variation. In: Internatinal Conference on Learning Representations (ICLR), Vancouver, Canada (2018)

24. Keller, T.A., Welling, M.: Topographic vaes learn equivariant capsules. In: Proceedings of Advances in Neural Information Processing Systems (NeurIPS), pp. 28585–28597. Virtual only (2021)

25. Kingma, D.P., Welling, M.: Auto-encoding variational bayes. In: International Conference on Learning Representations (ICLR), Banff, Canada (2014)

26. Kohonen, T.: Self-organized formation of topologically correct feature maps. Biol. Cybern. **43**(1), 59–69 (1982)

27. Kohonen, T.: Comparison of SOM point densities based on different criteria. Neural Comput. **11**(8), 2081–2095 (1999)

28. Kohonen, T.: Self-organizing Maps. Springer, Heidelberg (2001). https://doi.org/10.1007/978-3-642-56927-2

29. Krizhevsky, A.: Learning multiple layers of features from tiny images. Master's thesis, Computer Science Department, University of Toronto (2009)

30. Lee, D., Kim, C., Kim, S., Cho, M., Han, W.S.: Autoregressive image generation using residual quantization. In: Proceedings of IEEE Conference on Computer Vision and Pattern Recognition (CVPR), New Orleans, LA, USA, pp. 11523–11532 (2022)

31. Liska, A., Kruszewski, G., Baroni, M.: Memorize or generalize? searching for a compositional RNN in a haystack. In: AEGAP Workshop ICML, Stockholm, Sweden (2018)

32. Liu, D., Niehues, J.: Learning an artificial language for knowledge-sharing in multilingual translation. In: Proceedings of Conference on Machine Translation (WMT), Abu Dhabi, pp. 188–202 (2022)

33. Liu, D., et al.: Adaptive discrete communication bottlenecks with dynamic vector quantization. Preprint arXiv:2202.01334 (2022)

34. Liu, D., et al.: Discrete-valued neural communication. In: Proceedings of Advances in Neural Information Processing Systems (NeurIPS), pp. 2109–2121. Virtual only (2021)

35. Lloyd, S.: Least squares quantization in PCM. IEEE Trans. Inf. Theory **28**(2), 129–137 (1982)

36. MacQueen, J.: Classification and analysis of multivariate observations. In: Proceedings of Berkeley Symposium Mathematics Statistics Probability, pp. 281–297 (1967)

37. von der Malsburg, C.: Self-organization of orientation sensitive cells in the striate cortex. Kybernetik **14**(2), 85–100 (1973)

38. von der Malsburg, C., Willshaw, D.J.: How to label nerve cells so that they can interconnect in an ordered fashion. Proc. Natl. Acad. Sci. **74**(11), 5176–5178 (1977)

39. Manduchi, L., Hüser, M., Faltys, M., Vogt, J.E., Rätsch, G., Fortuin, V.: T-DPSOM: an interpretable clustering method for unsupervised learning of patient health states. In: Proceedings of Conference on Health, Inference, and Learning (CHIL), pp. 236–245. Virtual only (2021)

40. Martinetz, T., Schulten, K.: A "neural-gas" network learns topologies. In: Proceedings of International Conference on Artificial Neural Networks (ICANN), Espoo, Finland (1991)

41. Nasrabadi, N.M., Feng, Y.: Vector quantization of images based upon the Kohonen self-organizing feature maps. In: Proceedings of IEEE International Conference on Neural Networks (ICNN), vol. 1, pp. 101–105 (1988)

42. Oja, E.: Simplified neuron model as a principal component analyzer. J. Math. Biol. **15**(3), 267–273 (1982)

43. van den Oord, A., Vinyals, O., Kavukcuoglu, K.: Neural discrete representation learning. In: Proc. Advances in Neural Information Processing Systems (NIPS), Long Beach, CA, pp. 6306–6315 (2017)

44. Ozair, S., Li, Y., Razavi, A., Antonoglou, I., van den Oord, A., Vinyals, O.: Vector quantized models for planning. In: Proceedings of International Conference on Machine Learning (ICML), pp. 8302–8313. Virtual only (2021)

45. Ramesh, A., et al.: Zero-shot text-to-image generation. In: Proceedings of International Conference on Machine Learning (ICML), vol. 139, pp. 8821–8831. Virtual only (2021)

46. Razavi, A., van den Oord, A., Vinyals, O.: Generating diverse high-fidelity images with VQ-VAE-2. In: Proceedings of Advances in Neural Information Processing Systems (NeurIPS), Vancouver, Canada, pp. 14837–14847 (2019)

47. Rombach, R., Blattmann, A., Lorenz, D., Esser, P., Ommer, B.: High-resolution image synthesis with latent diffusion models. In: Proceedings of IEEE Conference on Computer Vision and Pattern Recognition (CVPR), New Orleans, LA, USA, pp. 10674–10685 (2022)

48. Roy, A., Vaswani, A., Parmar, N., Neelakantan, A.: Towards a better understanding of vector quantized autoencoders. OpenReview (2018)

49. Schlag, I., Irie, K., Schmidhuber, J.: Linear Transformers are secretly fast weight programmers. In: Proceedings of International Conference on Machine Learning (ICML). Virtual only (2021)

50. Schmidhuber, J.: Learning to control fast-weight memories: an alternative to recurrent nets. Technical Report. FKI-147-91, Institut für Informatik, Technische Universität München (March 1991)

51. Tirunagari, S., Bull, S., Kouchaki, S., Cooke, D., Poh, N.: Visualisation of survey responses using self-organising maps: a case study on diabetes self-care factors. In: Proceedings of IEEE Symposium Series on Computational Intelligence (SSCI), pp. 1–6 (2016)

52. Tjandra, A., Sakti, S., Nakamura, S.: Transformer VQ-VAE for unsupervised unit discovery and speech synthesis: zerospeech 2020 challenge. In: Proceedings of Interspeech, pp. 4851–4855. Virtual only (2020)

53. Träuble, F., et al.: Discrete key-value bottleneck. Preprint arXiv:2207.11240 (2022)

54. Vaswani, A., et al.: Attention is all you need. In: Proceedings of Advances in Neural Information Processing Systems (NIPS), Long Beach, CA, USA, pp. 5998–6008 (2017)

55. Walker, J., Razavi, A., Oord, A.V.d.: Predicting video with VQVAE. Preprint arXiv:2103.01950 (2021)

56. Willshaw, D.J., von der Malsburg, C.: How patterned neural connections can be set up by self-organization. Proc. Roy. Soc. Lond. Seri. B. Biol. Scie. **194**(1117), 431–445 (1976)

57. Willshaw, D.J., von der Malsburg, C.: A marker induction mechanism for the establishment of ordered neural mappings: its application to the retinotectal problem. Phil. Trans. Roy. Soc. Lond. B, Biol. Sci. **287**(1021), 203–243 (1979)

58. Yan, W., Zhang, Y., Abbeel, P., Srinivas, A.: VideoGPT: video generation using vq-vae and transformers. Preprint arXiv:2104.10157 (2021)

59. Yin, H.: The self-organizing maps: background, theories, extensions and applications. In: Computational Intelligence: A Compendium, pp. 715–762 (2008)
60. Yu, J., et al.: Vector-quantized image modeling with improved VQGAN. In: International Conference on Learning Representations (ICLR). Virtual only (2022)
61. Yu, J., et al.: Scaling autoregressive models for content-rich text-to-image generation. Preprint arXiv:2206.10789 (2022)
62. Zeghidour, N., Luebs, A., Omran, A., Skoglund, J., Tagliasacchi, M.: Soundstream: an end-to-end neural audio codec. IEEE/ACM Trans. Audio Speech Lang. Process. **30**, 495–507 (2021)

Neural Processes

Combined Global and Local Information Diffusion of Neural Processes

Jinyang Tai[1]([✉]) and Yike Guo[2]

[1] School of Computer Engineering and Science, Shanghai University,
No. 99, Shangda Road, Shanghai 200444, China
1203146411@shu.edu.cn
[2] School of Computer Engineering and Science, Hong Kong University of Science
and Technology, Clear Water Bay, Kowloon, Hong Kong 999077, China
yikeguo@hkbu.edu.hk

Abstract. Neural Processes (NPs) are a novel technique that combines neural networks and stochastic processes. NPs map the distribution function of target sample points from context sample points in input-output pairs. During the decoding process, context sample points are used as conditional inputs to generate distributional outputs for target sample points. Prior research has focused on issues within the NPs model structure, such as using only the average of context sample points or a single latent variable in the encoding module. However, this paper places emphasis on the data of context sample points within the NPs model and introduces a new model called Diffusion Neural Processes (DNPs). This model employs a controlled gradient sampling approach to ensure the quality of locally informative context sample point sampling. Additionally, the model introduces a pre-training noise method to capture global information from context sample points, thereby enhancing the complexity of their latent distribution representation. DNPs demonstrates excellent performance in 1D and 2D data, showcasing the model's potential across a range of applications.

Keywords: Neural Processes · Diffusion · Global Information

Neural Processes (NPs) are a new type of model that combines Neural Networks (NNs) [17] and Stochastic Processes. Unlike traditional Gaussian Processes (GPs) [12,22], NPs require fewer context sample points to influence the model, enabling rapid adaptation to new functions and unrestricted prediction of the original true distribution [20]. NPs are trained using a Meta-learning [2,18] framework, where observed context sample points are obtained from the real distribution. This allows NPs to fit all properties of the distribution using NNs, eliminating the need for domain knowledge and experience to select appropriate kernel functions, which can greatly impact computational complexity and model accuracy in GPs [5].

While NPs have shown promising progress compared to GPs, there are still limitations that need to be addressed. One such limitation is the problem of

M. Wand et al. (Eds.): ICANN 2024, LNCS 15016, pp. 365–380, 2024.
https://doi.org/10.1007/978-3-031-72332-2_24

underfitting context sample points in NPs model. To tackle this issue, Kim [10] proposes using attention in the encoder to assign weights to each context sample point. Additionally, the Stochastic Processes of NPs generate uncertainty dependent on a single latent variable, which reduces its ability to adapt to new functions. Juho [11] suggests a new extension of the bootstrap approach to the NPs family that can learn the stochasticity present in NPs without relying on a single latent variable. Despite these efforts, NPs face practical limitations, including their inability to model serial data during application. Tung [15] resolves this limitation by employing an autoregressive likelihood-based objective and constructing a new model based on the Transformer. Overall, while NPs have made progress, there are still bottlenecks that affect their performance.

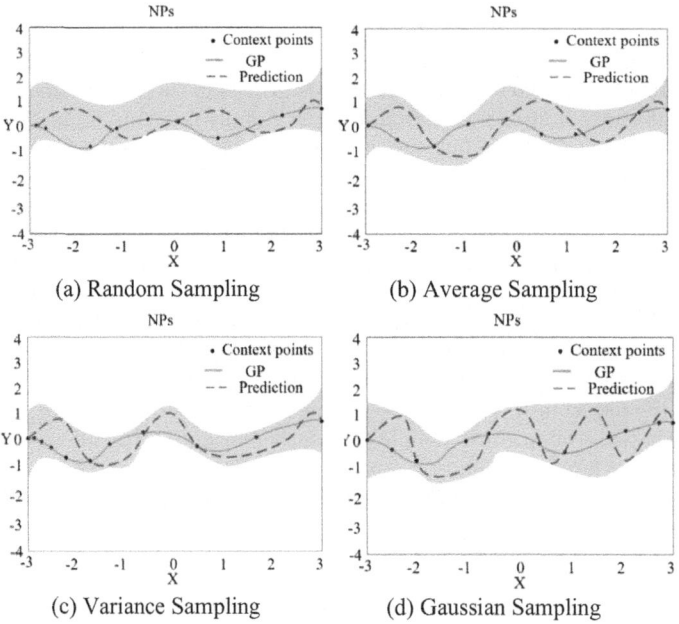

Fig. 1. We use different approaches to predict the sample point distribution by sampling 10 context sample points from the one-dimensional (1D) data distribution.

However, the studies mentioned above [10, 11, 15] mainly focus on improving the structure of NPs models without considering the effects of sampling local information in the context sample points and global information from the original data on NPs models. In Fig. 1, we present a comparison of various sampling methods (variance sampling, average sampling, random sampling, and Gaussian sampling) applied to different context sample points within the NPs model. This analysis aims to illustrate the importance of data quality in generating samples when utilizing 1D data [1].

To address the aforementioned issues, we were inspired by diffusion models. The Langevin sampler samples initial samples from a prior distribution and performs multi-step sampling using the inverse process to generate high-quality samples [9]. Similar to the Denoising Diffusion Probabilistic Models (DDPM), we include context sample point sampling in model training, but we introduce a gradient to enable sampling of different context sample points and to ensure local information (context sample point) richness. Based on the noise injection method of diffusion models, we introduce the noise-injected results as a pre-training method, endowing global information among context sample points, and enhancing the complexity of distributional representation. Based on the above analysis, We proposed a Diffusion of Neural Processes (DNPs) model to fit the context sample points. We evaluate the DNPs model on 1D data and 2D data results. The results of the DNPs model show our superiority in sampling context sample points and the ability to learn more complex representations in the enhanced context sample points.

1   Background

1.1   Neural Processes

The NPs model is a type of machine learning model that takes a sequence of inputs $x_{1:n} = (x_1, ..., x_n)$ with $X \in \mathbb{R}^d$ and corresponding output values $y_{1:n} = (y_1, ..., y_n)$ with $Y \in \mathbb{R}^d$ during the regression process of a given dataset D. And, there is a dataset $Q = \{x_1, ..., x_m\} \in X$ which does not have corresponding output values. These are referred to as the context data C and target data T, respectively. The conditional distribution for the output values of the target data, given the context sample points, can be expressed as $p(y_T|x_T, x_C, y_C) = (x_c, y_c)_{a \in C}$ with $|C| = n$. In this equation, the context sample points x_C and y_C are paired together, and their fusion is achieved by mapping them to a single value, denoted as r_C. This mapping relationship between x_C and y_C can be implemented using a Multi-Layer Perceptron (MLP), which is a type of neural network. The mean value r_C is used to represent the aggregated information of the context sample points in practice.

A global latent variable Z is included in the NPs module to account for the uncertainty of sampling out the context sample points (x_C, y_C) to predict the target sample points y_T. The path of the latent variable can be seen as a complement to the deterministic path (As shown in Fig. 2). Z is represented by the Gaussian parameter $s_C := s(x_C, y_C)$.

$$p(y_T \mid x_T, x_C, y_C) := \int p(y_T \mid x_T, r_C, z) \, q(z \mid s_C) \, dz \qquad (1)$$

The functions q, r, s are the encoder and the corresponding likelihood to the decoder result. The above decoding and encoding parameters are updated via the following ELBO:

$$\log p(y_T \mid x_T, x_C, y_C) \geq \mathbb{E}_{q(z|s_T)} \left[\log p(y_T \mid x_T, r_C, z) \right] - D_{\mathrm{KL}}(q(z \mid s_T) \| q(z \mid s_C)) \qquad (2)$$

The main role of NPs is to learn target reconstruction, while KL divergence regularization implements approximation between the context sample points and the target sample points.

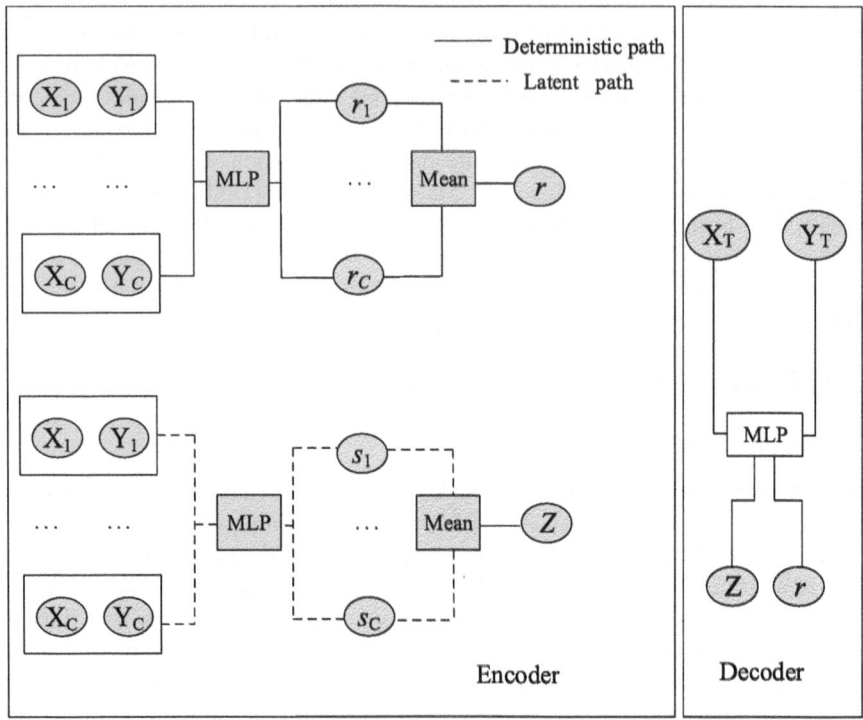

Fig. 2. Neural Processes model architecture. The left side indicates the process of encoding the context sample points. The right side indicates the process of decoding the context sample points

1.2   Related Work

Garnello et al. [4] address the issue of the high cost of Gaussian kernel computation in GPs, as well as the limitation that the distribution must satisfy the prior condition of the Gaussian distribution. In contrast, modern neural networks use gradient descent to fit curves, as shown in [7,13]. The combination of these two techniques results in Conditional Neural Processes (CNPs), which utilize a fixed dimensional encoding approach for the input context sample points, limiting the flexibility of the corresponding output values. To address this issue, NPs [5] introduce a latent global variable to enrich the representation of context sample points. However, this model uses aggregation by taking an average between sample points, which can lead to underfitting between context sample points. Attentive Neural Processes (ANPs) [10] address the issue of underfitting in NPs by aggregating context with multi-headed attention, assigning weights to different sample points. However, NPs still have serious flaws in sequential decision-making. To address this, the temporal function in the current transformer is combined with NPs to produce Transformer Neural Processes (TNPs) [15]. Additionally, Convolutional Conditional Neural Processes (CCNPs) [24] solve the problem of translation equivalence of NPs in data, such as time series and spatial data. This model extends data processing from finite to infinite dimensionality.

2   Diffusion of Neural Processes Model

This section examines the DNPs model, which is a combination of the DDPM and NPs models. The DNPs model aims to address two key issues in the NPs model: the low quality of sampled context sample points, and the inability to effectively integrate context information from these context sample points. The DDPM model is incorporated into the DNPs model to improve the quality of the context sample points by generating high-quality context sample points. These samples are then used to train the NPs model, which in turn allows for the effective integration of context information from these samples. Overall, the DNPs model provides a more robust and accurate approach to modeling complex data distributions by leveraging the strengths of both the DDPM and NPs models.

2.1   A Guided Continuous Sampling Approach

Random sampling in the NPs model does not ensure the quality of context sample points [19]. To address this issue, we integrate the sampling method into the model training process using the DDPM model. Specifically, we assume that the context sample points are obtained one time step at a time and solve the Ordinary Differential Equation (ODE) to extract the solution [19]. In general, conventional ODEs are solved using methods such as Euler's [14], Heun's [14], and Runge-Kutta's [16].

We propose using the control gradient method to guide the ODE sampling process. This method utilizes the same training procedure as no-parameter sampling (i.e., Euler's Method) and does not require any additional modifications. However, the sampling parameters are derived from additional gradient settings. In particular, the sampling process is represented as:

$$\hat{\epsilon} = \epsilon_\theta \left(x_{t-1} \right) - \sqrt{1 - \bar{\alpha}_{t-1}} \nabla_{x_{t-1}} \tag{3}$$

$$x_t = \sqrt{\bar{\alpha}_t} \left(\frac{x_{t-1} - \sqrt{1 - \bar{\alpha}_{t-1}} \hat{\epsilon}}{\sqrt{\bar{\alpha}_{t-1}}} \right) + \sqrt{1 - \bar{\alpha}_t} \hat{\epsilon} \tag{4}$$

where, x_t represents context sample points at each time step. The notation $\nabla_{x_{t-1}}$ indicates the gradient at time step $t - 1$. To increase the non-linearity during sampling, it's necessary to control the direction of the samples. This can be achieved by projecting the gradients at different time steps onto the normal plane (see Fig. 3). As shown in the figure, if the angle between the gradient update directions at the previous and next time steps is greater than 90 degrees (or the inner product is less than 0), the gradient direction of the next time step needs to be adjusted. The blue dashed line in Fig. 3 (b) indicates the projection of the gradient vector ∇_{x_t} onto the normal vector of $\nabla_{x_{t-1}}$.

$$\nabla_{x_t} = \nabla_{x_t} - \frac{\nabla_{x_t} \cdot \nabla_{x_{t-1}}}{\left\|\nabla_{x_{t-1}}\right\|^2} \nabla_{x_{t-1}} \tag{5}$$

The formula (26) is an orthogonal decomposition of ∇_{x_t} about the direction of $\nabla_{x_{t-1}}$, and then subtracting the reverse part of its co-linearity with $\nabla_{x_{t-1}}$. This leads to the projection of the final result onto the normal vector of $\nabla_{x_{t-1}}$. As a result, the gradient approach described above provides a means of controlling the sampling location, which in turn produces context sample points of high quality.

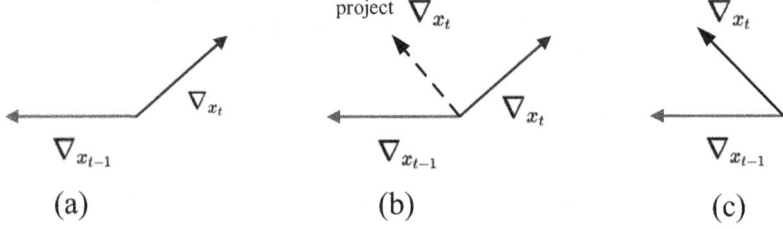

Fig. 3. Different time step gradient direction representation and project adjustment.

2.2   Diffusion Encoding and Decoding Approach

The DNPs model requires additional encoding beyond solving the context sample points sampling problem and controlling the gradient. Similar to the NPs model, the DNPs model encodes the same path into two separate paths. The first is the deterministic path that is used to express the locations of the context sample points. The second is the latent path that is used to express the latent distribution of the context sample points. Both paths need to be completed in order to fully encode the DNPs model.

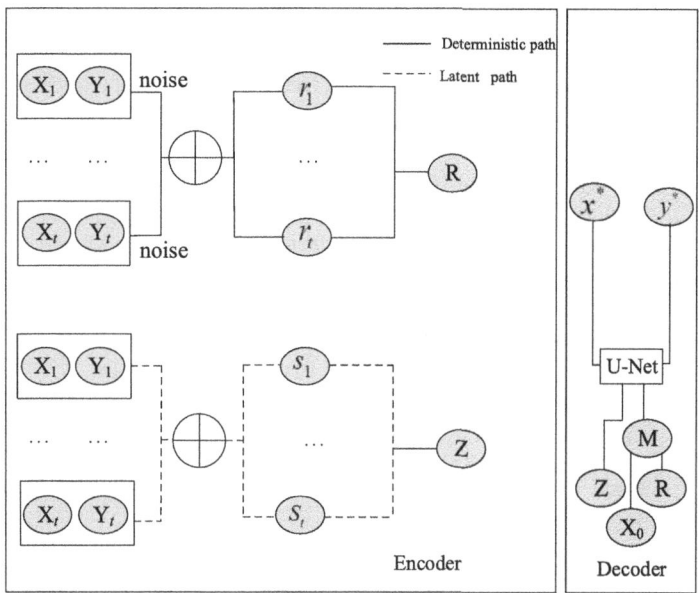

Fig. 4. The results of the DNPs model for the deterministic path implementing the sampled sample noise addition method are denoted by R. The latent paths are represented by the multivariate latent Gaussian distribution Z. In the decoding process, pre-training noise results and sampling noise are introduced to invert X_0, and the final results are fused with the sampling results to produce plus-noise M. Finally, the sample distribution prediction is completed by denoising.

The sampling time step and the noise addition time step are equal for the DNPs. The amount of noise at each time step is controlled by specifying the variance scheduling $\{\beta_t \in (0,1)\}_{t=1}^{T}$. We sample the context sample points according to Sect. 3.1 as (x_1, y_1), and the noise at each time step can be expressed as:

$$q\left(y_{1:T}, x_{1:T}\right) = \prod_{t=1}^{T} q\left(\begin{bmatrix} x_t \\ y_t \end{bmatrix} \mid \begin{bmatrix} x_{t-1} \\ y_{t-1} \end{bmatrix}, t\right) \tag{6}$$

where the joint Gaussian distribution is expressed as:

$$q\left(\begin{bmatrix} x_t \\ y_t \end{bmatrix} \mid \begin{bmatrix} x_{t-1} \\ y_{t-1} \end{bmatrix}, t\right) = \mathcal{N}\left(\begin{bmatrix} x_t \\ y_t \end{bmatrix} \mid a_t \begin{bmatrix} x_{t-1} \\ y_{t-1} \end{bmatrix}, \beta_t I\right) \tag{7}$$

where $a_t = \sqrt{1 - \beta_t}$. As shown in Fig. 4, we stitch together the context sample points after adding noise to get r. We calculate the process of adding noise at a forward arbitrary time step as follows:

$$x_t = \sqrt{\bar{\alpha}_t} x_0 + \sqrt{1 - \bar{\alpha}_t} \varepsilon_x \tag{8}$$

$$y_t = \sqrt{\bar{\alpha}_t} y_0 + \sqrt{1 - \bar{\alpha}_t} \varepsilon_y \tag{9}$$

where $\bar{\alpha}_t = \prod_{j=1}^{t} (1 - \beta_j)$, $\varepsilon_x \sim \mathcal{N}(0, I)$, and $\varepsilon_y \sim \mathcal{N}(0, I)$. We represent the results of the above noise addition through matrix R. Meanwhile, the latent path processing context sample points splicing results are represented by a latent Gaussian distribution $Z \sim \mathcal{N}(\mu_\theta(x_t, t), \Sigma_\theta(x_t, t))$.

The DNPs model introduces auxiliary information X_0 in addition to the data representation R and the latent distribution Z during decoding. The purpose of this is to introduce global information between context sample points, which enhances the complexity of learning the latent distribution. X_0 is the result of pre-training and involves adding noise through the original distribution, excluding the deterministic path sampling points. Essentially, the method involves pre-training noise but excluding sampling noise. The results of sampling added noise R and pre-training to inverse X_0 are fused to produce M.

$$M = \gamma \times R + (1 - \gamma) \times X_0 \tag{10}$$

where γ represents the weight parameter. Finally, the distribution is predicted by the above process (As shown in Fig. 5).

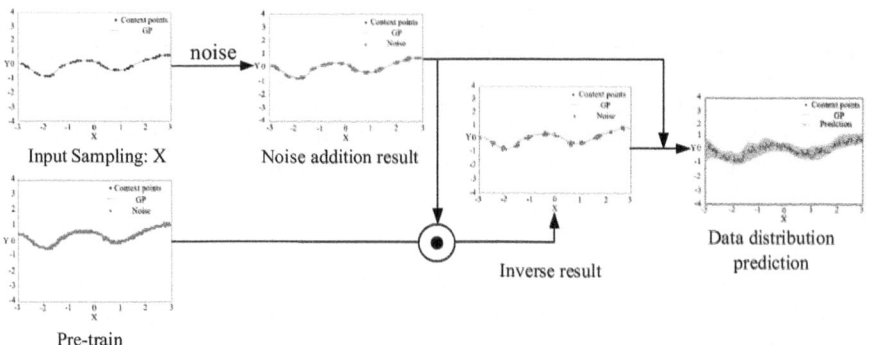

Fig. 5. DNPs predict the true distribution processes by sampling out context sample points from 1D GP data. Gaussian noise is added based on the context sample points sampled out. The pre-training noise is subtracted from the input sampling noise, thus forming a complementary relationship. We use weights to form a fusion between the two according to the position. Finally, the predicted true distribution is formed by denoising.

On the decoder side, the deterministic path involves searching the topological features matrix using the query r^*. The latent path is represented by a normal distribution Z across various dimensions of the topological feature matrix. During the decoding phase, the target sample points are generated based on the encoding results obtained earlier and the latent path. Specifically, the decoding phase uses the following probabilistic model: $p(Y_T | X_T, Z, r^*)$.

3   Experimental Result

3.1   1D Function for GP Data

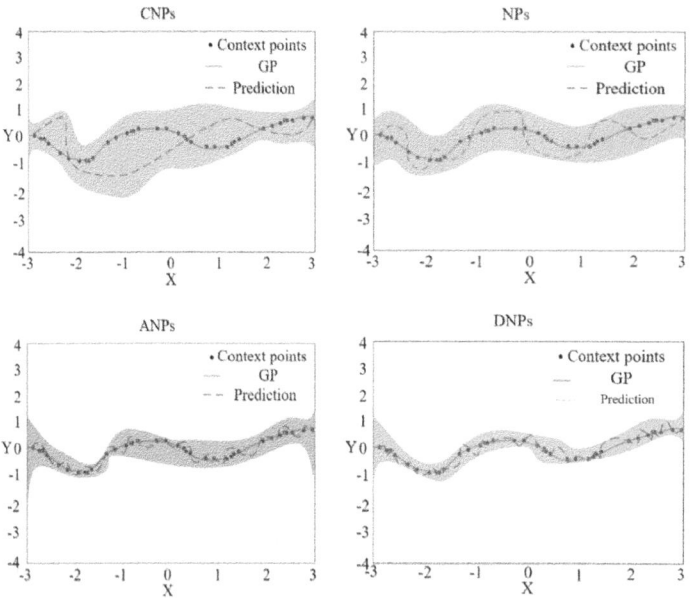

Fig. 6. Visualizing the effects of various sampling methods (CNPs, NPs, ANPs, and DNPs) on the prediction results of GPs using 40 sample points. CNPs, NPs, and ANPs represent random sampling methods, while DNPs represent guided continuous sampling. The true fit curve for GPs is depicted in blue, the context sample points in black, and the prediction curve in red. (Color figure online)

As an initial experiment, we tested the DNPs model on a regression task. For this experiment, we generated a distribution using a Gaussian kernel function for a 1D function. The training data was sampled from this distribution and consists of context sample points, while the associated target sample points were used for testing. Each time step in the experiment represents a sampled context sample point result. Assuming that t time steps were sampled, the sampled context sample points are represented as $(x_t, y_t)_C$, which means there are a total of t context sample points. Typically, this part of the sample points is used as the training dataset for the DNPs model. In addition to the training data, we also selected a certain number of context sample points for the testing data using the same method. The task of the model is to input context points $(x, y)_C$, target points x_T, and output predicted values y_T^* corresponding to the ground truth values y_T. The data x in the 1D GP data has a range of values restricted to $[-3, 3]$.

For this 1D data distribution, We explore the effects of the fusion of DNPs and diffusion approaches. In order to illustrate the superior performance of the DNPs model, the results of existing popular NPs [5][1], CNPs [4][2], and ANPs [10][3] are required to be compared on 1D GP data (see Fig. 6). The DNPs optimized loss is represented as follows:

$$L(\theta) = \mathbb{E}_{t,x_1,\epsilon,\nabla_t}[||\epsilon - \epsilon_\theta(\sqrt{\bar{\alpha}_t}x_1 + \sqrt{1 - \bar{\alpha}_t}\epsilon, x_0, t)||^2 + ||\epsilon - \epsilon_\theta(\nabla_t, x_1, t)||^2] \tag{11}$$

The gradient of sampling gradient is as follows:

$$\tilde{\epsilon}_\theta(\nabla_t, x_1, t) = \epsilon_\theta(\nabla_t, x_1, t) - w\nabla_t \log p_\theta(x_t, y_t, t) \tag{12}$$

where w is a parameter controlling the gradient at different time steps.

In Fig. 6, We provide 1D GP data to represent the visualization of CNPs, NPs, ANPs, and DNPs. DNPs model provides new sampling methods and a prior diffusion noise to solve the bottleneck of context sample points fitting. From Fig. 6, it is find that the DNPs model fitted the 1D GP data most effectively. Therefore, we visualize the data in 1D GP data to illustrate the superiority of the DNPs model.

We are conducting a study on training 1D GP data using various NPs such as CNPs, NPs, ANPs, and DNPs. The objective is to extract the minimum value in the GP prior by comparing the CNPs, NPs, ANPs, and DNPs to the original kernel values in a 1D oracle GP. We are using the best simple regret method, which measures the difference between the best observed solution and the global optimal solution in the 1D GP data. The model considers agents from the previous family of NPs for the objective function and extracts the waiting function using Thompson sampling from the agents and actions. We have chosen 100 objective functions, and the results are shown in Fig. 7. The simple regret is used to express the distance between the true oracle GP value and the predicted value, and the rate at which the slope in cumulative regret decreases is related to the utilization of the target function. The minimum difference between the DNPs and oracle GP values indicates that the function evaluation was well used for the subsequent prediction of the minimum value of the function. Based on our findings in Fig. 7, the DNPs result in better predicted minimum values than the CNPs, NPs, and ANPs. This suggests that the deep neural processes are more effective in training the 1D GP data compared to the other NPs.

---

[1] https://github.com/EmilienDupont/neural-processes.
[2] https://github.com/stratisMarkou/conditional-neural-processes.
[3] https://github.com/soobinseo/Attentive-Neural-Process.

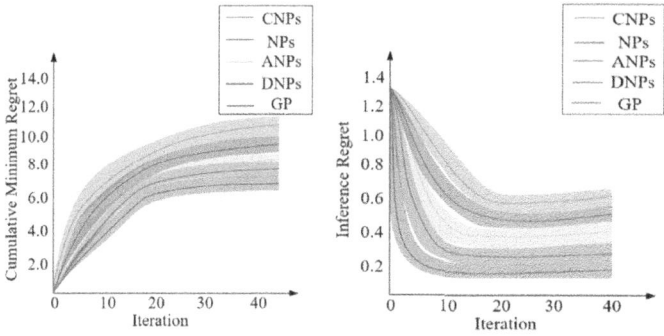

Fig. 7. Simple and cumulative regret for Bayesian optimized.

We require various kernel functions, including the Gaussian kernel, to account for the reliability of the DNPs model. To evaluate the performance of different models, we compare NPs [5], CNPs [4], ANPs [10], BNPs [11][4], CCNPs [24][5], TNPs [15][6], VNPs [6][7], and NDP [3][8], AENPs [21], and CAENPs [21], using metrics such as Mean Square Error (MSE). Additionally, we compare the results from various kernel types, such as RBF kernels, Matérn 5/2, and Periodic, using 1D GP data. Tables 1 demonstrate that the DNPs model, along with other models, showcased superior performance in terms of MSE measurements. Based on these results, we can confidently conclude that the DNPs model is highly reliable.

3.2   2D Function Data Images

To validate NPs models for processing high dimensional data, it is essential to test them on 2D data as well. In this study, we chose 2D data from common image data, including all two-dimensional data, not just images. We considered the dependence between pixel values in an image as the background value, represented by x_i, while predicting the target pixel y_i as the complete image problem. We compare the NPs [5], CNPs [4], ANPs [10], BNPs [11], CCNPs [24], TNPs [15], VNPs [6], NDP [3], AENPs [21], and CAENPs [21] on the MINIST dataset and CelebA dataset.

The results of visualizing CNPs, NPs, ANPs, and DNPs on this MNIST dataset (As shown in Fig. 8) and CelebA dataset (As shown in Fig. 8). We predict the pixels of the image $p(y^* \mid (M, Z, X_t, x^*))$. From Figs. 8 and 9, we find that the DNPs model predicts higher quality output samples than the results of the ANPs, NPs, and CNPs models. Observing from Fig. 8, we find that DNPs sample larger valid regions than the random sampling approach either on the

---

[4] https://github.com/juho-lee/bnp.
[5] https://github.com/cambridge-mlg/convcnp.
[6] https://github.com/tung-nd/TNP-pytorch.
[7] https://github.com/ZongyuGuo/Versatile-NP.
[8] https://github.com/vdutor/ndp.

Table 1. We experimented with different kernel functions in 1D GP. The mean and standard deviation of the five runs are reported (MSE Measures).

Method	RBF kernels	Matérn 5/2	Periodic
CNPs	0.278 ± 0.003	0.310 ± 0.003	0.652 ± 0.001
NPs	0.282 ± 0.003	0.315 ± 0.003	0.650 ± 0.002
ANPs	0.193± 0.001	0.230 ± 0.000	0.703 ± 0.002
BNPs	0.269± 0.003	0.301 ± 0.003	0.649 ± 0.002
CCNPs	0.254± 0.003	0.228 ± 0.001	0.681± 0.002
TNPs	0.177 ± 0.001	0.222 ± 0.000	0.670 ± 0.009
VNPs	0.174 ± 0.001	0.232 ± 0.000	0.601 ± 0.007
NDP	0.169 ± 0.000	0.210 ± 0.003	0.591± 0.005
AENPs	0.152 ± 0.003	0.217 ± 0.001	0.593 ± 0.005
CAENPs	0.149 ± 0.001	0.199 ± 0.003	0.524 ± 0.001
DNPs(Ours)	**0.146± 0.001**	**0.185 ± 0.000**	**0.518 ± 0.001**

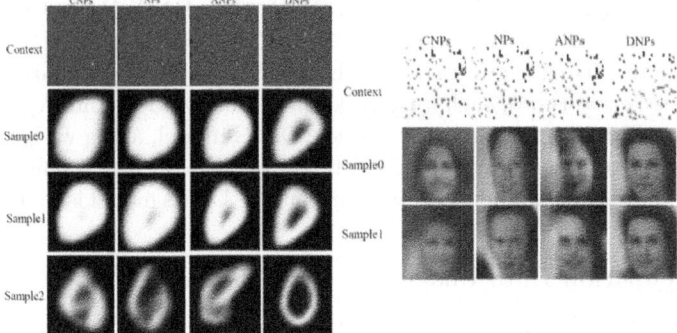

Fig. 8. The NPs, CNPs, and ANPs models utilize a random sampling of 10 context sample points (MNIST left) or 100 context sample points (CelebA right), while the DNPs model employs a guided continuous sampling approach for selecting 10 context sample points and 100 context sample points. These methods are used to predict images based on their learned distributions. Each model then selects three stage samples for presentation.

MNIST dataset or on the CelebA dataset. There is a large link between sampling and the performance of the model.

The superiority of the DNPs model is analysed from the visual qualitative approach in Figs. 8. We are required to evaluate the results of the complete images after sampling. The results are evaluated in quantitative form for the complete images. We use the Frechet Inception Distance (FID) [8] metric to measure the quality of the complete images. We have expressed the formula for the above metric as follows:

$$FID = ||\mu_r - \mu_g||_2^2 + Tr(\sum\nolimits_r + \sum\nolimits_g - 2(\sum\nolimits_r \sum\nolimits_g)^{1/2}) \qquad (13)$$

Table 2. The FID result of selecting a different number of pixel points from the images as context sample points to the complete images (MNIST dataset)

Methods	50	100	200	400
	FID	FID	FID	FID
CNPs	90.41	87.23	74.10	63.92
NPs	87.56	79.03	71.88	65.35
ANPs	79.62	72.15	60.84	52.71
BNPs	74.00	63.69	52.42	49.38
CCNPs	78.13	68.40	55.83	51.56
TNPs	67.42	54.51	47.29	39.50
VNPs	64.02	53.13	46.10	37.84
NDP	63.98	50.66	43.84	35.13
AENPs	58.16	43.57	34.83	19.00
CAENPs	57.43	40.64	32.48	**17.14**
DNPs(Ours)	**53.92**	**37.56**	**30.85**	15.30

Table 3. Experimental results on HDD

Methods	50	100	200	400
	FID	FID	FID	FID
CNPs	94.13	79.04	68.26	51.72
NPs	87.52	76.11	60.34	48.60
ANPs	80.00	73.89	55.91	37.45
BNPs	68.95	61.54	56.07	45.52
CCNPs	74.23	69.08	59.86	50.31
TNPs	53.18	50.24	46.96	40.99
VNPs	60.36	58.93	45.07	42.75
NDP	58.97	55.69	48.24	39.47
AENPs	60.84	55.72	41.09	31.55
CAENPs	55.47	43.30	39.17	30.70
DNPs(Ours)	**50.84**	**42.59**	**36.12**	**29.37**

where r represents the ground truth images, g represents the complete images, and μ represents the average of the distribution. The quality of complete images can be assessed using the FID metric, where a smaller FID indicates higher image quality. From Table 2 and 3, it can be observed that the DNPs model achieves the lowest FID values at sample points 50, 100, 200, and 400, indicating superior image quality compared to the NPs, CNPs, ANPs, BNPs, CCNPs, TNPs, VNPs, NDP, AENPs, and CAENPs models. These results suggest that DNPs are better at capturing the context sample points relationships that lead to high-quality complete images. This conclusion is supported by the MNIST and CelebA experiments.

3.3   Ablation Study

Sampling Approach: Analysis of the effect of sampling mode on the model. From Fig. 1, we have tried to illustrate its effect on the NPs model with different sampling methods. The DNPs model is chosen with Random sampling (DNPs-R), Alias sampling [23] (DNPs-A), and Markov Chain Monte Carlo sampling (DNPs-MCMC). It is also compared with some existing solution methods of ODE: Euler's Method (DNPs-Euler's), Heun's Method [14] (DNPs-Heun's), and Runge-Kutta Method [16] (DNPs-Runge-Kutta). In the MNIST dataset, their FID values were calculated separately using the combination of the sampling methods described above and the DNPs model (The results are shown in Table 4 and 5). From Table 4 and 5, it is found that the DNPs model gives the best results in the available control sampling methods to solve the ODE with an additional gradient orientation.

Table 4. A prior and different sampling methods are compared in the DNPs module for results (MNIST dataset).

Method	50	100	200	400
	FID	*FID*	*FID*	*FID*
DNPs-PT	94.46	83.47	72.56	69.03
DNPs-R	89.53	85.02	74.43	68.16
DNPs-A	88.09	74.29	67.18	60.49
DNPs-MCMC	80.31	79.01	64.28	59.95
DNPs-Euler's	**70.42**	**67.37**	**52.64**	**41.00**
DNPs-Heun's	**68.50**	**58.42**	**43.86**	**36.94**
DNPs-Runge-Kutta	**66.85**	**55.19**	**39.07**	**31.99**
DNPs(Ours)	**58.92**	**42.56**	**31.79**	**19.30**

Table 5. A priori and different sampling methods are compared in the DNPs module for results (CelebA dataset).

Method	50	100	200	400
	FID	*FID*	*FID*	*FID*
DNPs-PT	98.53	90.64	87.05	81.92
DNPs-R	95.44	86.45	79.01	68.16
DNPs-A	90.21	83.18	74.23	70.96
DNPs-MCMC	85.16	77.59	70.15	69.95
DNPs-Euler's	**81.88**	**78.43**	**69.16**	**65.18**
DNPs-Heun's	**77.00**	**69.51**	**58.32**	**45.23**
DNPs-Runge-Kutta	**67.43**	**58.92**	**51.11**	**42.17**
DNPs(Ours)	**50.84**	**44.00**	**36.09**	**31.70**

4 Conclusion

The traditional NPs model improvement is based on the network structure and does not consider the quality of context sample points. We propose the DNPs method, which combines the strengths of the DDPM and NPs approaches to address two key challenges in probabilistic modeling: the sampling of context sample points and the oversimplification of each context sample point to represent the latent distribution. To address the first challenge, we propose a method to control the gradient in the sampling process, ensuring high-quality context sample point sampling. Additionally, we introduce pre-training noise as auxiliary information in the decoding process. This auxiliary information helps the sample points learn global context information and facilitates each context sample point to learn more complex latent distributions.

References

1. Akuzawa, K., Iwasawa, Y., Matsuo, Y.: Information theoretic regularization for learning global features by sequential VAE. ArXiv, abs/2106.04804 (2021)
2. Bergman, A., Kellnhofer, P., Wetzstein, G.: Fast training of neural lumigraph representations using meta learning. In: Advances in Neural Information Processing Systems, pp. 172–186. Curran Associates, Inc. (2021)
3. Dutordoir, V., Saul, A., Ghahramani, Z., Simpson, F.: Neural diffusion processes. ArXiv, abs/2301.01475 (2023)
4. Garnelo, M.: Conditional neural processes. In: Proceedings of the 35th International Conference on Machine Learning, pp. 1704–1713. PMLR (2018)
5. Garnelo, M., et al.: Neural processes. In: ICML 2018 Workshop, abs/1807.01622 (2018)
6. Guo, Z., Lan, C., Zhang, Z., Lu, Y., Chen, Z.: Versatile neural processes for learning implicit neural representations. In: The Eleventh International Conference on Learning Representations (2023)
7. El Hanchi, A., Stephens, D., Maddison, C.: Stochastic reweighted gradient descent. In: Proceedings of the 39th International Conference on Machine Learning, pp. 8359–8374. PMLR (2022)
8. Heusel, M., Ramsauer, H., Unterthiner, T., Nessler, B., Hochreiter, S.: GANs trained by a two time-scale update rule converge to a local Nash equilibrium. In: Advances in Neural Information Processing Systems, vol. 30. Curran Associates, Inc. (2017)
9. Ho, J., Jain, A., Abbeel, P.: Denoising diffusion probabilistic models. In: Larochelle, H., Ranzato, M., Hadsell, R., Balcan, M., Lin, H. (eds.) Advances in Neural Information Processing Systems, vo. 33, pp. 6840–6851. Curran Associates, Inc. (2020)
10. Kim, H.: Attentive neural processes. abs/1901.05761 (2019)
11. Lee, J., Lee, Y., Kim, J., Yang, E., Hwang, S.J., Teh, Y.W.: Bootstrapping neural processes. In: Advances in Neural Information Processing Systems, pp. 6606–6615. Curran Associates, Inc. (2020)
12. Li, Z., Xi, T., Deng, J., Zhang, G., Wen, S., He, R.: GP-NAS: Gaussian process based neural architecture search. In: Proceedings of the IEEE/CVF Conference on Computer Vision and Pattern Recognition (CVPR), June 2020
13. Lin, W., Nielsen, F., Emtiyaz, K.M., Schmidt, M.: Tractable structured natural-gradient descent using local parameterizations. In: Proceedings of the 38th International Conference on Machine Learning, pp. 6680–6691. PMLR (2021)
14. Lu, C., Zhou, Y., Bao, F., Chen, J., Li, C., Zhu, J.: DPM-Solver: a fast ode solver for diffusion probabilistic model sampling in around 10 steps. In: Advances in Neural Information Processing Systems, vol. 35, pp. 25685–25698. Curran Associates, Inc. (2022)
15. Nguyen, T., Grover, A.: Transformer neural processes: uncertainty-aware meta-learning via sequence modeling. In: Proceedings of the 39th International Conference on Machine Learning, pp. 16569–16594. PMLR, July 2022
16. Nie, W., Guo, B., Huang, Y., Xiao, C., Vahdat, A., Anandkumar, A.: Diffusion models for adversarial purification. In: Proceedings of Machine Learning Research, vol. 162, pp. 16805–16827. PMLR (2022)
17. Pezeshki, M., Kaba, O., Bengio, Y., Courville, A.C., Precup, D., Lajoie, G.: Gradient starvation: a learning proclivity in neural networks. In: Advances in Neural Information Processing Systems, pp. 1256–1272. Curran Associates, Inc. (2021)

18. Rothfuss, J., Heyn, D., Chen, J., Krause, A.: Meta-learning reliable priors in the function space. In: Advances in Neural Information Processing Systems, pp. 280–293. Curran Associates, Inc. (2021)

19. Rothfuss, J., Heyn, D., Chen, J., Krause, A.: Meta-learning reliable priors in the function space. In: Advances in Neural Information Processing Systems, vol. 34, pp. 280–293. Curran Associates, Inc. (2021)

20. Sun, S., Shi, J., Wilson, A.G., Grosse, R.: Scalable variational Gaussian processes via harmonic kernel decomposition. In: International Conference on Machine Learning (ICML), abs/2106.05992 (2021)

21. Tai, J.: Global perception based autoregressive neural processes. In: Proceedings of the IEEE/CVF International Conference on Computer Vision (ICCV) (2023)

22. Titsias, M.K. Schwarz, J., Matthews, A.G.D.G., Pascanu, R., Teh, Y.W.: Functional regularisation for continual learning using gaussian processes. In: International Conference on Learning Representations (ICLR) (2020)

23. Vasconcelos, C., Larochelle, H., Roux, N.L., Romijnders, R., Goroshin, R., Dumoulin, V.: Impact of aliasing on generalization in deep convolutional networks. arXiv preprint (2021)

24. Ye, Z., Yao, L.: Contrastive conditional neural processes. In: 2022 IEEE/CVF Conference on Computer Vision and Pattern Recognition (CVPR), pp. 9677–9686. IEEE (2022)

Topology of Neural Processes

Jinyang Tai$^{1(\boxtimes)}$ and Yike Guo^2

^{1} ENE, No. 99, Shangda Road, Shanghai 200444, China
1203146411@shu.edu.cn
^{2} School of Computer Engineering and Science, Hong Kong University of Science and
Technology, Clear Water Bay, Kowloon 999077, Hong Kong, China
yikeguo@hkbu.edu.hk

Abstract. Neural networks and stochastic processes are combined to produce Neural processes (NPs). This method breaks away from traditional Gaussian Processes that are limited by kernels. However, both NPs and their subsequent variants ignore the intrinsic relationship between context sample points. This problem causes the NPs family model to require the latent distribution for each context sample point. Topology can mine the intrinsic relationships between sample points. We use the Vietoris-Rips complex of topology to compute topological features as a representation of the intrinsic relationships between sample points. The different-dimensional topological features are expressed separately in terms of the latent distributions. Based on this idea, we propose a new model called the Topology of Neural Processes. The regression problem from the 1D dataset, Bayesian optimization, and the complete image of the 2D dataset obtained better results compared to the results of the NPs family.

Keywords: Neural Processes · Topology · Vietoris-Rips Complex

1 Introduction

The primary objective of regression analysis is to establish a functional relationship between continuous input and output variables, such as stock prices and weather forecasts [23]. Gaussian Processes (GPs) are a statistical model that employs a Gaussian kernel function to map input samples onto an infinite polynomial space, enabling the estimation of covariance among the target sample points. The parameter update process in GPs aims to identify the feature space that best captures the covariance structure, thereby facilitating accurate estimation of the target samples from that space. However, selecting an appropriate kernel function in practice often requires domain knowledge and experience, as different kernel functions may result in varying levels of computational complexity and model accuracy. Therefore, choosing the right kernel function is crucial to optimizing model performance [21].

Nowadays, Neural Networks (NNs) have emerged as the prevailing approach for fitting data, outperforming other methods in fitting various distributions

© The Author(s), under exclusive license to Springer Nature Switzerland AG 2024
M. Wand et al. (Eds.): ICANN 2024, LNCS 15016, pp. 381–401, 2024.
https://doi.org/10.1007/978-3-031-72332-2_25

[18]. By integrating with combined stochastic processes, the selection of kernel functions in Gaussian Processes (GPs) no longer requires specialized expertise and experience. As a result, Neural Processes (NPs) have been developed to address this challenge [4]. NPs provide a useful way to model the sample distribution of continuous inputs in regression problems and enable the efficient testing of new distributions using a meta-learning framework [1,20]. Specifically, NPs are variational models that estimate conditional distributions and provide labels for the context sample points in advance, allowing for quick adaptation to new data distributions through the uncertainty in the latent variables [4]. To explain this more formally, suppose we have a dataset $D = \{(x_i, y_i)\} i = 0^{n-1}$ containing n input context sample points $x_i \in X$ and output context sample points $y_i \in Y$. Let F be the function of the stochastic process. For each $x1 : n = (x_1, ..., x_n)$ value of the function in dataset D, there corresponds a distribution $Y_{1:n}$, defined as $Y_{1:n} := (F(x_1), ..., F(x_n))$. Then, the probability marginal distribution can be formulated as $\rho_{x_{1:n}}(Y_{1:n}) = \rho_{x_1,x_2,...,x_n}(y_1, y_2, ..., y_n)$ [14].

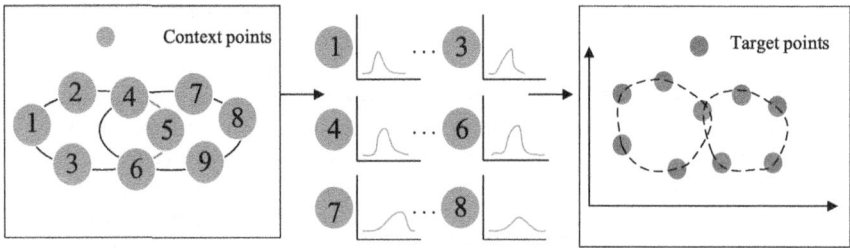

Fig. 1. We selected nine sample points from a concentric circle distribution as context sample points. To obtain predictions for the target sample points, the NPs model requires the generation of a latent distribution for each context sample point, which is achieved through the processing of the above results. Ultimately, the NPs model successfully predicts the target sample points.

There exist several variants of current NPs that address certain structural issues, such as taking an average value of multiple mapping relationships [12] or utilizing a single latent variable to indicate uncertainty [14]. However, existing NPs variants fail to consider the rationality of generating a normal distribution for each context sample point separately, which oversimplifies the learned distribution, as shown in Fig. 1. Traditional approaches ignored the intrinsic correlation between context sample points, which has been demonstrated both intuitively and empirically. To overcome this limitation, we suggest mining the global information between context sample points to achieve better curve fitting and reduce the bias in predicting the target sample points.

We propose a novel NPs model that combines geometric topological information with NPs to capture the complex structure of an object while preserving its invariant features during the transformation process. Our model achieves better results by reducing the number of sampling context sample points for downstream tasks. Context sample points are represented by different normal distributions, depending on the dimensionality of the topological classification, with the aim of enhancing the representation of complex distributions [19]. The Topology of TNPs is evaluated in 1D dimensional function regression, Bayesian optimization, and 2D image completion tasks [12]. The results demonstrate that TNPs outperform NPs and their variants in handling the intrinsic relationships between context sample points and representing the complexity of the latent normal distribution.

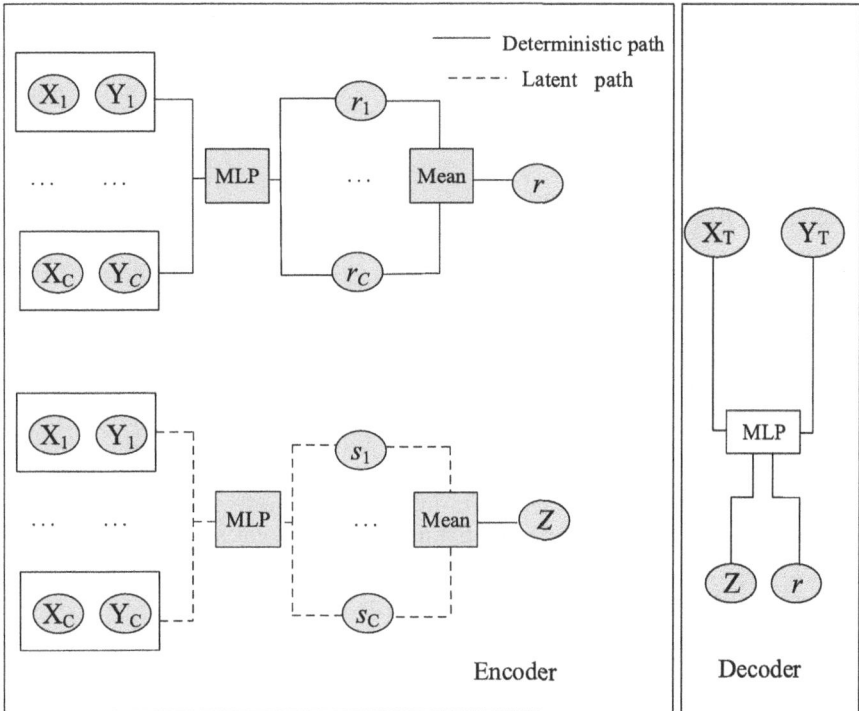

Fig. 2. Neural Processes model architecture. The left side indicates the process of encoding the context sample points. The right side indicates the process of decoding the context sample points.

2   Background

2.1   Neural Processes

The NPs model is a type of machine learning model that takes a sequence of inputs $x_{1:n} = (x_1, ..., x_n)$ with $X \in \mathbb{R}^d$ and corresponding output values $y_{1:n} = (y_1, ..., y_n)$ with $Y \in \mathbb{R}^d$ during the regression process of a given dataset D. And, there is a dataset $Q = \{x_1, ..., x_m\} \in X$ which does not have corresponding output values. These are referred to as the context data C and target data T, respectively.

The conditional distribution for the output values of the target data, given the context sample points, can be expressed as $p(y_T|x_T, x_C, y_C)$. In this equation, the context sample points x_C and y_C are paired together, and their fusion is achieved by mapping them to a single value, denoted as r_C. This mapping relationship between x_C and y_C can be implemented using a Multi-Layer Perceptron (MLP), which is a type of neural network. The mean value r_C is used to represent the aggregated information of the context sample points in practice.

A global latent variable Z is included in the NPs module to account for the uncertainty of sampling out the context sample point (x_C, y_C) to predict the variable y_T. The path of the latent variable can be seen as a complement to the deterministic path (As shown in Fig. 2). Z is represented by the Gaussian parameter $s_C := s(x_C, y_C)$.

$$p(y_T \mid x_T, x_C, y_C) := \int p(y_T \mid x_T, r_C, z) q(z \mid s_C) dz \tag{1}$$

The functions q, r, s are the encoder and the corresponding likelihood to the decoder result. The above decoding and encoding parameters are updated via the following ELBO:

$$\log p(y_T \mid x_T, x_C, y_C) \geq \mathbb{E}_{q(z|s_T)} [\log p(y_T \mid x_T, r_C, z)]$$
$$- D_{\mathrm{KL}}(q(z \mid s_T) \| q(z \mid s_C)) \tag{2}$$

The primary function of NPs is to acquire knowledge about target reconstruction, while regularization through KL divergence helps to establish an approximation between the context sample points and the target sample points.

2.2   Topology

Topological Data Analysis (TDA) is a powerful technique for studying dynamic data [2], such as structures [11] and point cloud data [25]. By exploiting the topological structure of the data, TDA can uncover unique geometric properties that are invariant under successive changes. This enables researchers to gain a deeper understanding of the data's underlying features and relationships. In essence, TDA represents a topology-driven approach to feature extraction, which offers distinct advantages over traditional data analysis methods. Overall, TDA has the potential to unlock new insights and applications across a wide range of scientific and engineering disciplines [17].

Simplicial Complex. We construct a topological space object T for which there is a chain complex C_T. This chain complex $C_0, ..., C_T$ are connected by homomorphisms $\partial_0 : C_0 \rightarrow C_1$. Homology groups $H_0 = \ker\partial_0/\mathrm{im}\partial_1$ are the focus of the study of this object. K is a finite set of the simplicial complex in the Euclidean space T. The simplicial complex of the connecting homomorphisms $\partial_0 : C_0(K) \rightarrow C_1(K)$.

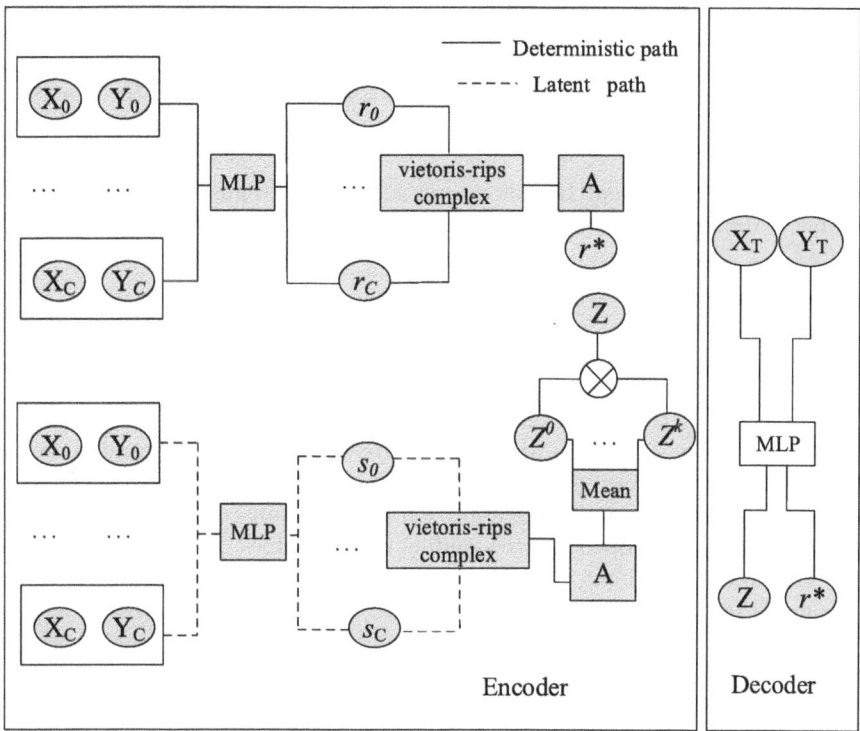

Fig. 3. The Topology of Neural Processes model has a structured architecture. The encoding process employs the Vietoris-Rips complex to generate topological features **A** from context sample points after mapping them r times. A deterministic path in r^* selects the corresponding value from the matrix of topological features **A** based on the distance. The latent path is divided into different dimensional topological features according to the Vietoris-Rips complex, which is represented by different normal distributions based on the dimensionality. The encoding results in Z and r^* are then decoded using a multilayer perceptron (MLP) through $p(Y_T \mid X_T, r^*, Z)$.

Persistent Homology . A persistent homology is a mathematical tool used to study topological features of data over different scales. In Persistent Homology, a topological space object T is represented by a family of simplicial complexes $(K^i)_{i=0}^d$ of different dimensions, where $K^0 \subset, ..., K^d = K$. Persistent Homology groups $H_d^{i,j}$ are then constructed using the boundary operator ∂ of the simplicial complexes, where d denotes the dimension, i denotes the scale parameter for the d-dimensional simplices, and j denotes the scale parameter for the $(d + 1)$-dimensional simplices. The Persistent Homology group $H_d^{i,j}$ is defined as the quotient group of the kernel of the boundary operator ∂_d^i by the subspace generated by the image of the boundary operator ∂_{d+1}^j and the kernel of the boundary operator ∂_d^i, i.e. $H_d^{i,j} = \ker\partial_d^i/(\mathrm{im}\partial_{d+1}^j \cup \ker\partial_d^i)$. The Betti number $\beta_d^{i,j}$ is the rank of the Persistent Homology group $H_d^{i,j}$, which indicates the number of homologous features in different dimensions. For example, β_0 represents the number of connected components in the data, while β_1 represents the number of 0-dimensional holes or tunnels in the data. Therefore, the Persistent Homology groups $H_d^{i,j}$ can be described in terms of Betti numbers as follows: $\beta_d^{i,j} = rank(H_d^{i,j})$

In summary, Persistent Homology provides a way to analyze the topological features of data over different scales using the concept of Betti numbers.

2.3   Related Work

Garnello et al. [5] address the issue of the high cost of Gaussian kernel computation in GPs, as well as the limitation that the distribution must satisfy the prior condition of the Gaussian distribution. In contrast, modern neural networks use gradient descent to fit curves, as shown in [8,15]. The combination of these two techniques results in Conditional Neural Processes (CNPs), which utilize a fixed dimensional encoding approach for the input context sample points, limiting the flexibility of the corresponding output values. To address this issue, NPs [4] introduce a latent global variable to enrich the representation of context sample points. However, this model uses aggregation by taking an average between sample points, which can lead to underfitting between context sample points. Attentive Neural Processes (ANPs) [12] address the issue of underfitting in NPs by aggregating context with multi-headed attention, assigning weights to different sample points. However, NPs still have serious flaws in sequential decision-making. To address this, the temporal function in the current transformer is combined with NPs to produce Transformer Neural Processes (TNPs) [16]. Additionally, Convolutional Conditional Neural Processes (CCNPs) [6] solve the problem of translation equivalence of NPs in data, such as time series and spatial data. This model extends data processing from finite to infinite dimensionality.

3   Topology of Neural Processes

In this section, our focus is on the model of NPs, where each context sample point produces a family of normal distributions. However, traditional NPs ignore the association between different context sample points, producing different normal distributions. To address this issue, our paper proposes to leverage topology to learn the association between context sample points, which enables us to achieve accurate predictions of target sample points.

3.1   Topological Feature Calculation

Given a dataset of context sample points $D = \{(x_i, y_i)\}_{i=0}^n$, where x_i belongs to the set X and y_i to Y, we can construct the Vietoris-Rips complex by mapping these points to a set of radii r_i. This mapping is done by setting r_i equal to the parameter ϵ (where $0 \leq \epsilon < \infty$) that allows the Vietoris-Rips complex to be

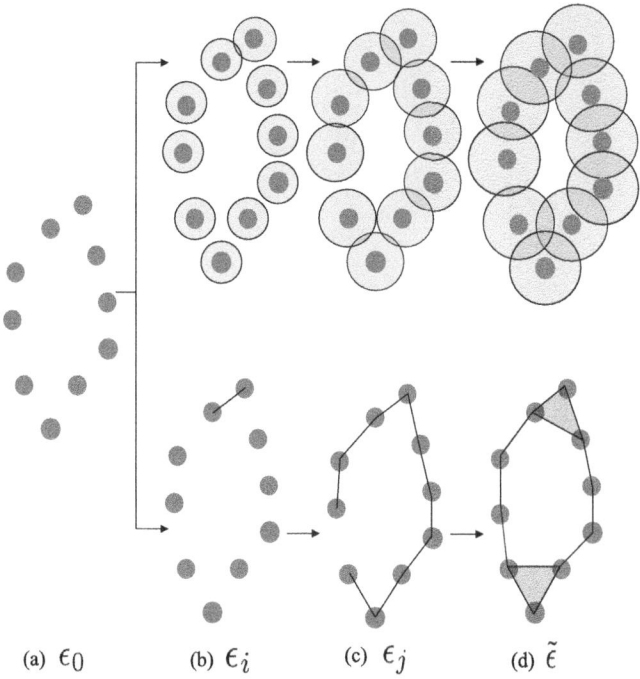

(a) ϵ_0 (b) ϵ_i (c) ϵ_j (d) $\tilde{\epsilon}$

Fig. 4. Context sample points calculation of continuous homotopy process diagram. (a) The circle is used as the prototype of the structure. (b) The evolution of the Vietoris-Rips complex when varying with the size of the parameter ϵ_i, which is expressed as the radius of the yellow circle around each data point. (c) The evolution of the Vietoris-Rips complex parameters from ϵ_i to ϵ_j. (d) The parameters of the Vietoris-Rips complex evolve $\tilde{\epsilon}$ to achieve the optimal results for the topological features. The 0-dimensional topology is represented by lines and the 1-dimensional topology by triangles and edges.

computed, as expressed by $\Re\epsilon(r)$. The elements $\{r_0, r_1, \ldots, r_C\}$ of r must satisfy the condition $dist\,(r_i, r_j) \leq \epsilon$, which means that any two points whose radii are closer than ϵ are connected by an edge in the Vietoris-Rips complex.

To measure the distances between context sample points and store them in a matrix $\mathbf{A}$, we use the Euclidean space S. The matrices $\Re_\epsilon(r)$ and $\Re_\epsilon(\mathbf{A})$ can be converted into each other by computing the distance between a pair of r values. Moreover, as we increase the value of ϵ (as shown in Fig. 4), we obtain a sequence of nested Vietoris-Rips complexes, where $\Re_{\epsilon_i}(r) \subseteq \Re_{\epsilon_j}(r)$ for $\epsilon_i \leq \epsilon_j$. This sequence enables us to determine the topological features of the data through the homology groups. By choosing an appropriate value of $\tilde{\epsilon}$, we can achieve the maximum stability of the topological features in the Vietoris-Rips complex. Thus, the complex $\Re_{\tilde{\epsilon}}(r)$ can include all the relevant topological information of the data.

We utilize the persistent homotopy approach, specifically the $\mathrm{PH}\,(\Re_\epsilon(X))$ method, to calculate the Vietoris-Rips complex. The resulting output is segre-

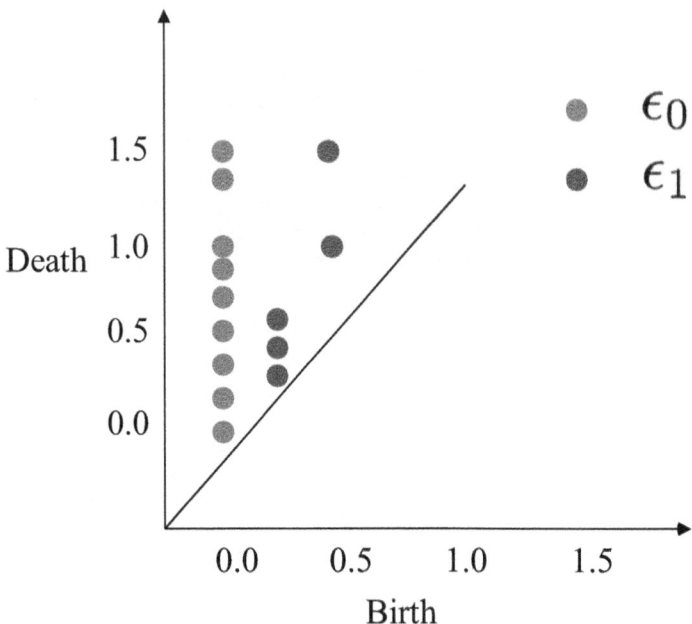

Fig. 5. Birth and death of topological features at different scales. The distance between points and diagonals is significant for topological features. The greater distance means that the topology is maintained for a longer period of time and the homologous topology is more stable.

gated into distinct k-dimensional components. We distinguish the k-dimensional components within $\Re_\epsilon(r)$ by analyzing the coordinate form (ϵ_0, ϵ_1) in the persistence diagram, as depicted in Fig. 5. Here, ϵ_0 represents the threshold ϵ at which topological features of the data are formed in the d-dimensional Vietoris-Rips complex, while ϵ_1 denotes the threshold $\tilde{\epsilon}$ at which these topological features are destroyed.

In the Vietoris-Rips complex, the 0-dimensional topological feature of the data is represented as lines, while the 1-dimensional topological feature is represented as triangles and edges (most of the topological information is stored in these two dimensions [17]). To retrieve the 0-dimensional topological feature, we use a distance measure for edges, where $dist\,(r_i, r_j) \leq \epsilon_0$. However, the 1-dimensional topological feature cannot be processed using this method due to the triangular and edge structures. To address this, we drew inspiration from the persistence diagram and mapped high-dimensional topological features of the data to two vertex distances (as shown in Fig. 5 'ϵ_1'). We use the longest side of the triangle as the representation of all points in that topological feature, with the condition that $Max\,(dist\,(r_i, r_j, r_o)) \leq \epsilon_1$. All of the above metrics are implemented in Euclidean space and stored in the matrix $\mathbf{A}$.

3.2 Fusion of the Latent Distributions

We calculate the topological features of the mapped sample points by using the Vietoris-Rips complex. These topological features are represented by the latent distribution Z for each of the differences in dimensionality (in general, most of the topological information is stored in these two dimensions [17]). We denote the 0-dimensional topological features as ϵ_0 and 1-dimensional topological features as ϵ_1. The above two types of topological dimensions are represented by the latent distribution $\epsilon_0 \sim \left(u_0, \delta_0^2\right), \epsilon_1 \sim \left(u_1, \delta_1^2\right)$. The probability distribution function of ϵ_0 is expressed as $g_1(x)$ and ϵ_1 is expressed as $g_1(x)$.

$$
\begin{aligned}
g_1(x) \cdot g_2(x) &= \frac{1}{\sqrt{2\pi}\delta_0} e^{-\frac{(x-u_0)^2}{2\delta_0^2}} \cdot \frac{1}{\sqrt{2\pi}\delta_1} e^{-\frac{(x-u_1)^2}{2\delta_1^2}} \\
&= \frac{1}{2\pi\sigma_0\sigma_1} e^{-\left(\frac{(x-\mu_0)^2}{2\sigma_0^2} + \frac{(x-\mu_1)^2}{2\sigma_1^2}\right)}
\end{aligned}
\tag{3}
$$

The exponential part of the formula can be dealt with first:

$$
\begin{aligned}
\beta &= \frac{(x-u_0)^2}{2\delta_0^2} + \frac{(x-u_1)^2}{2\delta_1^2} \\
&= \frac{\left(\delta_0^2+\delta_1^2\right)x^2 - 2\left(u_1\delta_0^2 + u_0\delta_1^2\right)x + \left(u_0^2\delta_1^2 + u_1^2\delta_0^2\right)}{2\delta_0^2\delta_1^2} \\
&= \frac{x^2 - 2\frac{u_1\delta_0^2+u_0\delta_1^2}{\delta_0^2+\delta_1^2}x + \frac{u_0^2\delta_1^2+u_1^2\delta_0^2}{\delta_0^2+\delta_1^2}}{\frac{2\delta_0^2\delta_1^2}{\delta_0^2+\delta_1^2}}
\end{aligned}
\tag{4}
$$

Constructing a new normal distribution $g_3(x)$:

$$
\begin{aligned}
g_3(x) &= \frac{\left(x - \frac{u_1\delta_0^2+u_0\delta_1^2}{\delta_0^2+\delta_1^2}\right)^2 + \frac{u_0^2\delta_1^2+u_1^2\delta_0^2}{\delta_0^2+\delta_1^2} - \left(\frac{u_1\delta_0^2+u_0\delta_1^2}{\delta_0^2+\delta_1^2}\right)^2}{\frac{2\delta_0^2\delta_1^2}{\delta_0^2+\delta_1^2}} \\
&= \underbrace{\frac{\left(x - \frac{u_1\delta_0^2+u_0\delta_1^2}{\delta_0^2+\delta_1^2}\right)^2}{\frac{2\delta_0^2\delta_1^2}{\delta_0^2+\delta_1^2}}}_{\gamma} + \underbrace{\frac{\frac{u_0^2\delta_1^2+u_1^2\delta_0^2}{\delta_0^2+\delta_1^2} - \left(\frac{u_1\delta_0^2+u_0\delta_1^2}{\delta_0^2+\delta_1^2}\right)^2}{\frac{2\delta_0^2\delta_1^2}{\delta_0^2+\delta_1^2}}}_{\lambda}
\end{aligned}
\tag{5}
$$

So, we can express the above process as $\beta = \gamma + \lambda$, where γ is the normal distribution, and λ is a constant. Continue the simplification of the representation of λ:

$$\lambda = \frac{\frac{u_0^2 \delta_1^2 + u_1^2 \delta_0^2}{\delta_0^2 + \delta_1^2} - \left(\frac{u_1 \delta_0^2 + u_0 \delta_1^2}{\delta_0^2 + \delta_1^2}\right)^2}{\frac{2\delta_0^2 \delta_1^2}{\delta_0^2 + \delta_1^2}}$$

$$= \frac{\left(u_0^2 \delta_1^2 + u_1^2 \delta_0^2\right)\left(\delta_0^2 + \delta_1^2\right) - \left(u_1 \delta_0^2 + u_0 \delta_1^2\right)^2}{2\delta_0^2 \delta_1^2 \left(\delta_0^2 + \delta_1^2\right)}$$ (6)

$$= \frac{\delta_0^2 \delta_1^2 \left(u_0^2 + u_1^2 - 2u_0 u_1\right)}{2\delta_0^2 \delta_1^2 \left(\delta_0^2 + \delta_1^2\right)}$$

$$= \frac{(u_0 - u_1)^2}{2\left(\delta_0^2 + \delta_1^2\right)}$$

Two multiplying normal distributions:

$$g_1(x) * g_2(x) = \frac{1}{2\pi\delta_0\delta_1} e^{-\beta} = \frac{1}{2\pi\delta_0\delta_1} e^{-(\gamma+\lambda)}$$

$$= \frac{1}{2\pi\delta_0\delta_1} e^{-\gamma} \cdot e^{-\lambda}$$ (7)

$$= \frac{1}{2\pi\delta_0\delta_1} e^{-\frac{(x-u)^2}{2\delta^2}} \cdot e^{-\frac{(u_0-u_1)^2}{2(\delta_0^2+\delta_1^2)}}$$

where $u = \frac{u_1 \delta_1^2 + u_0 \delta_2^2}{\delta_0^2 + \delta_1^2}$, $\delta 1 = \frac{\delta_0^2 \delta_1^2}{\delta_0^2 + \delta_1^2}$. We can express it directly as follows:

$$g_1(x) * g_2(x) = S_g \cdot \frac{1}{\sqrt{2\pi}\delta} e^{-\frac{(x-u)^2}{2\delta^2}}$$ (8)

where $S_g = \frac{1}{\sqrt{2\pi(\delta_0^2 + \delta_1^2)}} e^{-\frac{(u_0-u_1)^2}{2(\delta_0^2+\delta_1^2)}}$. Therefore, the distribution function after multiplication is a reduced or enlarged Gaussian distribution. S_g is the scaling factor.

3.3   Topology of Neural Processes Encoder and Decoder

In the following, we incorporate the above-mentioned topological features into the NPs model to generate a new Topology of Neural Processes (TNPs) model (as shown in Fig. 3).

The context sample points in Fig. 3 (left) are encoded by TNPs, which consist of two types of paths: deterministic and latent. In the deterministic path, we replace the NPs model's average value operation with the retrieval of topological features. For each mapped context sample point r in the latent path, the NPs generate a normal distribution. We obtain an average result based on different dimensions ϵ_k in the topological features and treat it as a normal distribution with an independent identical distribution Z^k (as shown in Fig. 6). In the latent path, we take an average of the context sample points for the model in each dimension ϵ_k from the topological feature matrix $\mathbf{A}$. In practice (taking

0-dimension and 1-dimension), the two types of dimensional topological features dimensionally normal are represented as $\epsilon_0 \sim \left(u_0, \delta_0^2\right), \epsilon_1 \sim \left(u_1, \delta_1^2\right)$. We multiply the above two types of dimensional normal distributions together to obtain Z. The result is as follows:

$$Z = \frac{1}{2\pi^2 \sigma_0 \sigma_1} \exp\left(-\frac{\sigma_1^2 \left(\bar{r}_0 - \mu_0\right)^2 + \sigma_0^2 \left(\bar{r}_1 - \mu_1\right)^2}{2\sigma_0^2 \sigma_1^2}\right) \tag{9}$$

The value of $\bar{r}0$ represents the average location of the context sample points that correspond to the 0-dimensional topological features. Similarly, the value of $\bar{r}1$ represents the average location of the sample points that correspond to the 1-dimensional topological features. In the computation of topological features of context sample points on TNPs, there exists inconsistency in the topological structure due to the varying number and location of randomly sampled context sample points. This is reflected in the absence of the 1-dimensional topological feature, which violates the consistency principle in the random processes. To address this particular issue, we represent Z^1 with a standard normal distribution, which ensures consistency in the distribution under different topological conditions.

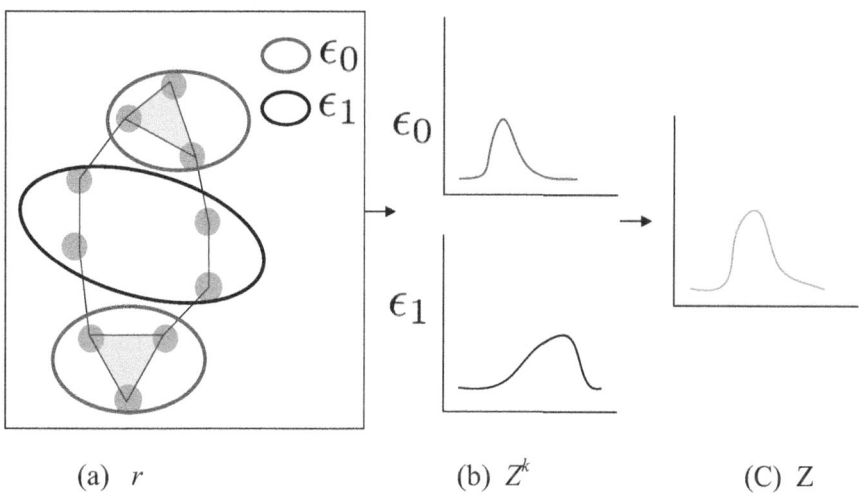

(a) r (b) Z^k (C) Z

Fig. 6. The calculation process of the Vietoris-Rips complex for the context sample point mapping result r. The same dimension is represented by the same latent normal distribution between the same dimensions. (a) Two type of topological dimensions within r: 0-dimension represent ϵ_0 and 1-dimension represent ϵ_1. (b) Latent normal distribution for different dimensions $Z^k = \left(Z^0, Z^1\right)$. (c) The result of multiplying and fusing two normal distributions Z.

On the decoder side, the deterministic path involves searching the topological features matrix using the query r^*. The latent path is represented by a normal distribution Z across various dimensions of the topological feature matrix. During the decoding phase, the target sample points are generated based on the encoding results obtained earlier and the latent path. Specifically, the decoding phase uses the following probabilistic model: $p\left(Y_T|X_T, Z, r^*\right)$.

In the past, the process of generating a normal distribution for each sample point used by TNPs has been changed. This model achieves a global linkage of geometric information between context sample points through the use of topological features. As a result, the actual distribution is represented by normal distributions of varying dimensions.

4 Experimental Result

TNPs are considered as one of the variants of NPs within a family. To validate their performance, we compared the model with other models using both 1D and 2D datasets.

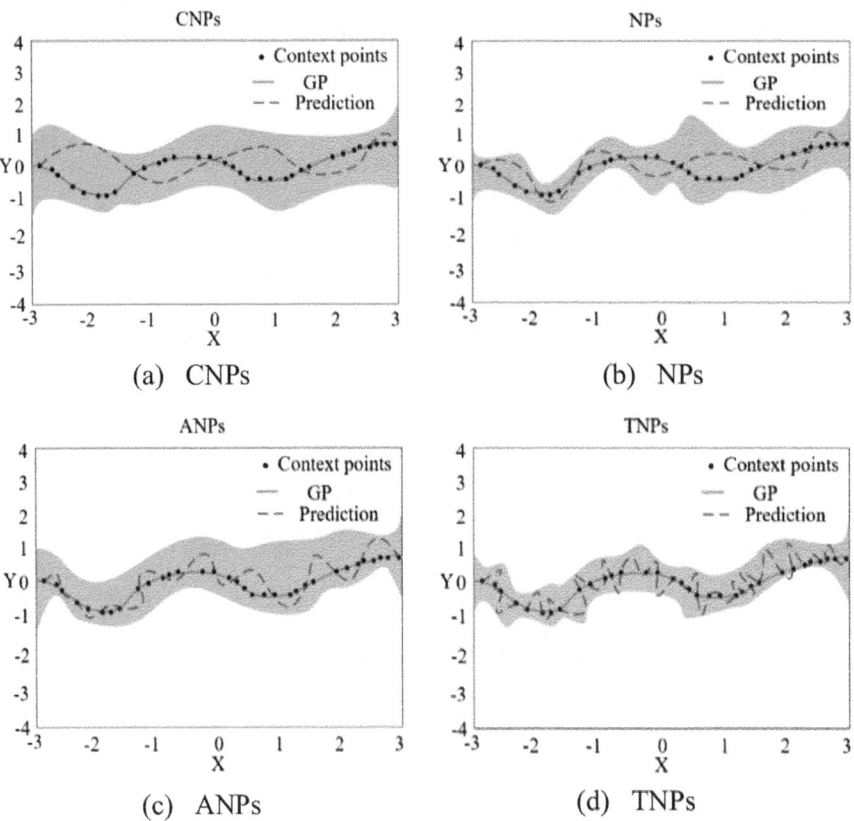

Fig. 7. Results of CNPs, NPs, ANPs, TNPs model fitting on GPs data

4.1   1D Function GP Data

We typically evaluate the accuracy of TNPs models by predicting target sample points. In this particular case, we are using a 1D GP data distribution to perform a regression task. TNPs generate GP data using Gaussian kernel functions, and we use the function f to randomly select n context sample points $(x, y)_C$ as the training dataset for the TNPs model. Additionally, we randomly select m target sample points $(x, y)_T$ as the testing dataset, with each sample point's value range restricted to $[-3,3]$. The model's task is to take in the context sample points $(x, y)_C$ and target sample points X_T and output predicted values Y_T^* corresponding to the ground truth values Y_T. We use the 1D distribution to explore the superiority of the TNPs model and compare its results with existing popular NPs [4][1], CNPs [5][2], and ANPs [12][3] on GP data (as shown in Fig. 7) to illustrate its superior performance. In the above process, we iterate using the following loss:

$$
\begin{aligned}
logp(y_T|x_T, x_C, y_C) \geq &\ \mathbb{E}_{q(Z|x_T, y_T)}[logp(y_T|x_T, r^*, Z)] \\
&- D_{\mathrm{KL}}(q(Z|s_T)||q(Z|s_C)))
\end{aligned}
\tag{10}
$$

In the above equation, r^* is the result obtained from $p(r^*|A(x_T, y_T))$. Z is the fusion of two types of dimensional topological features with a normal distribution.

We utilized a Bayesian approach to optimize the results obtained from the 1D GP data, as described previously. The optimization processes involved using the best simple regret method [24], which calculates the difference between the best observed solution and the globally optimal solution in the 1D GP data. The model takes into consideration agents from the previous family of NPs for the objective function, and we employed Thompson sampling to extract a simple function for optimizing the predicted minimum. We calculated the average value of the results obtained from testing 100 objective functions, and the results are presented in Fig. 8. As depicted in the figure, we observed that the TNPs of simple regret was the smallest and closest to the real GP. Additionally, the TNPs model had the steepest curve in cumulative regret (given the number of iterations), which indicates a better current functional function. Based on these results, we can conclude that the TNPs model is superior to the other models tested in this experiment. To ensure the reliability of the TNPs model, we evaluated its performance using various kernel functions, including NPs [4], CNPs [5], ANPs [12], BNPs [14][4], CCNPs [26][5], and TransNPs [16][6], VNPs [7][7], AENPs [22], CAENPs [22]. We compared the Mean Square Error (MSE)

---

[1] https://github.com/EmilienDupont/neural-processes.
[2] https://github.com/stratisMarkou/conditional-neural-processes.
[3] https://github.com/soobinseo/Attentive-Neural-Process.
[4] https://github.com/juho-lee/bnp.
[5] https://github.com/cambridge-mlg/convcnp.
[6] https://github.com/tung-nd/TNP-pytorch.
[7] https://github.com/ZongyuGuo/Versatile-NP.

and Calibration Error (CE) metrics of these models. Additionally, we compared the performance of different kernel functions, such as RBF kernels, Matérn 5/2, and Periodic, using data generated by the GP model. Our analysis revealed that the TNPs model and other models have the best performance in terms of both MSE and CE metrics. Specifically, Tables 1 and 2 show that TNPs outperformed other models. Therefore, based on these results, we can conclude that the TNPs model is a reliable choice for this study.

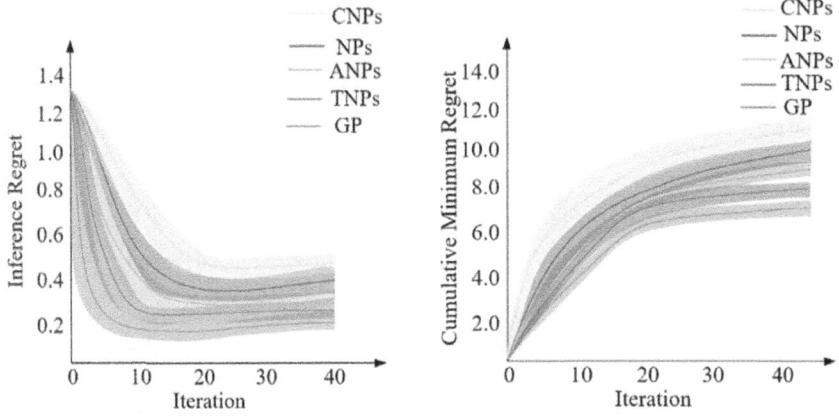

Fig. 8. Bayesian optimization results for GP prior functions

Table 1. We experimented with different kernel functions in 1D GP. The mean and standard deviation of the five runs are reported (MSE measures).

Method	RBF kernels	Matérn 5/2	Periodic
CNPs	0.278 ± 0.003	0.310 ± 0.003	0.652 ± 0.001
NPs	0.282 ± 0.003	0.315 ± 0.003	0.650 ± 0.002
ANPs	0.193 ± 0.001	0.230 ± 0.000	0.703 ± 0.002
BNPs	0.269 ± 0.003	0.301 ± 0.003	0.649 ± 0.002
CCNPs	0.254 ± 0.003	0.228 ± 0.001	0.681 ± 0.002
TransNPs	0.177 ± 0.001	0.222 ± 0.002	0.670 ± 0.007
VNPs	0.171 ± 0.002	0.219 ± 0.004	0.643 ± 0.009
AENPs	0.152 ± 0.003	0.217 ± 0.001	0.593 ± 0.005
CAENPs	0.149 ± 0.001	0.199 ± 0.003	0.524 ± 0.001
TNPs(Ours)	$\mathbf{0.134 \pm 0.001}$	$\mathbf{0.150 \pm 0.000}$	$\mathbf{0.424 \pm 0.001}$

Table 2. We experimented with different kernel functions in 1D GP. The mean and standard deviation of the five runs are reported (CE measures).

Method	RBF kernels	Matérn 5/2	Periodic
CNPs	0.078 ± 0.002	0.051 ± 0.000	0.143 ± 0.002
NPs	0.093 ± 0.002	0.056 ± 0.001	0.130 ± 0.007
ANPs	0.085 ± 0.001	0.169 ± 0.001	0.265 ± 0.002
BNPs	0.093 ± 0.003	0.054 ± 0.002	0.115 ± 0.004
CCNPs	0.094 ± 0.000	0.195 ± 0.001	0.162 ± 0.003
TransNPs	0.077 ± 0.001	0.50 ± 0.002	0.155 ± 0.007
VNPs	0.070 ± 0.002	0.096 ± 0.004	0.189 ± 0.003
AENPs	0.052 ± 0.002	0.047 ± 0.002	0.137 ± 0.000
CAENPs	0.049 ± 0.003	0.041 ± 0.000	0.122 ± 0.001
TNPs(Ours)	**0.040 ± 0.000**	**0.039 ± 0.000**	**0.111 ± 0.003**

4.2   2D Function Data Images

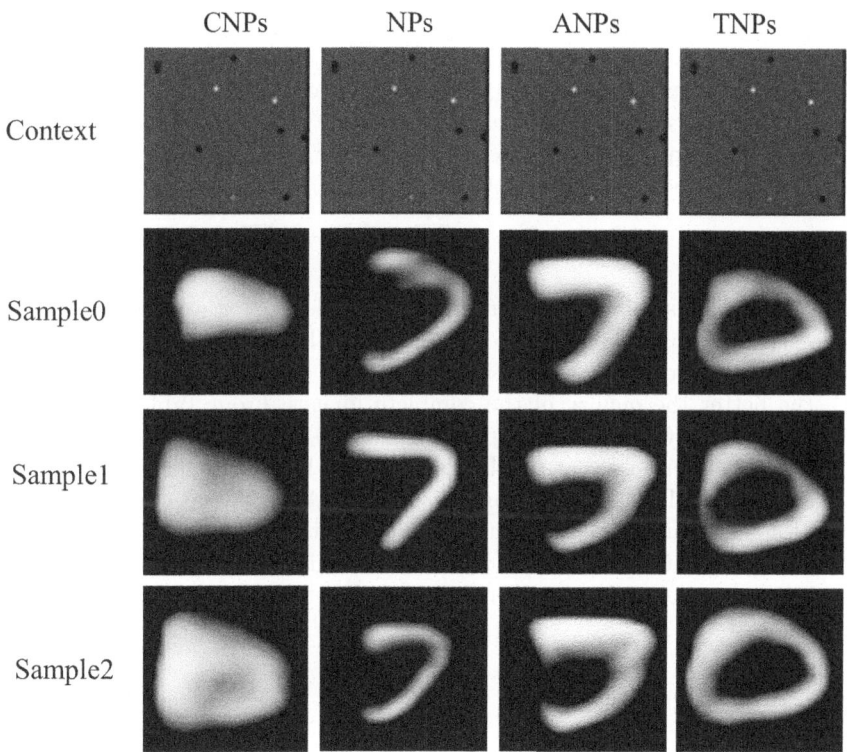

Fig. 9. We choose 10 pixel points as the context sample points. For the above 10 sample points, we use CNPs, NPs, ANPs, and TNPs methods to complement the images in the MNIST dataset. The results of 3 complete images for each model are selected for presentation

To further evaluate the TNPs model, it is important to test it on 2D datasets in addition to the 1D data experiments. In this paper, we have chosen to use images from 2D datasets as the subject of our research. We consider the dependency between pixel values in an image, where the background value is denoted as X_i and the target pixel value is denoted as Y_i. Our analysis is focused on the complete image problem, with the test and training datasets divided into various types of non-parametric priors, including NPs [4], CNPs [5], ANPs [12], BNPs [14][8], CCNPs [6][9], TransNPs [16][10], VNPs [7][11] and TNPs on the MNIST [13] dataset and CelebA [12] dataset.

The visualizations of CNPs, NPs, ANPs, and TNPs on the MNIST (as shown in Fig. 9) and CelebA (as shown in Fig. 10) datasets were evaluated, using 10 and 100 pixel points as context sample points to complete the images. The results indicate that TNPs are more effective than CNPs, NPs, ANPs, and TNPs in complementing pixel points in the images. The reason for this advantage is that TNPs use different dimensional topological features to express the latent distribution of each context sample point, allowing for correlations between them to be identified. This approach is distinct from existing NPs in which each context sample point generates a latent distribution. As a result, TNPs produce higher quality complementary images than CNPs, NPs, and ANPs.

While Figs. 9 and 10 provide a qualitative analysis of the results, it is also important to perform a quantitative analysis using different models. To achieve this, we use the number of image pixel points as context sample points, specifically 10, 50, 100, and 200. To measure the quality of the complete images, we use the Frechet Inception Distance (FID) metric [10]. We have expressed the formula for the above metric as follows:

$$FID = ||\mu_r - \mu_g||_2^2 + Tr(\sum_r + \sum_g - 2(\sum_r \sum_g)^{1/2}) \qquad (11)$$

where r represents the ground truth images, g denotes the complete images, and μ denotes the average of the distribution. The larger value of FID, the worse the quality of the complete images. On the contrary, the smaller value of FID, the better the quality of the complete images. From Table 5, it is found that the FID is lowest at sample points 10,50,100,200 for the complete images. The above results imply that the TNPs are of the highest quality than the NPs, CNPs, ANPs, BNPs, CCNPs, TransNPs, VNPs, AENPs, and CAENPs models of the complete images. The results of experiments on the CelebA and MNIST datasets show that the TNPs model achieves higher image completion quality by mining the global geometric relationships between context sample points (Tables 3 and 4).

---

[8] https://github.com/juho-lee/bnp.
[9] https://github.com/cambridge-mlg/convcnp.
[10] https://github.com/tung-nd/TNP-pytorch.
[11] https://github.com/ZongyuGuo/Versatile-NP.

Table 3. The FID result of selecting a different number of pixel points from the images as context sample points to the complete images (MNIST dataset).

Methods	10	50	100	200
	FID	FID	FID	FID
NPs [4]	87.56	79.03	71.88	65.35
CNPs [5]	90.41	87.23	74.10	63.92
ANPs [12]	79.62	72.15	60.84	52.71
BNPs [14]	74.00	63.69	52.42	49.38
CCNPs [6]	78.13	68.40	55.83	51.56
TransNPs [16]	67.42	54.51	47.29	39.50
VNPs [7]	66.31	49.00	42.13	28.94
AENPs [22]	58.16	43.57	34.83	19.00
CAENPs [22]	57.43	40.64	32.48	**17.14**
TNPs(Ours)	**50.84**	**35.00**	**30.09**	31.70

Table 4. The FID result of selecting a different number of pixel points from the images as context sample points to the complete images (CelebA dataset)

Methods	50	50	100	200
	FID	FID	FID	FID
CNPs [4]	95.18	89.03	80.57	70.14
NPs [5]	90.00	81.64	75.69	69.62
ANPs [12]	81.47	73.83	65.28	59.80
BNPs [14]	80.35	71.06	63.18	58.44
CCNPs [6]	78.13	75.32	69.21	53.78
TransNPs [16]	75.64	58.09	53.57	41.43
VNPs [7]	70.81	55.32	48.29	40.84
AENPs [22]	60.84	55.72	41.09	31.55
CAENPs [22]	55.47	43.30	39.17	30.70
TNPs(Ours)	**53.92**	**40.56**	**35.78**	**28.36**

4.3 Ablation Study

In our analysis of the TNPs model's performance, we examine the various factors that influence it. It is worth noting that topology plays a distinct and significant role in the TNPs model.

Metric Between Different Context Sample Points. To analyze the context sample point mapping results r, we applied different techniques such as k-means clustering [3], Gaussian Mixture Model (GMM) [9], and the Vietoris-Rips complex [19]. However, we only used the results obtained from these techniques to

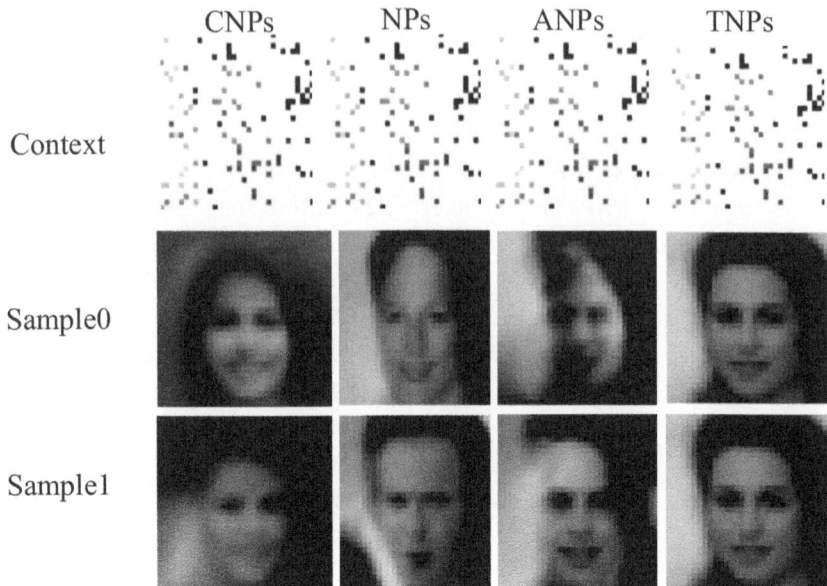

Fig. 10. We choose 100 pixel points as the context sample points. For the above 100 sample points, we use CNPs, NPs, ANPs, and TNPs methods to complement the images in the celebA dataset. The results of 2 complete images for each model are selected for presentation.

represent the latent distribution. To demonstrate the effectiveness of our experimental results, we applied two types of clustering methods, i.e., k-means and GMM, and divided the results into two regions. We used k-means to cluster the sample point r, and GMM to cluster the resulting data into two regions. Each region was then represented using a latent distribution. As shown in Table 5, our analysis revealed that topological features are the most effective in treating TNPs when compared to complete images.

Different Number of Context Sample Points. Explore the impact of different numbers of context sample points on the model. Here, we have chosen to select the number of context sample points according to the size of a ratio. We implement the above experiments on the MNIST dataset. We have chosen five different percentages of the ratio: 10%, 30%, 50%, 70%, and 90%. We also compare the results with the above NPs [4], CNPs [5], and ANPs [12], BNPs [14], CCNPs [6], TransNPs [16], VNPs [7], AENPs [22], and CAENPS [22] model as shown in Fig. 11. It is found that the TNPs outperform the results of the other models at any stage regardless of the number of context sample points. From Figure, it is found that the slope of TNPs changes more slowly for context sample points above their certain threshold (about two-thirds). Therefore, we can predict better FID results for target sample points based on fewer context sample points.

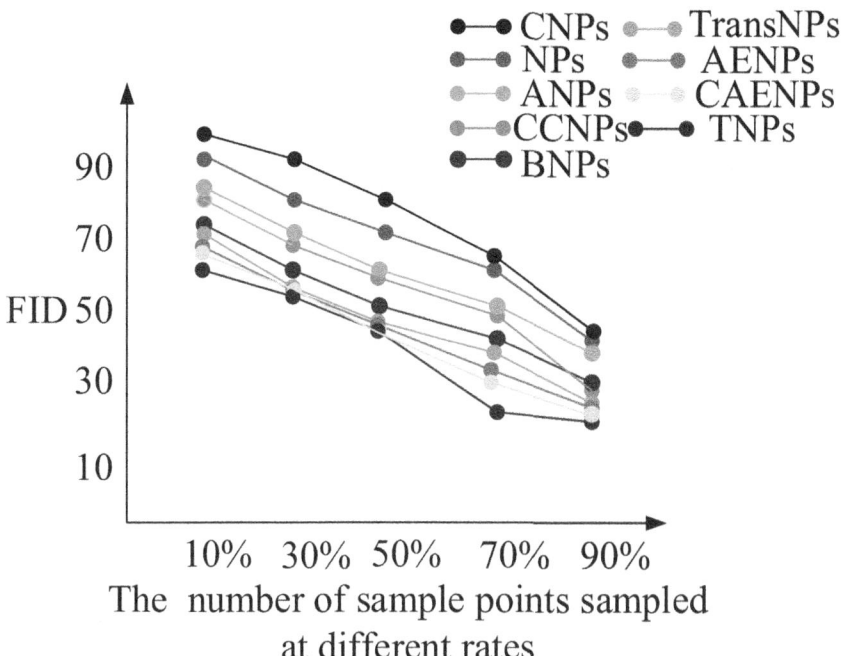

Fig. 11. We choose different ratios of sample points to compare the FID values of CNPs, NPs, ANPs, CCNPs, BNPs, TransNPs, and VNPs in the complete images (MNIST dataset). The lowest TNPs are found in the comparison of FID values for all models. It means that TNPs the complete images with the highest quality

Table 5. Context sample points mappings r respective k-means, GMM, and the Vietoris-Rips complexes were processed to represent the latent normal distribution after calculating the FID results (MNIST dataset).

Methods	10	50	100	200
	FID	*FID*	*FID*	*FID*
TNPs-k-means	70.22	65.14	54.28	42.60
TNPs-GMM	62.35	58.93	57.86	54.29
TNPs(Ours)	**59.92**	**42.56**	**31.79**	**19.30**

5   Conclusion

We propose a new method called TNPs, which aims to improve the accuracy of predicting target sample points by augmenting the expressiveness of NPs with topology. This is in response to the limitations of existing NPs, which rely on a latent distribution and do not take into account the intrinsic geometric relationships between context sample points. To address this issue, we use the Vietoris-Rips complex to calculate the topological features of context sample points, rather than simply taking an average value as in traditional NPs. This approach allows us to explore the geometric relationships between sample points more effectively. We also divide the topological features of the Vietoris-Rips complex into different dimensions and replace them with the latent distribution for each sample point. This helps to reduce computational costs and provides a good representation of the latent distribution. Our experiments with 1D and 2D datasets have shown that TNPs significantly improve the accuracy of predicting target sample points.

References

1. Bergman, A., Kellnhofer, P., Wetzstein, G.: Fast training of neural lumigraph representations using meta learning. In: Proceedings of the Conference on Neural Information Processing Systems (NeurIPS), pp. 172–186. Curran Associates, Inc. (2021)
2. Bergomi, M.G., Frosini, P., Giorgi, D., Querciaoli, N.: Towards a topological-geometrical theory of group equivariant non-expansive operators for data analysis and machine learning. Nat. Mach. Intell. **1**(9), 423–433 (2019)
3. Cho, M., Alizadeh-Vahid, K., Adya, S., Rastegari, M.: DKM: differentiable K-means clustering layer for neural network compression. In: Proceedings of the International Conference on Learning Representations (ICLR) (2022)
4. Garnelo, M., Rezende, D.J., Ali Eslami, S.M., Teh., Y.W.: Neural processes. In: Proceedings of the International Conference on Machine Learning (ICML) Workshop, vol. abs/1807.01622 (2018)
5. Garnelo, M., et al.: Conditional neural processes. In: Proceedings of the International Conference on Machine Learning (ICML). PMLR, July 2018
6. Gordon, J., Bruinsma, W.P., Foong, A.Y.K., Requeima, J., Dubois, Y., Turner, R.E.: Convolutional conditional neural processes. In: Proceedings of the International Conference on Learning Representations (ICLR) (2020)
7. Guo, Z., Lan, C., Zhang, Z., Lu, Y., Chen, Z.: Versatile neural processes for learning implicit neural representations. In: Proceedings of the Eleventh International Conference on Learning Representations (ICLR) (2023)
8. El Hanchi, A., Stephens, D., Maddison, C.: Stochastic reweighted gradient descent. In: Proceedings of the International Conference on Machine Learning (ICML), pp. 8359–8374. PMLR, July 2022
9. Hayashi, H., Uchida, S.: A discriminative gaussian mixture model with sparsity. In: Proceedings of the International Conference on Learning Representations (ICLR) (2019)
10. Heusel, M., Ramsauer, H., Unterthiner, T., Nessler, B., Hochreiter, S.: GANs trained by a two time-scale update rule converge to a local Nash equilibrium. In: Advances in Neural Information Processing Systems (NeurIPS) (2017)
11. Jing, B., Eismann, S., Suriana, P., Townshend, R.J.L., Dror, R.: Learning from protein structure with geometric vector perceptrons. In: Proceedings of the International Conference on Learning Representations (ICLR) (2021)
12. Kim, H., et al.: Attentive neural processes. arXiv preprint arXiv:1901.05761, abs/1901.05761 (2019)
13. LeCun, Y., Bottou, L., Bengio, Y., Haffner, P.: Gradient-based learning applied to document recognition. Proc. IEEE **86**(11), 2278–2324 (1998)

14. Lee, J., Yang, E., Hwang, S.J., Teh, Y.W.: Bootstrapping neural processes. In: Proceedings of the Conference on Neural Information Processing Systems (NeurIPS), pp. 6606–6615. Curran Associates, Inc. (2020)
15. Lin, W., Nielsen, F., Emtiyaz, K.M., Schmidt, M.: Tractable structured natural-gradient descent using local parameterizations. In: Proceedings of the International Conference on Machine Learning (ICML), pp. 6680–6691. PMLR, July 2021
16. Nguyen, T., Grover, A.: Transformer neural processes: uncertainty-aware meta-learning via sequence modeling. In: Proceedings of the International Conference on Machine Learning (ICML), pp. 16569–16594. PMLR, July 2022
17. Parker-Holder, J., et al.: Evolving curricula with regret-based environment design. In: Proceedings of the International Conference on Machine Learning (ICML). PMLR, July 2022
18. Pezeshki, M., Kaba, O., Bengio, Y., Courville, A.C., Precup, D., Lajoie, G.: Gradient starvation: a learning proclivity in neural networks. In: Proceedings of the Conference on Neural Information Processing Systems (NeurIPS) (2021)
19. Rosenkrantz, D.J., et al.: Efficiently learning the topology and behavior of a networked dynamical system via active queries. In: Proceedings of the International Conference on Machine Learning (ICML), pp. 18796–18808. PMLR, July 2022
20. Rothfuss, J., Chen, J., Krause, A.: Meta-learning reliable priors in the function space. In: Proceedings of the Conference on Neural Information Processing Systems (NeurIPS), pp. 280–293. Curran Associates, Inc. (2021)
21. Sun, S., Shi, J., Wilson, A.G., Grosse, R.: Scalable variational Gaussian processes via harmonic kernel decomposition. In: Proceedings of the International Conference on Machine Learning (ICML), vol. abs/2106.05992 (2021)
22. Tai, J.: Global perception based autoregressive neural processes. In: *Proceedings of the International Conference on Computer Vision (ICCV)* (2023)
23. Titsias, M.K., Schwarz, J., de G Matthews, A.G., Pascanu, R., Teh, Y.W.: Functional regularisation for continual learning using Gaussian processes. In: Proceedings of the International Conference on Learning Representations (ICLR) (2020)
24. Wong, C.-C., Vong, C.-M.: Persistent homology based graph convolution network for fine-grained 3D shape segmentation. In: Proceedings of the International Conference on Computer Vision (ICCV), pp. 7098–7107, October 2021
25. Yamada, R., Kataoka, H., Chiba, N., Domae, Y., Ogata, T.: Point cloud pre-training with natural 3D structures. In: Proceedings of the Conference on Computer Vision and Pattern Recognition (CVPR), pp. 21283–21293, June 2022
26. Ye, Z., Yao, L.: Contrastive conditional neural processes. In: Proceedings of the Conference on Computer Vision and Pattern Recognition (CVPR), pp. 9677–9686 (2022)

Novel Architectures for Computer Vision

DEEPAM: Toward Deeper Attention Module in Residual Convolutional Neural Networks

Shanshan Zhong[1], Wushao Wen[1(✉)], Jinghui Qin[2], and Zhongzhan Huang[1]

[1] Sun Yat-sen University, Guangzhou, China
{zhongshsh5,huangzhzh23}@mail2.sysu.edu.cn, wenwsh@mail.sysu.edu.cn
[2] Guangdong University of Technology, Guangzhou, China

Abstract. The efficacy of depth in boosting the performance of residual convolutional neural networks (CNNs) has been well-established through abundant empirical or theoretical evidences. However, despite the attention module (AM) being a crucial component for high-performance CNNs, most existing research primarily focuses on their structural design, overlooking a direct investigation into the impact of AM depth on performance. Therefore, in this paper, we explore the influence of AM depth under various settings in detail. We observe that (1) appropriately increasing AM depth significantly boosts performance; (2) deepening AM exhibits a higher cost-effectiveness compared to traditional backbone deepening. However, deepening AM introduces inherent challenges in terms of parameter and inference cost. To mitigate them while enjoying the benefit of deepening AM, we propose a novel AM called DEEPAM, leveraging mechanisms from recurrent neural networks and the design of lightweight AMs. Extensive experiments on widely-used benchmarks and popular attention networks validate the effectiveness of our proposed DEEPAM.

Keywords: Attention mechanism · Visual recognition

1 Introduction

For visual recognition, growing empirical and theoretical evidences [2,12,13,18, 22,27,36,37,45] highlights the critical role of **depth** in enhancing the effectiveness of residual convolutional neural networks (CNNs), which can usually [12] formulated as

$$x_{t+1} = x_t + f_t(x_t), 1 \leq t \leq m, \qquad (1)$$

where m is number of blocks, a.k.a. the depth of CNN, and x_t is the input of the t-th block and $f_t(\cdot)$ is the residual mapping. In general, as the value of m increases, neural networks have greater potential to achieve high performance [2]. However, for the attention modules (AMs) [16,19,40,42], as one of the most popular methods for enhancing CNNs' performance, which can be written as:

$$x_{t+1} = x_t + f_t(x_t) \otimes \sigma[\mathbf{A}(f_t(x_t); W_t^{(1)})], \qquad (2)$$

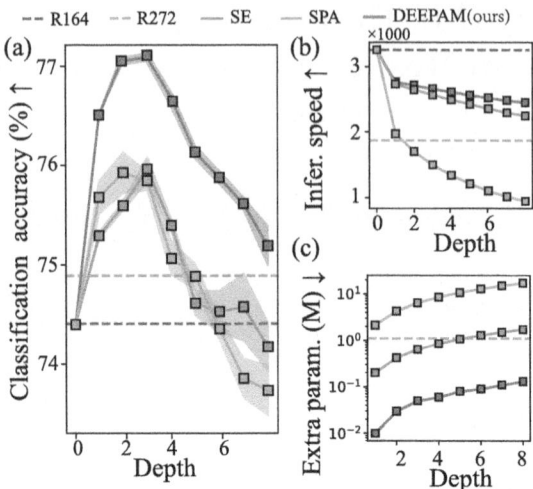

Fig. 1. The exploration of deepening AM depth d. R164 denotes ResNet164. All experiments of various AMs are based on ResNet164. (a) The classification accuracy (%): It shows a rapid increase followed by a gradual decrease. And a suitably increasing, like $d = 2$ or $d = 3$, can significantly boost the performance. (b) and (c) The extra number of parameters (million) and inference speed (frames per second): large d have a great impact on the computational cost.

where $\mathbf{A}(\cdot; W_t^{(1)})$ is the AM with learnable parameters $W_t^{(1)}$, σ is usually a Sigmoid function and $\otimes$ is the element-wise multiplication, existing research [16,17, 19,25,28,40,42,52] primarily focuses on the structural design of $\mathbf{A}(\cdot; W_t^{(1)})$, with little exploration into the impact of the "depth" d of the AM on performance. Therefore, we first conduct an in-depth investigation into d, which can be simply defined as the number of times the AM with the same structure is used in one block, e.g., $x_{t+1} = x_t + f_t(x_t) \otimes \sigma(a_d)$, where

$$a_i = \mathbf{A}(a_{i-1}; W_t^{(i)}), \quad 2 \leq i \leq d, \tag{3}$$

is the feature measured from i-th depth of AM and $a_1 = \mathbf{A}(f_t(x_t); W_t^{(1)})$. Taking ResNet164 on CIFAR100 dataset [24] as an example, we select two popular AMs as baseline, including SE [16] and SPA [10], to explore the influence of depth d. For each setting, we repeat the run under 5 different random seeds. The results is illustrated in Fig. 1, we can observe that:

(1) **Suitably increasing depth d significantly boosts performance.**
Figure 1 (a) illustrates the trend of the model's performance as the depth d of AM increases. The performance shows a rapid increase followed by a gradual decrease, reaching its optimal level at $d = 2$ or $d = 3$. This phenomenon may be attributed to the under-attention in shallow d and over-attention in deep d, and see Sect. 5 for details.

(2)**Deepening AM shows more cost-effective than traditional backbone deepening**. Notably, from Fig. 1, even when the depth d of AM is doubled or tripled, the associated increase in computational costs, including parameter and inference time, tends to be considerably smaller than that incurred by deepening ResNet164 (R164) to ResNet272 (R272). Moreover, the AM with these depth d also maintains a significant performance improvement advantage.

Based on the aforementioned observations and considering the inherent additional parameter and inference costs associated with deepening AM, as shown in Fig. 1 (b-c), in this paper, we draw inspiration from the mechanisms of recurrent neural networks [14] and the principles of lightweight AM design. We propose a novel attention mechanism named DEEPAM to allow CNNs to enjoy the performance benefits of deepening AM while mitigating the associated additional computational costs as much as possible. In Sect. 4, we conduct extensive experiments on widely-used benchmarks and popular attention networks to validate the effectiveness of our DEEPAM. Finally, in Sect. 5, we perform a detailed ablation study on the design of the proposed AM, analyzing the limitations of DEEPAM and examining the phenomenon of performance initially rising and then declining with the increase of depth d as shown in Fig. 1 (a).

We summarize **our contribution** as follows:

– Motivated by the profound impact of depth on CNNs, we explore the attention modules' (AMs) depth d. We find that suitably increasing d significantly enhances model performance, with a performance-to-cost ratio far superior to that achieved by deepening the network backbone.
– We propose a novel attention module named DEEPAM, which aims to enable CNNs to enjoy the performance benefits of deepening AMs while mitigating the extra computational cost associated with deepening.
– We further delve into a detailed discussion of the observed phenomenon of initial performance improvement followed by a decline as the depth d increases. Besides, we also analyze the structural design and limitations of DEEPAM.

2   Related Works

2.1   Attention Mechanism in CNNs

Attention, in a broad sense, can be conceptualized as a mechanism for directing processing resources towards the most salient aspects of an input signal [10,16,42]. Notably, visual attention integrated with Convolutional Neural Networks has gained prominence in various computer vision tasks, facilitating the identification of pertinent regions within images and capturing long-range structural dependencies among image components [1,3,7,8,29,55]. These methods encompass diverse areas such as image classification [21,44,54], image super-resolution [50], object detection [5,20], semantic segmentation [6,46], person re-identification [26], action recognition [41], image generation [9,47], and 3D vision [44].

Algorithm 1. Deep attention module (DEEPAM) in one block

Input: Input feature x_t. $d-1$ group normalizations $\{\mathbf{GN}_k\}_{k=1}^{d-1}$. Linear transformation parameter γ and β, residual mapping $f_t(\cdot)$.
Output: The feature x_{t+1}.

1:   ▷ Measuring Original Attention
2: Extract global average information $a_0 = \mathbf{GAP}(x_t) \in \mathbb{R}^{C \times 1 \times 1}$
3: Measure the initial attention $a_1 = a_0 \otimes \gamma + \beta$
4:   ▷ Generating Deeper Attetion
5: **for** k from 2 to d **do**
6:     $a_k = \mathbf{GN}_{k-1}(a_{k-1}) \otimes \gamma + \beta$
7: **end for**
8: **return** $x_{t+1} = x_t + f_t(x_t) \otimes \sigma(a_d)$

2.2 Enhanced Methods of Attention Mechanism

The purpose of attention modules is to capture dependencies, whether channel-wise or spatial-wise, through the utilization of global information embeddings. Researchers explore various methods to enhance attention modules from multiple perspectives. SENet [16] adaptively recalibrates channel-wise feature responses by explicitly modeling interdependencies between channels, which makes use of fully connected networks. CBAM [42] sequentially infers attention maps along two separate dimensions, channel and spatial, and then the attention maps are multiplied by the input feature map for adaptive feature refinement. ECA [40] can be efficiently implemented via 1D convolution. LSAS [54] enhances the discriminative ability of attention modules by incorporating sub-attention modules. ESA [52], taking into account the strengths of different attention modules, devises an automatic mechanism for integrating various excitation modules across different network layers. Unlike these studies, we introduce DEEPAM, which harnesses mechanisms from recurrent neural networks and employs lightweight attention modules. This approach enables CNNs to exploit the performance advantages of deep attention modules while alleviating the additional computational overhead associated with deepening them.

3 Deep Attention Module

In Fig. 1, we empirically showcase the enhanced performance resulting from an appropriately increased attention module depth (d). In this section, as illustrated in Fig. 2 and Algorithm 1, we present our innovative attention module, DEEPAM, which is specifically crafted to alleviate the extra computational burden associated with deepening attention modules, all while reaping the performance advantages of such deepening. This is achieved through a series of steps outlined below:

(1) **Lightweight module ensuring the efficiency.** To ensure the efficiency of the deepening AM process, it is essential for the employed $\mathbf{A}(\cdot; W_t^{(1)})$ to

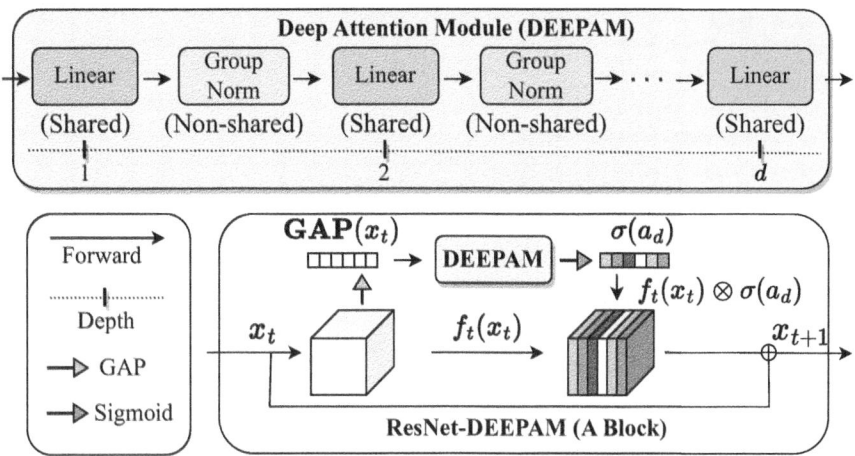

Fig. 2. The architecture details of DEEPAM. **GAP** is global average pooling and σ is Sigmoid activation function. Figure 3 shows a specific Pytorch implementation of DEEPAM.

be sufficiently lightweight [40]. Therefore, we consider a simple-yet-effective linear transformation, denoted as $g(\cdot)$, to serve as the foundational structure of DEEPAM. Some previous works [28,54] have shown that even simple linear transformation can also measure powerful attention value. Specifically, for a tensor $X \in \mathbb{R}^{C \times 1 \times 1}$, where C denotes the number of channels, we have:

$$g(X) = X \otimes \gamma + \beta, \qquad (4)$$

where the two learnable parameters $\gamma \in \mathbb{R}^{C \times 1 \times 1}$ and $\beta \in \mathbb{R}^{C \times 1 \times 1}$ take charge of scaling and shifting the tensor X to calibrate the representation power of feature map. The parameters of each $g(\cdot)$ is $2C$.

(2) **Sharing parameters to reduce cost.** As shown in Eq. (3), when the depth d in the t-th block is large, the extra number of parameters from $\{W_t^{(i)}\}_{i=2}^d$ become significant for both the model's inference and training phases. Therefore, even with the lightweight module illustrated in Eq. (4), we still need to consider further reducing the parameter cost. Motivated by recurrent neural networks [14], we consider reusing certain parameters within the AM. As shown in Fig. 2, all linear transformers $g(\cdot)$ are shared learnable parameters. Consequently, the total number of parameters introduced by $g(\cdot)$ in any block of DEEPAM is limited to only $2C$. Nevertheless, iteratively scaling and shifting features in a manner similar to Eq. (4) may result in certain performance drawbacks (refer to Sect. 5 for details). Hence, we will introduce additional normalization methods building upon this observation.

(3) **Using normalisation to stabilize attention.** Due to the introduction of a recurrent mechanism in AM, the previous works [14,15] on recurrent neural networks have revealed the potential risk of amplifying gradient instability. It may lead to training instability and poor performance, which is consistent with our

```
1    import torch
2    import torch.nn as nn
3
4    class LinearTranModule(nn.Module):
5        def __init__(self, planes):
6            super(LinearTranModule, self).__init__()
7            self.w = nn.Parameter(torch.Tensor(1, planes, 1, 1))
8            self.b = nn.Parameter(torch.Tensor(1, planes, 1, 1))
9            self.w.data.fill_(0)
10           self.b.data.fill_(-1)
11
12       def forward(self, x):
13           return x * self.w + self.b
14
15
16   class DeepAM(nn.Module):
17       def __init__(self, planes, deep=2, group=4):
18           super(DeepAM, self).__init__()
19           org_lt = LinearTranModule(planes)
20           self.att = nn.Sequential(
21               nn.AdaptiveAvgPool2d((1, 1)),
22               org_lt)
23           for _ in range(1, deep):
24               self.att.add_module(str(len(self.att)), nn.GroupNorm(int(planes/group), planes))
25               self.att.add_module(str(len(self.att)), org_lt)
26           self.att.add_module(str(len(self.att)), nn.Sigmoid())
27
28       def forward(self, x):
29           out = self.att(x)
30           out = out * x
31           return out
```

Fig. 3. The Pytorch implementation of DEEPAM. Following the setting of [28], the two learnable parameters $\gamma \in \mathbb{R}^{C \times 1 \times 1}$ (self.w) and $\beta \in \mathbb{R}^{C \times 1 \times 1}$ (self.b) are initialised as all-zero and all-minus one vectors.

observations in Sect. 5. Therefore, we further incorporate normalization between linear transformations $g(\cdot)$ to stabilize attention. In contrast to AMs that consider batch normalization [23], such as SRM [25], we utilize group normalization (GN) [43] as illustrated in Fig. 2. This choice is motivated by the idea that, for attention mechanisms, attention values should only depend on the current sample and not mix information from other samples in the batch, which could interfere with the focusing region of the attention module. In Sect. 5, we will provide a detailed performance comparison of different normalizations, where the experimental results align with our analysis, and see Sect. 5 for details. Additionally, it is worth noting that, similar to $g(\cdot)$, each GN has learnable parameters of $2C$, resulting in a total number of parameters of DEEPAM is $\mathcal{O}(C)$.

4   Experiments

Implementation Details. In this Section, to evaluate the efficacy of DEEPAM, extensive experiments are conducted in two tasks: image classification and instance segmentation. For image classification, ResNet serves as the backbone,

Table 1. The image classification for various attention modules, backbones, and datasets. #P(M) denotes the number of parameters and the frames per second (FPS) is used for measuring the inference speed.

Model	STL10			CIFAR100			CIFAR10		
	#P(M) ↓	FPS ↑	top-1 acc. ↑	#P(M) ↓	FPS ↑	top-1 acc. ↑	#P(M) ↓	FPS ↑	top-1 acc. ↑
ResNet83 [12]	0.87	844	80.96	0.89	6351	73.55	0.87	6361	93.62
+SE [16]	0.97	717	82.98	0.99	5345	74.62	0.97	5346	94.21
+CBAM [42]	0.97	442	82.31	0.99	3102	74.39	0.97	3103	93.99
+SRM [25]	0.89	660	82.09	0.91	4940	74.49	0.89	4943	94.55
+SPA [10]	1.93	584	77.54	1.96	3894	74.64	1.93	3896	94.43
+DIA [19]	1.09	708	84.70	1.11	4701	75.48	1.09	4705	94.40
+FCA [32]	0.98	671	84.94	1.01	3967	75.75	0.98	3968	94.27
+SA [49]	0.87	462	82.05	0.89	3670	74.66	0.87	3672	93.83
+SPEM [53]	0.89	498	83.68	0.91	3383	73.87	0.89	3385	94.03
+ESA [52]	0.99	710	84.46	1.01	5117	74.90	0.99	5122	94.20
+StepNet [17]	0.89	444	84.97	0.91	2600	75.31	0.89	2601	94.57
+DEEPAM ($d=2$)	0.89	716	**85.76** (↑ 4.80)	0.91	5306	75.71 (↑ 2.16)	0.89	5311	**94.70** (↑ 1.08)
+DEEPAM ($d=3$)	0.89	709	**85.80** (↑ 4.84)	0.92	5210	**76.01** (↑ 2.46)	0.89	5208	94.67 (↑ 1.05)
ResNet164 [12]	1.70	432	82.66	1.73	3253	74.30	1.70	3254	93.45
+SE [16]	1.91	368	83.96	1.93	2723	75.25	1.91	2723	94.26
+CBAM [42]	1.90	224	83.96	1.93	1562	74.68	1.90	1563	93.91
+SRM [25]	1.74	338	84.73	1.76	2513	75.49	1.74	2514	94.59
+SPA [10]	3.83	298	75.33	3.86	1974	75.68	3.83	1976	94.57
+DIA [19]	1.92	363	85.39	1.95	2417	76.59	1.92	2416	94.50
+FCA [32]	1.81	343	85.34	1.83	2010	76.52	1.81	2011	94.66
+SA [49]	1.70	235	83.81	1.73	1859	74.86	1.70	1858	94.13
+SPEM [53]	1.74	254	84.39	1.76	1710	76.31	1.74	1711	94.80
+ESA [52]	1.95	364	85.73	1.97	2607	76.49	1.95	2609	94.52
+StepNet [17]	1.74	219	86.08	1.77	1261	77.04	1.74	1263	95.14
+DEEPAM ($d=2$)	1.74	368	**87.28** (↑ 4.62)	1.76	2703	77.08 (↑ 2.78)	1.74	2707	**95.16** (↑ 1.71)
+DEEPAM ($d=3$)	1.75	359	86.66 (↑ 4.00)	1.78	2652	**77.23** (↑ 2.93)	1.75	2653	95.10 (↑ 1.65)

and a diverse set of popular and state-of-the-art attention modules are selected as baselines for performance comparison with DEEPAM. The evaluation metric for image classification is the top-1 accuracy and Frames Per Second (FPS) for inference speed. For FPS measurements, we conducted 100 repeated forward experiments on an RTX2080 GPU with a batch size of 64. Four widely used image recognition datasets are employed, namely, ImageNet [33], STL10 [4], CIFAR10/100 [24]. ImageNet comprises 224×224-sized images distributed across 1000 classes. STL10 consists of 5k training images and 8k test images with a resolution of 96×96, categorized into 10 classes. CIFAR10 and CIFAR100 include 50k training images and 10k test images of size 32 by 32, with 10 and 100 classes, respectively. Additionally, the instance segmentation performance of DEEPAM is evaluated on MS COCO 2017 [30] using Mask R-CNN [11]. MS COCO consists of 80 categories, and the average precision (AP) is employed as the metric to measure image segmentation performance.

Experimental Results. In the classification experiments, as shown in Tables 1 and 2, our proposed method, DEEPAM, demonstrates outstanding classification performance and surpasses numerous state-of-the-art AMs across four popular

Table 2. The image classification for various attention modules, backbones, and large scale dataset ImageNet. #P(M) denotes the number of parameters and the frames per second (FPS) is used for measuring the inference speed.

Model	#P(M) ↓	FPS ↑	top-1 acc. ↑
ResNet18 [12]	11.69	2778	69.98
+SE [16]	11.78	2624	70.57
+CBAM [42]	11.78	2101	70.77
+SPA [10]	12.65	2304	70.82
+DEEPAM ($d = 2$)	11.70	2595	71.31 (↑ 1.33)
+DEEPAM ($d = 3$)	11.70	2581	**71.45** (↑ 1.47)
ResNet34 [12]	21.80	1634	73.99
+SE [16]	21.96	1546	74.37
+CBAM [42]	21.96	1212	74.89
+SPA [10]	23.53	1341	74.59
+DEEPAM ($d = 2$)	21.81	1533	75.16 (↑ 1.17)
+DEEPAM ($d = 3$)	21.82	1526	**75.44** (↑ 1.45)
ResNet50 [12]	25.56	838	76.02
+SE [16]	28.09	757	76.62
+CBAM [42]	28.09	538	76.38
+SPA [10]	53.22	597	77.02
+DEEPAM ($d = 2$)	25.62	754	**77.64** (↑ 1.62)
+DEEPAM ($d = 3$)	25.65	750	77.36 (↑ 1.34)

datasets, including STL10, CIFAR10, CIFAR100, and ImageNet, as well as various network backbones. Remarkably, this performance improvement is achieved with a modest increase in inference speed and number of parameters. In instance segmentation experiments, as depicted in Table 3, DEEPAM exhibits consistent performance gains similar to the classification experiments. Additionally, we can also observe that, in both types of experiments, the performance at $d = 2$ and $d = 3$ is similar, with no single depth d consistently dominating the other.

5   Ablation Study and Analysis

(1) **Why do we use normalization in DEEPAM?** In Sect. 3, we mention that the primary reason for introducing GN is its ability to alleviate potential gradient instability risks arising from the recurrent mechanism, as empirically validated in Fig. 4(a). Specifically, the model's performance improvement significantly deteriorates when normalization is not applied, and instability is observed across different random seeds.

Furthermore, in Sect. 3, we discuss the rationale for not using classical batch normalization, emphasizing that attention values should only depend on the

Table 3. The instance segmentation for existing attention modules and DEEPAM on MS COCO 2017 and Mask R-CNN. "AP" is the average precision for the instance segmentation task. "AP_{50}" and "AP_{75}": AP at IoU as 50% and 75%.

Model	AP	AP_{50}	AP_{75}
ResNet50 [12]	34.5	54.8	36.5
+SE [16]	35.7	56.7	37.6
+DEEPAM ($d = 2$)	**36.1** (↑ 1.6)	57.8 (↑ 3.0)	**38.2** (↑ 1.7)
+DEEPAM ($d = 3$)	36.0 (↑ 1.5)	**58.1** (↑ 3.4)	37.9 (↑ 1.4)
ResNet101 [12]	35.9	56.9	38.0
+SE [16]	36.6	59.0	39.0
+DEEPAM ($d = 2$)	37.7 (↑ 1.8)	60.5 (↑ 3.6)	**40.1** (↑ 2.1)
+DEEPAM ($d = 3$)	**37.9** (↑ 2.0)	**60.6** (↑ 3.7)	39.9 (↑ 1.9)

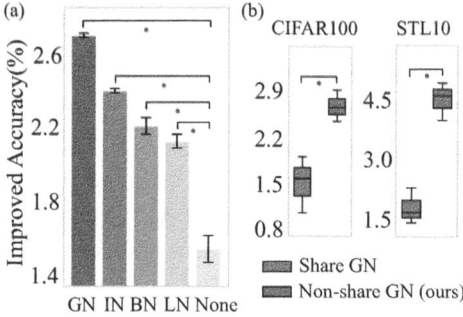

Fig. 4. The further analysis for normalization (norm) with $d = 2$. (a) The performance of other norm alternatives on CIFAR100. (b) The performance comparison under the settings of share GN and non-share GN (ours). IN, BN, and LN denote instance, batch, and layer norm, respectively. None means the DEEPAM without any norm. "*" denotes the p value $p < 0.05$ while using the Student's t-test [31] with the significance level (α) at 5%.

current sample and should not incorporate information from irrelevant samples in the batch. In practice, Instance Normalization (IN) [38], as a special case of GN, also considers only the information from the current sample. As evident from the results in Fig. 4 (a), IN demonstrates the second best performance, which is consistent with our analysis. However, since GN has been shown to exhibit more robust performance than IN across various tasks [43], we still recommend the use of GN.

(2) **Why does the performance curve exhibit an initial increase followed by a decrease?** In Fig. 1 (a), we illustrate a trend in the performance with an increase in depth d: initially rising and then declining. Taking ImageNet and STL10 as examples, we conduct a preliminary analysis of the potential causes of this phenomenon using Grad-CAM [34]. The visual results in Fig. 5

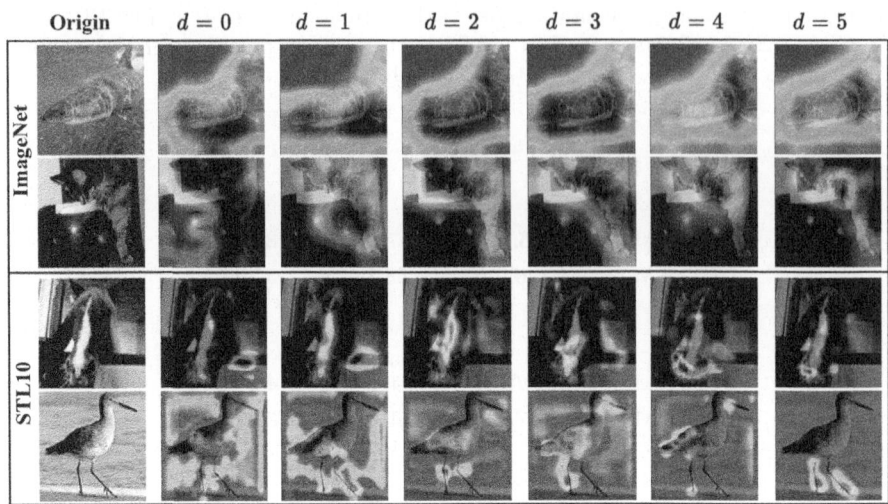

Fig. 5. Grad-CAM [34] visualization under various depth d. The red region indicates an essential place for a network to focus on while the blue one represents the opposite.

reveal that, compared to $d = 0$ (i.e., no use of AM), the model exhibits improved focus on the target at $d = 1$, yet it still lacks sufficient attention (a.k.a under-attention). This situation is alleviated at $d = 2$ or $d = 3$, where the model effectively concentrates attention, leading to a rapid performance improvement. However, with a further increase in depth d, the attention becomes overly confident (a.k.a over-attention), excessively concentrating on specific regions and deviating from the overall objective, resulting in a decline in performance. In fact, the iterative generation of attention is akin to a continuous refinement and calibration of attention values. Since there is a limit to the effectiveness of refinement [12,54], the transition from under-attention to over-attention is both natural and reasonable.

(3) **Why is normalization non-shared?** In our DEEPAM, the linear transformation is designed to reduce computational costs through shared parameters. A natural question arises: can we also implement parameter sharing for normalization (norm)? In Fig. 4 (b), we present a performance comparison between shared and non-shared norm, revealing a significant performance decline (p value $p < 0.05$ while using the Student's t-test [31]) when norm parameters are shared. This aligns with observations from prior work [35,39,48] introducing recurrence mechanisms in the backbone, which found that norm exhibits sensitivity and is not recommended to be involved in the recurrence. Instead, multiple norms are advised to be independently set during the recurrence process (see Fig. 2). Additionally, another intuitive explanation is that excluding certain parameters from the recurrence in AM contributes to the diversity of features during the attention generation process. Modules with entirely identical parameters yield

outputs with similar distributions, hindering the learning of more diverse features that could enhance performance.

(4) **Limitation and future work**. Despite the empirical success of $d = 2$ or $d = 3$ in enhancing the performance of AM, determining the optimal d still requires careful experimentation when designing new attention modules. Future investigations into the selection of d raise additional questions, such as whether different layers or blocks necessitate the same d and if there exist alternative definitions for the "depth" of AM. Addressing these questions will contribute to further optimizing the performance of AM. Additionally, in the current DEEPAM architecture, the computation of attention values involves multiple normalization steps, which incurs a certain computational cost. Exploring more specialized structures that allow for structural reparameterization [51] of the AM during the inference stage could significantly boost efficiency. However, these improvements require additional experiments to validate and represent a crucial avenue for future research.

6   Conclusion

In this paper, we empirically demonstrate that appropriately deepening the attention module (AM) significantly enhances model performance. This improvement exhibits greater cost-effectiveness when compared to deepening the network backbone. To simultaneously enjoy the performance gains from deepening the AM and alleviate the inherent costs in terms of parameters and inference speed, we introduce a novel attention module called DEEPAM. Extensive experiments and analysis conducted on widely-used benchmarks and popular AMs validate the effectiveness of our proposed DEEPAM.

Acknowledgments. This work was partly supported by the National Natural Science Foundation of China under Grant No. 62206314, No. 623B2099, and No. U1711264, GuangDong Basic and Applied Basic Research Foundation under Grant No. 2022A1515011835, Science and Technology Projects in Guangzhou under Grant No. 2024A04J4388.

References

1. Behera, A., Wharton, Z., Hewage, P.R., Bera, A.: Context-aware attentional pooling (cap) for fine-grained visual classification. In: Proceedings of the AAAI Conference on Artificial Intelligence, vol. 35, pp. 929–937 (2021)
2. Brahma, P.P., Wu, D., She, Y.: Why deep learning works: a manifold disentanglement perspective. TNNLS **27**(10), 1997–2008 (2015)
3. Chaudhari, S., Mithal, V., Polatkan, G., Ramanath, R.: An attentive survey of attention models. ACM Trans. Intell. Syst. Technol. (TIST) **12**(5), 1–32 (2021)
4. Coates, A., Ng, A., Lee, H.: An analysis of single-layer networks in unsupervised feature learning. In: JMLR Workshop, pp. 215–223 (2011)
5. Dai, J., et al.: Deformable convolutional networks. In: Proceedings of the IEEE International Conference on Computer Vision, pp. 764–773 (2017)

6. Fu, J., et al.: Dual attention network for scene segmentation. In: 2019 IEEE/CVF Conference on Computer Vision and Pattern Recognition (CVPR) (2020)
7. Gao, H., Pei, J., Huang, H.: Conditional random field enhanced graph convolutional neural networks. In: Proceedings of the 25th ACM SIGKDD International Conference on Knowledge Discovery and Data Mining, pp. 276–284 (2019)
8. Gao, Z., Xie, J., Wang, Q., Li, P.: Global second-order pooling convolutional networks. In: CVPR, pp. 3024–3033 (2019)
9. Gregor, K., Danihelka, I., Graves, A., Rezende, D., Wierstra, D.: Draw: a recurrent neural network for image generation. In: International Conference on Machine Learning, pp. 1462–1471. PMLR (2015)
10. Guo, J., Ma, X., et al: Spanet: spatial pyramid attention network for enhanced image recognition. In: ICME, pp. 1–6. IEEE (2020)
11. He, K., Gkioxari, G., Dollár, P., Girshick, R.: Mask r-cnn. In: ICCV, pp. 2961–2969 (2017)
12. He, K., Zhang, X., Ren, S., Sun, J.: Deep residual learning for image recognition. In: CVPR, pp. 770–778 (2016)
13. He, W., Huang, Z., Liang, M., Liang, S., Yang, H.: Blending pruning criteria for convolutional neural networks. In: Farkaš, I., Masulli, P., Otte, S., Wermter, S. (eds.) ICANN 2021. LNCS, vol. 12894, pp. 3–15. Springer, Cham (2021). https://doi.org/10.1007/978-3-030-86380-7_1
14. Hochreiter, S., Schmidhuber, J.: Long Short-Term Memory, vol. 9, pp. 1735–1780. MIT Press (1997)
15. Hopfield, J.J.: Neural networks and physical systems with emergent collective computational abilities. PNAS $79(8)$, 2554–2558 (1982)
16. Hu, J., Shen, L., Sun, G.: Squeeze-and-excitation networks. In: CVPR, pp. 7132–7141 (2018)
17. Huang, Z., Liang, M., Qin, J., Zhong, S., Lin, L.: Understanding self-attention mechanism via dynamical system perspective. In: ICCV, pp. 1412–1422 (2023)
18. Huang, Z., Liang, S., Liang, M., He, W., Yang, H., Lin, L.: The lottery ticket hypothesis for self-attention in convolutional neural network. arXiv preprint arXiv:2207.07858 (2022)
19. Huang, Z., Liang, S., Liang, M., Yang, H.: Dianet: dense-and-implicit attention network. In: AAAI, pp. 4206–4214 (2020)
20. Huang, Z., Shao, W., Wang, X., Lin, L., Luo, P.: Convolution-weight-distribution assumption: rethinking the criteria of channel pruning. arXiv preprint arXiv:2004.11627 (2020)
21. Huang, Z., Shao, W., Wang, X., Lin, L., Luo, P.: Rethinking the pruning criteria for convolutional neural network. Adv. Neural. Inf. Process. Syst. **34**, 16305–16318 (2021)
22. Huang, Z., Zhou, P., Yan, S., Lin, L.: Scalelong: towards more stable training of diffusion model via scaling network long skip connection. Adv. Neural Inf. Process. Syst. **36** (2024)
23. Ioffe, S., Szegedy, C.: Batch normalization: accelerating deep network training by reducing internal covariate shift. In: ICML, pp. 448–456. PMLR (2015)
24. Krizhevsky, A., Hinton, G., et al.: Learning multiple layers of features from tiny images (2009)
25. Lee, H., Kim, H.E., Nam, H.: SRM: a style-based recalibration module for convolutional neural networks. In: ICCV, pp. 1854–1862 (2019)
26. Li, W., Zhu, X., Gong, S.: Harmonious attention network for person re-identification. In: Proceedings of the IEEE Conference on Computer Vision and Pattern Recognition, pp. 2285–2294 (2018)

27. Liang, M., Zhou, J., Wei, W., Wu, Y.: Balancing between forgetting and acquisition in incremental subpopulation learning. In: European Conference on Computer Vision, pp. 364–380. Springer (2022)
28. Liang, S., Huang, Z., Liang, M., Yang, H.: Instance enhancement batch normalization: an adaptive regulator of batch noise. In: AAAI, vol. 34, pp. 4819–4827 (2020)
29. Liang, S., Huang, Z., Liang, M., Yang, H.: Instance enhancement batch normalization: an adaptive regulator of batch noise. In: Proceedings of the AAAI Conference on Artificial Intelligence, vol. 34, pp. 4819–4827 (2020)
30. Lin, T.Y., et al.: Microsoft coco: common objects in context. In: ECCV, pp. 740–755. Springer (2014)
31. Papoulis, A.: Probability and Statistics. Prentice-Hall, Inc. (1990)
32. Qin, Z., Zhang, P., Wu, F., Li, X.: Fcanet: frequency channel attention networks. In: ICCV, pp. 783–792 (2021)
33. Russakovsky, O., Deng, J., et al.: Imagenet large scale visual recognition challenge. IJCV **115**(3), 211–252 (2015)
34. Selvaraju, R.R., Cogswell, M., Das, A., Vedantam, R., Parikh, D., Batra, D.: Grad-cam: visual explanations from deep networks via gradient-based localization. In: ICCV, pp. 618–626 (2017)
35. Shen, Z., Liu, Z., Xing, E.: Sliced recursive transformer. In: ECCV, pp. 727–744. Springer (2022)
36. 44 Shi, C., Ni, H., Li, K., Han, S., Liang, M., Min, M.R.: Exploring compositional visual generation with latent classifier guidance. In: Proceedings of the IEEE/CVF Conference on Computer Vision and Pattern Recognition, pp. 853–862 (2023)
37. Sun, S., Chen, W., Wang, L., Liu, X., Liu, T.Y.: On the depth of deep neural networks: a theoretical view. In: AAAI, vol. 30 (2016)
38. Ulyanov, D., Vedaldi, A., Lempitsky, V.: Instance normalization: the missing ingredient for fast stylization. arXiv (2016)
39. Wang, J., Chen, Y., Yu, S.X., Cheung, B., LeCun, Y.: Compact and optimal deep learning with recurrent parameter generators. In: WACV, pp. 3900–3910 (2023)
40. Wang, Q., Wu, B., Zhu, P., Li, P., Hu, Q.: Eca-net: efficient channel attention for deep convolutional neural networks. In: CVPR (2020)
41. Wang, X., Girshick, R., Gupta, A., He, K.: Non-local neural networks. In: CVPR, pp. 7794–7803 (2018)
42. Woo, S., Park, J., Lee, J.Y., Kweon, I.S.: Cbam: convolutional block attention module. In: ECCV, pp. 3–19 (2018)
43. Wu, Y., He, K.: Group normalization. In: ECCV, pp. 3–19 (2018)
44. Xie, S., Liu, S., Chen, Z., Tu, Z.: Attentional shapecontextnet for point cloud recognition. In: Proceedings of the IEEE Conference on Computer Vision and Pattern Recognition, pp. 4606–4615 (2018)
45. Xing, X., Liang, M., Wu, Y.: Toa: Task-oriented active VGA. In: Thirty-Seventh Conference on Neural Information Processing Systems (2023)
46. Yuan, Y., Huang, L., Guo, J., Zhang, C., Chen, X., Wang, J.: Ocnet: object context network for scene parsing. arXiv preprint arXiv:1809.00916 (2018)
47. Zhang, H., Goodfellow, I., Metaxas, D., Odena, A.: Self-attention generative adversarial networks. In: International Conference on Machine Learning, pp. 7354–7363. PMLR (2019)
48. Zhang, J., et al.: Minivit: compressing vision transformers with weight multiplexing. In: CVPR, pp. 12145–12154 (2022)
49. Zhang, Q.L., Yang, Y.B.: Sa-net: shuffle attention for deep convolutional neural networks. In: ICASSP, pp. 2235–2239. IEEE (2021)

50. Zhang, Y., Li, K., Li, K., Wang, L., Zhong, B., Fu, Y.: Image super-resolution using very deep residual channel attention networks. In: ECCV, pp. 286–301 (2018)
51. Zhong, S., Huang, Z., Wen, W., Qin, J., Lin, L.: ASR: attention-alike structural re-parameterization. arXiv (2023)
52. Zhong, S., Huang, Z., Wen, W., Yang, Z., Qin, J.: ESA: excitation-switchable attention for convolutional neural networks. Neurocomputing **557**, 126706 (2023)
53. Zhong, S., Wen, W., Qin, J.: SPEM: self-adaptive pooling enhanced attention module for image recognition. In: Dang-Nguyen, D.-T., et al. (eds.) MMM 2023, Part II, pp. 41–53. Springer, Cham (2023). https://doi.org/10.1007/978-3-031-27818-1_4
54. Zhong, S., Wen, W., Qin, J., Chen, Q., Huang, Z.: LSAS: lightweight sub-attention strategy for alleviating attention bias problem. arXiv (2023)
55. Zhu, W., Yeh, W., Chen, J., Chen, D., Lin, Y.: Evolutionary convolutional neural networks using ABC. In: ICMLC 2019: International Conference on Machine Learning and Computing (2019)

Differentiable Largest Connected Component Layer for Image Matting

Xinshuang Liu[1(✉)] and Yue Zhao[2]

[1] UC San Diego, La Jolla, CA 92093, USA
xinsliu01@gmail.com
[2] University of Pennsylvania, Philadelphia, PA 19104, USA
yz2000@upenn.edu

Abstract. In the image matting task, a foreground object typically corresponds to a connected component in an image. Recent work has extracted the largest connected component in the raw object alpha matte to form the cleaned object alpha matte. However, existing works only consider connected components after training. We argue that including it in training can introduce prior knowledge that the object corresponds to a whole region, which can further improve the model's performance. To achieve this, we implement a differentiable largest connected component (LCC) layer, which finds the largest connected component and calculates the backward propagation result for the gradient. This LCC layer can be added to a model just like other PyTorch layers. Furthermore, we designed a gradient backward propagation method for our LCC layer, which facilitates the model to learn to output foreground pixels as a connected component. Its effectiveness has been comprehensively validated on different models and datasets. Finally, although our LCC layer significantly improves the models' performance on average on general datasets, we further conduct a study of which sorts of objects in the image matting task will benefit from the LCC layer, to give guidance on when our LCC layer should be used. The code is available at https://github.com/XinshuangL/LCCLayer.

Keywords: Image Matting · Connected Component Labeling

1 Introduction

Image matting is one of the fundamental tasks in computer vision, which can be seen as an extension of the foreground segmentation task [6,17,24]. The image matting task aims at extracting the foreground object from the image and estimating its transparency instead of just a binary mask. The objective of the image matting task is to solve for the transparency or alpha value for each pixel in the image, as shown in this equation:

$$I_{i,j} = \alpha_{i,j} F_{i,j} + (1 - \alpha_{i,j}) B_{i,j}, \quad \alpha_{i,j} \in [0,1], \tag{1}$$

M. Wand et al. (Eds.): ICANN 2024, LNCS 15016, pp. 419–431, 2024.
https://doi.org/10.1007/978-3-031-72332-2_27

where (i, j) is the position of the pixel, $I_{i,j}$, $F_{i,j}$, and $B_{i,j}$ are the RGB colors of the input image, the foreground, and the background image, and $\alpha_{i,j}$ is the alpha value (or the transparency).

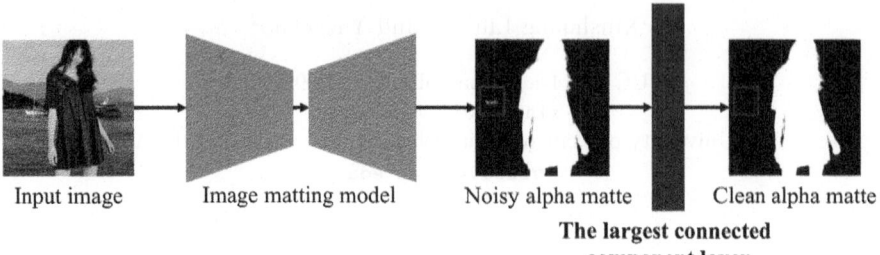

Input image   Image matting model   Noisy alpha matte   Clean alpha matte

The largest connected component layer

Fig. 1. Illustration of the functionality of the largest connected component layer. It can be added to the image matting model as an additional neural network layer, to enforce the output alpha mattes to contain only one connected component.

In real-world applications, a foreground object in the image matting task typically corresponds to a connected component in an image. For example, pixels of a person in an image are generally connected, since the human body is an integrated object. Based on this observation, a recent image matting method [25] has extracted the largest connected component in the raw object alpha matte to form the cleaned object alpha matte for the portrait matting task. However, existing works only consider connected components after training. We argue that including it in training can introduce prior knowledge that the predicted alpha matte should extract a whole and connected region of the object, not scattered pixels. This prior knowledge can further improve the model's performance.

To include finding the largest connected component during training, we implement a differentiable largest connected component (LCC) layer, which can be added to a model (as shown in Fig. 1) just like other PyTorch layers [21]. Instead of only using the LCC layer as postprocessing, we use it in both training and testing, letting the whole model (the image model combined with the LCC layer) be trained in an end-to-end way. Both the forward and backward propagation of the LCC layer are implemented by parallel computing. Thus, the computation time of the LCC layer is negligible compared to the forward propagation or the backward propagation of the existing image matting models [12,14,20].

We note that the LCC layer has no parameters to be trained, but it can propagate the gradient backward to train the image matting model. However, calculating the backward propagation of the gradient for the LCC layer remains difficult, because the pixels are not independent during calculating the largest connected component (the forward propagation). Moving a pixel away (setting the alpha value as 0) from a connected component may not only influence that pixel but also influence the pixels connected to it, making those connected pixels not belong to the original connected components. Thus, we propose a method to approximately calculate the gradient, which facilitates the model to learn to output foreground pixels as a connected component. The effectiveness of our approximate gradient backward propagation is validated on different models and datasets.

Furthermore, enforcing an object to be the largest connected component in the object alpha matte is just a constraint based on prior knowledge. Like the other prior knowledge in computer vision, this prior knowledge also has limitations. Thus, although our LCC layer significantly improves the models' performance on average on general datasets, we further conduct a study of which sorts of objects in the image matting task will benefit from the LCC layer, to give guidance on when our LCC layer should be used.

In experiments, we first evaluate the effectiveness of our approximate backward propagation of the gradient for model training. On some typical masks, our LCC layer enables the training while using a naive backward propagation of the gradient could not. Then, we evaluate the effectiveness of our LCC layer in improving the performance of image matting models. Our LCC layer improves the performance of different typical models [12,14,20] on different general image matting datasets [23,27]. Significant improvements introduced by our LCC layer are observed. Finally, we conduct an in-depth study of which sorts of objects actually benefit from the LCC layer and which do not. We analyze the failure objects and the successful objects, to find the sorts of objects that the LCC layer is suitable for. Further experiments on those sorts of objects validated these findings.

Our main contributions are summarized as follows:

- We implement a differentiable largest connected component (LCC) layer, which can be added to a model just like other PyTorch layers.
- We propose a method to approximately calculate the gradient, which facilitates the model to learn to output foreground pixels as a connected component and has been comprehensively validated as effective on different models and datasets.
- Finally, we further conduct a study of which sorts of objects in the image matting task will benefit from the LCC layer, to give guidance on when our LCC layer should be used.

2   Related Works

2.1   Image Matting

Traditional image matting methods can be divided into two categories: the propagation-based [4,13,26] and the sampling-based [9–11]. Due to the limitation of the handcrafted rules in these traditional methods, they cannot effectively learn from the training data. On the contrary, deep-learning-based image matting methods depend less on pre-defined rules and can effectively fit the patterns in the training data.

The first deep convolutional neural network for image matting is proposed by Cho *et al.* [7]. Then, Xu *et al.* [27] proposed an image matting network in the encoder-decoder architecture. The follow-up works include adding using a learnable indexing mechanism for downsampling and upsampling to process fine details [18], performing image matting on image patches to achieve high-resolution results [28], and disentangling the image matting task into trimap adaptation task and alpha prediction task to make the model robust to trimap quality [3]. Recently, some works proposed to automatically estimate the alpha mattes to save the users' time [5,12,16,23,29].

2.2   Connected Component Labeling

The connected component is a critical concept in computer vision, which has been adopted in the image matting task [25] and the video object segmentation task [2]. The computation of connected components can be efficiently implemented in parallel [22], which makes this concept practical in real-world applications. However, existing works only consider finding the connected components for postprocessing. We design a differentiable largest connected component layer to include it in training and further improve the model's performance.

3   The Largest Connected Component Layer

As shown in Fig. 1, our largest connected component (LCC) layer can be added to a model just like other PyTorch layers [21]. With this LCC layer, the predicted alpha matte will be a whole object region (which is a connected component of the foreground pixels), instead of scattered pixels. During training, the layer supports forward propagation to compute the largest connected component and backward propagation to approximately compute the gradient. The details of these two parts will be introduced in this section.

3.1   The Forward Propagation

We use the 4-connectivity in this work, where the pixels $(i-1,j)$, $(i+1,j)$, $(i,j-1)$, and $(i,j+1)$ are considered to be the neighbors of the pixel (i,j). For the image matting task, we consider two neighbors to be connected if they both belong to the foreground object, *i.e.*, their alpha values are greater than a threshold. We use 0.5 as the threshold in this work. Based on the definition of connectivity, a connected component is defined as a set of all pixels that are connected.

The connected components can be computed in parallel using the union-find algorithm [1]. The basic idea is to merge the incomplete connected components to gradually form complete connected components. After getting the connected components, we only keep the largest connected component in the alpha matte as the output and remove the other connected components. This is because there is generally only one foreground object in an image in the image matting task.

3.2   The Backward Propagation

The LCC layer has no parameters to be trained. For backward propagation, it only needs to propagate the gradient backward to train the image matting model. However, the pixels are not independent during the forward propagation, which makes calculating the backward propagation of the gradient difficult. For example, if we move a pixel away from a connected component (set the alpha value as 0), it may not only influence that pixel but also influence the pixels connected to it, making those connected pixels not belong to the original connected components. Thus, the computation of the gradient is extremely difficult.

To tackle this issue, we aim at approximately computing a gradient that can facilitate the model to output foreground pixels as a connected component. Based on the above analysis, our intuition is the pixels close to the largest connected component are more important because they can affect the regions farther away. Thus, we rescale the gradient that is passed to the LCC layer and use the rescaled gradient as the backward propagation result. When the gradient is letting the alpha value decrease, which means the prediction does not underestimate the alpha value, we do not do the rescaling. Otherwise, we use the scale:

$$s_{i,j} = (1 - \alpha_{i,j})e^{max(0,\, 1-d_{i,j}/L)}, \tag{2}$$

where (i, j) is the position of the pixel, $s_{i,j}$ the scale for the gradient at that pixel, $\alpha_{i,j}$ the original alpha value, $d_{i,j}$ the distance of that pixel to the largest connected component, and L a hyper-parameter denoting the maximum distance. In this work, we set $L = 32$.

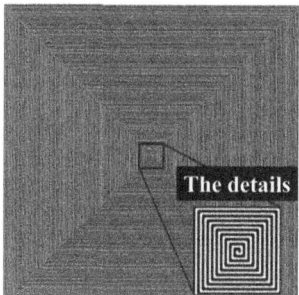

The sphere mask   The spiral mask

Fig. 2. The typical masks to test our largest connected component layer. The sphere mask corresponds to the easy case, the spiral mask corresponds to the difficult case.

3.3   The Adjustment During Evaluation

We observed that some tiny parts of the objects were wrongly removed by the original LCC layer because they were too small or too transparent. A mistake

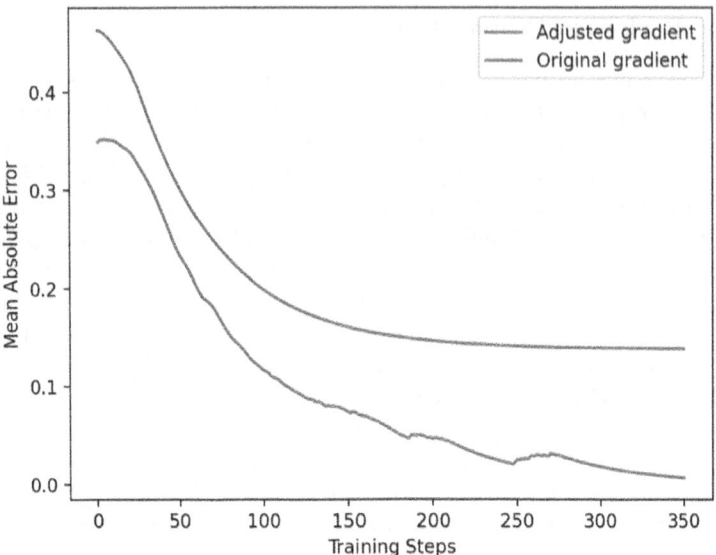

Fig. 3. The experimental results of training the model to map random noises to typical masks.

could probably cause an object separated into two connected components. To tackle this issue, we propose to first dilate all connected components by 32 pixels and then compute the updated largest connected component during evaluation. In this way, we are not likely to separate an object due to mistakes. Finally, we remove the other connected components by applying a soft mask. This mask's value decreases exponentially based on the distance from the edge of the largest component and is truncated to zero at a distance of 32. We note that the output is not the dilated alpha matte, but the original alpha matte masked by the updated largest connected component.

4   Experiments on Typical Masks

In this session, we evaluate our largest connected component (LCC) layer on two typical masks, as shown in Fig. 2. The spiral mask is a difficult case also used by the previous work [19].

4.1   Validation of the Backward Propagation

To validate the effectiveness of our proposed approximate backward propagation of the gradient, which rescales the gradient based on the importance of the pixels, we train an image matting model [14] to map random noises to the two masks and observe whether rescaling brings improvements. The batch size is 32, where each typical mask corresponds to 16 random noises. The optimizer is Adam with a learning rate of 0.01.

Figure 3 shows the experimental results. We observe that the gradient without rescaling leads to a suboptimal solution and fails to further reduce the training loss given enough epochs. In contrast, our rescaled gradient leads to the global optimal solution and achieves faster convergence. This indicates that our approximate gradient facilitates the model to learn to output foreground pixels as a connected component.

Table 1. The computation time of the largest connected component layer.

Batch size	Forward (ms)	Backward (ms)
2	0.3	0.2
8	0.5	0.6
32	1.8	2.4
128	6.9	9.3
512	27	37
2048	110	150

Table 2. Experimental results on the entire AMD dataset and Distinctions-646 dataset. Testing LCC denotes using our LCC layer during the test. Training LCC denotes using our LCC layer during training. '†' indicates we use their pre-trained model for weight initialization.

Method	Testing LCC	Training LCC	AMD		Distinctions-646	
			MSE ↓	MAD ↓	MSE ↓	MAD ↓
GCA Matting [14]			0.069	0.102	0.043	0.071
	✓		0.070	0.103	0.043	0.071
	✓	✓	**0.031**	**0.073**	**0.036**	**0.068**
MatteFormer [20]			0.090	0.140	0.049	0.083
	✓		0.091	0.140	0.050	0.083
	✓	✓	**0.071**	**0.134**	**0.038**	**0.078**
MODNet [12]			0.061	0.091	0.029	**0.054**
	✓		0.061	0.090	0.029	**0.054**
	✓	✓	**0.040**	**0.071**	**0.028**	0.057

4.2 Computation Time

We measure the computation time of the forward propagation and backward propagation of our LCC layer on a single NVIDIA A100. The input batch of data contains the same number of the two typical masks. The experimental results are shown in Table 1 The computation time of both the forward propagation and backward propagation of our LCC layer is negligible compared to the computation time of existing deep neural networks.

5 Experiments on the Image Matting Task

5.1 Dataset

We use two typical image matting datasets, the Adobe Image Matting dataset [27] and the Distinctions-646 dataset [23]. The Adobe Image Matting dataset contains 431 foreground images for training and 50 foreground images for test. The Distinctions-646 dataset contains 596 foreground images for training and 50 foreground images for testing. The synthetic images are generated by combining the foreground images with background images. For training, background images are randomly sampled from the MS COCO dataset [15]. For testing, each foreground image is combined with 20 background images in the Pascal VOC dataset [8]. To prevent interference with the evaluation, we eliminate background images that contain clearly visible humans.

5.2 Models and Evaluation Metrics

To evaluate the effectiveness of our proposed largest connected component (LCC) layer, we combine it with the following state-of-the-art image matting models:

– Guided Contextual Attention Matting (GCA Matting) [14]: A typical convolution image matting model to test the compatibility of our LCC layer with the convolution neural networks.
– Transformer-based Image Matting (MatteFormer) [20]: A typical transformer-based image matting model, to test the compatibility of our LCC layer with the transformers.
– Matting Objective Decomposition Network (MODNet) [12]: A typical image matting model trained by using multiple objectives to test the compatibility of our LCC layer with multi-objective learning.

The choice of these typical models makes a comprehensive evaluation of our LCC layer. For the evaluation metrics, we follow the previous work [12] and use Mean Square Error (MSE) and Mean Absolute Difference (MAD).

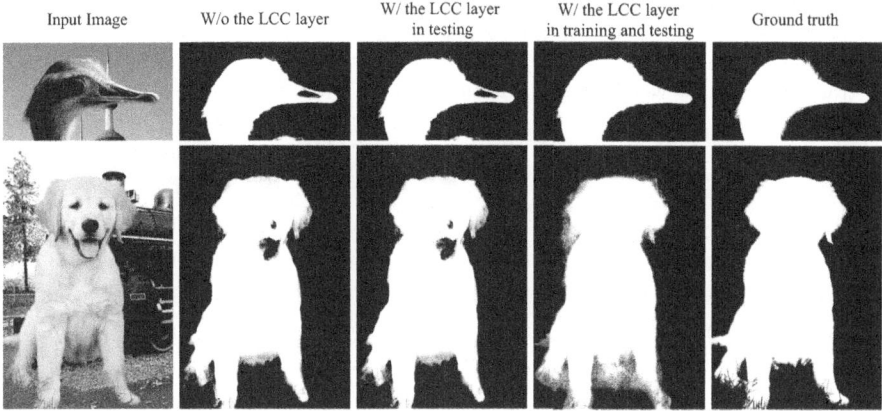

Fig. 4. Successful results on the image matting task.

5.3 Implementation Details

The Adam optimizer is used with an initial learning rate of 1×10^{-3}. The learning rate is adjusted by Warmup for the first 10 and exponential learning rate scheduler throughout the training process. In each epoch, the learning rate is multiplied by 0.998. The models are trained for 100 epochs with their official loss functions. To combine our LCC layer, we use the average of the loss functions with and without using our LCC layer. We initialize the model weights with their pre-trained official network backbone weights (they were not pre-trained on the image matting task). Since our task is trimap-free image matting, we

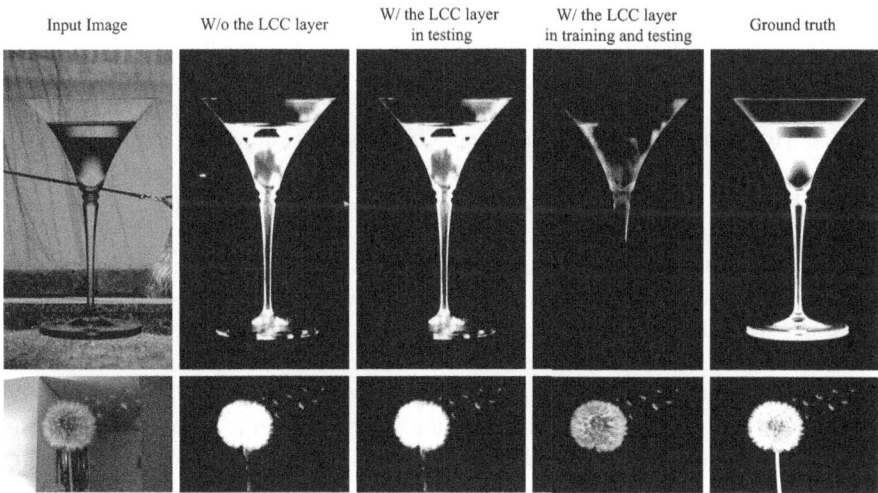

Fig. 5. Failure results on the image matting task.

Table 3. Experimental results on the human and animal subjects of AMD dataset and Distinctions-646 dataset. Testing LCC denotes using our LCC layer during the test. Training LCC denotes using our LCC layer during training. '†' indicates we use their pre-trained model for weight initialization.

Method	Testing LCC	Training LCC	AMD		Distinctions-646	
			MSE ↓	MAD ↓	MSE ↓	MAD ↓
GCA Matting [14]			0.126	0.141	0.028	0.037
	✓		0.126	0.140	0.028	0.038
	✓	✓	**0.092**	**0.116**	**0.015**	**0.028**
MatteFormer [20]			0.138	0.164	0.025	0.039
	✓		0.138	0.164	0.025	0.039
	✓	✓	**0.070**	**0.104**	**0.017**	**0.033**
MODNet [12]			0.040	0.053	0.028	0.037
	✓		0.039	0.052	0.028	0.037
	✓	✓	**0.028**	**0.042**	**0.019**	**0.028**

remove the parts that are designed for trimaps in the models. During training, data augmentation settings are adopted for better generalizability, including randomly combining two foreground images, random affine transformation, random cropping to the size of 512×512, and random color jitters.

5.4 Experimental Results

To comprehensively evaluate the effectiveness of our LCC layer, we use three experimental settings: 1) without our LCC layer; 2) with our LCC layer during testing; and 3) with our LCC layer during both training and testing. The experimental results of different methods on different datasets are shown in Table 2. We observe that only using the LCC layer during the test cannot effectively improve the performance. However, using our LCC layer during both training and testing significantly improves the performance. This observation supports our intuition that including the LCC layer in training can introduce prior knowledge that an object should correspond to a whole region, which can further improve the model's performance.

We further study the successful results and the failure results, to identify which sorts of objects benefit from the LCC layer. The successful results are shown in Fig. 4. The failure results are shown in Fig. 5. We observe that the successful results are general humans or animals since their bodies are not transparent and are connected. The failure results are some glasses or scattered objects, which do not correspond to one single connected component. Based on this observation, we recommend using the LCC layer on humans and animals and avoiding using it on scattered or mostly transparent objects. Based on this observation, we further conduct experiments on the human and animal subjects in the AMD and Distinctions-646 datasets, where the models are re-trained. The experimental results are shown in Table 3. We observe that our LCC significantly improves the models' performance on these human and animal subjects, which validated our findings.

6   Conclusions

In this work, we proposed a differentiable largest connected component layer, which can be added to a model just like other PyTorch layers. An approximation of the backward gradient propagation is proposed to facilitate the model to learn to output foreground pixels as a connected component. Experimental results on different models and datasets validated the effectiveness of the LCC layer. We conduct a study of which sorts of objects in the image matting task will benefit from the LCC layer, to give guidance on when our LCC layer should be used.

Acknowledgements. The authors gratefully acknowledge the ICANN 2024 reviewers for their valuable feedback.

References

1. Allegretti, S., Bolelli, F., Grana, C.: Optimized block-based algorithms to label connected components on GPUs. IEEE Trans. Parallel Distrib. Syst. **31**(2), 423–438 (2020)
2. Appiah, K., Hunter, A., Dickinson, P., Meng, H.: Accelerated hardware video object segmentation: from foreground detection to connected components labelling. Comput. Vision Image Understand **114**(11), 1282–1291 (2010)
3. Cai, S., et al.: Disentangled image matting. In: International Conference on Computer Vision, pp. 8818–8827. IEEE (2019)
4. Chen, Q., Li, D., Tang, C.: KNN matting. In: Proceedings of the IEEE/CVF Conference on Computer Vision and Pattern Recognition, pp. 869–876. IEEE Computer Society (2012)
5. Chen, Q., Ge, T., Xu, Y., Zhang, Z., Yang, X., Gai, K.: Semantic human matting. In: ACM Multimedia, pp. 618–626. ACM (2018)
6. Cheng, B., Misra, I., Schwing, A.G., Kirillov, A., Girdhar, R.: Masked-attention mask transformer for universal image segmentation. In: Proceedings of the IEEE/CVF Conference on Computer Vision and Pattern Recognition, pp. 1280–1289. IEEE (2022)
7. Cho, D., Tai, Y.-W., Kweon, I.: Natural image matting using deep convolutional neural networks. In: Leibe, B., Matas, J., Sebe, N., Welling, M. (eds.) ECCV 2016. LNCS, vol. 9906, pp. 626–643. Springer, Cham (2016). https://doi.org/10.1007/978-3-319-46475-6_39
8. Everingham, M., Gool, L.V., Williams, C.K.I., Winn, J.M., Zisserman, A.: The pascal visual object classes (VOC) challenge. Int. J. Comput. Vision **88**(2), 303–338 (2010)
9. Feng, X., Liang, X., Zhang, Z.: A cluster sampling method for image matting via sparse coding. In: Leibe, B., Matas, J., Sebe, N., Welling, M. (eds.) ECCV 2016. LNCS, vol. 9906, pp. 204–219. Springer, Cham (2016). https://doi.org/10.1007/978-3-319-46475-6_13

10. Gastal, E.S.L., Oliveira, M.M.: Shared sampling for real-time alpha matting. Comput. Graph. Forum **29**(2), 575–584 (2010)
11. He, K., Rhemann, C., Rother, C., Tang, X., Sun, J.: A global sampling method for alpha matting. In: Proceedings of the IEEE/CVF Conference on Computer Vision and Pattern Recognition, pp. 2049–2056. IEEE Computer Society (2011)
12. Ke, Z., Sun, J., Li, K., Yan, Q., Lau, R.W.H.: Modnet: real-time trimap-free portrait matting via objective decomposition. In: Association for the Advancement of Artificial Intelligence, pp. 1140–1147. AAAI Press (2022)
13. Levin, A., Lischinski, D., Weiss, Y.: A closed-form solution to natural image matting. IEEE Trans. Pattern Anal. Mach. Intell. **30**(2), 228–242 (2007)
14. Li, Y., Lu, H.: Natural image matting via guided contextual attention. In: Association for the Advancement of Artificial Intelligence, pp. 11450–11457. AAAI Press (2020)
15. Lin, T.-Y., et al.: Microsoft COCO: common objects in context. In: Fleet, D., Pajdla, T., Schiele, B., Tuytelaars, T. (eds.) ECCV 2014. LNCS, vol. 8693, pp. 740–755. Springer, Cham (2014). https://doi.org/10.1007/978-3-319-10602-1_48
16. Liu, J., et al.: Boosting semantic human matting with coarse annotations. In: Proceedings of the IEEE/CVF Conference on Computer Vision and Pattern Recognition, pp. 8560–8569. Computer Vision Foundation/IEEE (2020)
17. Long, J., Shelhamer, E., Darrell, T.: Fully convolutional networks for semantic segmentation. In: Proceedings of the IEEE/CVF Conference on Computer Vision and Pattern Recognition, pp. 3431–3440. IEEE Computer Society (2015)
18. Lu, H., Dai, Y., Shen, C., Xu, S.: Indices matter: learning to index for deep image matting. In: International Conference on Computer Vision, pp. 3265–3274. IEEE (2019)
19. Oliveira, V.M., Lotufo, R.A.: A study on connected components labeling algorithms using gpus. In: Workshop of Undergraduate Works, XXIII Sibgrapi, Conference on Graphics, Patterns and Images (2010)
20. Park, G., Son, S., Yoo, J., Kim, S., Kwak, N.: Matteformer: transformer-based image matting via prior-tokens. In: Proceedings of the IEEE/CVF Conference on Computer Vision and Pattern Recognition, pp. 11686–11696. IEEE (2022)
21. Paszke, A., et al.: Pytorch: an imperative style, high-performance deep learning library. In: Advances in Neural Information Processing Systems, pp. 8024–8035 (2019)
22. Playne, D.P., Hawick, K.: A new algorithm for parallel connected-component labelling on GPUs. IEEE Trans. Parallel Distrib. Syst. **29**(6), 1217–1230 (2018)
23. Qiao, Y., et al.: Attention-guided hierarchical structure aggregation for image matting. In: Proceedings of the IEEE/CVF Conference on Computer Vision and Pattern Recognition, pp. 13673–13682. Computer Vision Foundation/IEEE (2020)
24. Qiu, Y., et al.: SATS: self-attention transfer for continual semantic segmentation. Pattern Recogn. **138**, 109383 (2023)
25. Sengupta, S., Jayaram, V., Curless, B., Seitz, S.M., Kemelmacher-Shlizerman, I.: Background matting: The world is your green screen. In: Proceedings of the IEEE/CVF Conference on Computer Vision and Pattern Recognition, pp. 2288–2297. Computer Vision Foundation/IEEE (2020)
26. Sun, J., Jia, J., Tang, C., Shum, H.: Poisson matting. ACM Trans. Graph. **23**(3), 315–321 (2004)
27. Xu, N., Price, B.L., Cohen, S., Huang, T.S.: Deep image matting. In: Proceedings of the IEEE/CVF Conference on Computer Vision and Pattern Recognition, pp. 311–320. IEEE Computer Society (2017)

28. Yu, H., Xu, N., Huang, Z., Zhou, Y., Shi, H.: High-resolution deep image matting. In: Association for the Advancement of Artificial Intelligence, pp. 3217–3224. AAAI Press (2021)
29. Zhang, Y., et al.: A late fusion CNN for digital matting. In: Proceedings of the IEEE/CVF Conference on Computer Vision and Pattern Recognition, pp. 7469–7478. Computer Vision Foundation/IEEE (2019)

Enhancing Generalization in Convolutional Neural Networks Through Regularization with Edge and Line Features

Christoph Linse[(✉)] [iD], Beatrice Brückner, and Thomas Martinetz [iD]

Institute for Neuro- and Bioinformatics, University of Lübeck,
23562 Lübeck, Germany
{c.linse,thomas.martinetz}@uni-luebeck.de,
beatrice.brueckner@t-online.de
https://www.inb.uni-luebeck.de/en/home.html

Abstract. This paper proposes a novel regularization approach to bias Convolutional Neural Networks (CNNs) toward utilizing edge and line features in their hidden layers. Rather than learning arbitrary kernels, we constrain the convolution layers to edge and line detection kernels. This intentional bias regularizes the models, improving generalization performance, especially on small datasets. As a result, test accuracies improve by margins of $5 - 11$ percentage points across four challenging fine-grained classification datasets with limited training data and an identical number of trainable parameters. Instead of traditional convolutional layers, we use Pre-defined Filter Modules, which convolve input data using a fixed set of 3×3 pre-defined edge and line filters. A subsequent ReLU erases information that did not trigger any positive response. Next, a 1×1 convolutional layer generates linear combinations. Notably, the pre-defined filters are a fixed component of the architecture, remaining unchanged during the training phase. Our findings reveal that the number of dimensions spanned by the set of pre-defined filters has a low impact on recognition performance. However, the size of the set of filters matters, with nine or more filters providing optimal results.

Keywords: Convolutional neural networks · Pre-defined filters · Edge and line features · Regularization

1 Introduction

Deep Convolutional Neural Networks (CNNs) exhibit strong generalization capabilities on unseen data, especially in image recognition [1,6,11]. Various methods have emerged to enhance generalization further, including leveraging the power of additional data, transfer learning, or regularization. Our research proposes a new regularization technique to improve the generalization of CNNs by making them utilize edge and line information, two prominent feature types in computer

M. Wand et al. (Eds.): ICANN 2024, LNCS 15016, pp. 432–446, 2024.
https://doi.org/10.1007/978-3-031-72332-2_28

vision. In images, edges are boundaries where intensity values change sharply. Therefore, the image gradient conveys information to detect edges. Traditionally, edge detection employs convolution with first-order derivative kernels of various orientations. Similarly, the convolution with second-order derivative kernels can detect thin lines. While it is known that CNNs can develop such kernels during training, it remains unclear how much they rely on specific features in practice [2]. The training data might provide incentives to use other types of information. We demonstrate that the intentional processing of edge and line features in the convolutional layers of CNNs can enhance generalization. We implement this regularization technique by combining understandable pre-defined filters with CNNs. A convolution operation can be described as:

$$(f * g)[m, n] = \sum_{c=1}^{C} \sum_{i,j} f_c[i, j] g_c[m - i, n - j]. \tag{1}$$

Here, $f \in \mathbb{R}^{C \times k \times k}$ is the filter, and $g \in \mathbb{R}^{C \times W \times H}$ is the input feature map with the number of channels C, the size of the filter k, the width of the image W, and its height H. m and n index the pixels. We express the filters f_c using k^2 pre-defined filters $h_1, ... h_{k^2} \in \mathbb{R}^{1 \times k \times k}$ and weights $w \in \mathbb{R}^{k^2 \times C}$:

$$f_c[i, j] = \left(\sum_{l=1}^{k^2} w_{l,c} \cdot h_l \right)[i, j]. \tag{2}$$

The convolution becomes:

$$\text{PFM}_{\text{noReLU}}[m, n] = (f * g)[m, n] = \sum_{c=1}^{C} \sum_{l=1}^{k^2} w_{l,c} \cdot (h_l * g_c)[m, n]. \tag{3}$$

If the set of h_l has a full rank, the network can learn all possible kernels by adjusting the weights $w_{l,c}$. This changes by adding a ReLU [13].

$$\text{PFM}[m, n] = \sum_{c=1}^{C} \sum_{l=1}^{k^2} w_{l,c} \cdot \text{ReLU}(h_l * g_c)[m, n] \tag{4}$$

The additional ReLU nullifies negative values. It removes the information unrelated to the specific pre-defined filter, leading to a well-structured and comprehensible data representation. The output channels of the ReLU $\text{ReLU}(h_l * g_c)$ form distinct features, each containing positive filter responses only. Later, we choose the h_l as edge and line detectors in different orientations. A subsequent linear combination with the weights $w_{l,c}$ combines the distinct features. For implementation, we use the Pre-defined Filter Module (PFM), which was initially proposed in the paper [10] to reduce the number of trainable parameters within CNNs. It employs a depthwise 3×3 convolution, a batch normalization layer (here not shown), a subsequent ReLU, and a pixel-wise 1×1 convolution.

This research identifies edge and line filters as suitable pre-defined filters for a broad range of vision problems. First, we demonstrate the effectiveness of our regularization method using our own toy dataset for binary image classification. The dataset is available on github.com/Criscraft/Oriented_Dashes_Classification_Dataset. The task requires counting dashes in different orientations. We show that one PFM with pre-defined edge and line filters can solve the problem with only two trainable parameters, while a fully convolutional network with 3×3 kernels and ReLUs would require at least two layers and 36 parameters. Therefore, PFMs seem to be well-suited for problems where the orientations of edges and lines are relevant.

Second, we show that the generalization abilities of ResNet [5] and DenseNet [8] improve on the Fine-Grained Visual Classification of Aircraft dataset (FGVC Aircraft) [12], StanfordCars [9], Caltech-UCSD Birds-200-2011 (CUB) [18], and the 102 Category Flower dataset (Flowers) [14]. In additional experiments, we extract random features, achieved through randomly generated pre-defined filters drawn from a uniform distribution around zero. This setting does not harm the performance and sometimes increases the test accuracy. The results align with findings in the literature where the power of random filters is discussed [3,16].

Third, in a comprehensive overview, we study how many pre-defined filters are needed to reach high recognition rates. Furthermore, we show that the number of dimensions spanned by the set of filters has a minor effect on performance.

The paper is structured as follows. Section 2 describes the related work. Section 3 presents the toy dataset and shows how our regularized architecture solves the dataset with very few parameters. Implementation details are presented in Sect. 4. We show the performance metrics of our approach in Sect. 5. In Sect. 6, we argue why adding linearly dependent edge kernels to the filter set can further increase performance. Section 7 studies how the number of pre-defined filters affects performance. Section 8 discusses the results.

2 Related Work

Wimmer et al. [19] expressed 3×3 convolution kernels through various spanning vectors, as detailed in (3). Their approach, termed interspace pruning, aims to reduce the number of trainable parameters. Our work also employs spanning vectors, but we utilize pre-defined filters that are not adjusted to the training data. Another distinction lies in the intermediate ReLU function in (4) that eliminates patterns to which the filters do not respond positively.

A related study [10] used pre-defined filters to decrease the number of parameters in CNNs. While our approach shares similarities with the related study, our primary objective differs. We aim to regularize CNNs to improve their generalization capabilities rather than only focusing on parameter reduction. We incorporate their PFMs to improve the performance on four benchmark datasets. The module parameter f describes the internal number of copies of input channels. The previous study used $f = 1$ to save trainable parameters. Here, we choose f as the number of pre-defined filters in the filter pool. For $f = 9$ with nine edge- and line filters, the module convolves each input channel using each pre-defined filter as shown in (4). This setting does not save any parameters when compared to the baseline model.

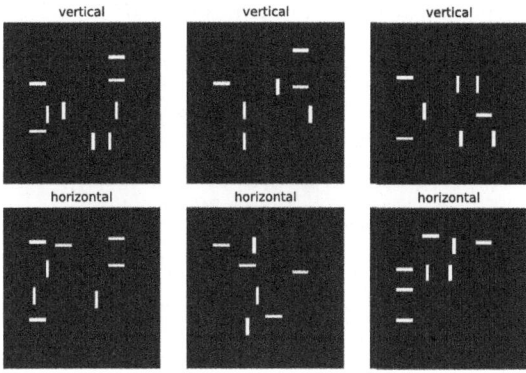

Fig. 1. Samples from the toy dataset with the classes *vertical* and *horizontal*.

The PFM is similar to depthwise separable convolution [7], which also has depthwise and pointwise convolution parts. However, our method does not adjust the filters in the depthwise part during training. Instead, our approach focuses on learning linear combinations of pre-defined filter outputs.

3   Toy Dataset

We study a simplified dataset for binary image classification that requires the processing of gradient information. We demonstrate that pre-defined filters are well-suited to tackle such problems using a minimal number of parameters. The dataset contains 1024 images featuring various horizontal and vertical dashes. The task is determining whether the image has more horizontal than vertical dashes, requiring the network to identify the orientations. Thus, effective utilization of gradient information is essential for solving this problem. Figure 1 shows some example images. The grayscale images have 48×48 pixels. The dashes are one pixel thick and five pixels long. Scenarios with an equal number of horizontal and vertical dashes do not occur. The function $f : \mathbb{R}^{H \times W} \to \mathbb{R}$

$$f(\mathbf{x}) = \sum_{i,j} \left(\text{ReLU}(H * \mathbf{x}) - \text{ReLU}(V * \mathbf{x}) \right) [i, j] \tag{5}$$

solves the problem with the pre-defined edge kernels

$$H = \begin{pmatrix} -1 & -1 & -1 \\ 2 & 2 & 2 \\ -1 & -1 & -1 \end{pmatrix}, \quad V = \begin{pmatrix} -1 & 2 & -1 \\ -1 & 2 & -1 \\ -1 & 2 & -1 \end{pmatrix}. \tag{6}$$

A non-negative output refers to the *horizontal* class, while a negative one refers to the *vertical* class. Indeed, (5) correctly classifies all images in the dataset.

Table 1. Different implementations of (5). p denotes the number of trainable parameters. Orange boxes contain trainable parameters. Blue boxes contain pre-defined filters. A dashed line corresponds to the weight value zero. I^+ denotes the identity.

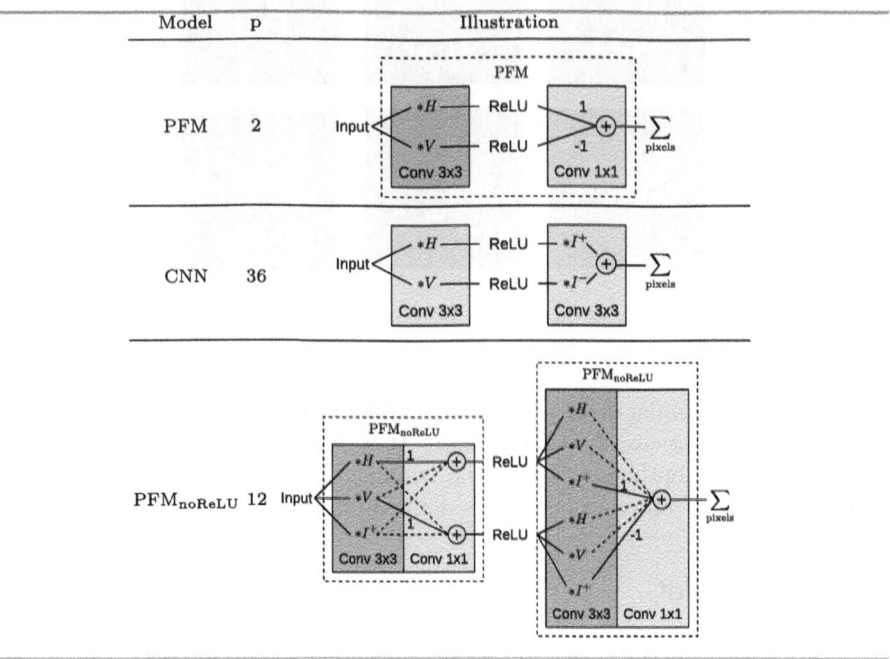

Table 1 presents three variants to implement (5) as a CNN. For the sake of simplicity, we ignore padding in our examples. The first variant uses our approach. It employs only one PFM consisting of two pre-defined edge kernels, a ReLU, and a trainable 1×1 convolution. Our variant implements (5) with only two trainable parameters. The second variant uses two common convolutional layers connected by a ReLU. The first convolutional layer has one input and two output channels. The second layer has two input channels and one output channel. The architecture needs a total of 36 trainable parameters. The third variant employs two $\text{PFM}_{\text{noReLU}}$ without the intermediate ReLU function. It needs a third pre-defined kernel (the identity) to implement (5). Here, all layers share the same set of pre-defined filters. The third variant has 12 trainable parameters.

The first variant using PFM offers the implementation with the fewest trainable parameters. It appears well-suited for image recognition problems where image gradients are relevant. The module enables the network to directly utilize gradient information, effectively filtering out other types of information. In the subsequent section, we apply PFM in deep CNN architectures and demonstrate that they provide a beneficial bias for challenging real-world image recognition problems.

Table 2. Number of trainable parameters of PFNet18 in millions. Below, the time for the forward pass (FP) and backward pass (BP) is measured for input tensors of shape $48 \times 3 \times 224 \times 224$ in milliseconds on an NVIDIA GeForce RTX 4090 GPU.

# Filters:	2	4	8	9	13	18	ResNet18
# Parameters:	2.7	5.2	10.1	11.3	16.2	22.4	11.3
Time FP:	7	11	21	24	33	46	6
Time BP:	14	21	36	40	54	73	14

4 Architectures and Sets of Filters

Starting from ResNet18 as the baseline, all convolutional layers are substituted with PFMs [10]. The resulting network is denoted PFNet18. Additionally, ResNet18 contains three skip connections with a 1×1 convolution and stride two. Following the recommendations of the paper [10], the skip connections are enhanced by smoothing their inputs using a Gaussian filter. This step addresses aliasing issues and increases the performance of PFNets.

The number of trainable parameters of PFNet18 depends on the number of pre-defined filters as seen in (4). Table 2 shows the model sizes for different filter sets. PFNet with nine filters exhibits the same number of parameters as the baseline ResNet18. Table 2 also demonstrates the time needed for the forward and backward pass. The times were measured for input tensors of shape $48 \times 3 \times 224 \times 224$ on an NVIDIA GeForce RTX 4090 GPU. Our regularization method tends to be slower compared to the baseline. One reason is that each convolutional layer in the original network is replaced by a PFM with two convolution steps, leading to more nodes in the computational graph.

To further study the applicability of our regularization method, we introduce PFMs to DenseNet121 [8] as an additional backbone architecture. Similar to ResNet, DenseNet is a widely used architecture in vision. It consists of 121 layers and contains 12.7 million trainable parameters. DenseNet, or *Densely Connected Convolutional Network*, connects all layers within a dense block in a feedforward manner. Each dense layer in a block receives the feature maps of all preceding layers as input, enhancing feature reuse and improving gradient flow.

All PFM modules in our experiments share the same pre-defined kernels, which remain unchanged throughout the training process. The experiments involve edge and line detectors in various orientations, as shown in Fig. 2. The filters are mean-free, and the sum of their absolute elements is one. We also employ random filters where the filter elements are drawn from a uniform distribution $[-1, 1]$ without normalization. All PFM modules of the same network share the same random filters. Nevertheless, distinct random seeds lead to different filters. We also utilize translating filters that have one element being one and the other elements being zero.

438 C. Linse et al.

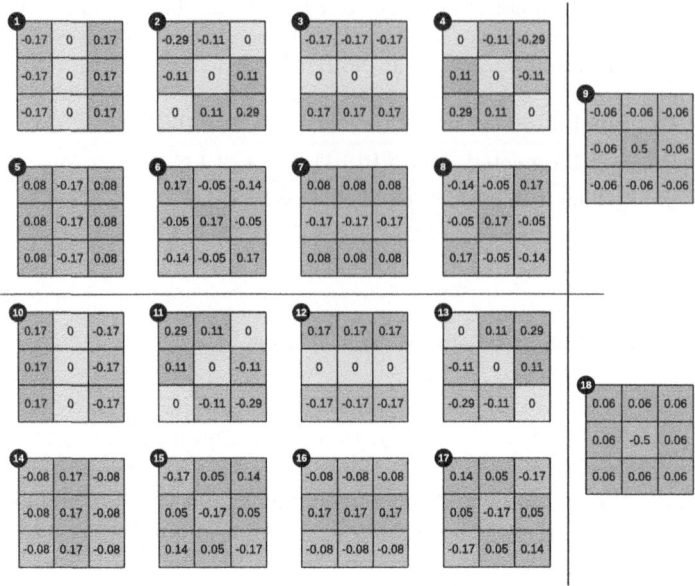

Fig. 2. Set of pre-defined filter kernels used in the experiments.

5 Performance on Benchmark Datasets

The models are trained and tested on the ImageNet Large Scale Visual Recognition Challenge (ILSVRC) [17], Fine-Grained Visual Classification of Aircraft dataset (FGVC Aircraft) [12], StanfordCars [9], Caltech-UCSD Birds-200-2011 (CUB) [18], and the 102 Category Flower dataset (Flowers) [14]. The CUB dataset has 5994 training and 5794 test images of 200 bird species. The images show birds in natural habitats, posing challenges like variations in lighting and backgrounds. The FGVC Aircraft dataset includes 6667 training and 3333 test images of 100 airplane models. The classes have notable intra-class variation due to various factors such as advertisement, airlines, and perspective. The Flowers dataset has 102 blossom types with strong intra-class variations. Regarding the Flowers dataset, we use the union of the official training and validation sets for training consisting of 2040 images. Testing occurs on the official test set with 6149 images. StanfordCars (2013) has 8144 training and 8041 test images of 196 car models, with images depicting single cars in various environments. The ILSVRC features 1281167 training images and 50000 validation images across 1000 classes. The images were extracted from various platforms, including Flickr, and were manually labeled with exactly one category.

The trainable network weights are initialized using Kaiming initialization [4]. The training hyperparameters for the ILSVRC are summarized in the Appendix 10.1. We use the training hyperparameters from the paper [10] for the remaining datasets. Table 3 presents the average test performance of five models trained with different seeds. The filter type 'edge, line' uses the edge and line detectors

Table 3. Average test accuracy. The pre-defined filters are not adjusted to the training data.

Backbone	Filter type	# Filters	Flowers	CUB	FGVC Aircraft	StanfordCars
ResNet18	Translating	9	73.01 ± 0.61	55.59 ± 0.62	73.72 ± 0.39	77.47 ± 0.44
ResNet18	Random	9	78.82 ± 1.99	60.29 ± 0.62	74.59 ± 3.74	79.80 ± 2.64
ResNet18	Random	18	81.16 ± 1.80	62.66 ± 1.01	77.60 ± 1.59	81.96 ± 1.45
ResNet18	Edge, line	9	84.28 ± 0.23	61.28 ± 0.28	79.66 ± 0.20	82.64 ± 0.17
ResNet18	Edge, line	18	**85.16 ± 0.15**	**63.01 ± 0.50**	**81.66 ± 0.23**	**83.66 ± 0.20**
ResNet18 [10]	–	–	73.4 ± 0.34	58.51 ± 0.53	73.32 ± 1.06	77.9 ± 0.37
DenseNet121	Edge, line	9	81.46 ± 0.38	60.43 ± 0.18	78.94 ± 0.29	80.62 ± 0.32
DenseNet121	Edge, line	18	**81.74 ± 0.35**	**62.10 ± 0.37**	**80.44 ± 0.17**	**81.04 ± 0.34**
DenseNet121	–	–	75.19 ± 0.66	58.26 ± 0.60	74.03 ± 0.38	77.43 ± 0.26

from Fig. 2 starting from index one. The sets with 9 and 18 detectors both span 9 dimensions.

The edge and line detectors exhibit performance improvements, achieving margins of $5 - 11$ percentage points compared to the baseline. This enhancement is consistent across all four datasets and is attributed to processing edge and line features in the convolutional layers, contributing to the regularization of the models. The experiments with DenseNet121 as a backbone show similar results. Interestingly, having more than 9 filters enhances the test performance further, even though the additional filters are linearly dependent. Section 6 studies this phenomenon in detail.

Experiments with pre-defined filters, randomly drawn from a uniform distribution around zero, are also shown in Table 3. Occasionally, PFNet18 with random filters surpasses the baseline model. However, the networks exhibit a high standard deviation in test accuracy, reaching up to 3.74% for the FGVC Aircraft dataset. Apparently, some random filter sets are more or less suited to the specific recognition tasks.

In alternative experiments, we employ translating filters instead of edge, line, or random filters. Referring to (4), the translating filters mimic the learning of the convolutional filters in the canonical basis. Compared to ResNet18, the performance drops up to 3%. We attribute this decline to the ReLU in the first layer, which sets approximately half of the pixels of the original input image to zero. The impact appears to depend on the dataset.

Table 4 presents the test accuracies on the ILSVRC (ImageNet). Our regularized model with nine filters exhibits a performance that is 2% below the baseline. We attribute the observed decline in performance to the similarity between training and test accuracy. The similarity implies that regularization might not be necessary in this scenario. Instead, the model's capacity should ideally increase. Indeed, when we augment the network with more filters, thereby expanding its capacity, the model shows a slight improvement over the baseline.

Table 4. Accuracy on the validation set of the ILSVRC (ImageNet). The pre-defined filters are not adjusted to the training data.

Filter type	# Filters	Top 1	Top 5
ResNet18 [10]	–	69.60	89.13
Edge, line	9	67.59	88.13
Edge, line	18	**70.16**	**89.49**

Table 5. Average test accuracy. The first column describes the initialization of the pre-defined filters. The filters are adjusted to the training data.

Filter type	# Filters	Flowers	CUB	FGVC Aircraft	StanfordCars
Translating	9	74.22 ± 0.92	59.35 ± 0.80	74.56 ± 0.38	79.29 ± 0.70
Random	9	79.61 ± 1.96	60.92 ± 1.58	75.19 ± 3.25	80.08 ± 2.31
Random	18	80.65 ± 1.35	62.94 ± 0.89	78.01 ± 1.89	81.81 ± 1.61
Edge, line	9	84.29 ± 0.33	61.93 ± 0.49	78.09 ± 0.36	81.54 ± 0.45
Edge, line	18	$\mathbf{84.62 \pm 0.41}$	$\mathbf{63.17 \pm 0.48}$	$\mathbf{79.54 \pm 0.55}$	$\mathbf{82.74 \pm 0.27}$
ResNet18 [10]	–	73.4 ± 0.34	58.51 ± 0.53	73.32 ± 1.06	77.9 ± 0.37

Furthermore, we adjust the pre-defined filters to the train data. We allow each PFM module to learn its own set with nine or 18 filters. However, these experiments do not improve the performance metrics, as shown in Table 5. The training process struggled to identify filters that outperformed our pre-defined edge and line detectors, underscoring their good generalization abilities.

6   Limited Impact of the Spanned Dimensions on Performance

This section investigates how the number of dimensions spanned by the set of pre-defined filters affects the recognition performance. The number of dimensions means how many of the nine possible dimensions are spanned by the set of pre-defined filters. It is obtained by counting the number of linearly independent filters. PFNet18 is trained using nine pre-defined kernels that span a varying number of dimensions. To get four dimensions, we choose the filters 1, 3, 5, 7, 10, 12, 14, and 16, and the sum of 14 and 16 from Fig. 2. Figure 3 presents the average test accuracies from five runs with different seeds. The results suggest that the dimensionality spanned by the pre-defined filters has minimal influence on performance, with four dimensions already yielding satisfactory results.

Interestingly, Table 3 unveils a notable performance gain between using 9 and 18 pre-defined kernels by margins of $1 - 3\%$. The nine 3×3 pre-defined filters already span all 9 dimensions. We suggest that the additional ReLU in the PFM can make the features linearly independent, even if the filter kernels

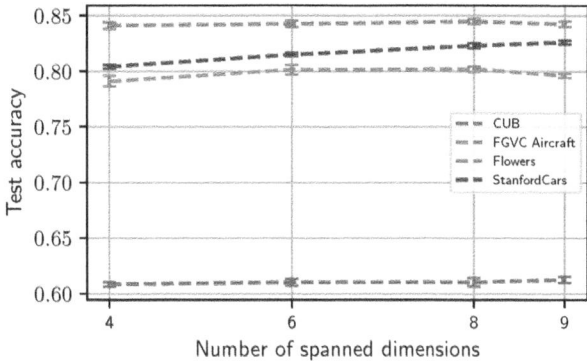

Fig. 3. Average test accuracy when using nine edge and line detectors that span a variable number of dimensions.

are linearly dependent. A module with one input channel and two pre-defined filters $\tilde{w}_1, \tilde{w}_2 \in \mathbb{R}^{k \times k}$ contains the two functions

$$f^{(\tilde{w}_1, m, n)}(\mathbf{x}) = \text{ReLU}(\tilde{w}_1 * \mathbf{x})[m, n]$$
$$f^{(\tilde{w}_2, m, n)}(\mathbf{x}) = \text{ReLU}(\tilde{w}_2 * \mathbf{x})[m, n] \tag{7}$$
$$f^{(\tilde{w}_1, m, n)}, f^{(\tilde{w}_2, m, n)} : \mathbb{R}^{M \times N} \to \mathbb{R}$$

with pixel coordinates $m, n \in N$. As shown in the appendix 10.2, if $a\tilde{w}_1 = \tilde{w}_2$ with $a < 0$, then the functions are linearly independent. The PFM module benefits from having a negative copy of a pre-defined filter because the rectified convolution outputs become linearly independent. The entire PFM module (ignoring normalization layers) can be written as

$$\text{PFM}[m, n] = \sum_{c=1}^{c_{\text{in}}} \sum_{l=1}^{2} q_{cl} \text{ReLU}(\tilde{w}_l * \mathbf{x})[m, n]$$
$$= q_{11} \text{ReLU}(\tilde{w}_1 * \mathbf{x})[m, n]$$
$$+ q_{12} \text{ReLU}(a\tilde{w}_1 * \mathbf{x})[m, n] \tag{8}$$
$$\text{Case 1: } (\tilde{w}_1 * \mathbf{x})[m, n] \geq 0 : q_{11}(\tilde{w}_1 * \mathbf{x})[m, n]$$
$$\text{Case 2: } (\tilde{w}_1 * \mathbf{x})[m, n] < 0 : aq_{12}(\tilde{w}_1 * \mathbf{x})[m, n].$$

The convolution output is either multiplied with q_{11} or aq_{12}. Here, ReLU acts like a switch, deciding which weight to apply.

We conclude that the set of pre-defined filters should incorporate pairs of filter kernels with inverted signs. For instance, the filters $\tilde{w}_1$ and $\tilde{w}_2 = -\tilde{w}_1$ could represent two edge detectors of opposing directions (see filters one and ten in Fig. 2). If the input contains an edge aligned with $\tilde{w}_1$, the first output channel has a positive activation while the second channel remains inactive. If the input contains the opposing edge, the second channel activates while the first channel remains inactive.

Table 6. Average test performance on the benchmark datasets. All experiments but the baseline use PFM$_{\text{noReLU}}$ modules without the additional ReLU. The pre-defined filters are not adjusted to the training data.

Filter type	# Filters	Flowers	CUB	FGVC Aircraft	Stanford Cars
Translating	9	74.93 ± 0.72	59.62 ± 0.52	74.51 ± 0.58	79.84 ± 0.40
Random	9	78.51 ± 1.64	59.28 ± 1.55	73.62 ± 4.21	79.11 ± 2.53
Random	18	78.54 ± 2.27	$\mathbf{60.55 \pm 1.02}$	$\mathbf{75.05 \pm 2.13}$	$\mathbf{80.04 \pm 1.47}$
Edge, line	9	78.29 ± 0.38	50.01 ± 0.69	71.25 ± 0.40	72.56 ± 0.28
Edge, line	18	$\mathbf{79.38 \pm 0.35}$	50.41 ± 0.46	72.43 ± 0.59	73.68 ± 0.49
ResNet18 [10]	–	73.4 ± 0.34	58.51 ± 0.53	73.32 ± 1.06	77.9 ± 0.37

Table 7. Filter subsets of different sizes and types.

Filter type	# Filters	Filter selection (see Fig. 2)
even	2	5, 7
even	4	5, 7, 14, 16
even	8	5–8, 14–17
uneven	2	1, 3
uneven	4	1, 3, 10, 12
uneven	8	1–4, 10–13
even-uneven	9	1–9
even-uneven	13	1–9, 11, 13, 15, 17
even-uneven	18	1–18

To better understand the benefit of linearly dependent filters, we repeat the experiments conducted in Sect. 5 in an ablation study. This time, we employ PFM$_{\text{noReLU}}$ modules without the additional ReLU function as described in (3). Since the pre-defined filters span all nine dimensions, the network retains its ability to learn all convolution kernels. Regularization does not occur. As expected, the results presented in Table 6 exhibit a notable performance drop for edge and line filters and a weak drop for random kernels. The baseline model ResNet18 outperforms the edge and line filters on the CUB, FGVC Aircraft, and StanfordCars datasets.

Furthermore, Table 6 shows that translating filters achieve test accuracies comparable to the baseline. This complements the prior experiments where an additional ReLU function decreased the recognition rates by 3%. The additional ReLU decreases the performance by setting dark input pixels to zero, erasing half of the original image's information.

7   Nine or More Filters are Needed for Optimal Results

This section studies the effect of the number of pre-defined filters on the recognition performance. We train PFNet18 on the CUB and the Flowers dataset using the filter subsets in Table 7. As illustrated in Fig. 4, the best results are obtained when utilizing all 18 filters. The test accuracies drop when choosing four or fewer filters. It is worth noting that a reduction in the number of pre-defined filters not only limits the information transferred to the subsequent layer but also leads to a decrease in trainable parameters, thereby diminishing the model's capacity, as shown in Table 2.

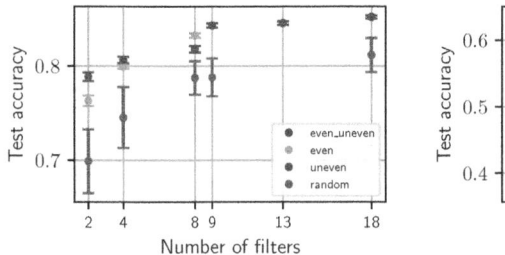 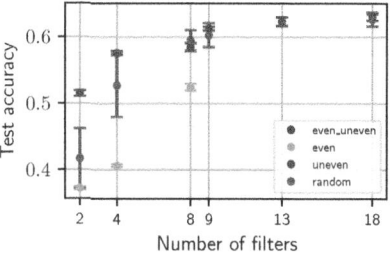

Fig. 4. Average test accuracy when using a variable number of filters. Left: Flowers dataset. Right: CUB dataset.

Interestingly, the edge filters (green color) often outperform the line filters (yellow color) or the random filters (red color). Given the abundance of edges in images, we hypothesize that edge filters convey more information than lines.

8   Discussion

Across four fine-grained classification datasets, we observed a notable test accuracy improvement ranging between $5 - 11$ percentage points using nine edge and line filters while maintaining the same number of parameters as the baseline model, ResNet18. Doubling the number of pre-defined filters resulted in further performance improvements. We also applied our regularization method to DenseNet121 with similar results. The experiments demonstrate the beneficial bias introduced by our regularization method. Notably, regularization was not achieved by reducing the number of trainable parameters but by biasing the CNN to process understandable edge and line features.

The ReLU in PFMs can remove specific information from the incoming feature maps. A question for future research is which pre-defined filters allow or do not allow the learning of the identity mapping. Consequently, we applied our regularization method to models with residual connections (ResNet) and densely connected layers (DenseNet), where the identity mapping is a fundamental part of the architecture.

The number of dimensions spanned by the set of pre-defined filters appears to have a low impact on recognition performance. This observation is attributed to the nature of the ReLU activation function. When applied to the outputs of convolution operations with linearly dependent filter kernels, ReLU can produce linearly independent results. The experiments indicate that four dimensions achieve comparable recognition values to those obtained with nine dimensions.

However, the number of pre-defined filters matters with optimal results obtained with nine or more 3×3 filters. Unfortunately, using more filters is bound to higher computational costs, as seen in Table 2, limiting the attractiveness for platforms with sensitive energy and speed requirements.

9   Conclusion

Processing edge and line features within CNNs improves their generalization abilities in image recognition. For ResNet18 and DenseNet121, we observed a noteworthy increase in test accuracy from $5 - 11$ percentage points across four classification benchmark datasets with the same number of trainable parameters. The results imply that pre-defined edge and line filters add a suitable bias to many image recognition problems. We demonstrated that the number of dimensions spanned by the set of pre-defined filters has a minimal impact on performance. However, the number of pre-defined filters matters. Using fewer than nine pre-defined 3×3 filters reduces test accuracy while using more than nine filters improves recognition performance but increases computational costs.

We believe pre-defined filters in CNNs are an underestimated area, offering improved generalization and the potential to save parameters. However, determining the optimal set of pre-defined filters for specific image recognition tasks remains challenging. Better problem-specific filters might exist. Adjusting the filters to the training data did not yield further improvements. Applying pre-defined filters to diverse tasks beyond image recognition, such as medical image, sound, or video analysis, is left for future research. Specialized features may offer significant benefits in these domains. Furthermore, investigating the compatibility of our regularization method with architectures beyond ResNet and DenseNet requires more experiments. Assessing PFM in different CNN architectures will help to determine their generalizability and effectiveness. Future research should also explore transfer learning with pre-defined filters, potentially reducing the need for extensive retraining.

Acknowledgment. The work of Christoph Linse was supported by the Bundesministerium für Wirtschaft und Klimaschutz through the Mittelstand-Digital Zentrum Schleswig-Holstein Project.

10   Appendix

10.1   Training Hyperparameters for ImageNet

The networks are trained with a batch size of 48 on five NVIDIA GeForce RTX 4090 GPUs. The remaining training hyperparameters are taken from the training reference of PyTorch [15]. The cross-entropy loss is minimized using stochastic gradient descent for 90 epochs with a momentum of 0.9 and weight-decay 0.0001. The initial learning rate of 0.1 is reduced by a factor of 0.1 every 30 epochs.

10.2   Linear Independency of ReLU-Based Functions

Two functions $f_1, f_2 : X \to Y$ are linearly independent if

$$(\forall \mathbf{x} \in X : c_1 f_1(\mathbf{x}) + c_2 f_2(\mathbf{x}) = 0) \Leftrightarrow c_1 = c_2 = 0. \tag{9}$$

Consider the functions from (7) that occur in the PFM module. The functions are linearly dependent if the pre-defined filters $\tilde{w}_1$ and $\tilde{w}_2$ are linearly dependent and $a\tilde{w}_1 = \tilde{w}_2, a \geq 0$.

Proof. Choose some arbitrary $\mathbf{x} \in \mathbb{R}^{M \times N}$. Choose $c_1 \in \mathbb{R} \backslash \{0\}$ and $c_2 = -c_1/a$. Then,

$$\begin{aligned} &c_1 f^{(\tilde{w}_1, m, n)}(\mathbf{x}) + c_2 f^{(\tilde{w}_2, m, n)}(\mathbf{x}) \\ &= c_1 \text{ReLU}(\tilde{w}_1 * \mathbf{x})[m, n] - c_1 \frac{a}{a} \text{ReLU}(\tilde{w}_1 * \mathbf{x})[m, n] = 0. \end{aligned} \tag{10}$$

$\square$

The functions in (7) are linearly independent if $\tilde{w}_1$ and $\tilde{w}_2$ are linearly dependent and $a\tilde{w}_1 = \tilde{w}_2, a < 0$.

Proof. The $\leftarrow$ direction is clear. To show the $\rightarrow$ direction, let $\mathbf{x} \in \mathbb{R}^{M \times N}$.

$$\begin{aligned} &c_1 f^{(\tilde{w}_1, m, n)}(\mathbf{x}) + c_2 f^{(\tilde{w}_2, m, n)}(\mathbf{x}) = 0 \\ &\Leftrightarrow c_1 \text{ReLU}(\tilde{w}_1 * \mathbf{x})[m, n] + c_2 \text{ReLU}(a\tilde{w}_1 * \mathbf{x})[m, n] = 0 \\ &\Leftrightarrow c_1 \text{ReLU}(\tilde{w}_1 * \mathbf{x})[m, n] - ac_2 \text{ReLU}(-\tilde{w}_1 * \mathbf{x})[m, n] = 0 \\ &\text{Case1} : (\tilde{w}_1 * \mathbf{x})[m, n] \geq 0 \implies c_1 = 0 \\ &\text{Case2} : (\tilde{w}_1 * \mathbf{x})[m, n] < 0 \implies c_2 = 0 \end{aligned} \tag{11}$$

The sum has to be zero for all $\mathbf{x} \in \mathbb{R}^{M \times N}$. This means that both coefficients c_1 and c_2 have to be zero. $\square$

References

1. Belkin, M., Hsu, D., Ma, S., Mandal, S.: Reconciling modern machine-learning practice and the classical bias-variance trade-off. Proc. Natl. Acad. Sci. **116**(32), 15849–15854 (2019)
2. Gavrikov, P., Keuper, J.: CNN filter DB: an empirical investigation of trained convolutional filters. In: 2022 IEEE/CVF Conference on Computer Vision and Pattern Recognition (CVPR), pp. 19044–19054. IEEE, New Orleans (2022)
3. Gavrikov, P., Keuper, J.: Rethinking 1x1 convolutions: can we train CNNs with frozen random filters? (2023). arXiv preprint arXiv:2301.11360 [cs]

4. He, K., Zhang, X., Ren, S., Sun, J.: Delving deep into rectifiers: surpassing human-level performance on ImageNet classification. In: 2015 IEEE International Conference on Computer Vision (ICCV), pp. 1026–1034. IEEE, Santiago (2015)

5. He, K., Zhang, X., Ren, S., Sun, J.: Deep residual learning for image recognition. In: 2016 IEEE Conference on Computer Vision and Pattern Recognition (CVPR), pp. 770–778. IEEE, Las Vegas (2016)

6. Hertel, L., Barth, E., Käster, T., Martinetz, T.: Deep convolutional neural networks as generic feature extractors. arXiv preprint arXiv:1710.02286 [cs] (2017)

7. Howard, A.G., et al.: Mobilenets: efficient convolutional neural networks for mobile vision applications. arXiv preprint arXiv:1704.04861 (2017)

8. Huang, G., Liu, Z., Van Der Maaten, L., Weinberger, K.Q.: Densely connected convolutional networks. In: Proceedings of the IEEE Conference on Computer Vision and Pattern Recognition, pp. 4700–4708 (2017)

9. Krause, J., Stark, M., Deng, J., Fei-Fei, L.: 3D Object rpresentations for fine-grained categorization. In: 2013 IEEE International Conference on Computer Vision Workshops, pp. 554–561. IEEE, Sydney (2013)

10. Linse, C., Barth, E., Martinetz, T.: Convolutional neural networks do work with pre-defined filters. In: 2023 International Joint Conference on Neural Networks (IJCNN), pp. 1–8. IEEE (2023)

11. Linse, C., Martinetz, T.: Large neural networks learning from scratch with very few data and without explicit regularization. In: Proceedings of the 2023 15th International Conference on Machine Learning and Computing (ICMLC 2023), pp. 279-283. Association for Computing Machinery, New York (2023)

12. Maji, S., Rahtu, E., Kannala, J., Blaschko, M., Vedaldi, A.: Fine-grained visual classification of aircraft. arXiv preprint arXiv:1306.5151 [cs] (2013)

13. Nair, V., Hinton, G.E.: Rectified linear units improve restricted Boltzmann machines. In: Proceedings of the 27th International Conference on International Conference on Machine Learning (ICML 2010), pp. 807–814. Omnipress, Madison (2010), event-place: Haifa, Israel

14. Nilsback, M.E., Zisserman, A.: Automated flower classification over a large number of classes. In: 2008 Sixth Indian Conference on Computer Vision. Graphics and Image Processing, pp. 722–729. IEEE, Bhubaneswar (2008)

15. Paszke, A., et al.: PyTorch: an imperative style, high-performance deep learning library. In: Wallach, H., Larochelle, H., Beygelzimer, A., AlchÃ-Buc, F.d., Fox, E., Garnett, R. (eds.) Advances in Neural Information Processing Systems, vol. 32, pp. 8024–8035. Curran Associates, Inc. (2019)

16. Ramanujan, V., Wortsman, M., Kembhavi, A., Farhadi, A., Rastegari, M.: What's hidden in a randomly weighted neural network? arXiv preprint arXiv:1911.13299 [cs] (2020)

17. Russakovsky, O., et al.: ImageNet large scale visual recognition challenge. Int. J. Comput. Vision **115**(3), 211–252 (2015)

18. Wah, C., Branson, S., Welinder, P., Perona, P., Belongie, S.: The Caltech-UCSD Birds-200-2011 Dataset. Tech. Rep. CNS-TR-2011-001, California Institute of Technology (2011)

19. Wimmer, P., Mehnert, J., Condurache, A.: Interspace pruning: using adaptive filter representations to improve training of sparse CNNs. In: Proceedings of the IEEE/CVF Conference on Computer Vision and Pattern Recognition, pp. 12527–12537 (2022)

Transformer Tracker Based on Multi-level Residual Perception Structure

Zhenhai Wang[1]([✉]), Hui Chen[2], Lutao Yuan[1], Ying Ren[1], and Hongyu Tian[3]

[1] College of Information Science and Engineering, Linyi University, Linyi 276000, China
wangzhenhai@lyu.edu.cn
[2] Computer and Artificial Intelligence, Jiangnan University, Wuxi 214222, China
[3] School of Physics and Electronic Engineering, Linyi University, Linyi 276005, China

Abstract. Recently, Transformer networks have been used for feature extraction and calculation of similarity in object tracking. This new structure is called one stream structure, and has achieved good results. However, the one stream structure of the Transformer tracker has too many network parameters, which limits the tracking speed of the network. For this reason, this work has designed a one stream Transformer structure that uses soft split operations to significantly reduce model parameters and computational complexity. To further improve the accuracy of tracking, this work proposes a multi-level residual perception structure to enhance the feature information of the target and reduce the background feature information, thereby enhancing the fore-ground and background discrimination ability of the model. To prove the speed and accuracy of this method, this work not only compared it with algorithms using deep neural network models, but also compared it with UAV tracking algorithms using shallow networks. Experimentally, the UAV123 dataset reached the level of SOTA; the inference speed can reach 130 FPS.

Keywords: Transformer · Object tracking · One stream structure · multi-level residual perception

1 Introduction

Visual object tracking—an area of increasing interest in the field of computer vision—determines an object's precise location, then tracks the object using a video sequence. The practical application of object tracking technology in the industry has received increasing attention, such as unmanned aerial vehicle object technology, un-manned sports live streaming, and other application scenarios.

Currently, the mainstream object tracking network still uses Siamese Network [1–4]; however, most backbones have changed from CNN to Transformer structure [5, 6]. With the Transformer structure, a small part of the object tracking network will splice extracted features part and relational modeling part together to use the Transformer, such as SimTrack [7], and OStrack [8]. They have achieved SOTA on multiple datasets, proving their methods' effectiveness.

M. Wand et al. (Eds.): ICANN 2024, LNCS 15016, pp. 447–460, 2024.
https://doi.org/10.1007/978-3-031-72332-2_29

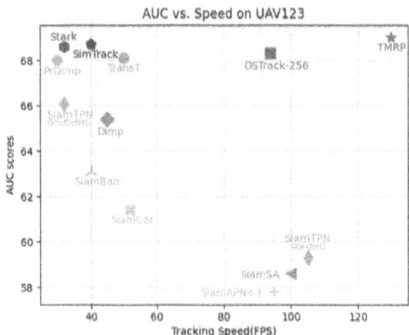

Fig. 1. A comparison of the quality and the speed of tracking methods on UAV123 test.

However, Transformer [9] was originally designed to process machine translation tasks in natural language. Transformers' long-distance dependence and attention mechanisms have always been required by computer vision, the feature dimension accepts and processes in 1D (even though computers mostly store images in 2D or 3D). This is a hallmark of the field of computer vision, as with image classification [10–12], target detection [13–15], semantic segmentation [16], et cetera.

In addition to taking advantage of Transformer's long-distance dependence and the advantages of parallel computing, these works also take advantage of the essence of Transformer's attention mechanism. For this reason, ViT [17] introduced Transformer into computer vision. It used CNN to divide the input image into 16*16 patches. These patches are then flattened through linear mapping and sent these patch dimensions into the Transformer to calculation. From patch embedding to subsequent calculations, there is no interaction between non-overlapping adjacent patches. The method used by Swin Transformer to generate patches is the same as that used by ViT and feed patches into Transformer. Then, it uses a moving windows method to facilitate the interaction between adjacent windows to achieve a global modeling effect. Some have attempted to use these two networks to build object tracking algorithms. OSTrack and SimTrack use ViT, and the SwinTrack [18] uses Swin-Transformer [19]. The T2T Transformer [10] idea is by proposing a method called T2T, replacing the simple tokenization of images in ViT, which is not CNN. In the tokenization of ViT, the images in simply divided into patches, and a linear transformation is applied to the patches. In the tokenization of T2T Transformers, a layer-wise T2T transformation is incorporated to progressively structure images to tokens by recursively aggregating neighboring Tokens into one Token. This work has improved the T2T structure so that it can be applied to target tracking algorithms, and subsequent calculations can improve the effectiveness of the model.

The simple Transformer's attention mechanism is insufficient for images. Consequently, this work designed a novel multi-level residual perception structure based on the Transformer block in ViT, and performs weighted attention to target, which is effective for tracker to focus on the target positioning because it weakens the influence of redundant features on tracker. To preserve 1D vector data calculated by Transformer, this work used the Multilayer Perceptron (MLP) to generate the target's regression box information to be tracked. This work also added a branch for accurate prediction of the target's center information.

To increase the tracking speed, this work deleted many complex structures, and finally compared the tracking speed with the current SOTA algorithm and UAV tracking algorithm. As shown in Fig. 1, the algorithm proposed in this work is practical and efficient. The contributions of this work are summarized as follows:

(1) A novel Transformer tracker named TMRP (Transformer Tracker based on Multi-level Residual Perception Structure) is proposed. TMRP does not use any local feature extractor for preserve the feature's global information as much as possible to enhance the model's feature expression.
(2) A novel multi-level residual perception structure is proposed to increase the network's attention to the target and reduce the focus on the background. This will help TMRP lock the target's position when tracking the object.
(3) TMRP achieves 69.0% AUC performance in the UAV123 dataset. More importantly, TMRP has fewer parameters. The tracking speed can reach 130 FPS.

2 Related Work

2.1 High-Speed Object Tracking Tracker

In the past decade, correlation filters and Siamese networks have been two of the most prominent examples in the field of object tracking. The object tracking method based on deep learning mainly utilizes the strong representation ability of depth features to achieve tracking. SiamFC [1] algorithm utilizes the Siamese network to train a similarity metric function offline. Due to the AlexNet [20] shallow neural network used by SiamFC, the extraction of feature information was significantly insufficient. Until later, others [3, 21, 22] used the ResNet [23] deep neural network as a feature extractor, although the tracking accuracy was high, the tracking speed significantly decreased. The Transformer tracking algorithm of One Stream, use the Transformer network to complete feature fusion when performing feature extraction. Accelerated by cuda, the tracking speed of SimTrack is 40 FPS and OSTrack is 89 FPS. For this reason, this work wondered whether the tracking speed could be improved if this structure had the same parameters as TransT [5], OSTrack and SimTrack. Therefore, this work used the T2T structure in T2T-Transformer to improve it, enabling it to be applied to the field of target tracking, and achieved significant results, with a tracking speed of up to 130 FPS.

2.2 Transformer

Transformer first appeared within the NLP (Natural Language Processing) task and can perform parallel machine translation. Later, networks like TransT and Stark used Transformer for object tracking and achieved remarkable tracking effects. Xing et al. proposed SiamTPN [24] for UVA tracking. Unlike the Siamese network that first extracts the features of different inputs and then fuses them to determine similarity, these one-stream jobs both concatenate and fuse features (i.e., relational modeling) while performing feature extraction. In this work, TMRP continue this one-stream working mechanism. TMRP have stopped using CNN and propose a pure Transformer tracking network based on the enhanced attention module and soft segmentation.

2.3 Patch-Based Approaches

In ViT, the image is divided into several patches, which are processed as a sequence of marks, and the position marks are added at the same time, which are sent to the encoder for feature learning. In 2021, Yuan et al. proposed token-to-token Vision Converter [10], uses a fixed-size and iterative approach to generate patches. It utilizes the proposed Token-to-Token module to iteratively generate patches from images. The generated patch is then fed to the T2T-ViT network for a final prediction. Inspired by this, this work have improved the T2T structure so that it can be applied to target tracking algorithms.

3 Method

As shown in Fig. 2, TMRP consists of a T2T block, ViT Transformer using multi-level residual perception structure, and a MLP to generate corner prediction heads.

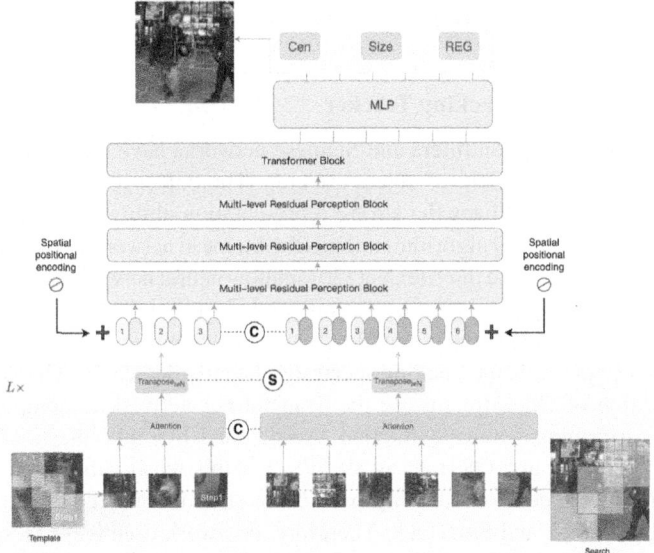

Fig. 2. An overview of the proposed offline regression network. "C" represents the concatenation operation, "S" represents the separation operation. The bottom is a T2T block used for processing image information, the middle is a three-level multi-layer residual perception module and Transformer module used for feature fusion, and the upper part is a bounding box prediction head.

3.1 T2T Block

The T2T block generates feature vectors that the Transformer can accept and process image-based information (bottom of Fig. 2). The T2T block consists of two parts. 1.) The soft split operation divides the picture into needed patches in sub-window form. 2.) Send the generated patches to the Transformer module for feature enhancement and dimensionality reduction.

Soft Split. The image pair of the input model, the template image $Z \in \mathbb{R}^{H_z \times W_z \times 3}$ and the search image $X \in \mathbb{R}^{H_x \times W_x \times 3}$ needs to be divided and flatten operation generates a *vector* $\in \mathbb{R}^{N \times dim}$ that Transformer can accept, here N is the number of patches, and *dim* is the dimension of each patch. Traditional Transformer uses convolution or linear mapping to divide the image into multiple independent and non-overlapping patches, which will lose the localization of the image and may destroy the semantics of the image. In the soft split module, the soft split aggregates the adjacent patches into a single patch, and the overlapping of the patches allows the model to capture the local information of the image, which makes up for the lack of local information in the Transformer's modeling. The soft split is depicted in Fig. 3.

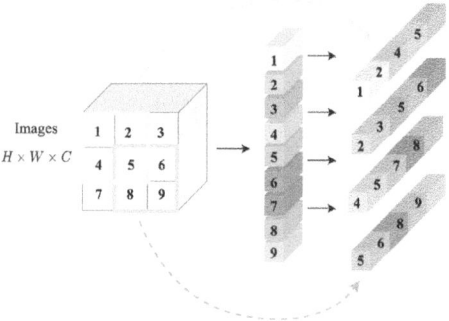

Fig. 3. Neighboring patches are aggregated into a single patch, and the aggregated patches overlap each other.

Using the Soft split to generate patch for the input image $X \in \mathbb{R}^{b \times h \times w \times c}$. After the soft split, the length of the output patch can be calculated as in (1).

$$l = \left\lfloor \frac{h + 2p - k + 1}{s} + 1 \right\rfloor \times \left\lfloor \frac{w + 2p - k + 1}{s} + 1 \right\rfloor \qquad (1)$$

Here, h and w are the width and height of the input image, p is the filling size, k is the sliding window size, and s is the sliding step. The input image dimension changes from $X \in \mathbb{R}^{b \times h \times w \times c}$ to $X \in \mathbb{R}^{b \times (c \times k^2) \times l}$. Where b is batch number, c is channel of image.

Attention Operation. The picture information was divided through the fully connected layer to generate queries Q, keys K, and values V. Then, TMRP performed the attention calculation according to the following formula:

$$\text{Attention}(Q, K, V) = \text{softmax}\left(\frac{QK^T}{\sqrt{d}}\right)V \qquad (2)$$

Here, d in the formula represents the key dimensionality. The purpose of this operation is to replace the multiplication and addition operation in convolution.

3.2 Multi-level Residual Perception Structure

Multi-level Residual Perception Structure. Part (a) in Fig. 4 shows the three-level multiple residual perception block in Fig. 2. Each layer of multi-level Residual Perception Block is composed of four layers of Transformers, as shown in Part (b) of Fig. 4. The features of the third layer are enhanced (Enh) and the features of the fourth layer are weakened (Wea) according to the attention score of the first layer of Transformer. In this way, not only feature constraints are strengthened and differentiated spatial representations are captured, but also the representation ability of spatial context is improved. Part (c) in Fig. 4 is a traditional Transformer block. This work enhances or weaken features based on attention scores in the dotted line section, as shown in Part (d) of Fig. 4. Please note that the Top-k and Low-k operations in Part (d) of Fig. 4 are not performed in the same Transformer.

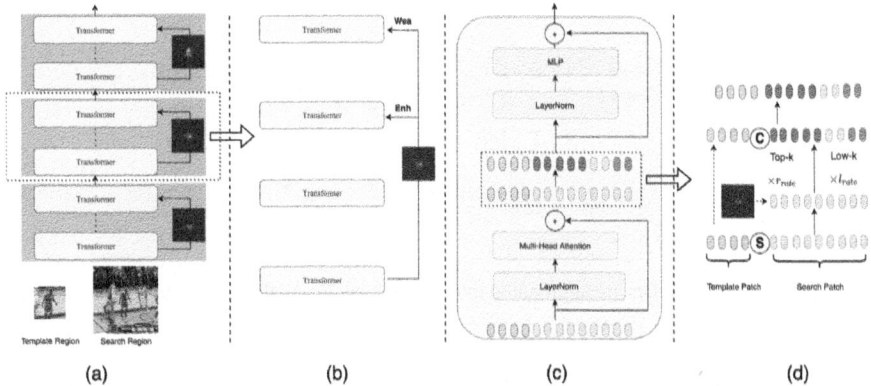

<table>
<tr><td>(a)</td><td>(b)</td><td>(c)</td><td>(d)</td></tr>
</table>

Fig. 4. Multi-level residual perception structure. (a) Three-level multi-layer residual structure. (b) Each level of multi-layer residual structure consists of four layers of Transformers. (c) A traditional Transformer structure that enhances and weakens features after multi-head attention. (d) Enhances the weight of the Top-K features and weakens the weight of the Low-K features in search. Note that the enhancement and weakening are performed in different Transformers.

3.3 Linear Predict Head

TMRP also improved the traditional corner prediction [25]. First, TMRP used a the MLP instead of the Fully Convolutional Network (FCN) to generate the predicted target information. The predicted target information includes the center point (Cen), offset (REG) and size (Size). Among them, the center point includes the target center's position information and the target classification prediction score $Center \in [0,1]$. Offset refers to the target's offset from the target's center. TMRP stride was set to 16, the distance between the two adjacent positions of the final feature map mapping the original map is 16, thus the center error is < 16. To reduce the center error, TMRP corrected the predicted center's coordinates using offset. And the Size score predicts the target box's width and height.

3.4 Loss

The loss of TMRP includes two parts: classification loss and regression loss. Positive and negative samples were established in an anchor-free manner. Since the network stride was set to 16, if the width or height of the sample ground truth (GT) was < 16, it would cause serious sample loss or offset. If the network stride is set to 8, it will consume too much memory. Therefore, this work used the weighted focal loss [26] to determine classification loss. Like OStrack, TMRP use a Gaussian kernel to generate a real ground truth heatmap, as follows (3):

$$\widehat{P}_{XY} = \exp\left(-\frac{(x - \tilde{p}_x) + (y - \tilde{p}_y)}{2\sigma_p^2}\right) \tag{3}$$

Here, $\tilde{p}_x$, $\tilde{p}_y$ is the GT target's center p_{gt}. Its corresponding low-resolution equivalent value, σ is an object size-adaptive standard deviation. So, classification loss as follows (4):

$$L_{cls} = -\Sigma \begin{cases} (1 - P_{xy})^2 \log(P_{xy}), & \text{if}\widehat{P}_{xy} = 1 \\ (1 - \widehat{P}_{xy})^4 (P_{xy})^2 \log(1 - P_{xy}), & \text{otherwise} \end{cases} \tag{4}$$

The regression loss is similar to TransT, OSTrack, SimTrack using l_1 loss and GIOU loss [27]. Regression loss is included as formula (5):

$$L_{reg} = \lambda_{iou} L_{iou} + \lambda_1 L_1 \tag{5}$$

Here, this work set the hyperparameters $\lambda_{iou} = 2$, $\lambda_1 = 5$. As a result, the total loss was (6):

$$L_{all} = L_{cls} + L_{reg} \tag{6}$$

4 Experiment

4.1 Implementation Details

Experiment Platform. The algorithm training in this paper was conducted using the Ubuntu18.04 operating system, computer hardware configuration: 1) CPU is Inter(R) Xeon(R) silver 4214R @2.40 GHz; 2) GPU is 4 Nvidia GeForce RTX 3080, memory 10G. TMRP tracker tested on Ubuntu18.04 operating system, and the following computer hardware configuration: 1) CPU is Inter i7-11700F @2.5 GHz; 2) GPU is 1 Nvidia GeForce RTX 3060Ti, memory 8G.

Model. TMRP is based on the pytracking toolkit and uses python 3.7 and pytorch 1.8. TMRP's backbone network used the T2T-ViT model for feature extraction and relationship modeling. The weights trained by T2T-ViT on the ImageNet-24K [28] dataset were used to initialize the weights of the backbone. The multi-level residual perception structure improved the Transformer model, and the implementation details are like those described in Sect. 3.2. The Head part uses 3 layers of fully connected layers to input a result, a total of 3 results, so a total of 3 × 3 fully connected layers, and the Head part uses a random initialization method to initialize the parameters of the network.

Train. The algorithm in this paper trains for 300 epochs and reduces the learning rate at the 220th epoch. The initial learning rate was $1e - 4$, using the AdamW optimizer [29]. TMRP was trained using GOT-10k [30], Trackingnet [31], LaSOT [32], and COCO2017 [33].

4.2 Compared with the SOTA Algorithms

TMRP is compared with multiple algorithms on UAV123 [34], UAV20L, UAV123@10FPS and LaSOT datasets. The FPS of TMRP is 130 under the Nvidia 3060TI GPU (Desktop), 129 TOPs, and the FPS was 33 under the 1050ti GPU (PC), which is lower than the RK3588 drone chip with a computing power of 6 TOPS or NVIDIA Jetson AGX Xavier with a computing power of 32 TOPs. The computing power is 5TOPs. RK3588 can be used as the master control of UAV equipment.

UAV123, UAV20L, UAV123@10FPS. The benchmark UAV123 dataset tests drone aerial photography object tracking algorithms. It consists of 123 high-resolution video sequences shot by drones, including nine target categories and 12 common challenges. The average video length is 915 frames. UAV20L is a subset of UAV123 for long-term object tracking. The UAV123@10FPS dataset is sampled at intervals of 10FPS; conequently, tracking is more difficult than with UAV123.

Table 1. Comparison with state-of-the-arts on three UAV benchmarks.

Method	Params	UAV123		UAV123@10fps		UAV20L	
		AUC	P	AUC	P	AUC	P
SiamFC [4]	2.3M	49.4	72.5	47.3	68.0	49.6	73.3
SiamRPN [2]	22.6M	53.7	75.3	52.4	70.6	55.0	74.9
SiamRPN++ [3]	53.7M	57.9	76.9	55.1	73.5	55.8	75.1
ATOM [35]	10.3M	65.0	82.2	–	–	63.5	82.3
DiMP [36]	43.1M	65.4	84.9	–	–	63.1	81.9
SiamCAR [21]	51.4M	61.4	76.0	–	–	60.6	75.3
SiamBAN [22]	53,7M	63.1	83.3	–	–	64.2	84.5
TrDiMP [37]	43.7M	67.0	87.6	–	–	67.4	88.2
TransT [5]	23.0M	68.1	87.6	64.3	84.1	68.2	88.9
SiamAPN++ [38]	14.7M	57.8	72.1	58.0	76.4	56.0	73.6
SiamSA [39]	14.8M	58.6	76.3	59.2	67.2	58.2	76.1
SiamTPN [24]	10.8M	65.3	84.9	64.3	84.6	64.5	83.1
SimTrack [7]	88.6M	**68.7**	**89.8**	*67.5*	*89.1*	*69.7*	*91.5*
OSTrack [8]	92.0M	68.3	89.1	64.6	84.7	68.3	89.3
TMRP	21.7M	*69.0*	*90.3*	**66.8**	**87.8**	**69.3**	**91.1**

Table 2. Comparison with state-of-the-arts on LaSOT benchmarks.

Method	SiamFC	SiamRPN++	SiamDW	ATOM	SiamMask [40]	SiamCar	SiamBan	TrDiMP	TransT	**TMRP**
AUC	33.6	49.5	34.7	49.9	46.7	51.6	51.4	63.9	**64.9**	*65.5*
P_{Norm}	42.0	57.0	43.7	57.0	55.2	61.0	59.8	73.0	**73.8**	*75.1*

UAV123, UAV20L and UAV123@10fps comparison results are shown in Table 1. In Table 1, red represents the best result and blue represents the second best result. Among these algorithms, SiamAPN++, SiamTPN and SiamSA are trackers for UAV tracking, and the others are general scene trackers. Although the evaluation result of TMRP on the UAV123@10fps and UAV20L dataset is slightly lower than that of and SimTrack, TMRP's tracking speed is 90FPS higher than them. Compared with the trackers [24, 38, 39] for UAV tracking, TMRP is far ahead of the others in terms of tracking speed. In addition, the parameters of TMRP are a quarter of those of the same type of structural algorithm [7, 8].

LaSOT. Lasot is an established, large-scale dataset that includes 1400 sequences distributed over 14 attributes. The various methods were ranked according to AUC, precision, and normalized precision (P_{Norm}). The evaluation results of the compared tracking algorithms are shown in Table 2. TMRP outperforms the second ranked algorithm TransT by 0.6% in Success, P_{Norm} leads by 1.3%. Figure 5 show the performance of TMRP and other trackers on a variety of tracking properties in the LaSOT test set. TMRP excels in all challenging attributes.

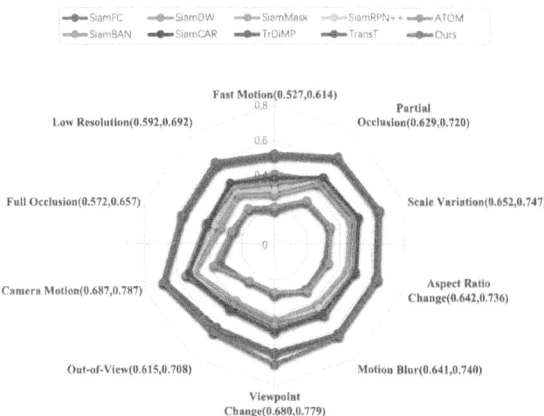

Fig. 5. AUC scores of different attributes on the LaSOT dataset.

4.3 Visual Comparison

To prove TMRP's effectiveness, this work visually compared multiple SOTA algorithms. For convenience, this work only visualized the algorithm using Transformer, as shown in Fig. 6.

In the first row of the Bird1_1 video sequence in Fig. 6, the target tracked in the second image completely disappears, but in the third image, TMRP can relock the target. In the second line of the building2 video sequence, there is a challenge of large-scale changes, and TMRP can still cope with this challenge. In the third row of the car7 video sequence, there are not only occlusion challenges but also similar object interference challenges. In frame 888 of the video sequence, TMRP still successfully locks the target. In the fourth line of the Person16 video sequence, it can be observed that TMRP can always lock onto the target itself, while other algorithms may experience tracking loss issues. In the last line of the uav4 video sequence, the main challenges are fast motion and the challenge of small targets. When other algorithms fail to track, TMRP can still stably lock the target.

4.4 Ablation Experiment

Using of ViT Without T2T. To verify the effectiveness of the T2T block, TMRP deleted this module to observe performance changes on the UAV123 and UAV20L datasets. Regarding the application of T2T block, this work mainly conducted two experiments: one is that the yellow part of the Attention module in Fig. 2 does not use concatenating operations, and the other is that concatenating operations are used. The results show that using splicing operations in T2T blocks is more effective. As shown in Table 3.

Table 3. The Ablation Study for T2T on UAV123, UAV20L Dataset

	Cat	Split	UAV123		UAV20L	
			AUC	P	AUC	P
TMRP	✓		69.0	90.3	69.4	91.1
TMRP-Split		✓	67.2	85.9	67.5	86.3
TMRP-nT			68.3	88.5	68.9	89.8

NOTE: TMRP-Split is a model that does not use concatenating operations in T2T blocks. TMRP-nT is a model that does not use T2T blocks. Cat means concatenate

As showed in Table 3, not using T2T (TMRP-nT) has a small effect on the accuracy of model testing, while not being able to use the T2T structure correctly (TMRT-Split) has a large impact on the results. The more obvious gap between TMRP and TMRP-nT is in the inference speed. The parameters of the model and multiply-accumulate computations (MAC) affect the inference speed of the tracker. Due to GPU hardware limitations, in TMRP-nT, this work only use the ViT-12 model for testing, which has 20G MAC, twice that of TMRP (10G), and parameters of 83M, four times that of TMRP (21M). The final FPS is 68, much lower than final model (130 FPS).

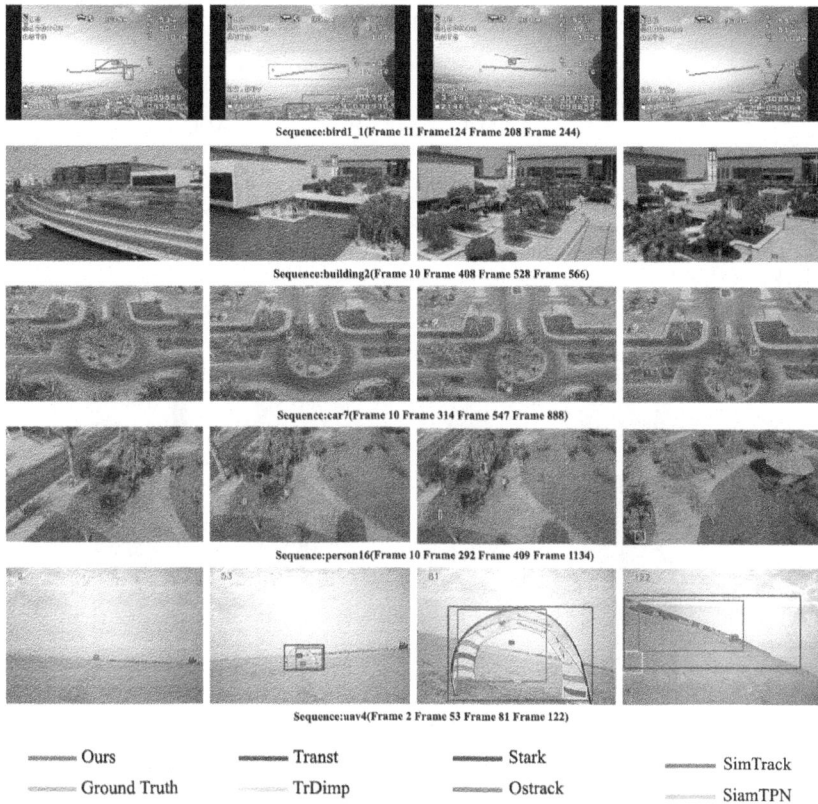

Sequence:bird1_1(Frame 11 Frame124 Frame 208 Frame 244)

Sequence:building2(Frame 10 Frame 408 Frame 528 Frame 566)

Sequence:car7(Frame 10 Frame 314 Frame 547 Frame 888)

Sequence:person16(Frame 10 Frame 292 Frame 409 Frame 1134)

Sequence:uav4(Frame 2 Frame 53 Frame 81 Frame 122)

| ▬ Ours | ▬ Transt | ▬ Stark | ▬ SimTrack |
| ▬ Ground Truth | TrDimp | ▬ Ostrack | SiamTPN |

Fig. 6. A comparison of the quality and the speed of tracking methods on UAV123 test set.

Application Multi-level Residual Perception Structure in Transformer. To demonstrate the effectiveness of the multi-level residual perception structure, this work deleted this module and conducted ablation experiments, and compared it with the early candidate elimination module used by OSTrack for visual attention score visualization, as shown in Fig. 7. To ensure fairness in the comparison, this work only delete this module, leaving the remaining model structures and hyperparameters unchanged. This work has presented the comparison results on the UAV123 and UAV20L datasets, as shown in Table 4. It is obvious that using the multi-level residual perception structure is more effective for the overall model.

To further demonstrate the effectiveness of TMRP, in addition to the data in Table 4. This work also conducted a visual attention score ablation experiment, as shown in Fig. 7, where row a represents the attention score map using the multi-level residual perception structure, row b represents the attention score map not using the structure, and row c represents the attention score map obtained by OSTrack. It is worth noting that for the fairness of the experiment, the above attention score map is taken from the last layer of Transformer structure. Through comparison, it can be intuitively observed that the

Table 4. The Ablation Study for MRP on UAV123, UAV20L Dataset

Method	MRP	UAV123		UAV20L	
		AUC	P	AUC	P
TMRP		69.0	90.3	69.4	91.1
TMRP-nMRP	✓	67.6	86.3	66.5	84.1

NOTE: TMRP-nMRP is a model that does not use c multi-level residual perception structure in T2T blocks

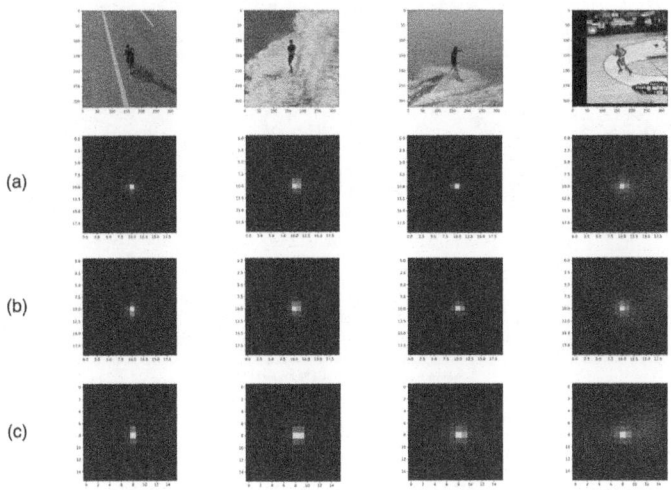

Fig. 7. Visualization of attention scores. (a) represents the attention score map using a multi-level residual perception structure (MRP), (b) represents the attention score map without using MRP, and (c) represents the attention score map using OSTrack.

multi-level residual perception structure effectively strengthens feature constraints and captures differentiated spatial representations.

CNN Head vs Linear Head. In order to prove that the combination of MLP and Transformer features can achieve better results in classification regression, this work conducted a ablation experiment using CNN Head. To verify its effectiveness, this work conducted validation on the UAV123 and UAV20L datasets. From Table 4, the use of MLP head improves the performance of AUC by about 1%.

5   Conclusion

This work proposes a pure Transformer Tracker. Given the advantages of Transformer parallel computing and less parameter quantity, the tracking speed of TMRP reached 130 FPS. This work uses the T2T structure in the T2T transformer and make improvements to make it applicable to visual tracking tasks, while significantly reducing network parameters. This work designs a multi-level residual perception structure to enhance the network's attention to the target, enabling it to have better feature expression capabilities. In addition, compared with the latest one stream Transformer visual tracking algorithm, TMRP parameter quantity is far lower than theirs, and compared with the deep CNN visual tracking algorithm, this work also far exceeds them in tracking speed and tracking accuracy. TMRP uses deep networks and achieves fast tracking speed, TMRP has important value for future visual object tracking in practical applications. In short, TMRP combines tracking speed and accuracy, so it is efficient, simple and practical.

Acknowledgments. This work was supported by Key Research and Development Program of Linyi City (Grant No.2022028) and The Natural Science Foundation of Shandong Province (Grant No. ZR2019MA030).

Disclosure of Interests. It is now necessary to declare any competing interests or to specifically state that the authors have no competing interests.

References

1. Bertinetto, L., Valmadre, J., Henriques, J.F., Vedaldi, A., Torr, P.H.: Fully-Convolutional Siamese Networks for Object Tracking, pp. 850–865. Springer (2016)
2. Li, B., Yan, J., Wu, W., Zhu, Z., Hu, X.: High Performance Visual Tracking with Siamese Region Proposal Network, pp. 8971–8980 (2018)
3. Li, B., Wu, W., Wang, Q., Zhang, F., Xing, J., Yan, J.: 'Siamrpn++: Evolution of Siamese Visual Tracking with Very Deep Networks, pp. 4282–4291 (2019)
4. Xu, Y., Wang, Z., Li, Z., Yuan, Y., Yu, G.: Siamfc++: Towards Robust and Accurate Visual Tracking with Target Estimation Guidelines, pp. 12549–12556 (2020)
5. Chen, X., Yan, B., Zhu, J., Wang, D., Yang, X., Lu, H.: Transformer Tracking, pp. 8126–8135 (2021)
6. Yan, B., Peng, H., Fu, J., Wang, D., Lu, H.: Learning Spatio-temporal Transformer For Visual Tracking, pp. 10448–10457 (2021)
7. Chen, B., et al.: Backbone is All Your Need: A Simplified Architecture for Visual Object Tracking, pp. 375–392. Springer (2022)
8. Ye, B., Chang, H., Ma, B., Shan, S., Chen, X.: Joint Feature Learning and Relation Modeling for Tracking: A One-Stream Framework, pp. 341–357. Springer (2022)
9. Vaswani, A., et al.: Attention is all you need. Adv. Neural Inf. Process. Syst. **30** (2017)
10. Yuan, L., et al.: Tokens-to-Token Vit: Training Vision Transformers From Scratch on Imagenet, pp. 558–567 (2021)
11. Yue, X., et al.: Vision Transformer with Progressive Sampling, pp. 387–396 (2021)
12. Hatamizadeh, A., et al.: FasterViT: Fast Vision Transformers with Hierarchical Attention (2023)
13. Zhu, X., Su, W., Lu, L., Li, B., Wang, X., Dai, J.: Deformable detr: deformable transformers for end-to-end object detection. arXiv preprint arXiv:2010.04159 (2020)
14. Carion, N., Massa, F., Synnaeve, G., Usunier, N., Kirillov, A., Zagoruyko, S.: 'End-to-End Object Detection with Transformers, pp. 213–229. Springer (2020)
15. Lv, W., et al.: Detrs beat yolos on real-time object detection (2023)

16. Xie, E., Wang, W., Yu, Z., Anandkumar, A., Alvarez, J.M., Luo, P.: SegFormer: simple and efficient design for semantic segmentation with transformers. Adv. Neural. Inf. Process. Syst. **34**, 12077–12090 (2021)
17. Dosovitskiy, A., et al.: An image is worth 16x16 words: transformers for image recognition at scale. arXiv preprint arXiv:2010.11929 (2020)
18. Lin, L., Fan, H., Zhang, Z., Xu, Y., Ling, H.: Swintrack: a simple and strong baseline for transformer tracking. Adv. Neural. Inf. Process. Syst. **35**, 16743–16754 (2022)
19. Liu, Z., et al.: Swin Transformer: Hierarchical Vision Transformer Using Shifted Windows, pp. 10012–10022 (2021)
20. Krizhevsky, A., Sutskever, I., Hinton, G.E.: Imagenet classification with deep convolutional neural networks. Commun. ACM **60**(6), 84–90 (2017)
21. Guo, D., Wang, J., Cui, Y., Wang, Z., Chen, S.: SiamCAR: Siamese Fully Convolutional Classification and Regression for Visual Tracking, pp. 6269–6277 (2020)
22. Chen, Z., et al.: SiamBAN: target-aware tracking with siamese box adaptive network. IEEE Trans. Pattern Anal. Mach. Intell. (2022)
23. He, K., Zhang, X., Ren, S., Sun, J.: Deep Residual Learning for Image Recognition, pp. 770–778 (2016)
24. Xing, D., Evangeliou, N., Tsoukalas, A., Tzes, A.: Siamese Transformer Pyramid Networks for Real-Time UAV Tracking, pp. 2139–2148 (2022)
25. Law, H., and Deng, J.: 'Cornernet: Detecting objects as paired keypoints', (Eds.): 'Book Cornernet: Detecting objects as paired keypoints' (2018, edn.), pp. 734–750
26. Lin, T.-Y., Goyal, P., Girshick, R., He, K., Dollár, P.: Focal Loss for Dense Object Detection, pp. 2980–2988 (2017)
27. Rezatofighi, H., Tsoi, N., Gwak, J., Sadeghian, A., Reid, I., Savarese, S.: Generalized Intersection Over Union: A Metric and a Loss for Bounding Box Regression, pp. 658–666 (2019)
28. Deng, J., Dong, W., Socher, R., Li, L.-J., Li, K., Fei-Fei, L.: Imagenet: A Large-Scale Hierarchical Image Database, pp. 248–255. IEEE (2009)
29. Loshchilov, I., Hutter, F.: Decoupled weight decay regularization. arXiv preprint arXiv:1711.05101 (2017)
30. Huang, L., Zhao, X., Huang, K.: Got-10k: a large high-diversity benchmark for generic object tracking in the wild. IEEE Trans. Pattern Anal. Mach. Intell. **43**(5), 1562–1577 (2019)
31. Muller, M., Bibi, A., Giancola, S., Alsubaihi, S., Ghanem, B.: Trackingnet: A Large-Scale Dataset and Benchmark for Object Tracking in the Wild, pp. 300–317 (2018)
32. Fan, H., et al.: Lasot: A High-Quality Benchmark for Large-Scale Single Object Tracking, pp. 5374–5383 (2019)
33. Lin, T.-Y., et al.: Microsoft Coco: Common Objects in Context, pp. 740–755. Springer (2014)
34. Mueller, M., Smith, N., Ghanem, B.: A Benchmark and Simulator for UAV Tracking, pp. 445–461. Springer (2016)
35. Danelljan, M., Bhat, G., Khan, F.S., Felsberg, M.: Atom: Accurate Tracking by Overlap Maximization, pp. 4660–4669 (2019)
36. Bhat, G., Danelljan, M., Gool, L.V., Timofte, R.: Learning Discriminative Model Prediction for Tracking, pp. 6182–6191 (2019)
37. Wang, N., Zhou, W., Wang, J., Li, H.: Transformer Meets Tracker: Exploiting Temporal Context for Robust Visual Tracking, pp. 1571–1580 (2021)
38. Cao, Z., Fu, C., Ye, J., Li, B., Li, Y.: SiamAPN++: Siamese Attentional Aggregation Network for Real-Time UAV Tracking, pp. 3086–3092. IEEE (2021)
39. Zheng, G., Fu, C., Ye, J., Li, B., Lu, G., Pan, J.: Scale-aware siamese object tracking for vision-based UAM approaching. IEEE Trans. Indust. Inf. (2022)
40. Wang, Q., Zhang, L., Bertinetto, L., Hu, W., Torr, P.H.: Fast Online Object Tracking and Segmentation: A Unifying Approach, pp. 1328–1338 (2019)

Fairness in Machine Learning

CFP: A Reinforcement Learning Framework for Comprehensive Fairness-Performance Trade-Off in Machine Learning

Simiao Zhang[1], Jitao Bai[2], Menghong Guan[3], Yueling Zhang[1(✉)], Jun Sun[4], Yihao Huang[5], Jiaping Wang[1], ChengCheng Wan[1], Ting Su[1], and Geguang Pu[1(✉)]

[1] National Trusted Embedded Software Engineering Technology Research Center, East China Normal University, Shanghai, China
{ylzhang,ggpu}@sei.ecnu.edu.cn
[2] Tianjin University, Tianjin, China
[3] East China Normal University, Shanghai, China
[4] Singapore Management University, Singapore, Singapore
[5] Nanyang Technological University, Singapore, Singapore

Abstract. Machine learning models are increasingly used for impactful decisions, such as loan approval, criminal sentencing, and resume filtering, raising concerns about ensuring fairness without sacrificing performance. However, fairness has multiple definitions, and existing techniques targeting specific metrics have limitations in improving multiple notions of fairness simultaneously. In this work, we establish a comprehensive measurement to simultaneously consider multiple fairness notions as well as performance, and propose new metrics through an in-depth analysis of the relationship between different fairness metrics. Based on the comprehensive measurement and new metrics, we present CFP, a reinforcement learning-based framework, to efficiently improve the fairness-performance trade-off in machine learning classifiers. We conduct extensive experiments to evaluate CFP on 6 tasks, 3 machine learning models, and 15 fairness-performance measurements. The results demonstrate that CFP can improve the classifiers on multiple fairness metrics without sacrificing its performance.

Keywords: Fairness in machine learning · Fairness-performance trade-off · Reinforcement learning · Ethics of AI

1 Introduction

Machine learning (ML) techniques have been widely used in many areas, such as criminal sentencing [26], medical imaging [24], and credit risk evaluation [17]. As a common task for ML, classification is drawing special attention in helping people make decisions. While garnering high accuracy, it has also raised public concerns about its fairness. For instance, it is found that a neural network

M. Wand et al. (Eds.): ICANN 2024, LNCS 15016, pp. 463–477, 2024.
https://doi.org/10.1007/978-3-031-72332-2_30

trained to predict recidivism is more likely to label male or non-Caucasian with higher risks [1]. Similarly, credit rating models may favor males or younger individuals [10]. These biases are unacceptable and can lead to ethical and legal issues. Therefore, ensuring fairness in ML techniques is crucial.

To address the issue of fair classification, several bias mitigation techniques have been developed [11,16,28]. But in general, they are plagued by two challenges: ① considering multiple fairness notions comprehensively, and ② the trade-off between fairness and performance. The first challenge for handling fairness issues lies in the diverse interpretations of fairness from different perspectives and the conflicts between their corresponding measurements. There are at least 21 mathematical definitions of fairness in current literature [2], such as Statistical Parity [11], Equalized Odds [13], and Equal Opportunity [13]. To quantitatively measure fairness, over 70 metrics [2] have been proposed based on these definitions. A comprehensive set of metrics is usually necessary for evaluation of fairness [3], as no current legal frameworks have appointed which metric to use [17,26] and there is no one best metric applicable across all contexts [2]. However, it is hard to combine these metrics due to the inevitable conflicts among them [30]. Another issue concerns the trade-off between fairness and performance. Many studies believe that an increase in fairness cannot be achieved without sacrificing accuracy. An empirical study [30] has shown that many existing bias mitigation techniques often lead to significant accuracy loss. Despite some studies have tried to cover the challenges [27], solutions often lack generality. Therefore, it is necessary to develop a generalized technique that can simultaneously improve multiple notions of fairness, while balancing the trade-off between performance and fairness.

In this work, we propose a reinforcement learning (RL) framework named CFP (Comprehensive Fairness-Performance) to improve the fairness-performance trade-off of ML classifier. The comprehensive measurement employed in CFP can simultaneously consider multiple metrics with different scales and monotonicity. Besides, we conduct both theoretical analysis and empirical study to understand the relationships among some fairness metrics, based on which we propose new independent metrics CFD_p and CFD_n that are more effective in improving multiple fairness simultaneously. The experiments on several commonly used datasets and ML models show that CFP can enhance the fairness of the model effectively while maintaining its acceptable performance. Specifically, we make the following contributions:

- We propose CFP, a RL framework, which can improve fairness-performance trade-off of ML models across multiple metrics.
- We establish a comprehensive measurement for multiple metrics regardless of their scale and monotonicity.
- We reveal the coupling effects between different fairness metrics and propose two new metrics, CFD_p and CFD_n, to simultaneously improve multiple fairness measures.
- We conduct a comprehensive evaluation of CFP on 6 tasks and 3 ML algorithms, and compare it with 4 state-of-the-art methods. The results show that CFP effectively improves the fairness-performance trade-off of models.

2 Preliminary

We begin with introducing the relevant terminology from the field of ML fairness. *sensitive attribute* (e.g., race, sex, age, religion) is an input feature associated with a protected characteristic, which is application-specific. Based on the value of sensitive attributes, all individuals are divided into two groups, i.e. *privileged groups* (have advantages in getting favorable labels) and *unprivileged groups* (have disadvantages in getting favorable labels). Let $D = \{X, Y, Z\}$ be a dataset where X is the training features, Y is the binary classification label ($Y = 1$ for favorable label and $Y = 0$ for unfavorable label), and Z is the binary sensitive attribute ($Z = 1$ for privileged group (abbreviated as p) and $Z = 0$ for unprivileged group (abbreviated as u)). Define $\hat{Y}$ as the predicted label ($\hat{Y}=1$ for favorable label and $\hat{Y} = 0$ for unfavorable label), and then some probability terms are given in Table 1. Fairness generally means no bias against individuals or groups based on their inherent or acquired traits. We briefly introduce the following 5 group fairness metrics as they are widely used [2,3,8].

Disparate Impact (DI): DI is measured by the ratio of probabilities for favorable predictions in the two subgroups [11].

$$DI = min(\frac{P[\hat{Y} = 1|Z = 0]}{P[\hat{Y} = 1|Z = 1]}, \frac{P[\hat{Y} = 1|Z = 1]}{P[\hat{Y} = 1|Z = 0]}) \tag{1}$$

Statistical Parity Difference (SPD): SPD is measured by the difference of the probabilities for favorable predictions in the two subgroups [5].

$$SPD = |P[\hat{Y} = 1|Z = 0] - P[\hat{Y} = 1|Z = 1]| \tag{2}$$

Equal Opportunity Difference (EOD): EOD is measured by the difference of the true positive rate (TPR) for two subgroups [13].

$$EOD = TPR_u - TPR_p \tag{3}$$

Average Odds Difference (AOD): AOD is measured by the mean value of false positive rate (FPR) difference and TPR difference in the two subgroups [13].

$$AOD = \frac{1}{2}[(FPR_u - FPR_p) + (TPR_u - TPR_p)] \tag{4}$$

Error Rate Difference (ERD): ERD is measured by the difference in the sum of the false negative rate (FNR) and FPR between two subgroups [9].

$$ERD = (FPR_u + FNR_u) - (FPR_p + FNR_p) \tag{5}$$

Table 1. Notations of probability terms used

Terms	Definitions	Terms	Definitions
FPR_u	$P[\hat{Y} = 1 \mid Y = 0, Z = 0]$	FPR_p	$P[\hat{Y} = 1 \mid Y = 0, Z = 1]$
FNR_u	$P[\hat{Y} = 0 \mid Y = 1, Z = 0]$	FNR_p	$P[\hat{Y} = 0 \mid Y = 1, Z = 1]$
TPR_u	$P[\hat{Y} = 1 \mid Y = 1, Z = 0]$	TPR_p	$P[\hat{Y} = 1 \mid Y = 1, Z = 1]$

3   Related Work

Bias mitigation techniques can be classified into three categories, including pre-processing, in-processing, and post-processing techniques [20].

Pre-processing Techniques are developed based on the assumption that the root cause of unfairness comes from the biased features in training data [4]. These techniques works on the dataset before training so that models can produce fairer predictions. A representative technique is reweighing, which adjusts input class weights according to their degree of being favored [3]. Others include modifying feature values [19] or feature representations [6]. However, modifying training data may have unexpected impacts on model fairness [29], and there is a huge controversy in the community on whether such action is acceptable.

In-processing Techniques modify the ML model to mitigate the bias in predictions. One common technique is adversary training. Usually, adversary training is performed on model level [7], but that on individual neuron level has also been developed [12]. Meanwhile, supervised and semi-supervised learning are also applied in some in-processing techniques [32]. Most of these methods require the design of specific fairness constraints, making it difficult to adapt them for new fairness metrics.

Post-processing Techniques directly modify predictions to reduce bias. They remove discrimination by adjusting some predictions of a classifier [13]. Multiple post-processing techniques have been developed based on post-hoc disparity [25], Gaussian process [22], group-aware threshold adaptation [15], etc. In particular, Sikdar et al. [27] employed RL techniques and developed a framework called GetFair, which can get competitive accuracy fairness trade-offs on tested models. It can also simultaneously improve statistical parity and equal opportunity by directly averaging the two fairness scores in the reward function. However, it ignores the relationships between different fairness notions, and the performance is unclear when more fairness metrics are involved simultaneously.

4   Methodology

In this section, we introduce our bias mitigation method CFP. The overall framework of CFP is shown in Fig. 1, in which the training process is shown on the left, and the construction of the reward function on the right. The training process is developed based on RL, and its reward function is constructed based on our comprehensive measurement and new metrics.

4.1   Comprehensive Measurement

In this section, we present how to take multiple metrics into consideration. The most straightforward way is to put them together. For metrics with positive values, we can directly add them up. While for those with negative values, we can add up their absolute values. However, different metrics may have different

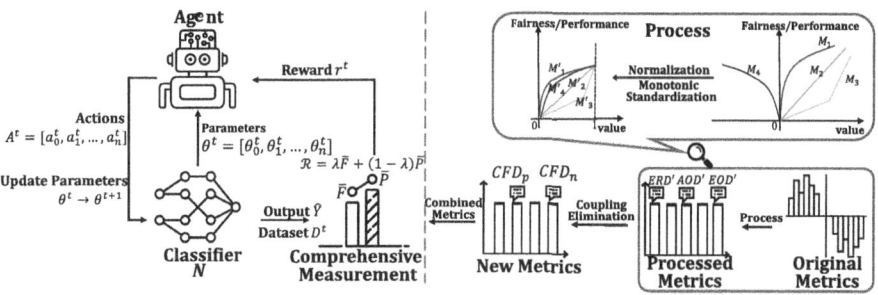

Fig. 1. Overview of CFP. The left part shows the training process. At step t, the agent inputs parameter θ^t and outputs the update directions A^t. θ^{t+1} is computed from θ^t by taking a step towards A^t. Then the classifier $N_{\theta^{t+1}}$ and the tuning dataset D^t are used to calculate the reward r^t. The right part shows the comprehensive measurement-based reward function from right to left. First, multiple fairness and performance metrics are processed to the same monotonicity and scales. Subsequently, we eliminate coupling effects among metrics by introducing CFD_p and CFD_n. Lastly, the values of processed fairness (performance) metrics are averaged to obtain $\bar{F}$ ($\bar{P}$).

scales, and the value change in metrics with small scales would have little influence on the sum. In other words, the effect of these metrics has been diluted. Therefore, the metrics should be normalized first. In addition, different metrics can have different monotonicity. That is, a larger/smaller value for metrics does not necessarily mean better fairness (performance). Thus, we cannot directly add the metrics up even if they are in the same scale, and the manipulation of monotonicity standardization is necessary.

Metrics can be classified into three categories, including monotonically increasing metrics, monotonically decreasing metrics, and non-monotonic metrics. If a larger value in a metric indicates better fairness or higher performance, then the metric is defined as a monotonically increasing metric. On the contrary, if a smaller value in a metric indicates better fairness or higher performance, then the metric is defined as a monotonically decreasing metric. While for those whose values have no monotonic correspondence to fairness or performance, they are defined as non-monotonic metrics.

Metrics in each category should be processed accordingly so that all the processed metrics share the same monotonicity. We process all the metrics to be monotonically increasing with a value range of $[0, 1]$. The manipulation for metrics with different monotonicity is presented as follows.

Monotonically Increasing Metrics. Denote monotonically increasing metrics by x_A, and the processed metric can be expressed as Eq. (6), where $x_{A_{min}}$ is the minimum of the metric, and $x_{A_{max}}$ is the maximum of the metric.

$$X_A = \frac{x_A - x_{A\,\min}}{x_{A\,\max} - x_{A\,\min}} \qquad (6)$$

Monotonically Decreasing Metrics. Denote monotonically decreasing metrics by x_B, and the processed metric can be expressed as Eq. (7), where $x_{B_{min}}$ is the minimum of the metric, and $x_{B_{max}}$ is the maximum of the metric.

$$X_B = 1 - \frac{x_B - x_{B\,min}}{x_{B\,max} - x_{B\,min}} \tag{7}$$

Non-monotonic Metrics. Denote a non-monotonic metric by x_C, the minimum of the metric by $x_{C_{min}}$, and the maximum of the metric by $x_{C_{max}}$, then there exists a metric value $x_0 \in (x_{C_{min}}, x_{C_{max}})$ that indicates the best fairness (performance). Take non-monotonic fairness metrics as an example, $x_C = x_0$ indicates the best fairness, $x_C < x_0$ indicates bias towards privileged group (unprivileged group), while $x_C > x_0$ indicates bias towards unprivileged group (privileged group). The processed metric can be calculated as Eq. (8).

$$X_C = \begin{cases} \frac{x_C - x_{C\,min}}{x_0 - x_{C\,min}} & x_C \le x_0 \\ \frac{x_C - x_{C\,max}}{x_0 - x_{C\,max}} & x_C > x_0 \end{cases} \tag{8}$$

In particular, when $x_{C_{min}}$ and $x_{C_{max}}$ are symmetrical about (i.e., $x_0 = \frac{1}{2}(x_{C_{min}} + x_{C_{max}})$), the processed metric X_C can be readdressed as Eq. (9).

$$X_C = 1 - 2\frac{|x_C - x_0|}{x_{C\,max} - x_{C_{min}}} \tag{9}$$

Let X_{A_i} $(i = 1, \ldots, I)$, X_{B_j} $(j = 1, \ldots, J)$, and X_{C_k} $(k = 1, \ldots, K)$ be the processed monotonic increasing, monotonic decreasing, and non-monotonic metrics, respectively. We can obtain the mean value of the processed metrics $\bar{X} = \frac{1}{N}\left(\sum_{i=1}^{I} X_{Ai} + \sum_{j=1}^{J} X_{Bj} + \sum_{k=1}^{K} X_{Ck}\right)$, where $N = I + J + K$ is the total number of processed metrics.

We define the mean value of the processed fairness metrics (denoted by $\bar{F}$) as comprehensive fairness measurement, and that of the processed performance metrics (denoted by $\bar{P}$) as comprehensive performance measurement.

4.2   Coupling Elimination via CFD_p and CFD_n

Section 4.1 has provided an effective method for taking multiple metrics simultaneously into account. However, it should be noted that not all metrics are suitable to be combined together. We assume that the coupling effects between metrics have a negative effect on the training process, and using metrics that are equivalent but more independent is better. We have demonstrated this through the theoretical analysis in this section and the empirical results in Sect. 5.2.

Consider the commonly used EOD, AOD, and ERD as an example, the combination of the these three metrics can make the result degenerate into a specific metric.

Proposition 1. *The combination of EOD, AOD and ERD can always be expressed with a certain one metric of the three.*

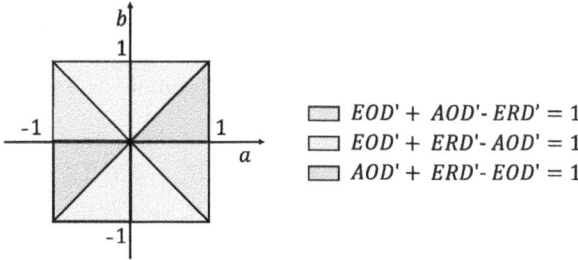

Fig. 2. Correlations of EOD', AOD', and ERD' on different domains.

Proof. AOD, EOD, and ERD (Eqs. 3 to 5) are non-monotonic metrics and are symmetrical about zero. Thus, they can be treated with Eq. (9). Let $a = FPR_u - FPR_p$, $b = FNR_u - FNR_p$, and $c = TPR_u - TPR_p$ $(a, b, c \in [-1, 1])$, the processed metrics are presented as:

$$\begin{cases} EOD' = 1 - |TPR_u - TPR_p| = 1 - |c| \\ AOD' = 1 - \frac{1}{2}|(FPR_u - FPR_p) + (TPR_u - TPR_p)| = 1 - \frac{1}{2}|a + c| \\ ERD' = 1 - \frac{1}{2}|(FPR_u + FNR_u) - (FPR_P + FNR_P)| = 1 - \frac{1}{2}|a + b| \end{cases}$$
(10)

According to Table 1, there is Eq. (11), or in other words, $c = -b$.

$$b + c = (FNR_u + TPR_u) - (FNR_p + TPR_p) = 1 - 1 = 0 \qquad (11)$$

While variables a and b are mutually independent. Substitute Eq. (11) into Eq. (10), we then have:

$$\begin{cases} EOD' = 1 - |c| = 1 - |b| \\ AOD' = 1 - \frac{1}{2}|a + c| = 1 - \frac{1}{2}|a - b| \\ ERD' = 1 - \frac{1}{2}|a + b| \end{cases}$$

The correlations of EOD', AOD', and ERD' are discussed on different domains of a and b, as shown in Fig. 2. Therefore, the sum of EOD', AOD', and ERD' always satisfies one of the three equations of Eq. (12) depending on different domains of a and b.

$$\begin{cases} EOD' + AOD' + ERD' = 1 + 2ERD' \\ EOD' + AOD' + ERD' = 1 + 2AOD' \\ EOD' + AOD' + ERD' = 1 + 2EOD' \end{cases}$$
(12)

That means the combination of the three metrics can always be expressed with a certain one metric and cannot play a combined role as expected (as expressed by Eq. (13), in which X_i represents any one of the three metrics).

$$EOD' + AOD' + ERD' = 1 + 2X_i \qquad (13)$$

In this case, two new metrics named Complementary False Positive Rate Difference (denoted by CFD_p) and Complementary False Negative Rate Difference (denoted by CFD_n) are developed herein to eliminate the negative effects of direct combination.

$$\begin{cases} CFD_p = 1 - |a| = 1 - |FPR_u - FPR_p| \\ CFD_n = 1 - |b| = 1 - |FNR_u - FNR_p| \end{cases}$$

We conduct empirical experiments to investigate the performance of various combinations of metrics. The results are presented in Sect. 5.2. Since CFD_p and CFD_n are mutually independent and have no negative coupling effects, they are expected to be able to achieve better results compared with the direct combination of EOD', AOD', and ERD'. So, can we say the combination of EOD', AOD', and ERD' is actually good-for-nothing? The answer is no. Although the combination of EOD', AOD', and ERD' can always be expressed with a certain one of the three metrics, it cannot be always expressed with a fixed one. With variables a and b falling on different domains, as illustrated in Fig. 2, X_i in Eq. (13) can change from one metric to another. In this sense, we can still say the combination has covered all three metrics, merely the metrics are covered one by one, rather than simultaneously. Therefore, the combination of EOD', AOD', and ERD' can still outperform any one of the three metrics that acts as an individual, even if it may not match the newly developed combination of CFD_p and CFD_n.

4.3   Bias Mitigation Through CFP Framework

Training Process. As a RL framework, the structure of CFP is composed of the agent and environment, i.e., the classifier N_θ with parameters $\theta = [\theta_0, \theta_1, ..., \theta_n]$ that we want to improve for better fairness-performance trade-off. We adopt the standard notations [31] for the state s^t, action a^t, reward r^t, and policy π. At the time step t, state is the parameters $\theta^t = [\theta_0^t, \theta_1^t, ..., \theta_n^t]$ of the classifier, action $a_i^t \in \{-1, 1\}$ is the update direction of the θ_i^t, and reward r^t is calculated by the reward function $\mathcal{R}$ defined in Eq. (14). Policy π_ϕ is a neural network called meta-optimizer with the parameters ϕ, which takes in the parameter sequence θ^t and outputs an action sequence $A^t = [a_0^t, a_1^t, ..., a_n^t]$.

In the beginning, we get the classifier N_{θ^0} trained for performance. At a time step t in an episode, π_ϕ takes the parameters θ^t as input and generates the action A^t. The next parameter θ^{t+1} is computed from θ^t by taking a step towards the direction A^t as $\theta^{t+1} = \theta^t + A^t \cdot lr \cdot c(t)$, where lr is a predetermined step size similar to the learning rate used in gradient descent methods, and $c(t)$ is an optional scaling parameter that reduces the step size when approaching the optimum. The classifier with the parameter θ^{t+1} is then used to calculate the reward on tuning dataset $D_t \subset D$. This process is repeated to generate an episode, and a trajectory is obtained: $traj = (\theta^0, A^0, r^0, \theta^1, A^1, r^1, ..., \theta^t, A^t, r^t, ...)$. The episode ends when a predetermined number of steps is reached or when the performance of N_{θ^t} reaches a threshold. ϕ is updated at the end of each episode based on the rewards collected during the entire episode using the REINFORCE algorithm.

Reward Function. The reward function $\mathcal{R}$ is constructed with the comprehensive fairness measurement $(\bar{F})$ and comprehensive performance measurement $(\bar{P})$ defined in Sect. 4.1, with the weights adjusted by parameter λ $(0 \leq \lambda \leq 1)$, as shown in Eq. (14).

$$\mathcal{R} = \lambda\bar{F} + (1 - \lambda)\bar{P} \tag{14}$$

Get Updated Models. During the training process, we obtain different parameters θ and corresponding models N_θ. Meanwhile, we record the scores of N_θ on comprehensive fairness measurement and comprehensive performance measurement, record as $\bar{F}_\theta$ and $\bar{P}_\theta$ respectively. Using $\bar{F}_\theta$ and $\bar{P}_\theta$ of each model as the horizontal and vertical coordinates respectively, we can obtain their scatters. The set of models that form the convex hull of these points is denoted by θ^*, which represents the final models obtained from the training process.

5   Experimental Evaluation

5.1   Experimental Setup

Baselines. We compare our approach with 4 types of bias mitigation techniques described in Sect. 3, including pre-processing method Disparate Impact Remover (DIR) [11], in-processing method Adversarial Debiasing (ADV) [28], post-processing method Reject Option Classification (ROC) [16], and RL-based method GetFair [27].

Datasets. We conduct experiments on four commonly used fairness-related datasets: Adult [10], Compas [18], German [10], and Bank [21]. For German and Bank datasets, the sensitive attributes are sex and age, respectively. For Adult and Compas datasets, there are two sensitive attributes: sex and race. In each task, only one sensitive attribute is considered, leading to 6 tasks in total (e.g., Compas-Sex, Compas-Race, etc.).

Machine Learning Models. We improve fairness-performance trade-off for 3 ML algorithms including Neural Network (NN), Logistic Regression (LR), and Support Vector Machine (SVM), which are widely adopted in fairness literature [3,8,30]. Following previous work [30], our NN contains five hidden layers, each consisting of 64, 32, 16, 8, and 4 units respectively. For SVM and LR [8], we employed the default configuration provided by the scikit-learn library [23].

Evaluation Metrics. We measure the performance of the model using 3 metrics: accuracy (ACC), F1-Score of the positive class (F1) and area under the ROC curve (AUC). We evaluate the fairness of the model using 5 fairness metrics, including DI, SPD, EOD, AOD, and ERD, which have been described in Sect. 2. SPD, EOD, AOD and ERD are processed to SPD', EOD', AOD', and ERD', respectively, as shown in Eqs. (10 and 15).

$$SPD' = 1 - |P[\hat{Y} = 1|Z = 0] - P[\hat{Y} = 1|Z = 1]| \tag{15}$$

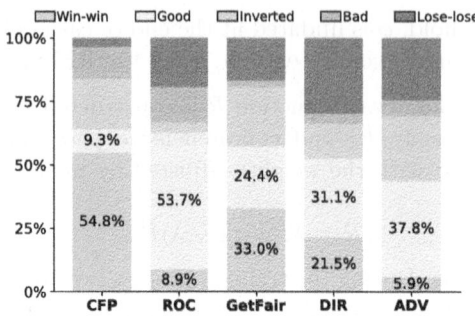

Fig. 3. The distribution of effectiveness levels for CFP and other baseline methods. CFP achieves the best trade-off, with 54.8% cases falling in *win-win*.

Fairness-Performance Trade-Off Evaluation Method. We employ Fairea [14] to assess the effectiveness of bias mitigation methods. According to Fairea, the performance of bias mitigation techniques is classified into five levels: *win-win*, *good*, *inverted*, *bad*, and *lose-lose*. A higher proportion of the *win-win* and *good* levels indicates a more effective bias mitigation method.

Implementation. The default reward function of CFP is $0.5(CFD_p + CFD_n) + 0.5AUC$. Each experiment setting is repeated 10 times, and we calculated the average of ML performance, fairness and the trade-off effectiveness (indicated as the proportion of five levels) of the corresponding setting over the 10 runs.

5.2 Experimental Results

Effectiveness and Efficiency of CFP. Figure 3 shows results for all methods across different models and tasks. We can observe that CFP outperforms baseline methods. CFP (54.8%) have achieved the highest proportion for *win-win* cases. In contrast, ROC (8.9%), GetFair (33%), DIR (21.5%), and ADV (5.9%) have decreased by 45.9%, 21.8%, 33.3% and 48.9% in proportions respectively, with an average of 37.5% compared with CFP. This shows that CFP can effectively improve the fairness-performance trade-off for ML classifiers.

Generality and Stability. CFP shows wide effectiveness in improving the fairness-performance trade-off of various ML algorithms on most tasks and is more stable than compared methods. The results on different ML algorithms are presented in Fig. 4 (a). It can be found that CFP achieves the highest proportion of cases falling in *win-win* for all algorithms, with 54% for LR, 42% for SVM, and 54% for NN. The results on different tasks are illustrated in Fig. 4 (b). For CFP, it outperforms all other methods in most tasks, except for Bank-Age. In addition, the stability of CFP is also investigated. Compared with other methods, CFP has exhibited much higher stability. The proportions of cases falling in *win-win* for CFP are all higher than 17% regardless of the tasks. While for other methods, much greater fluctuations in such proportion can be observed.

For instance, ROC has more than 15% cases falling in *win-win* on Adult-Sex and Adult-Race, but such proportion drops to 0% on Bank-Age. The case is worse for DIR, which has 100% cases falling in *win-win* on Bank-Age but 0% on Adult-Sex, Compas-Sex, and Compas-Race. In contrast, CFP has at least 31% cases falling in *win-win* on these 3 tasks while still reaching 18% on Bank-Age.

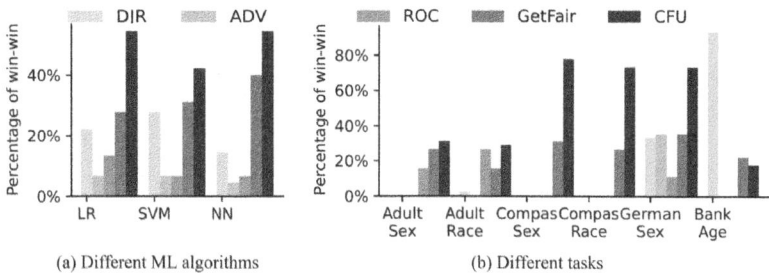

(a) Different ML algorithms   (b) Different tasks

Fig. 4. The percentage of *win-win* organized by different models (a) and tasks (b). CFP achieves a larger proportion of *win-win* than baseline methods across all models and most tasks.

Efficiency. CFP can improve multiple fairness for ML classifiers simultaneously while maintaining its performance. The scatter diagram of the original model and models processed by 5 bias mitigation methods are plotted in Fig. 5, with the performance value of the model as abscissa and the fairness value as ordinate. Due to limited space, only the results for the LR algorithm on Compas-Race task are presented, and those for other tasks are similar. Almost all the points for CFP are located at the upper right of the points for the original model, indicating that the model processed by CFP has better fairness and performance and can achieve a better fairness-performance trade-off. Moreover, CFP works well for all fairness metrics. The model processed by CFP reaches ((c), (h), (m), (e), (j), and (o) subgraphs) or close to (other subgraphs) the highest fairness scores of all models. In contrast, most points for other bias mitigation methods are located in the upper-left of the points for the original model, indicating that they have sacrificed performance while improving fairness. In addition, they also restricted to improving only one specific fairness notion. For instance, DIR is good at improving statistical parity, with the highest score on both DI and SPD metrics, while GetFair and ADV work well at equal opportunity and equalized odds respectively. This suggests that CFP can efficiently improve the fairness-performance trade-off across multiple metrics.

In summary, CFP is superior and has outperformed existing methods in improving fairness-performance trade-off across multiple metrics. In addition, it has superior generality and stability over existing methods on various ML algorithms and tasks.

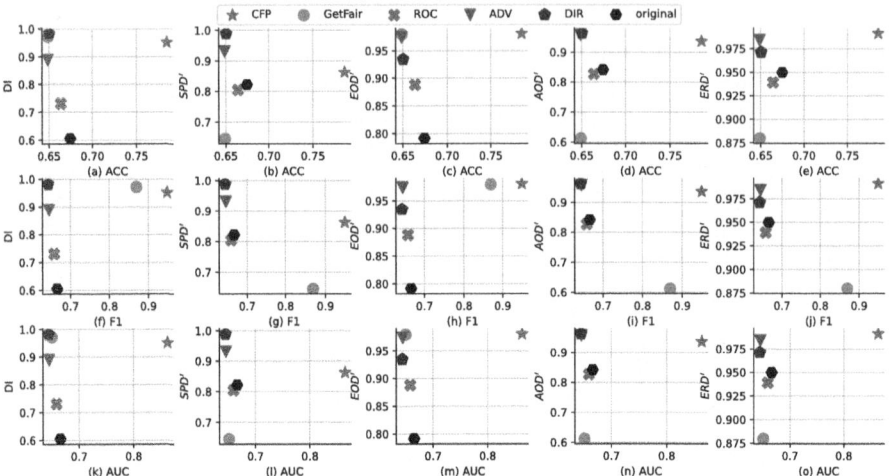

Fig. 5. The fairness-performance trade-off was evaluated by three performance metrics and five fairness metrics on LR model for Compas-Race task. The original points correspond to the original LR model, while the points of bias mitigation methods correspond to the models processed by these methods. Higher scores indicate a better degree of performance or fairness. It can be observed that the model processed by CFP achieves higher scores across various metrics, compared to the original model and other bias mitigation methods.

Analysis of Different Metrics Combination. We test numerous reward functions of CFP, which are constructed with different fairness and performance metrics. The results are evaluated by Fairea and the proportions of *win-win* cases are recorded, as presented in Table 2. The experiments consist of 3 groups. Firstly, we employ 1 existing fairness metric paired with 1 performance metric to form 15 combinations. Secondly, we bundle EOD, AOD, and ERD fairness metrics together (select all or none) and name them as Triplet. By using 3 fairness metrics (SPD, DI, Triplet) and 2 performance metrics (ACC, AUC), we explore 15 combinations. Finally, we combine our two new metrics (CFD_p, CFD_n) with 2 performance metrics (ACC, AUC) to form 9 combinations. We use each combination to construct the reward function, and conduct 10 repeated experiments on 3 algorithms with 6 tasks, resulting in 7020 experiments.

Characteristics of Existing Metrics. In the first experimental group, it can be found that the combinations containing F1 (denoted by a grey background) have all experienced a decrease in proportions of *win-win* cases. In contrast, those containing AUC (denoted by a green background) can achieve higher proportions for *win-win* cases. These findings suggest that F1 is ineffective in CFP, while AUC is an excellent performance metric. Additionally, using EOD, AOD, and ERD simultaneously can still perform well in *win-win* cases even though it can not match default CFP. The observation aligns with the theoretical analysis presented in Sect. 4.2.

Table 2. Effectiveness of CFP with different metrics combination.

1 Fairness × 1 Performance (%)

Performance	DI	SPD	EOD	AOD	ERD
ACC	30.0	33.0	26.3	29.7	24.8
AUC	32.2	33.7	30.4	33.7	31.9
F1	23.7	23.7	23.7	18.2	17.4

Multiple Fairness × Multiple Performance (%)

Performance	SPD	DI	SPD Triplet	DI Triplet	SPD DI Triplet	CFD_p	CFD_n	$CFD_p CFD_n$
ACC	30.0	30.7	29.3	25.6	30.4	41.1	45.9	46.3
AUC	40.0	35.6	34.1	33.3	34.4	50.4	54.4	54.8
ACC AUC	34.1	36.3	34.4	37.0	33.7	53.7	50.4	44.4

Effectiveness of Comprehensive Measurement and New Metrics. The superiority of CFP comes from the comprehensive measurement and new metrics. In Sect. 5.2, we can see that GetFair cannot competes with CFP when it comes to *win-win* cases. In this experiment, GetFair has three kinds of combinations, including SPD+ACC, EOD+ACC, and AOD+ACC. The highest proportion of *win-win* cases that are achieved by these combinations is 33%, only 60.14% of default CFP. Since the only difference between CFP and GetFair lies in the reward function, which has involved the comprehensive measurement and newly proposed metrics for CFP, it is reasonable to believe that the comprehensive measurement and new metrics do have played a vital role.

The Efficiency of CFD_p and CFD_n. The highest proportion of *win-win* cases (54.8%, denoted by a red background) is achieved when both CFD_p and CFD_n are used as fairness metrics and AUC as performance metric (the default setting of reward function), while the combinations without CFD_p and CFD_n seem to be not effective: the *win-win* cases for combinations including CFD_p or CFD_n are all higher than those without them. The highest proportion of *win-win* cases that can be achieved without CFD_p and CFD_n (denoted by a pink background) is 40%, which is around 72.75% of default CFP. On average, the cases falling in *win-win* for combinations using newly proposed metrics have been improved by 61.58% compared with the combinations without using it. Therefore, it is concluded that CFD_p and CFD_n are superior in improving the trade-off between fairness and performance.

In summary, CFD_p and CFD_n are superior in improving the trade-off between fairness and performance. The best combination of CFP consists of CFD_p, CFD_n and AUC, which has outperformed all other metric combinations.

6 Conclusion

Fairness is an important measure in ML classification models, as it enhances the credibility of the models and safeguards the rights of vulnerable groups. This paper proposes CFP framework to optimize the fairness-performance trade-off in ML classifiers. It also establishes a comprehensive measurement that considers multiple fairness and performance measures, along with introducing two new metrics based on analyzing the relationship between different fairness metrics. We evaluate the CFP framework with a variety of experiments to prove its capability to improve the classifier's fairness without sacrificing performance. In future work, we aim to generalize the framework to a wider range of ML tasks beyond classification.

Acknowledgement. This work was partially supported by the Shanghai Pujiang Program (No. 22PJD021), National Trusted Embedded Software Engineering Technology Research Center (East China Normal University), National Natural Science Foundation of China (NSFC 62202166), Shanghai Industrial Collaborative Innovation Leading Group Office (No. XTCX-KJ-2022-2-10) and Chenguang Program of Shanghai Education Development Foundation and Shanghai Municipal Education Commission (No. 23CGA33).

References

1. Angwin, J., Larson, J., Mattu, S., ProPublica, L.K.: Machine bias: there's software used across the country to predict future criminals, and it's biased against blacks (2016). https://www.propublica.org/article/machine-bias-risk-assessments-in-criminal-sentencing
2. Bellamy, R.K.E., et al.: Ai fairness 360: an extensible toolkit for detecting and mitigating algorithmic bias. IBM JResearch Develop. **63**(4/5), 4:1–4:15 (2019)
3. Biswas, S., Rajan, H.: Do the machine learning models on a crowd sourced platform exhibit bias? an empirical study on model fairness. In: Proceedings of the 28th ACM Joint Meeting on European Software Engineering Conference and Symposium on the Foundations of Software Engineering, pp. 642–653 (2020)
4. Biswas, S., Rajan, H.: Fair preprocessing: towards understanding compositional fairness of data transformers in machine learning pipeline. In: Proceedings of the 29th ACM Joint Meeting on European Software Engineering Conference and Symposium on the Foundations of Software Engineering, pp. 981–993 (2021)
5. Calders, T., Verwer, S.: Three Naive Bayes approaches for discrimination-free classification. Data Mining Knowl. Discov. (2010)
6. Calmon, F., Wei, D., Vinzamuri, B., et al.: Optimized pre-processing for discrimination prevention. Adv. Neural Inf. Process. Syst. **30** (2017)
7. Celis, L.E., Mehrotra, A., Vishnoi, N.: Fair classification with adversarial perturbations. Adv. Neural. Inf. Process. Syst. **34**, 8158–8171 (2021)
8. Chen, Z., Zhang, J.M., Sarro, F., Harman, M.: Maat: a novel ensemble approach to addressing fairness and performance bugs for machine learning software. In: Proceedings of the 30th ACM Joint European Software Engineering Conference and Symposium on the Foundations of Software Engineering, pp. 1122–1134 (2022)
9. Chouldechova, A.: Fair prediction with disparate impact: a study of bias in recidivism prediction instruments. arXiv, Applications (2016)
10. Dua, D., Graff, C.: UCI machine learning repository (2017). http://archive.ics.uci.edu/ml. Accessed 17 Feb 2023
11. Feldman, M., Friedler, S.A., Moeller, J., Scheidegger, C., Venkatasubramanian, S.: Certifying and removing disparate impact. arXiv, Machine Learning (2014)
12. Gao, X., Zhai, J., Ma, S., Shen, C., Chen, Y., Wang, Q.: Fairneuron: improving deep neural network fairness with adversary games on selective neurons. In: Proceedings of the 44th International Conference on Software Engineering, pp. 921–933 (2022)
13. Hardt, M., Price, E., Srebro, N.: Equality of opportunity in supervised learning. Adv. Neural Inf. Process. Syst. **29** (2016)
14. Hort, M., Zhang, J.M., Sarro, F., Harman, M.: Fairea: a model behaviour mutation approach to benchmarking bias mitigation methods. In: Proceedings of the 29th ACM Joint Meeting on European Software Engineering Conference and Symposium on the Foundations of Software Engineering, pp. 994–1006 (2021)

15. Jang, T., Shi, P., Wang, X.: Group-aware threshold adaptation for fair classification. In: Proceedings of the AAAI Conference on Artificial Intelligence. vol. 36, pp. 6988–6995 (2022)

16. Kamiran, F., Karim, A., Zhang, X.: Decision theory for discrimination-aware classification. In: 2012 IEEE 12th International Conference on Data Mining, pp. 924–929 (2012)

17. Khandani, A.E., Kim, A.J., Lo, A.W.: Consumer credit-risk models via machine-learning algorithms. J. Bank. Financ **34**(11), 2767–2787 (2010)

18. Larson, J., Mattu, S., Kirchner, L., Angwin, J.: How we analyzed the compas recidivism algorithm (2016). https://www.propublica.org/article/how-we-analyzed-the-compas-recidivism-algorithm

19. Li, Y., et al.: Training data debugging for the fairness of machine learning software. In: Proceedings of the 44th International Conference on Software Engineering, pp. 2215–2227 (2022)

20. Mehrabi, N., Morstatter, F., Saxena, N., Lerman, K., Galstyan, A.: A survey on bias and fairness in machine learning. ACM Comput. Surv. **54**(6) (2021)

21. Moro, S., Cortez, P., Rita, P.: A data-driven approach to predict the success of bank telemarketing. Decis. Support Syst. **62**, 22–31 (2014)

22. Nguyen, D., et al.: Fairness improvement for black-box classifiers with gaussian process. Inf. Sci. **576**, 542–556 (2021)

23. Pedregosa, F., Varoquaux, G., Gramfort, A., et al.: Scikit-learn: machine learning in Python. J. Mach. Learn. Res. **12**, 2825–2830 (2011)

24. Ricci Lara, M.A., Echeveste, R., Ferrante, E.: Addressing fairness in artificial intelligence for medical imaging. Nat. Commun. **13**(1), 4581 (2022)

25. Rodolfa, K.T., Lamba, H., Ghani, R.: Empirical observation of negligible fairness-accuracy trade-offs in machine learning for public policy. Nat. Mach. Intell. **3**(10), 896–904 (2021)

26. Ruoss, A., et al.: Learning certified individually fair representations. In: Proceedings of the 34th International Conference on Neural Information Processing Systems (NIPS 2020). Curran Associates Inc., Red Hook (2020)

27. Sikdar, S., Lemmerich, F., Strohmaier, M.: Getfair: generalized fairness tuning of classification models. In: Proceedings of the 2022 ACM Conference on Fairness, Accountability, and Transparency, pp. 289–299 (2022)

28. Zhang, B.H., Lemoine, B., Mitchell, M.: Mitigating unwanted biases with adversarial learning. In: Proceedings of the 2018 AAAI/ACM Conference on AI, Ethics, and Society, pp. 335–340 (2018)

29. Zhang, J.M., Harman, M.: Ignorance and prejudice in software fairness. In: 2021 IEEE/ACM 43rd International Conference on Software Engineering (ICSE), pp. 1436–1447. IEEE (2021)

30. Zhang, M., Sun, J.: Adaptive fairness improvement based on causality analysis. In: Proceedings of the 30th ACM Joint European Software Engineering Conference and Symposium on the Foundations of Software Engineering, pp. 6–17 (2022)

31. Zhang, S., et al.: Figcps: effective failure-inducing input generation for cyber-physical systems with deep reinforcement learning. In: 2021 36th IEEE/ACM International Conference on Automated Software Engineering (ASE), pp. 555–567. IEEE (2021)

32. Zhang, T., Zhu, T., Han, M., Chen, F., et al.: Fairness in graph-based semi-supervised learning. In: Knowledge and Information Systems, pp. 1–28 (2022)

Author Index

SPRINGER NATURE

GPSR Compliance

The European Union's (EU) General Product Safety Regulation (GPSR) is a set of rules that requires consumer products to be safe and our obligations to ensure this.

If you have any concerns about our products, you can contact us on ProductSafety@springernature.com

In case Publisher is established outside the EU, the EU authorized representative is:

Springer Nature Customer Service Center GmbH
Europaplatz 3
69115 Heidelberg, Germany

The manufacturer's authorised representative in the EU is Springer
Nature Customer Service Centre GmbH, Europaplatz 3, 69115 Heidelberg,
Germany. If you have any concerns regarding our products, please
contact ProductSafety@springernature.com

Printed and bound by CPI Group (UK) Ltd, Croydon, CR0 4YY
24/04/2026
02096351-0012